Sustainable Development Goals Series

The **Sustainable Development Goals Series** is Springer Nature's inaugural cross-imprint book series that addresses and supports the United Nations' seventeen Sustainable Development Goals. The series fosters comprehensive research focused on these global targets and endeavours to address some of society's greatest grand challenges. The SDGs are inherently multidisciplinary, and they bring people working across different fields together and working towards a common goal. In this spirit, the Sustainable Development Goals series is the first at Springer Nature to publish books under both the Springer and Palgrave Macmillan imprints, bringing the strengths of our imprints together.

The Sustainable Development Goals Series is organized into eighteen subseries: one subseries based around each of the seventeen respective Sustainable Development Goals, and an eighteenth subseries, "Connecting the Goals", which serves as a home for volumes addressing multiple goals or studying the SDGs as a whole. Each subseries is guided by an expert Subseries Advisor with years or decades of experience studying and addressing core components of their respective Goal.

The SDG Series has a remit as broad as the SDGs themselves, and contributions are welcome from scientists, academics, policymakers, and researchers working in fields related to any of the seventeen goals. If you are interested in contributing a monograph or curated volume to the series, please contact the Publishers: Zachary Romano [Springer; zachary.romano@springer.com] and Rachael Ballard [Palgrave Macmillan; rachael.ballard@palgrave.com].

Srijan Goswami • Moharana Choudhury
Palas Samanta
Editors

Micro-Nano Plastics Exposure, Environmental Degradation and Public Health Crisis

Perspectives and Concerns for Sustainable Development

Springer

Editors
Srijan Goswami
Department of Biotechnology
University of North Bengal
Darjeeling, West Bengal, India

Moharana Choudhury
Environmental Research and
Management Division
Voice of Environment (VoE)
Guwahati, Assam, India

Palas Samanta
Department of Environmental Science
Sukanta Mahavidyalaya
University of North Bengal
Dhupguri, West Bengal, India

ISSN 2523-3084 ISSN 2523-3092 (electronic)
Sustainable Development Goals Series
ISBN 978-981-96-5046-0 ISBN 978-981-96-5047-7 (eBook)
https://doi.org/10.1007/978-981-96-5047-7

This Springer imprint is published by the registered company Springer Nature Singapore Pte Ltd.
The registered company address is: 152 Beach Road, #21-01/04 Gateway East, Singapore 189721, Singapore

Contents

Micro- and Nano-plastics, the Globally Emerging Environmental and Public Health Crisis

1

Srijan Goswami, Palas Samanta, and Manab Deb Adhikari

1.1 Introduction

The development and widespread adoption of plastics can be traced back to the early twentieth century when scientific advancements and industrial innovation converged. Plastics, primarily synthetic polymers, have revolutionized several sectors, such as consumer products, constructing, automobiles, and packing, largely due to their versatility, low cost, and durability. However, their rise in production and consumption has also been accompanied by significant environmental and health issues, including their environmental persistence and potential for toxicity (Plastics—American Chemistry Council 2024; Baca 2022). The origins of plastic production can be traced to the development of Bakelite, the first fully synthetic plastic, invented by Belgian chemist Leo Baekeland in 1907. Bakelite was a phenolic resin that could be molded into various shapes and sizes, marking a key turning point in material science. It was initially used for electrical insulators, automotive parts, and household items due to its heat resistance and electrical insulating properties (Baekeland 1909). This early development laid the foundation for the growth of the synthetic polymer industry. Following Baekeland's innovation, other significant synthetic plastics were developed. In the 1920s, polymethyl methacrylate (PMMA), or acrylic, was introduced, offering transparency and weather resistance. In the 1930s, the development of polyvinyl chloride (PVC) and polyethylene expanded the range of materials available, each possessing unique properties that catered to different industrial needs (Plastics—American Chemistry Council 2024; Baca 2022). The post-World War II era saw an explosion in the use and production of plastics. This was largely driven by advancements in polymer chemistry, along with the increased availability of cheap oil and natural gas, which provided the raw materials needed for large-scale plastic production. The introduction of low-density polyethylene (LDPE) and high-density polyethylene (HDPE) in the 1950s, along with polypropylene (PP), contributed to plastics' widespread adoption due to their ease of production and wide range of applications (Plastics—American Chemistry Council 2024; Baca 2022). During the 1950s and 1960s, the mass production

S. Goswami · M. D. Adhikari (✉)
Department of Biotechnology, University of North Bengal, Darjeeling, West Bengal, India

P. Samanta
Department of Environmental Science, Sukanta Mahavidyalaya, University of North Bengal, Dhupguri, West Bengal, India

S. Goswami et al. (eds.), *Micro-Nano Plastics Exposure, Environmental Degradation and Public Health Crisis*, Sustainable Development Goals Series, https://doi.org/10.1007/978-981-96-5047-7_1

of plastics was further facilitated by the growth of consumerism in developed countries. Plastics began to replace materials such as wood, metal, and glass in many everyday products, including packaging, household goods, and toys. As plastics became integral to modern life, production volumes increased dramatically, fueled by innovations in polymerization techniques and the development of new plastic forms such as polystyrene (PS), polyethylene terephthalate (PET), and polycarbonate (PC) (Plastics—American Chemistry Council 2024; Baca 2022).

The latter half of the twentieth century saw an unprecedented rise in global plastic consumption. By the 1970s and 1980s, plastics had become ubiquitous in packaging, medical devices, electronics, and automotive components. The rise of disposable plastic products, such as packaging materials, single-use bags, and food containers, marked a significant shift toward convenience-driven consumption. The global plastic production grew rapidly, with major producers like ExxonMobil, Dow Chemical, and BASF leading the charge in the chemical manufacturing sector (Geyer et al. 2017; Plastics Europe 2022). Global manufacturing of plastic accelerated from two million metric tons (MMT) in 1950 to approximately 368 MMT in 2019, and projections indicate continued growth in the coming decades. The world's top manufacturers and users of plastics is now Asia, especially China, followed by North America and Europe. In 2019, China accounted for over 30% of global plastic production, and the country is expected to remain a dominant player in the industry due to its extensive manufacturing capabilities and large consumer market (Plastics Europe 2022). A worsening environmental disaster has been brought on by the quick increase in plastic manufacture and usage. Plastics, being highly durable and resistant to degradation, accumulate in the natural environment, causing widespread pollution. The vast majority of plastic products are not recycled, with less than 10% of all plastics produced globally being recycled into new products. As a result, plastic waste has become a significant contributor to land and marine pollution, with an estimated 8 MMT of plastic entering the ocean annually (Jambeck et al. 2015). In addition to waste accumulation, plastics also pose potential health risks due to the migration of toxic substances, such as bisphenol A (BPA) and phthalates, from plastics into the surrounding environment including human body. These chemicals have been linked to endocrine disruption and various other health issues, raising concerns about the safety of widespread plastic use, particularly in food packaging and medical applications (Rochman et al. 2013a). The origin and rise of plastics in global production and consumption can be attributed to significant technological advancements, industrial demand, and the economic appeal of plastic materials. From Bakelite's invention to the current era of ubiquitous plastic usage, plastics have played a key role in shaping modern society. However, the environmental and health challenges linked with plastic waste and chemical leachates require urgent attention and innovative solutions. The future of plastics will likely depend on advancements in recycling technologies, the development of biodegradable alternatives, and stronger regulatory frameworks to mitigate the environmental impact.

1.2 Understanding Microplastics (MPs) and Nanoplastics (NPs): Size, Composition, and Prevalence

Microplastics (MPs) are defined as plastic particles that measure less than 5 mm in size, and they are usually divided into two types: primary and secondary. Primary MPs are developed at small sizes for specific applications, such as in cosmetics, cleaning products, and personal care products (PCPs). The disintegration of bigger plastic objects, like bottles, fishing nets, and bags, produces secondary MPs, which degrade into smaller fragments over time due to physical, chemical, or biological processes (Barnes et al. 2009). Nanoplastics (NPs), on the other hand, refer to plastic particles that are smaller than 100 nanometers (nm). Due to their even smaller size, NPs can exhibit unique properties, such as increased reactivity and mobility, compared to their microplastic counterparts. NPs are often the result of further degradation of MPs, or they can be intentionally manufactured for use in nanotechnologies, coatings, and consumer goods (Eerkes-Medrano et al. 2015). The size of MPs and NPs plays a crucial role in their interaction with environmental media and living organisms. MPs, with a size range from a few micrometers to 5 mm, are large enough to be consumed by a diverse organism, spanning from plankton to marine mammals. NPs, in contrast, are so small that they can cross biological membranes and penetrate cells, allowing them to interact with cellular structures, including organelles and proteins. This property makes NPs a particular concern for potential toxicity (Bouwmeester et al. 2015). Both MPs and NPs are primarily composed of synthetic polymers, which are the building blocks of most plastics. The popular polymeric types found in MPs and NPs include polyethylene (PE), PP, polyvinyl chloride (PVC), PS, and PET (Rochman et al. 2015). These polymers are made from petrochemicals and are intended to be long lasting and resistant to deterioration. As a result, they endure for extended periods of time in the environment, contributing to the accumulation of plastic waste. Additives, such as plasticizers, stabilizers, and colorants, are often incorporated into plastics to enhance their properties. These substances have the potential to gradually seep out of the plastic matrix and may have harmful effects on both wildlife and humans. For example, phthalates and BPA, common additives in plastic products, have been linked to endocrine disruption and other health concerns (Bouwmeester et al. 2015; Rochman et al. 2015). Numerous habitats, including rivers, lakes, sediments, seas, and even far-flung regions like the deep sea and the Arctic, have been shown to have higher MPs and NPs levels. MPs are widespread and ubiquitous, with reports of their presence in almost every ecosystem and environmental medium (Cózar et al. 2014). In marine environments, MPs have been found in large concentrations, often in the form of floating debris or as small particles within the water column. Due to the buildup of plastic debris in the oceans, sizable "garbage patches," like the Great Pacific Trash Patch, have developed areas that contain elevated levels of MPs (Lebreton et al. 2012). Microplastics are also prevalent in freshwater ecosystems, where they are transported through rivers and lakes, often entering the ocean through river outflows. In addition to water bodies, MPs have been detected in soils, agricultural fields, and even in the atmosphere. The movement of MPs through atmospheric processes has been linked to their transport over long distances, with reports of microplastics found in remote locations far from their sources (Dris et al. 2015). Nanoplastics, being even smaller than MPs, are more challenging to detect and quantify. Due to their size, NPs can be dispersed more easily through air and water, potentially reaching environments that are less accessible for larger MPs. Studies have indicated that NPs are present in surface waters, sediments, and soils, though their abundance is less understood compared to MPs (Eerkes-Medrano et al. 2015). The environmental impacts of MPs and NPs are wide ranging. Many creatures, such as plankton, fish, birds, and marine mammals, can consume such particles, which can cause physical injury, decreased feeding efficiency, and in rare circumstances, mortality. MPs and NPs can be carriers of harmful substances, such as heavy metals (HMs) and persistent organic pollutants (POPs), which can stick to the plastic surface in addition to their physical effects. After being consumed by creatures, these substances might trigger bioaccumulation alongside biomagnification in the food chain (Teuten et al. 2009). The potential health risks to humans are also a growing concern. As MPs and NPs are found in food products, drinking water, and even the air, there is increasing evidence that human exposure to these particles is inevitable. While the health effects of NPs are not fully understood, studies have suggested that these particles may cause an inflammatory response, oxidative stress, and even cellular damage upon entering the human body (Smith et al. 2018). The ability of NPs to cross the blood–brain barrier and interact with brain tissue further highlights their toxicity to human health (Bouwmeester et al. 2015).

1.3 Sources of Micro- and Nano-plastics

1.3.1 Primary Sources: Microbeads, Industrial Abrasives, and Cosmetic Additives

Primary origins of MPs and NPs include microbeads, industrial abrasives, and cosmetic additives, all of which contribute to the environmental contamination of plastics. Microbeads, typically made from polyethylene, polypropylene, or polymethyl methacrylate, are commonly utilized in PCPs such as exfoliating scrubs, toothpaste, and facial cleansers due to their small size and abrasive properties (Fig. 1.1). These tiny plastic particles, which are too tiny for wastewater treatment facilities to filter efficiently, wind up in aquatic ecosystems, and greatly increase the amount of MPs pollution in freshwater and marine ecosystems (Rochman et al. 2013b). Industrial abrasives, including plastic grit used for sandblasting and polishing, are another important source of primary MP. These abrasive particles, when used in industries like construction, shipbuilding, and automotive manufacturing, are released into the environment, either as waste or runoff during cleaning processes (Ziani et al. 2023). Additionally, plastics are often added as additives in cosmetics to enhance texture, stability, or appearance. These plastic-based additives, including small plastic particles added to products like foundation, sunscreen, and hair care items, are washed off during use and contribute to MPs buildup in wastewater and natural water bodies (Guerranti et al. 2019). Together, these sources represent significant contributors to the growing prevalence of MPs and NPs in the environment, further exacerbating the challenges of plastic pollution.

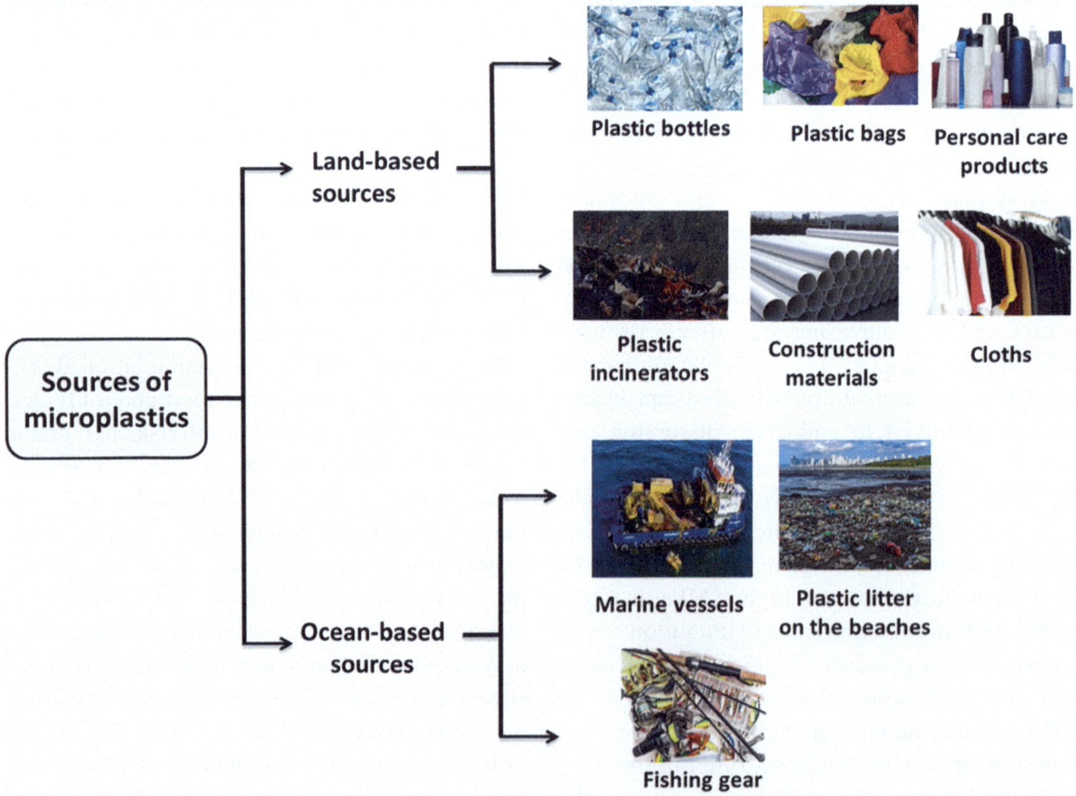

Fig. 1.1 MPs sources into the natural environment. (Adapted from Osman et al. (2023); open access)

1.3.2 Secondary Sources: Fragmentation of Larger Plastics Due to Environmental Factors

Secondary MPs and NPs are primarily generated through larger plastic particle disintegration as a result of factors such as UV radiation, mechanical abrasion, and temperature fluctuations. Over time, plastics exposed to sunlight undergo photodegradation, where UV rays break down the polymer chains, causing the material to weaken and fragment into smaller pieces. This process is particularly pronounced in marine environments, where plastics exposed to both UV radiation and salt water can degrade faster, leading to the formation of MPs (Andrady 2011). Additionally, mechanical forces, such as waves, wind, and friction with other materials, further break down larger plastic objects like bottles, fishing nets, and packaging into smaller particles. These smaller fragments can range from MPs (less than 5 mm) to NPs (less than 100 nm) and are widely distributed across ecosystems, contributing to persistent plastic pollution (Cózar et al. 2014). Temperature variations also play a role in the fragmentation process, as they can cause the plastic to expand and contract, weakening the structure of the material and accelerating its breakdown (Thompson et al. 2009). These secondary MPs and NPs are pervasive in aquatic and terrestrial environments, making it difficult to mitigate their environmental impact.

1.4 Environmental Pathways

1.4.1 Microplastics Transport Through Terrestrial, Freshwater, and Marine Environments

MPs are ubiquitously transported across terrestrial, freshwater, and marine environments through a variety of physical, biological, and chemical pathways. In terrestrial environments, MPs are primarily introduced through waste disposal, agricultural practices, and atmospheric deposition. These particles can be transported by wind, water runoff, and even through the movement of soil, leading to their widespread distribution in fields, forests, and urban areas (Du et al. 2021). After entering freshwater environments like rivers and lakes, MPs can travel great distances via water currents before exiting the system and reaching the ocean. In freshwater environments, MPs are often carried along the surface or in the sediment, depending on their size and density (Dris et al. 2015). Marine environments are particularly vulnerable to MP contamination due to the high volume of plastic waste entering the oceans through rivers, shipping, and coastal activities. Once in the ocean, MPs can float on the surface, become suspended in the water column, or settle in the sediment depending on their size and composition. The movement of MPs in the marine environment is further influenced by ocean currents, wind patterns, and biological activities, such as ingestion by marine organisms, which can transport the particles across various trophic levels (Cózar et al. 2014). These environmental pathways contribute to the global spread of MPs, highlighting the challenges associated with mitigating plastic pollution across different ecosystems.

1.4.2 Role of Atmospheric Deposition in Spreading Microplastics Globally

Atmospheric deposition regulates the global spreading of MPs, facilitating their transport over long distances and their deposition in remote and previously uncontaminated areas. Microplastics are released into the atmosphere through diverse origins, including larger plastic disintegration, industrial activities, and the wear and tear of synthetic textiles (Dris et al. 2015). These tiny particles could also be transported by wind currents, often traveling hundreds or even thousands of kilometers, before eventually settling onto land, water bodies, and other surfaces. Studies have shown that atmospheric deposition of MPs is widespread, with particles detected in both urban and rural environments, as well as in pristine regions such as the Arctic and remote mountain ranges (Huang et al. 2021). Once in the atmosphere, MPs can become part of dust particles or remain suspended for extended periods, depending on their size, shape, and chemical makeup (Wright et al. 2020). The ability of MPs to be dispersed by atmospheric processes highlights the global nature of plastic pollution, demonstrating that even the most remote ecosystems are not immune to its effects. Atmospheric deposition of MPs has become a critical pathway in the spread of this pervasive environmental contaminant, underlining the complexity of addressing plastic pollution on a global scale.

1.4.3 Plastic Pollution in Remote Ecosystems: Polar Regions, Deep Sea, and Mountains

Plastic pollution has infiltrated some of the most remote ecosystems on Earth, including polar regions, the deep sea, and high-altitude mountain ranges, highlighting the widespread nature of plastic contamination. In the polar regions, particularly the Arctic and Antarctic, MPs have been found in snow, ice, and water samples, demonstrating that these areas are receiving plastics through atmospheric deposition and ocean currents (Peeken et al. 2018). The harsh conditions of these environments, combined with increasing global shipping activities and ice melt due to climate change, contribute to the dispersal of plastics, which then accumulate in the region's fragile ecosystems. Similarly, in the deep sea, plastic debris has been discovered at great depths, often in areas previously thought to be pristine. MPs and larger plastic items are transported by ocean currents to the deepest parts of the ocean, where these could be consumed by marine organisms, impacting biodiversity and food chains (Woodall et al. 2014). High-altitude mountain ecosystems, such as the Himalayas and the Alps, have also been identified as sinks for microplastics, with plastic particles being deposited through atmospheric transport and precipitation (Bergmann et al. 2017). The plastic accumulation in these remote locations underscores the ubiquitous character of plastic pollution, raising concerns about its effects on wildlife and human populations, even in the most isolated and pristine environments.

1.5 Fate and Transport in the Environment

1.5.1 Degradation and Fragmentation Mechanisms—Photodegradation, Chemical Degradation, and Mechanical Breakdown

Microplastics undergo several degradation processes, including photodegradation, chemical degradation, and mechanical breakdown, which contribute to their fragmentation and transformation in the environment. Photodegradation is one of the most significant processes affecting plastics exposed to sunlight, particularly ultraviolet (UV) radiation. UV rays cause the polymer chains in plastics to break down, leading to the production of free radicals and the eventual breakdown of the material into smaller particles (Andrady 2011). This process is especially prevalent in marine and terrestrial environments, where plastics are exposed to sunlight for extended periods. Chemical degradation, on the other hand, occurs through interactions with environmental chemicals, such as acids, bases, and oxidative agents. Over time, these chemicals can alter the chemical structure of plastics, facilitating their breakdown into microplastics and smaller compounds. Mechanical breakdown, driven by physical forces such as abrasion, wave action, and collisions with other objects, also contributes significantly to the fragmentation of larger plastic debris into microplastics. This process is common in marine environments where plastics are subjected to constant movement and stress (Cózar et al. 2014). Together, these degradation mechanisms accelerate the formation and distribution of microplastics, making them persistent environmental pollutants that are difficult to mitigate.

1.5.2 Degradation and Fragmentation Mechanisms—Factors Influencing Plastic Degradation (UV Exposure, Temperature, and Wave Action)

Plastic degradation in the environment is primarily influenced by a combination of factors, such as ultraviolet (UV) exposure, temperature, and wave action, which collectively accelerate the breakdown and fragmentation of plastic materials. UV exposure, particularly from sunlight, is a leading factor in the degradation of plastics, initiating photodegradation. UV radiation breaks the chemical bonds within the polymer chains, creating free radicals and leading to the formation of smaller, more brittle particles (Andrady 2017). Temperature also plays a significant role in plastic disintegration, with higher temperatures promoting chemical reactions that degrade plastics more quickly. The thermal breakdown of plastics is often accelerated in environments with significant temperature fluctuations, as this increases the mobility of polymer chains and makes them more vulnerable to chemical and mechanical stresses (Thompson et al. 2009). Additionally, wave action in marine environments causes mechanical stress and abrasion, further fragmenting plastics into smaller pieces. Repeated physical impacts from waves, tides, and currents break down larger plastic debris, contributing to the widespread presence of MPs in oceans (Cózar et al. 2014). These combined factors such as UV exposure, temperature, and mechanical forces are known to create an environment in which plastics are continually degraded and fragmented, leading to the persistence of plastic pollution in various ecosystems.

1.5.3 Long-Term Environmental Persistence—Durability of Plastic Materials

Plastic materials are renowned for their durability, which, while beneficial for many applications, also contributes to their long-term persistence in the environment. Plastics are highly durable to natural degradation processes due to their stable polymeric structures, which do not easily break down in the presence of water, microorganisms, or other environmental factors (Andrady 2017). This durability means that plastics can persist in ecosystems for decades or even centuries, accumulating in terrestrial, freshwater, and marine environments. Plastic disintegration is a slow process, and even after significant fragmentation into microplastics, the particles remain in the environment, continuing to pose environmental and ecological risks (Thompson et al. 2009). Many common plastics, such as PE and PP, are particularly resistant to biodegradation because their chemical composition is not readily broken down by natural processes (Geyer et al. 2017). As a result, plastics accumulate in landfills, waterways, and oceans, where they can cause long-lasting damage to ecosystems, wildlife, and human health. The persistence of plastics in the environment highlights the need for more sustainable materials and waste management practices to mitigate their impact.

MPs are increasingly found in sedimentary environments, where they accumulate and interact with organic matter, leading to complex ecological consequences. As MPs are denser than water, they often sink to the bottom of aquatic environments, accumulating in sediments where they can persist for long periods (Wright et al. 2020). In these sediments, MPs can become coated with organic matter, such as proteins, lipids, and polysaccharides, which form biofilms on their surface. Biofilm formation enhances the adhesion of microplastics to sediments and increases their persistence in the environment (Zettler et al. 2013). These biofilms not only alter the physical properties of the MPs, making them more difficult to degrade, but also provide a substrate for microorganisms, including bacteria and algae, which can colonize the plastic particles (Disasanayakek et al. 2022). Additionally, the interaction of MPs with organic matter can affect their role as vectors for the transport of harmful substances, such as POPs and HMs, which may be adsorbed onto the particle's surface (Teuten et al. 2009). This complex interaction between microplastics, organic matter, and biofilms contributes to the environmental impact of plastic pollution by increasing the mobility and toxicity of associated pollutants, affecting sediment quality, and influencing benthic ecosystems.

1.5.4 Bioavailability and Trophic Transfer

1.5.4.1 Uptake of Microplastics by Aquatic Organisms (Filter Feeders, Plankton, and Fish)

The uptake of MPs by aquatic organisms is a growing concern, as these particles are easily consumed by a wide variety of species, spanning from plankton to larger fish and filter feeders. Filter feeders, such as bivalves and certain species of krill, are particularly vulnerable to MPs ingestion due to their feeding mechanism, which involves filtering large volumes of water for food particles, including MPs (Browne et al. 2010). Similarly, plankton, which forms the base of aquatic food webs, can ingest MPs that are similar in size to their natural food particles (Cole et al. 2013). Once consumed, MPs can accumulate in the digestive tracts of these organisms, where they may cause physical harm or affect feeding and growth. Fish, particularly those at higher trophic levels, may also ingest MPs through their diet, either directly or through consuming contaminated prey (Lusher et al. 2013). The MPs' ingestion by fish can lead to bioaccumulation, potentially introducing plastics and associated toxicants into the food chain, which can have long-term ecological and human health implications (Rochman et al. 2013b). The transfer of MPs through aquatic food webs raises concerns about the impacts of MPs pollution on biodiversity, ecosystem functioning, and the safety of seafood for human consumption.

MP pollution poses significant threats to food webs and trophic-level transfer in aquatic ecosystems. As microplastics are ingested by organisms at various trophic levels, from plankton to large fish, they have the potential to disrupt biological processes and impact ecosystem health. The ingestion of microplastics by primary producers such as plankton can result in bioaccumulation, with plastics and associated toxicants transferring up the food chain as predators consume contaminated prey (Lusher et al. 2013). This trophic-level transfer can lead to the accumulation of microplastics in higher trophic organisms, including fish and marine mammals, affecting their health, reproduction, and survival (Rochman et al. 2013a). Furthermore, microplastics can alter feeding behaviors, reduce growth rates, and cause physical harm to organisms, disrupting their role in the food web and potentially leading to shifts in population dynamics (Browne et al. 2008). The transfer of MPs and the associated harmful chemicals, such as POPs and HMs, through the food web raises concerns not only for biodiversity but also for human health, as these pollutants may eventually reach seafood consumed by humans (Teuten et al. 2009). The impact of MPs on food webs and trophic interactions underscores the need for urgent action to mitigate MP pollution and protect both aquatic ecosystems and human food security.

1.6 Ecological Impacts of Micro- and Nano-plastics

1.6.1 Effects on Marine and Freshwater Organisms

Plastic pollution, including macroplastics and MPs, poses a severe danger to marine and freshwater organisms, with ingestion, entanglement, and physical harm being among the most documented impacts. Many aquatic animals confuse plastic waste for food, which can result in swallowing and hinder development and reproduction by inducing internal damage, physical obstructions in the digestive tract, decreased feeding efficiency, and false satiation (Lusher et al. 2013). In freshwater systems, species such as fish and amphibians are similarly affected, with the ingestion of microplastics often resulting in reduced nutrient uptake and energy storage (Du et al. 2021). Entanglement in larger plastic debris, such as fishing nets and packaging materials, is another common hazard, especially for marine mammals, sea turtles, and birds. This can lead to restricted movement, starvation, and even death (Kühn et al. 2015). Physical harm extends to abrasion injuries caused by sharp plastic edges, leading to infections or increased vulnerability to predators. Further, plastics can act as carriers for harmful pathogens and invasive species, amplifying their impact on aquatic ecosystems (Gregory 2009). These effects not only threaten individual organisms but also disrupt population dynamics and ecosystem stability, emphasizing the immediate action for addressing plastic pollution in aquatic environments. MPs and associated chemical pollutants exert sub-lethal effects on marine and freshwater organisms, significantly impacting their reproductive, developmental, and behavioral functions. Ingestion of microplastics has been shown to interfere with energy allocation, often diverting resources away from reproduction and growth, leading to reduced fecundity in species such as fish and invertebrates (Cole et al. 2015). Developmental effects are also prominent, as exposure to microplastics and their associated toxicants, such as endocrine-disrupting chemicals, can impair embryonic development and result in morphological abnormalities in larvae (Wright et al. 2013). Behavioral alterations, including reduced feeding efficiency, impaired predator avoidance, and changes in social interactions, have been found in organisms that range from zooplankton to fish (Mattsson et al. 2015). These behavioral shifts are often linked to neurotoxicity or physical discomfort caused by microplastics, which can alter survival outcomes and affect population dynamics. Such sublethal effects may accumulate over time, leading to cascading impacts on food web structures and ecosystem functions. These findings emphasize the pervasive nature of microplastic pollution and its subtle yet significant influence on aquatic life (Fig. 1.2).

1.6.2 Microplastics as Vectors for Toxic Substances

1.6.2.1 Adsorption of Persistent Organic Pollutants (pops) and Heavy Metals

MPs act as effective vectors for toxic substances in aquatic environments due to their higher surface dimension and hydrophobic properties, enabling the adsorption of POPs and HMs. POPs, including polychlorinated biphenyls (PCBs) and polycyclic aromatic hydrocarbons (PAHs),

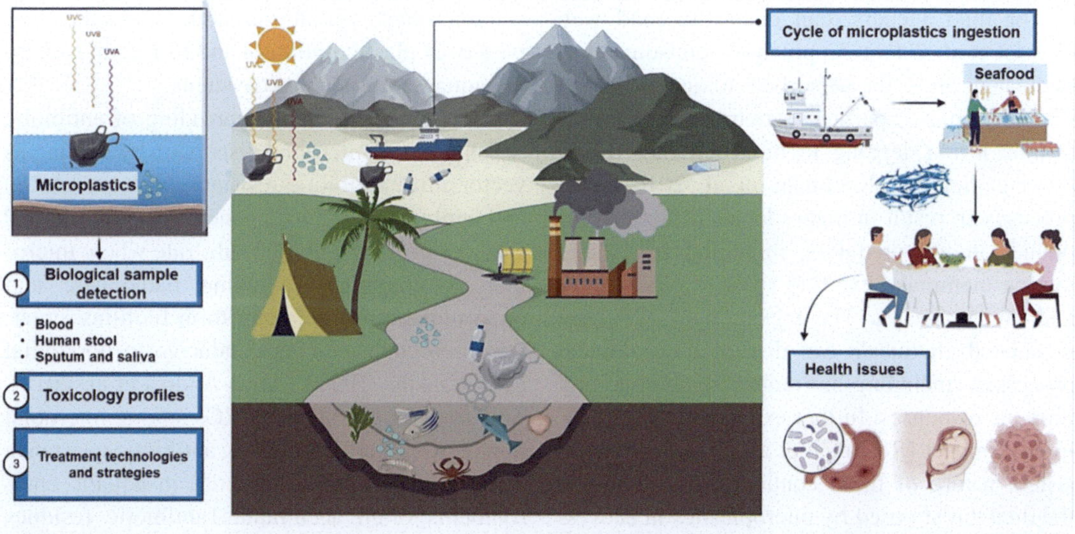

Fig. 1.2 Impacts and routes of microplastics on humans and the natural world. (Adapted from Osman et al. (2023); open access)

adhere to the surface of microplastics from surrounding water, concentrating these hazardous substances far beyond their ambient levels (Teuten et al. 2009). Similarly, heavy metals such as lead, cadmium, and mercury bind to the surface of microplastics through ionic interactions and complexation processes, potentially exacerbating their bioavailability and toxicity (Rochman et al. 2013a). When ingested by aquatic organisms, microplastics can serve as a direct pathway for the transfer of these contaminants into biological systems, leading to bioaccumulation and biomagnification within food webs. This dual threat of physical plastic particles and adsorbed toxicants amplifies the environmental and ecological risks of microplastic pollution, raising concerns about their long-term impacts on biodiversity and ecosystem health, as well as the safety of seafood for human consumption (Koelmans et al. 2016).

1.6.2.2 Transfer of Chemical Contaminants Through the Food Chain

MPs facilitate the movement of chemical substances through the food chain by serving as carriers for harmful substances, including POPs and HMs. These contaminants adsorb onto the MPs' surface in aquatic environments and, when consumed by lower trophic organisms such as plankton or filter feeders, can enter the food web (Teuten et al. 2009). As predators consume contaminated prey, the adsorbed toxicants, along with microplastic particles, are transferred up the trophic levels, leading to bioaccumulation and biomagnification (Koelmans et al. 2016). This process can result in higher levels of toxic compounds in top predators, including fishes and marine mammals, exerting risks to biodiversity and human health. The ingestion of MPs and associated chemicals can disrupt physiological processes, including hormonal regulation and immune responses, further exacerbating the ecological impact (Rochman et al. 2013a). The persistent nature of these contaminants highlights the dual threat posed by microplastics in ecosystems, emphasizing the implementation of strategies to reduce their prevalence in the environment and mitigate their impact on food web dynamics.

1.6.3 Interaction with Microbial Communities

MPs provide novel and persistent surfaces for the colonization and growth of microbial communities, leading to the formation of biofilms. These biofilms, often termed "plastispheres," consist of diverse microorganisms, including bacteria, algae, and fungi, which attach to the surface of plastics in aquatic environments (Zettler et al. 2013). The unique chemical and physical properties of plastics, such as their hydrophobicity and longevity, make them suitable habitats for microbial colonization. Over time, these biofilms can alter the buoyancy and transport of microplastics, facilitating their movement across ecosystems (Amaral-Zettler et al. 2020). The microbial communities within plastispheres may include both beneficial and harmful organisms, such as pathogenic bacteria, which can be a threat to aquatic life and human health if they are transported via contaminated plastics (Kirstein et al. 2018). Additionally, biofilm formation can influence the adsorption of toxic substances onto MPs, further amplifying their environmental and ecological impact. The interaction between microplastics and microbial communities underscores the complexity of plastic pollution and its role in reshaping aquatic microbial ecosystems.

MPs contribute to the spreading of antibiotic resistance and microbial dispersion by serving as vectors for antibiotic-resistant bacteria (ARB) and antibiotic-resistance genes (ARGs). The MP surfaces provide a unique substrate where microbial communities, including pathogenic and opportunistic bacteria, can form biofilms, creating microenvironments conducive to horizontal gene transfer (HGT) (Arias-Andres et al. 2018). This transfer can spread ARGs among microbial populations, perhaps making antibiotic resistance a worldwide problem. Plastics in aquatic environments often accumulate antibiotic residues and heavy metals, which can act as selective pressures, promoting the proliferation of resistant strains (Lu et al. 2019). Moreover, as MPs are transported across different ecosystems by currents or ingestion by mobile organisms, they facilitate the dispersion of ARB and ARGs to new environments, including pristine habitats and human-influenced areas. This capability of plastics to act as a reservoir and vector for antibiotic resistance highlights their role in amplifying both environmental and public health challenges linked to plastic pollution (Beans 2023).

1.7 Human Health Risks

1.7.1 Pathways of Human Exposure

1.7.1.1 Ingestion Through Contaminated Water, Food (Seafood, Salt), and Air

Humans are exposed to MPs through multiple pathways, with ingestion being one of the most significant routes, primarily via contaminated water, food, and air. Drinking water, both bottled and tap, has been widely reported to contain MP particles, potentially due to contamination during production or distribution (Koelmans et al. 2019). Seafood, particularly filter feeders like mussels and oysters, is a significant dietary source of MPs, as these organisms accumulate MPs directly from their environment (Cox et al. 2019). Table salt, especially sea salt, has also been identified as a source of MP ingestion due to the concentration of particles during the evaporation process (Karami et al. 2018). Inhalation of airborne MPs is another emerging concern, as these particles are present in indoor and outdoor air, especially in places where plastic garbage and synthetic textile usage is prevalent. Once ingested or inhaled, MPs can interact with human tissues, potentially causing physical and chemical harm, though the long-term health implications remain a focus of ongoing research.

Dermal exposure and inhalation are emerging pathways of human exposure to MPs and NPs, particularly in environments with high plastic particulate loads (Fig. 1.3). Dermal exposure occurs when microplastics come into contact with human skin through PCPs such as scrubs, exfoliants, and cosmetics, which often contain microbeads (Tsochatzis et al. 2024). While the intact skin acts as a natural barrier, concerns persist about the potential penetration of smaller nanoplastics through micro-abrasions or interaction with hair follicles, although further study is needed to confirm these pathways (Vethaak and Legler 2021). Inhalation represents a significant exposure route, especially in urban areas with high levels of airborne MPs derived from the degradation of synthetic textiles, tire wear particles, and industrial emissions (Allen et al. 2019). Once inhaled, these particles can deposit in the respiratory tract, potentially leading to inflammation, oxidative stress, or other respiratory issues, although the long-term health effects remain under investigation. Together, dermal exposure and inhalation highlight the pervasive dominance of MPs and NPs and their potential implications for human health.

1.7.2 Potential Health Effects

Plastic materials often contain additives such as BPA and phthalates, which are used to improve their durability, flexibility, or performance but can leach into the environment and pose significant toxicological risks. BPA, a monomer used in polycarbonate plastics, is a known endocrine disruptor that can interfere with hormonal signaling, leading to developmental, reproductive, and metabolic disorders in humans and wildlife (Rochester 2013). Phthalates, commonly used as plasticizers, are similarly associated with endocrine disruption and have been linked to reduced fertility, altered neurodevelopment, and increased risks of certain cancers (Hauser and Calafat 2005). In addition to intrinsic additives, MPs also adsorb environmental pollutants, including POPs like PCBs and PAHs, as well as HMs, from surrounding water and sediments. These adsorbed contaminants can desorb in biological systems when MPs are ingested, acting as vectors for toxic substances and amplifying exposure risks through bioaccumulation and trophic transfer (Koelmans et al. 2016). The combined toxicity of plastic additives and adsorbed pollutants under-

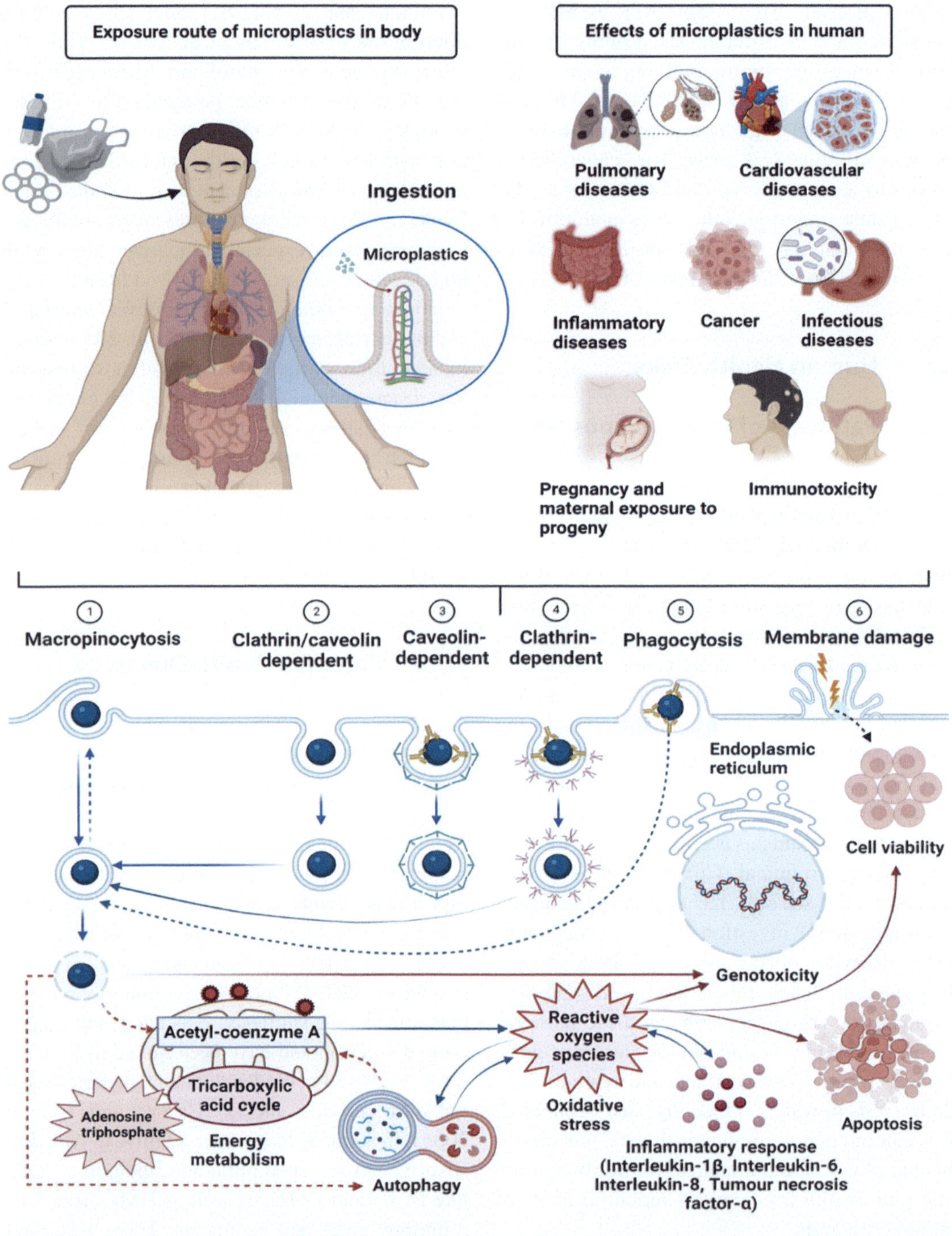

Fig. 1.3 MPs' effects on human health and associated mechanism. (Adapted from Osman et al. (2023); open access)

scores the multifaceted environmental and health impacts of plastic pollution.

Plastic additives, such as BPA, phthalates, and flame retardants, associated with MPs can trigger adverse biological responses, including inflammation, oxidative stress, and potentially carcinogenic effects. Upon exposure, these chemicals can disrupt cellular homeostasis by generating reactive oxygen species (ROS), leading to oxidative injury to lipids, proteins, and DNA (Wright and Kelly 2017). This oxidative stress can activate pro-inflammatory signaling, such as the nuclear factor-kappa B (NF-κB) pathway, resulting in chronic inflammation, a key driver of various diseases, including cancer (Prata et al. 2020). Additionally, some additives like BPA and certain phthalates are classified as endocrine disruptors, with evidence linking them to increased risks of hormone-dependent cancers, such as breast and prostate cancer (Rochester 2013). Persistent exposure to these chemicals, particularly in vulnerable populations namely pregnant women and such as infants, raises concerns about their long-term health impacts. The cumulative effects of inflammatory responses, oxidative stress, and potential carcinogenicity emphasize the need for stricter regulations on plastic additive use and an enhanced understanding of their interactions within biological systems.

1.8 Conclusion

MPs and NPs are ubiquitous in our environment and represent a multifaceted and degraded global environmental problem and public health concern. These molecules, so small they are invisible to the human eye, have entered ecosystems, food chains, and human bodies, posing major threats to environmental health and as well public health. Small in size, micro- and nano-plastics have a huge collective impact, harming natural habitats, contaminating drinking water, and potentially destroying biodiversity, wildlife, and possibly human health worldwide. Since MPs and NPs are already affecting ecosystems, upsetting habitats, and getting into food chains, their effects on the environment are already clear. Plastics' tenacity and bioaccumulation in the environment make these issues even worse, posing a long-term obstacle to preservation and environmentally sound management initiatives. More worrisome is the mounting evidence that these tiny particles may be responsible for human health issues including toxic buildup and inflammation as well as physical harm to living things. This increases the need to apply mitigation, prevention, and remediation strategies on local, national, and international scales as our understanding of those risks expands. A multifaceted strategy is needed to address this expanding challenge, one that includes cutting back on the manufacturing and utilization of plastic, enhancing waste management procedures, and creating new materials and technologies to reduce plastic pollution. Robust laws, international cooperation, and public education about the risks of plastic pollution are also essential in addressing this problem. In order to stop additional environmental deterioration and protect ecological and human health for future generations, worldwide action is imperative in response to the issue posed by MPs and NPs. In conclusion, it is evident that the danger posed by MPs and NPs is both immediate and complicated, even though the entire scope of their effects is still being determined. To lessen the harm, a quick and persistent action is required at all levels scientific, legislative, industrial, and social. By taking a comprehensive approach to this problem through innovation, legislative change, and international collaboration—we may endeavor to protect the planet's and its people's health for future generations.

References

Allen S, Allen D, Phoenix VR et al (2019) Atmospheric transport and deposition of microplastics in a remote mountain catchment. Nat Geosci 12:339–344. https://doi.org/10.1038/s41561-019-0335-5

Amaral-Zettler LA, Zettler ER, Mincer TJ (2020) Ecology of the plastisphere. Nat Rev Microbiol 18(3): 139–151. https://doi.org/10.1038/s41579-019-0308-0

Andrady AL (2011) Microplastics in the marine environment. Mar Pollut Bull 62(8):1596–1605. https://doi.org/10.1016/j.marpolbul.2011.05.030. ISSN 0025-326X

Andrady AL (2017) The plastic in microplastics: a review. Mar Pollut Bull 119(1):12–22. https://doi.org/10.1016/j.marpolbul.2017.01.082

Arias-Andres M, Klümper U, Rojas-Jimenez K, Grossart HP (2018) Microplastic pollution increases gene exchange in aquatic ecosystems. Environ Pollut (Barking, Essex: 1987) 237:253–261. https://doi.org/10.1016/j.envpol.2018.02.058

Baca J (2022) The story of plastics (and ACC) – American Chemistry Council. American Chemistry Council. https://www.americanchemistry.com/chemistry-in-america/news-trends/blog-post/2022/the-story-of-plastics-and-acc

Baekeland L (1909) The synthesis, constitution, and uses of bakelite. J Ind Eng Chem 1:149–161

Barnes DK, Galgani F, Thompson RC, Barlaz M (2009) Accumulation and fragmentation of plastic debris in global environments. Philos Trans R Soc Lond Ser B Biol Sci 364(1526):1985–1998. https://doi.org/10.1098/rstb.2008.0205. PMID: 19528051; PMCID: PMC2873009
Beans C (2023) Are microplastics spreading infectious disease? Proc Natl Acad Sci USA 120(31):e2311253120. https://doi.org/10.1073/pnas.2311253120. Epub 2023 Jul 26. PMID: 37494399; PMCID: PMC10401022
Bergmann M, Wirzberger V, Krumpen T, Lorenz C, Primpke S, Tekman MB, Gerdts G (2017) High quantities of microplastic in arctic deep-sea sediments from the HAUSGARTEN observatory. Environ Sci Technol 51(19):11000–11010. https://doi.org/10.1021/acs.est.7b03331
Bouwmeester H, Hollman PC, Peters RJ (2015) Potential health impact of environmentally released micro- and nanoplastics in the human food production chain: experiences from nanotoxicology. Environ Sci Technol 49(15):8932–8947. https://doi.org/10.1021/acs.est.5b01090
Browne MA, Dissanayake A, Galloway TS, Lowe DM, Thompson RC (2008) Ingested microscopic plastic translocates to the circulatory system of the mussel, Mytilus edulis (L). Environ Sci Technol 42(13):5026–5031. https://doi.org/10.1021/es800249a
Browne MA, Galloway TS, Thompson RC (2010) Spatial patterns of plastic debris along Estuarine shorelines. Environ Sci Technol 44(9):3404–3409. https://doi.org/10.1021/es903784e
Cole M, Lindeque P, Fileman E, Halsband C, Goodhead R, Moger J, Galloway TS (2013) Microplastic ingestion by zooplankton. Environ Sci Technol 47(12):6646–6655. https://doi.org/10.1021/es400663f
Cole M, Lindeque P, Fileman E, Halsband C, Galloway TS (2015) The impact of polystyrene microplastics on feeding, function and fecundity in the marine copepod Calanus helgolandicus. Environ Sci Technol 49(2):1130–1137. https://doi.org/10.1021/es504525u
Cox KD, Covernton GA, Davies HL, Dower JF, Juanes F, Dudas SE (2019) Human consumption of microplastics. Environ Sci Technol 53(12):7068–7074. https://doi.org/10.1021/acs.est.9b01517
Cózar A, Echevarría F, González-Gordillo JI, Irigoien X, Ubeda B, Hernández-León S, Palma AT, Navarro S, García-de-Lomas J, Ruiz A, Fernández-de-Puelles ML, Duarte CM (2014) Plastic debris in the open ocean. Proc Natl Acad Sci USA 111(28):10239–10244. https://doi.org/10.1073/pnas.1314705111
Dissanayake PD, Kim S, Sarkar B, Oleszczuk P, Sang MK, Haque MN, Ahn JH, Bank MS, Ok YS (2022) Effects of microplastics on the terrestrial environment: a critical review. Environ Res 209:112734. https://doi.org/10.1016/j.envres.2022.112734
Dris R, Gasperi J, Rocher V, Saad M, Renault N, Tassin B (2015) Microplastic contamination in an urban area: a case study in Greater Paris. Environ Chem 12:592–599. https://doi.org/10.1071/EN14167
Du S, Zhu R, Cai Y, Xu N, Yap P, Zhang Y, He Y, Zhang Y (2021) Environmental fate and impacts of microplastics in aquatic ecosystems: a review. RSC Adv 11(26):15762–15784. https://doi.org/10.1039/d1ra00880c
Eerkes-Medrano D, Thompson RC, Aldridge DC (2015) Microplastics in freshwater systems: a review of the emerging threats, identification of knowledge gaps and prioritisation of research needs. Water Res 75:63–82. https://doi.org/10.1016/j.watres.2015.02.012
Geyer R, Jambeck JR, Law KL (2017) Production, use, and fate of all plastics ever made. Sci Adv 3(7):e1700782. https://doi.org/10.1126/sciadv.1700782
Gregory MR (2009) Environmental implications of plastic debris in marine settings—entanglement, ingestion, smothering, hangers-on, hitch-hiking and alien invasions. Philos Trans R Soc Lond Ser B Biol Sci 364(1526):2013–2025. https://doi.org/10.1098/rstb.2008.0265. PMID: 19528053; PMCID: PMC2873013
Guerranti C, Martellini T, Perra G, Scopetani C, Cincinelli A (2019) Microplastics in cosmetics: environmental issues and needs for global bans. Environ Toxicol Pharmacol 68:75–79. https://doi.org/10.1016/j.etap.2019.03.007. ISSN 1382-6689
Hauser R, Calafat AM (2005) Phthalates and human health. Occup Environ Med 62(11):806–818. https://doi.org/10.1136/oem.2004.017590
Huang Y, He T, Yan M, Yang L, Gong H, Wang W, Qing X, Wang J (2021) Atmospheric transport and deposition of microplastics in a subtropical urban environment. J Hazard Mater 416:126168. https://doi.org/10.1016/j.jhazmat.2021.126168. ISSN 0304-3894
Jambeck JR et al (2015) Plastic waste inputs from land into the ocean. Science 347:768–771. https://doi.org/10.1126/science.1260352
Karami A, Golieskardi A, Choo CK, Larat V, Karbalaei S, Salamatinia B (2018) Microplastic and mesoplastic contamination in canned sardines and sprats. Sci Total Environ 612:1380–1386. https://doi.org/10.1016/j.scitotenv.2017.09.005. ISSN 0048-9697
Kirstein IV, Wichels A, Krohne G, Gerdts G (2018) Mature biofilm communities on synthetic polymers in seawater – specific or general? Mar Environ Res 142:147–154. https://doi.org/10.1016/j.marenvres.2018.09.028. ISSN 0141-1136
Koelmans AA, Bakir A, Burton GA, Janssen CR (2016) Microplastic as a vector for chemicals in the aquatic environment: critical review and model-supported reinterpretation of empirical studies. Environ Sci Technol 50(7):3315–3326. https://doi.org/10.1021/acs.est.5b06069
Koelmans AA, Mohamed Nor NH, Hermsen E, Kooi M, Mintenig SM, De France J (2019) Microplastics in freshwaters and drinking water: critical review and assessment of data quality. Water Res 155:410–422. https://doi.org/10.1016/j.watres.2019.02.054
Kühn S, Bravo Rebolledo EL, van Franeker JA (2015) Deleterious effects of litter on marine life. In:

Bergmann M, Gutow L, Klages M (eds) Marine anthropogenic litter. Springer, Cham. https://doi.org/10.1007/978-3-319-16510-3_4

Lebreton LC, Greer SD, Borrero JC (2012) Numerical modelling of floating debris in the world's oceans. Mar Pollut Bull 64(3):653–661. https://doi.org/10.1016/j.marpolbul.2011.10.027. PMID: 22264500

Lu L, Luo T, Zhao Y, Cai C, Zhengwei F, Jin Y (2019) Interaction between microplastics and microorganism as well as gut microbiota: a consideration on environmental animal and human health. Sci Total Environ 667:94–100. https://doi.org/10.1016/j.scitotenv.2019.02.380. ISSN 0048-9697

Lusher AL, McHugh M, Thompson RC (2013) Occurrence of microplastics in the gastrointestinal tract of pelagic and demersal fish from the English Channel. Mar Pollut Bull 67(1–2):94–99. https://doi.org/10.1016/j.marpolbul.2012.11.028

Mattsson K, Ekvall MT, Hansson LA, Linse S, Malmendal A, Cedervall T (2015) Altered behavior, physiology, and metabolism in fish exposed to polystyrene nanoparticles. Environ Sci Technol 49(1):553–561. https://doi.org/10.1021/es5053655

Osman AI, Hosny M, Eltaweil AS et al (2023) Microplastic sources, formation, toxicity and remediation: a review. Environ Chem Lett 21:2129–2169. https://doi.org/10.1007/s10311-023-01593-3

Peeken I, Primpke S, Beyer B, Gütermann J, Katlein C, Krumpen T, Bergmann M, Hehemann L, Gerdts G (2018) Arctic sea ice is an important temporal sink and means of transport for microplastic. Nat Commun 9(1):1505. https://doi.org/10.1038/s41467-018-03825-5

Plastics – American Chemistry Council (2024) American Chemistry Council. https://www.americanchemistry.com/chemistry-in-america/chemistry-in-everyday-products/plastics

Plastics Europe (2022) Plastics – the facts 2020. Plastics Europe. https://plasticseurope.org/knowledge-hub/plastics-the-facts-2020/

Prata JC, da Costa JP, Lopes I, Duarte AC, Rocha-Santos T (2020) Environmental exposure to microplastics: an overview on possible human health effects. Sci Total Environ 702:134455. https://doi.org/10.1016/j.scitotenv.2019.134455

Rochester JR (2013) Bisphenol A and human health: a review of the literature. Reprod Toxicol (Elmsford, NY) 42:132–155. https://doi.org/10.1016/j.reprotox.2013.08.008

Rochman CM, Browne MA, Halpern BS, Hentschel BT, Hoh E, Karapanagioti HK, Rios-Mendoza LM, Takada H, Teh S, Thompson RC (2013a) Policy: classify plastic waste as hazardous. Nature 494(7436):169–171. https://doi.org/10.1038/494169a

Rochman CM, Hoh E, Kurobe T, Teh SJ (2013b) Ingested plastic transfers hazardous chemicals to fish and induces hepatic stress. Sci Rep 3:3263. https://doi.org/10.1038/srep03263

Rochman CM, Tahir A, Williams SL, Baxa DV, Lam R, Miller JT, Teh FC, Werorilangi S, Teh SJ (2015) Anthropogenic debris in seafood: plastic debris and fibers from textiles in fish and bivalves sold for human consumption. Sci Rep 5:14340. https://doi.org/10.1038/srep14340. PMID: 26399762; PMCID: PMC4585829

Smith M, Love DC, Rochman CM, Neff RA (2018) Microplastics in seafood and the implications for human health. Curr Environ Health Rep 5(3):375–386. https://doi.org/10.1007/s40572-018-0206-z. PMID: 30116998; PMCID: PMC6132564

Teuten EL, Saquing JM, Knappe DR, Barlaz MA, Jonsson S, Björn A, Rowland SJ, Thompson RC, Galloway TS, Yamashita R, Ochi D, Watanuki Y, Moore C, Viet PH, Tana TS, Prudente M, Boonyatumanond R, Zakaria MP, Akkhavong K, Ogata Y et al (2009) Transport and release of chemicals from plastics to the environment and to wildlife. Philos Trans R Soc Lond Ser B Biol Sci 364(1526):2027–2045. https://doi.org/10.1098/rstb.2008.0284

Thompson RC, Moore CJ, vom Saal FS, Swan SH (2009) Plastics, the environment and human health: current consensus and future trends. Philos Trans R Soc Lond Ser B Biol Sci 364(1526):2153–2166. https://doi.org/10.1098/rstb.2009.0053

Tsochatzis ED, Gika H, Theodoridis G, Maragou N, Thomaidis N, Corredig M (2024) Microplastics and nanoplastics: exposure and toxicological effects require important analysis considerations. Heliyon 10(11):e32261. https://doi.org/10.1016/j.heliyon.2024.e32261. PMID: 38882323; PMCID: PMC11180319

Vethaak AD, Legler J (2021) Microplastics and human health. Science (New York, NY) 371(6530):672–674. https://doi.org/10.1126/science.abe5041

Woodall LC, Sanchez-Vidal A, Canals M, Paterson GL, Coppock R, Sleight V, Calafat A, Rogers AD, Narayanaswamy BE, Thompson RC (2014) The deep sea is a major sink for microplastic debris. R Soc Open Sci 1(4):140317. https://doi.org/10.1098/rsos.140317

Wright SL, Kelly FJ (2017) Plastic and human health: a micro issue? Environ Sci Technol 51(12):6634–6647. https://doi.org/10.1021/acs.est.7b00423

Wright SL, Thompson RC, Galloway TS (2013) The physical impacts of microplastics on marine organisms: a review. Environ Pollut (Barking, Essex: 1987) 178:483–492. https://doi.org/10.1016/j.envpol.2013.02.031

Wright SL, Ulke J, Font A, Chan KLA, Kelly FJ (2020) Atmospheric microplastic deposition in an urban environment and an evaluation of transport. Environ Int 136:105411. https://doi.org/10.1016/j.envint.2019.105411. ISSN 0160-4120

Zettler ER, Mincer TJ, Amaral-Zettler LA (2013) Life in the "plastisphere": microbial communities on plastic marine debris. Environ Sci Technol 47(13):7137–7146. https://doi.org/10.1021/es401288x

Ziani K, Ioniță-Mîndrican CB, Mititelu M, Neacșu SM, Negrei C, Moroșan E, Drăgănescu D, Preda OT (2023) Microplastics: a real global threat for environment and food safety: a state of the art review. Nutrients 15(3):617. https://doi.org/10.3390/nu15030617. PMID: 36771324; PMCID: PMC9920460

2 Environmental Degradation of Plastics: Understanding the Mechanisms

Thong Nguyen Tan, Palas Samanta, Moharana Choudhury, Kaleeswari Kalimuthu, Subhankar Dhar, and Janitha Kularajasingam

2.1 Introduction

Plastics are widely used across consumer and industrial sectors, including construction, automotive manufacturing, electronics, and agriculture, due to their durability, versatility, and cost-effectiveness (Zhang et al. 2021). It is estimated that approximately 438 million tons of plastic are produced yearly (Tsakona et al. 2021). However, the widespread use of plastic has led to a concerning accumulation of plastic waste. Improper disposal of plastic waste leads to significant environmental pollution, contributing to biodiversity loss, human health risks, food and water contamination, economic strain, and the exacerbation of climate change due to its persistent presence in ecosystems (Lin et al. 2022; Tsakona et al. 2021). Plastic pollution is prevalent in terrestrial and marine environments, accumulating and posing long-term ecological risks (Sutkar et al. 2023). Microplastics (MPs, 5 mm > MPs > 100 nm in size) and nanoplastics (NPs; <100 nm in size), which result from the breakdown of larger plastic debris or macroplastics (MaPs; >5 mm in size), have been identified in almost all ecosystems, including natural ecosystems, from ocean and coast to mountains, cities, and rural areas (Lin et al. 2022; Zhang et al. 2021). MPs and NPs are ingested by marine and terrestrial organisms, entering the food chain, and they also act as carriers for harmful pollutants, exacerbating contamination (Shashoua et al. 2024; Tsakona et al. 2021).

Plastics degrade slowly, often taking decades to centuries, depending on polymer composition and environmental conditions. Their chemical structure enables resistance to natural degradation

T. N. Tan
Research Institute for Sustainable Development, Ho Chi Minh City University of Natural Resources and Environment (HCMUNRE), Tan Binh District, Ho Chi Minh City, Vietnam

P. Samanta
Department of Environmental Science, Sukanta Mahavidyalaya, University of North Bengal Dhupguri, West Bengal, India

M. Choudhury (✉)
Environmental Research and Management Division, Voice of Environment (VoE), Guwahati, Assam, India

K. Kalimuthu
Department of Chemistry, Faculty of Science, Mahidol University, Bangkok, Thailand

S. Dhar
Chemical Engineering, Mingchi University of Technology, Taipei, Taiwan

J. Kularajasingam
School of Management and Marketing; Sustainable Tourism Impact Lab, Taylor's University, Selangor, Malaysia

S. Goswami et al. (eds.), *Micro-Nano Plastics Exposure, Environmental Degradation and Public Health Crisis*, Sustainable Development Goals Series, https://doi.org/10.1007/978-981-96-5047-7_2

processes, causing them to break down primarily through physical, chemical, and mechanical means (Lin et al. 2022; Shashoua et al. 2024). This persistence poses significant risks to biodiversity, ecosystem health, and human well-being, hindering progress toward the United Nations Sustainable Development Goals (SDGs). Despite global recognition of the plastic waste crisis, progress toward effective solutions still needs to be made. In 2020, bioplastics represented just 0.5% of global production, limiting their impact on reducing plastic pollution (EEA 2020).

To address the plastic waste crisis, research is focused on exploring solutions such as biodegradable polymers, advanced catalysts for accelerated degradation, and microbial systems capable of breaking down conventional plastics. Moreover, a deep understanding of the degradation mechanisms of nonbiodegradable plastics is essential for mitigating the growing threat of plastic pollution (Roy and Choudhury 2022). Accordingly, this chapter explores the fundamental mechanisms of plastic degradation, including physical, chemical, and biological processes, and explores the factors influencing these processes. Besides, this chapter reviews recent studies on plastic degradation methods, highlighting progress and challenges in the degradation of plastics.

2.2 Degradation Mechanism of Plastic Fragments

Plastic degradation mechanisms can be categorized into physical processes (e.g., cracking and flaking), chemical processes (e.g., oxidation and polymer chain scission), and biological processes (Chamas et al. 2020). Plastics break down into fragments, such as MaPs, MPs, and NPs (Sutkar et al. 2023). These PFs are then further degraded through biotic and abiotic processes, ultimately converting into components like CO_2, H_2O, and inorganic compounds (Li et al. 2023; Sutkar et al. 2023).

2.2.1 Biotic Degradation Mechanisms of Plastics

Biotic degradation involves microorganisms that modify and consume polymers or polymeric monomers, resulting in the production of low molecular weight molecules, such as acids, aldehydes, terpenes, and water H_2O, as well as gases like carbon dioxide (CO_2), methane (CH_4), and nitrogen (N_2) (Lim and Thian 2022; Nabi et al. 2023). The biodegradation of plastics occurs in four primary stages: biodeterioration, depolymerization, assimilation, and mineralization (Nabi et al. 2023; Wu et al. 2023), as presented in Fig. 2.1. The process may cease at any stage during plastic biodegradation; however, the verification of plastic biodegradation requires evidence of mineralization.

The initial stage of biotic degradation, known as bio-deterioration, involves the breakdown of large plastics into smaller fragments, such as MaPs, MPs, and NPs, by microorganisms and/or abiotic factors (Ali et al. 2021; Cai et al. 2023). This surface-level degradation alters the plastic's mechanical, physical, and chemical properties and is visible to the naked eye. This stage is also characterized by deformations, including cracks, roughness, scratches, holes, or weight reduction (Cai et al. 2023).

In the second stage, known as bio-fragmentation/depolymerization, microorganisms excrete catalytic agents, such as enzymes and free radicals, which reduce the molecular weight of polymers (Anjana et al. 2020). This process cleaves the polymers into smaller molecules, including oligomers, dimers, or monomers, without altering the chemical composition or structure of the monomer units (Cai et al. 2023). Microbial extracellular and intracellular depolymerase enzymes are crucial in breaking down polymers into smaller, simpler intermediates. These intermediates dissolve quickly in water and are subsequently absorbed by microbial cells. The effectiveness of depolymerization can be assessed through enzymatic assays (Amobonye et al. 2021).

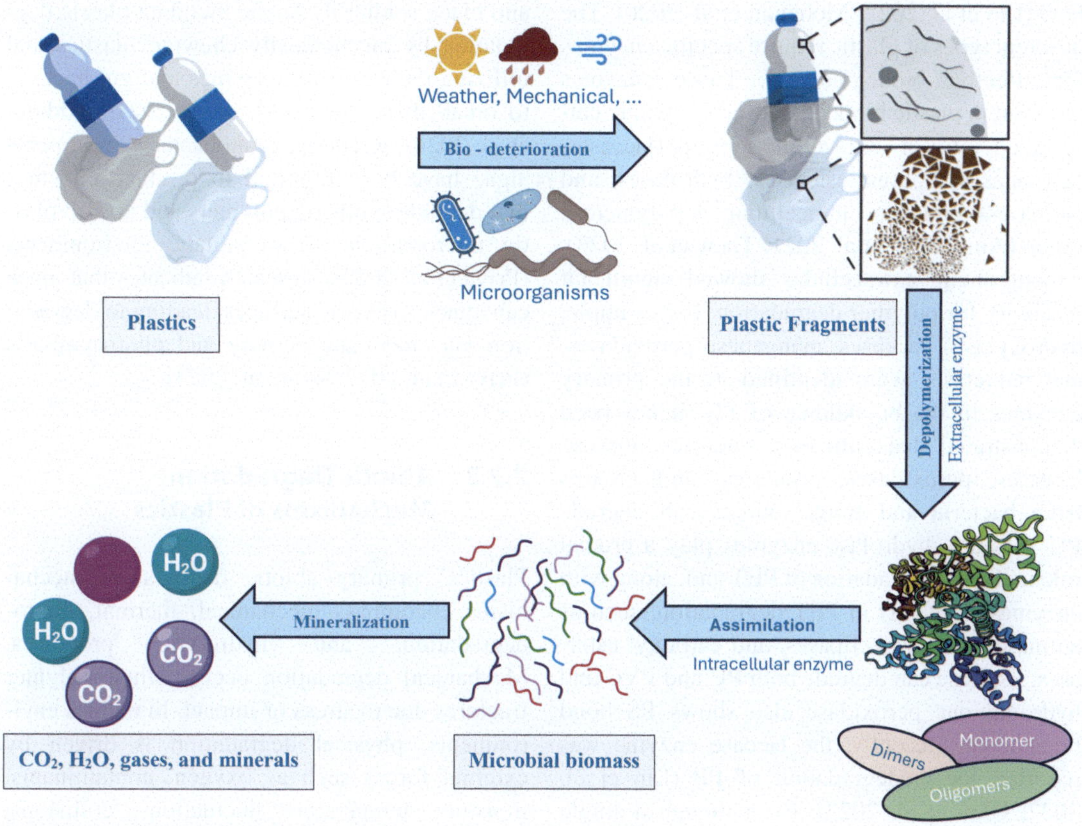

Fig. 2.1 Primary biodegradation pathway of plastics based on microorganisms

During the assimilation stage, small molecules generated from depolymerization/biofragmentation are incorporated into microbial metabolism within the cytoplasm (Ali et al. 2021). These molecules are utilized to produce energy, biomass, and primary or secondary metabolites, with microorganisms using polymers as carbon and nitrogen sources (Cai et al. 2023; Wu et al. 2023). Assimilation occurs through metabolic processes such as aerobic respiration, anaerobic respiration, or fermentation, producing ATP, which supports microbial growth (Lin et al. 2022).

The final stage of plastic biodegradation, mineralization, occurs alongside assimilation and involves converting organic substances into safer compounds like CO_2, H_2O, and minerals (Nabi et al. 2023). Microorganisms carry out this process, some breaking down complex polymers into CO_2 while others convert them into simpler monomers used for energy. The fate of molecules depends on oxygen availability: aerobic degradation results in the formation of CO_2 and H_2O, while anaerobic metabolism produces CO_2, H_2O, CH_4, and H_2S (Cai et al. 2023). Mineralization efficiency, a key biodegradation indicator, is assessed through respirometry methods measuring gas evolution or oxygen consumption (Cai et al. 2023). The process ends when microorganisms consume all biodegradable compounds, completing the breakdown of plastic pollution in air, land, and water ecosystems (Hebner and Maurer-Jones 2020; Nabi et al. 2023).

Microbial enzymes play a leading role in the enzymatic pathways involved in the biodegradation of plastic. These enzymes are generated by various microorganisms such as actinomycetal, algae, bacteria, and fungi. It stimulates the chemical bonds on the surface of plastics for breaking into smaller chains through oxidation and hydrolysis, which are entirely utilized by microbial cells as the carbon source to generate CO_2 and

H_2O (Lin et al. 2022; Mohanan et al. 2020). The different types of plastic require specific enzymes for effective biodegradation. Those enzymes involved in plastic biodegradation are mainly categorized into an extracellular group (laccases, peroxidases, lipases, esterases, hydrolases, and cutinases) and the intracellular depolymerase group (Amobonye et al. 2021; Thew et al. 2023). Among them, extracellular showed significant reactivity for polymer degradation. For example, hydroxylases, laccases, manganese peroxidases, and reductases were identified as the primary enzymes in the breakdown of PE. It has been demonstrated that cutinases, esterases, lipases, laccases, peroxidases, proteases, and ureases from bacteria and fungi sources can degrade PU. Besides, hydrolase enzymes play a crucial role in the biodegradation of PET and, along with enzymes, take part in PET degradation, such as cutinases, esterases, lipases, and carboxyl esterases. Esterase can degrade both PU and PVC and hydroquinone peroxidase also shows PS bond breakdown. Recently, the laccase enzyme was reported for biodegradation of PP (Lin et al. 2022; Okal et al. 2023). Even though a single enzyme was used in targeting plastic depolymerization, frequently combined enzymes showed an effective way. In addition, new enzymes that can degrade plastic are still being found and characterized in this field of study. Plastics are naturally exposed to many bacteria, contributing to their breakdown (Thew et al. 2023). While mixed microorganism strains with the ability to break down plastics have been discovered, almost all studies have concentrated on individual strains (Li et al. 2023; Zeenat et al. 2021). Table 2.1 shows the biodegradation of polymers by microorganisms for several plastics.

In addition to biotic plastic degradation by organisms, factors such as algae, worms, and insects contribute to plastic degradation through bioaccumulation, bioaugmentation, and biodegradation (Siddiqui et al. 2024). Worms, specifically mealworms and superworms, have shown the capacity to consume and degrade plastics such as PP, PE, PS, PVC, HDPE, and LDPE (Pham et al. 2023; Wang et al. 2024). Their gut microbiota plays a significant role in depolymerizing plastics into smaller molecules. Waxworms and black soldier fly larvae facilitate plastic degradation by mechanically chewing plastics and utilizing their salivary enzymes and gut bacteria to break them down (Ali et al. 2021; Siddiqui et al. 2024). Recently, cyanobacteria and microalgae have been reported to produce enzymes that degrade synthetic polymers and absorb plastic microparticles from aquatic environments (Priya et al. 2022). Research indicates that algae can attach to plastic surfaces, leading to degradation via enzymatic activity and photosynthesis (Priya et al. 2022; Wu et al. 2023).

2.2.2 Abiotic Degradation Mechanisms of Plastics

Plastics' primary abiotic degradation mechanisms encompass mechanical, thermal, photodegradation, and hydrolytic processes. Mechanical degradation occurs when polymer fractures due to stress or impact. In natural environments, physical degradation is driven by external forces such as oxygen, contaminants, moisture, temperature fluctuations, collisions, and abrasions, including interactions with sand and waves in marine settings (Shashoua et al. 2024; Xi et al. 2022). Mechanical degradation offers simplicity, low cost, and shorter duration than other degradation methods (Xi et al. 2022). Mechanical degradation is often induced by mechanical stresses, such as compression, tension, and shear, reducing tensile strength, flexural strength, hardness, and impact resistance (Chamas et al. 2020). These factors are frequently used as supplementary indicators for abiotic and biotic degradation. The deterioration of fibers into MPs results from clothing wear and household cleaning tools subjected to abrasion and force (Zhang et al. 2021). Figure 2.2 presents the abiotic degradation pathways of plastic.

Thermal and hydrolytic degradation processes also contribute to the degradation of plastics. Thermal degradation occurs when polymers are exposed to elevated temperatures, resulting in changes to their molecular structure. This process involves random scission, depolymerization, and side-group elimination, which alter molecular weight and properties such as color, mechanical

Table 2.1 Biodegradation pathways of plastics

Types of plastics	Degradation pathway	Primary factors of the degradation process	Degradation efficiency (%)	Time	References
PP	Organisms	*Bacillus* sp. and *Rhodococcus*	6.4% by *Rhodococcus* sp. and 4.0% by *Bacillus* sp.	40 days	Auta et al. (2018)
PE	Organisms	*Bacillus* sp. and *Paenibacillus* sp.	Particle dry weight decreased by 14.7% and mean diameter by 22.8%	60 days	Park and Kim (2019)
PET	Organisms	*Gordonia* sp. CN2K	40.43% of the PET was degraded	45 days	Chandramouli Swamy et al. (2024)
PEMPs (polyethylene MPs) and NMPs (nylon 6,6 MPs)	Organisms	*Achromobacter xylosoxidans* and mixed culture species	*Achromobacter xylosoxidans* wt % by 26.7% (PEMPs) and 21.3% (NMPs). The mixed culture reduced wt % by 19.3% (PEMPs) and 20% (NMPs)	40 days	Tiwari et al. (2024)
PE, PS, PVC, and PP	Worms	Superworms (larvae of *Zophobas atratus)*	44.16 mg PE/100 larvae for PE-bran-fed and 39.61 mg PS/100 larvae for PS-bran-fed.	30 days	Wang et al. (2024)
PU, PE, and tire rubber	Insects	18 fungal strains	*Fusarium, Penicillium, Botryotinia cinerea EN41*, and *Trichoderma* showed high potential to degrade PU, rubber, and PE, consuming over 90% of O_2 in less than 14 days.	16 weeks	Ibrahim et al. (2024)
PS	Algae	*Chlorella vulgaris*	The release rate of dissolved organic matter from PS was elevated from 0.120 mg/L/day to 0.577 mg/L/day. Algal extracellular polymeric substances accelerated the aging process of PS.	20 days	Li et al. (2024)

strength, and impact resistance (Dirauf et al. 2022). Thermal degradation occurs via radical, ionic, or mixed pathways and can be monitored through mass changes, molecular weight analysis, and reaction enthalpy measurements (Dirauf et al. 2022; Mahapatra et al. 2024). At 450 °C, both PE and PP accumulated in sludge degraded into MPs, while at 500 °C, these MPs were eliminated (Ni et al. 2020). Furthermore, hydrolytic degradation involves the reaction of water molecules with polymer bonds, forming new chain ends and causing the plastic to become brittle and disintegrate over time. The rate of hydrolytic degradation is influenced by factors such as the plastic type, temperature, pH, and the presence of catalysts (Liu et al. 2022).

Photodegradation happens as plastic is exposed to artificial or natural light, which leads to polymer bond breakdown. This process leads to cracking, color changes, and a loss of physical properties. Additives, pigments, metallic catalysts, and residues in polymers can absorb solar radiation more effectively than the polymer, exacerbating the degradation process (Du et al. 2021). Chemical degradation of plastics is often driven by environmental interactions with oxygen, water, heat, ultraviolet radiation, and chemical substances (Lin et al. 2020; Nabi et al. 2020). Among these, oxygen (oxidation) and water

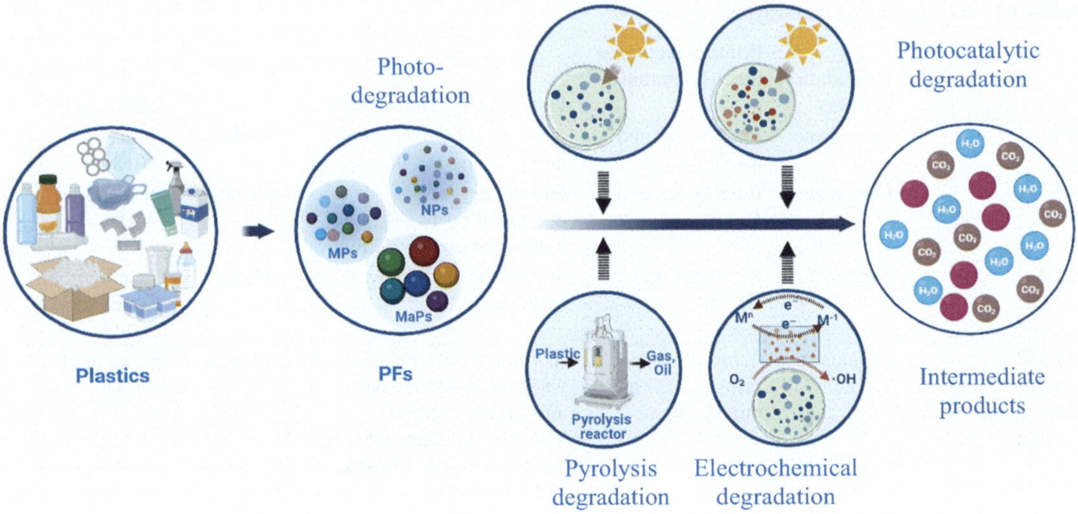

Fig. 2.2 Primary abiotic degradation pathways of plastic

(hydrolysis) are the most significant agents. Advanced oxidation processes (AOPs), which generate reactive oxygen species (ROS), are particularly effective in plastic degradation under aqueous conditions (Kim et al. 2022). Due to their high reactivity, ROS attack organic materials and destabilize chemical bonds. AOPs involve various reactions, including photocatalytic oxidation, electrochemical oxidation, and photodegradation (Du et al. 2021). Photodegradation, especially under UV radiation, is a critical step in plastic breakdown, causing the formation of environmentally persistent free radicals (EPFRs) and structural changes such as fragmentation and fissures (Hebner and Maurer-Jones 2020). Recent advancements in the abiotic degradation of plastic under different approaches can be found in Table 2.2.

2.3 Factors Controlling the Degradation Pathway of Plastic

Abiotic degradation of plastics involves nonbiological factors such as chemical reactions, UV radiation, temperature, and mechanical stress, leading to the breakdown of polymer structures (Chamas et al. 2020; Kim et al. 2022; Ni et al. 2020). The chemical composition plays a crucial role, as polymers with long carbon chains are generally more resistant to degradation, whereas introducing heteroatoms or unsaturation increases susceptibility to processes like thermal oxidation. Besides, factors such as molecular weight, hydrophobicity, and the presence of additives like pro-oxidants or fillers also significantly influence the rate of degradation (Mahapatra et al. 2024; Sutkar et al. 2023). External conditions like exposure to oxygen, light, temperature, and mechanical stress further accelerate the process by inducing chemical changes in the polymer matrix, leading to the formation of lower molecular weight compounds and structural weakening (Chandramouli Swamy et al. 2024; Shashoua et al. 2024).

Moreover, plastic's biotic degradation is primarily driven by microorganisms such as bacteria and fungi that metabolize polymers, especially those modified to include functional groups like carbonyls (Amobonye et al. 2021). Hydrophilic polymers or those blended with natural substances like starch are more prone to microbial attack, as they provide a conducive environment for microbial growth (Zeenat et al. 2021). Microorganisms initiate the degradation process by breaking down polymer chains at susceptible sites, leading to a decrease in molecular weight and physical properties over time (Li et al. 2023;

Table 2.2 Physio-chemical degradation of plastics

Degradation pathway	Types of plastics	Mechanism of the degradation process	Degradation efficiency (%)	Degradation products	References
Physical degradation	Single-use plastic products (SUPs) include PS, PP, PLA, PE, and PET	SUPs exposed to natural weather and marine conditions underwater for 3 years	Faster degradation of lighter-colored SUPs. Translucent PP food containers degraded to 31.8% per year; white PE bottles at 19.3% per year	Byproducts of plastic degradation from SUPs after exposure to tests for 3 years.	Shashoua et al. (2024)
Photocatalytic	PS, PE	TiO_2 nanoparticle films under UV light	99.99% of PS after 24 h	CO_2	Nabi et al. (2020)
Electrocatalytic	PVC	TiO_2/graphite cathode	56 wt. % removal and 75% dechlorination efficiency for 6 h	CO_2 and H_2O	Miao et al. (2020)
Pyrolysis	HDPE	Heating plastic to 600 °C in a nitrogen-purged reactor	Volatile content is 99.33% at 550 °C and 5 mm particle size (HDPE) for 30 min	The wax consists of alkanes, alkenes, and dienes, with carbon chains from C7 to C44, including 1-tridecene (9.39%) and 5-tetradecene (8.87%); oil; gases	Rodriguez-Luna et al. (2021)
Wet air oxidation (WAO)	PE, PS, and PET	Heating an MPs–water mixture under high pressure and injecting oxygen	43.6% of PE, 68.2% of PS, and 0.9% of PET were mineralized to CO_2 after 120 min WAO treatment	Acetic acid, short-chain ketones, and aldehydes are the primary intermediates CO_2, volatile acids, and ketones are in the gas emissions	Wang et al. (2023)

Tiwari et al. 2024). Furthermore, additives that promote microbial colonization or chemical changes can enhance biodegradability, whereas synthetic polymers with high molecular weight and hydrophobic character resist microbial action (Lim and Thian 2022; Thew et al. 2023).

2.4 Current Challenges in Plastic Degradation

The degradation of plastics by microorganisms and enzymes poses several challenges due to the inherent properties of plastic. Firstly, the existence of recalcitrant polymers causes incomplete degradation by microorganisms even when pretreated by physical and chemical factors (Mohanan et al. 2020; Wu et al. 2023). Besides, plastics are macromolecules and long-chain molecules, making enzymes challenging to break down, while biodegradable plastics contain functional groups with sensitive ester linkages. Secondly, releasing MPs leads to an awareness of the toxicity of the biodegradation process. Both phthalates (PAEs) and bisphenol A (BPA) are widely used in the plastic product industry, and they cause serious problems that inhibit microorganisms and enzymes (Amobonye et al. 2021; Park and Kim 2019). Furthermore, the biodegradation process is affected by abiotic and biological factors and other variables, such as roughness and surface characteristics, which facilitate the adherence of microorganisms to plastics. Finally, the cost and degradation times by biotic pathways differ from the large amount of plastic released daily.

Several factors, such as waves, light, wind, and sand friction, cause the degradation of plastics in the environment into smaller particle shapes (Zeenat et al. 2021; Zhang et al. 2021). UV combined with sand particles accelerates plastic degradation and fragmentation, while mechanical abrasion alone increases the reaction contact area (Hebner and Maurer-Jones 2020; Lin et al. 2020). Thus, mechanical forces significantly promote plastic degradation. Increased temperature may enhance the process of plastic breakdown (Ammala et al. 2011). The thermal degradation mechanism for plastics is similar to photodegradation, using heat instead of heat and light. The degradation of plastics via the photodegradation method is greatly influenced by temperature, both directly and indirectly, alongside catalysts based on the mechanisms of activation energy, quantum yield, and radical diffusion (Lin et al. 2022). High temperatures promote natural plastic degradation but do not differ from photocatalyst processes. While the existence of a catalyst in the degradation of the plastic process leads to the activation energy of the reaction, heating does not contribute to reducing the activation energy and promoting the degradation rate. As a result, controlling temperature should be considered when evaluating different degradation mechanisms.

2.5 Future Research Directions

Efforts to tackle plastic degradation in the environment should concentrate on enhancing existing methods and creating novel technologies with vital future directions. Firstly, research should prioritize identifying microorganisms and enzymes to improve their efficiency and specificity in breaking down plastics. By utilizing genetic engineering and synthetic biology, it is possible to develop microbial communities capable of degrading a broader spectrum of plastics in various environments. Additionally, a deeper understanding of physical and electrochemical processes such as photodegradation, thermal degradation, and AOPs leads to the development of more efficient and environmentally friendly techniques. Developing catalysts and materials that improve these processes while reducing the production of harmful by-products is crucial. Notably, integrating biotic and abiotic methods should enhance and clarify the effectiveness of degradation processes. For example, applying abiotic methods to plastics can increase their susceptibility to microbial attack, significantly accelerating the degradation process.

Secondly, developing plastics naturally degrade while retaining promising functionality, potentially involving novel polymers or engineered enzymes. Furthermore, it is crucial to address MPs and NPs formation during the degradation of larger plastics by improving their durability to minimize fragmentation or creating effective methods for the soft recovery of PFs.

Finally, enhancing plastic production, use, and disposal regulations is crucial for mitigating plastic pollution. Implementing policies that encourage using biodegradable plastics, enhance recycling efforts, and improve waste management practices is essential. Simultaneously, increasing public consciousness regarding the environmental consequences of plastic and promoting sustainable consumption can stimulate the desire for environmentally friendly products.

2.6 Conclusions

Plastic has primarily affected modern society because of its exceptional durability, flexibility, and cost-effectiveness. The severe problem of plastic degradation in the environment requires several efficient approaches. Plastic products are generally divided into two categories: biodegradable plastic and nonbiodegradable plastic. Specific methods are necessary for the degradation process in each category. Nonbiodegradable plastics pose a significant environmental threat due to their widespread use and resistance to decomposition. Plastic degradation mechanisms can be classified into two main categories: biotic pathways, which encompass biological processes, and abiotic pathways, which encompass physical and chemical processes. While biotic degradation is a process based on the action of microbial enzymes and microorganisms, abiotic degradation is the process where plastics are broken down by physical and chemical processes related to UV radiation, mechanical abrasion, and chemical reactions. However, there are still ongoing challenges and limitations. There is a pressing requirement for improved and enduring solutions, as evidenced by the incomplete degradation of plastics, the formation of MPs, and the release of EPFRs during degradation processes. Future research should focus on optimizing degradation pathways, developing new strategies, and minimizing the environmental impact of degradation by-products. Addressing the degradation of plastics through an integrated approach that combines biotic and abiotic mechanisms is crucial for achieving sustainable development goals.

References

Ali SS, Elsamahy T, Al-Tohamy R, Zhu D, Mahmoud YA, Koutra E, Metwally MA, Kornaros M, Sun J (2021) Plastic wastes biodegradation: mechanisms, challenges and future prospects. Sci Total Environ 780:146590. https://doi.org/10.1016/j.scitotenv.2021.146590

Ammala A, Bateman S, Dean K, Petinakis E, Sangwan P, Wong S, Yuan Q, Yu L, Patrick C, Leong KH (2011) An overview of degradable and biodegradable polyolefins. Prog Polym Sci 36(8):1015–1049. https://doi.org/10.1016/j.progpolymsci.2010.12.002

Amobonye A, Bhagwat P, Singh S, Pillai S (2021) Plastic biodegradation: frontline microbes and their enzymes. Sci Total Environ 759:143536

Anjana K, Hinduja M, Sujitha K, Dharani G (2020) Review on plastic wastes in marine environment–biodegradation and biotechnological solutions. Mar Pollut Bull 150:110733

Auta HS, Emenike CU, Jayanthi B, Fauziah SH (2018) Growth kinetics and biodeterioration of polypropylene microplastics by Bacillus sp. and Rhodococcus sp. isolated from mangrove sediment. Mar Pollut Bull 127:15–21. https://doi.org/10.1016/j.marpolbul.2017.11.036

Cai Z, Li M, Zhu Z, Wang X, Huang Y, Li T, Gong H, Yan M (2023) Biological degradation of plastics and microplastics: a recent perspective on associated mechanisms and influencing factors. Microorganisms 11(7):1661

Chamas A, Moon H, Zheng J, Qiu Y, Tabassum T, Jang JH, Abu-Omar M, Scott SL, Suh S (2020) Degradation rates of plastics in the environment. ACS Sustain Chem Eng 8(9):3494–3511

Chandramouli Swamy TM, Nagarathna SV, Reddy PV, Nayak AS (2024) Efficient biodegradation of Polyethylene terephthalate (PET) plastic by Gordonia sp. CN2K isolated from plastic contaminated environment. Ecotoxicol Environ Saf 281:116635. https://doi.org/10.1016/j.ecoenv.2024.116635

Dirauf M, Muljajew I, Weber C, Schubert US (2022) Recent advances in degradable synthetic polymers for biomedical applications – beyond polyesters. Prog Polym Sci 129:101547. https://doi.org/10.1016/j.progpolymsci.2022.101547

Du H, Xie Y, Wang J (2021) Microplastic degradation methods and corresponding degradation mechanism: research status and future perspectives. J Hazard Mater 418:126377. https://doi.org/10.1016/j.jhazmat.2021.126377

EEA (2020) Biodegradable and compostable plastics—challenges and opportunities. European Environment Agency, Copenhagen

Hebner TS, Maurer-Jones MA (2020) Characterizing microplastic size and morphology of photodegraded polymers placed in simulated moving water conditions. Environ Sci Process Impacts 22(2):398–407. https://doi.org/10.1039/c9em00475k

Ibrahim SS, Ionescu D, Grossart H-P (2024) Tapping into fungal potential: biodegradation of plastic and rubber by potent Fungi. Sci Total Environ 934:173188. https://doi.org/10.1016/j.scitotenv.2024.173188

Kim S, Sin A, Nam H, Park Y, Lee H, Han C (2022) Advanced oxidation processes for microplastics degradation: a recent trend. Chem Eng J Adv 9:100213

Li S, Yang Y, Yang S, Zheng H, Zheng Y, Jun M, Nagarajan D, Varjani S, Chang J-S (2023) Recent advances in biodegradation of emerging contaminants-microplastics (MPs): feasibility, mechanism, and future prospects. Chemosphere 331:138776

Li F, Bai X, Ji Y, Kang M (2024) Understanding microplastic aging driven by photosensitization of algal extracellular polymeric substances. J Hazard Mater 469:133949

Lim BKH, Thian ES (2022) Biodegradation of polymers in managing plastic waste – a review. Sci Total Environ 813:151880. https://doi.org/10.1016/j.scitotenv.2021.151880

Lin J, Yan D, Fu J, Chen Y, Ou H (2020) Ultraviolet-C and vacuum ultraviolet inducing surface degradation of microplastics. Water Res 186:116360. https://doi.org/10.1016/j.watres.2020.116360

Lin Z, Jin T, Zou T, Xu L, Xi B, Xu D, He J, Xiong L, Tang C, Peng J, Zhou Y, Fei J (2022) Current progress on plastic/microplastic degradation: fact influences and mechanism. Environ Pollut 304:119159. https://doi.org/10.1016/j.envpol.2022.119159

Liu G, Wang J, Wang M, Ying R, Li X, Hu Z, Zhang Y (2022) Disposable plastic materials release microplastics and harmful substances in hot water. Sci

Total Environ 818:151685. https://doi.org/10.1016/j.scitotenv.2021.151685
Mahapatra PM, Aech S, Mandal PK, Panda AK (2024) Pyrolysis of multilayered plastic packets to hydrocarbon fuel feedstock: thermal degradation behaviour, kinetics and thermodynamic analysis, and semi-batch studies. J Energy Inst 115:101690. https://doi.org/10.1016/j.joei.2024.101690
Miao F, Liu Y, Gao M, Yu X, Xiao P, Wang M, Wang S, Wang X (2020) Degradation of polyvinyl chloride microplastics via an electro-Fenton-like system with a TiO(2)/graphite cathode. J Hazard Mater 399:123023. https://doi.org/10.1016/j.jhazmat.2020.123023
Mohanan N, Montazer Z, Sharma PK, Levin DB (2020) Microbial and enzymatic degradation of synthetic plastics. Front Microbiol 11:580709. https://doi.org/10.3389/fmicb.2020.580709
Nabi I, Bacha AU, Li K, Cheng H, Wang T, Liu Y, Ajmal S, Yang Y, Feng Y, Zhang L (2020) Complete photocatalytic mineralization of microplastic on TiO(2) nanoparticle film. Iscience 23(7):101326. https://doi.org/10.1016/j.isci.2020.101326
Nabi I, Zaheer M, Jin W, Yang L (2023) Biodegradation of macro-and micro-plastics in environment: a review on mechanism, toxicity, and future perspectives. Sci Total Environ 858:160108
Ni B-J, Zhu Z-R, Li W-H, Yan X, Wei W, Xu Q, Xia Z, Dai X, Sun J (2020) Microplastics mitigation in sewage sludge through pyrolysis: the role of pyrolysis temperature. Environ Sci Technol Lett 7(12):961–967
Okal EJ, Heng G, Magige EA, Khan S, Wu S, Ge Z, Zhang T, Mortimer PE, Xu J (2023) Insights into the mechanisms involved in the fungal degradation of plastics. Ecotoxicol Environ Saf 262:115202
Park SY, Kim CG (2019) Biodegradation of micro-polyethylene particles by bacterial colonization of a mixed microbial consortium isolated from a landfill site. Chemosphere 222:527–533. https://doi.org/10.1016/j.chemosphere.2019.01.159
Pham TQ, Longing S, Siebecker MG (2023) Consumption and degradation of different consumer plastics by mealworms (Tenebrio molitor): effects of plastic type, time, and mealworm origin. J Clean Prod 403:136842. https://doi.org/10.1016/j.jclepro.2023.136842
Priya AK, Jalil AA, Dutta K, Rajendran S, Vasseghian Y, Karimi-Maleh H, Soto-Moscoso M (2022) Algal degradation of microplastic from the environment: mechanism, challenges, and future prospects. Algal Res 67:102848. https://doi.org/10.1016/j.algal.2022.102848
Rodriguez-Luna L, Bustos-Martinez D, Valenzuela E (2021) Two-step pyrolysis for waste HDPE valorization. Process Saf Environ Prot 149:526–536
Roy P, Choudhury M (2022) Bioeconomy and green plastic production. In: Baskar C, Ramakrishna S, Daniela La Rosa A (eds) Encyclopedia of green materials. Springer, Singapore. https://doi.org/10.1007/978-981-16-4921-9_189-1
Shashoua Y, Peydaei A, Mortensen MN, Kanstrup AB, Gregory DJ (2024) Physio-chemical degradation of single-use plastics in natural weather and marine environments. Environ Pollut 357:124414. https://doi.org/10.1016/j.envpol.2024.124414
Siddiqui SA, Abdul Manap AS, Kolobe SD, Monnye M, Yudhistira B, Fernando I (2024) Insects for plastic biodegradation – a review. Process Saf Environ Prot 186:833–849. https://doi.org/10.1016/j.psep.2024.04.021
Sutkar PR, Gadewar RD, Dhulap VP (2023) Recent trends in degradation of microplastics in the environment: a state-of-the-art review. J Hazard Mater Adv 11:100343. https://doi.org/10.1016/j.hazadv.2023.100343
Thew CXE, Lee ZS, Srinophakun P, Ooi CW (2023) Recent advances and challenges in sustainable management of plastic waste using biodegradation approach. Bioresour Technol 374:128772. https://doi.org/10.1016/j.biortech.2023.128772
Tiwari N, Santhiya D, Sharma JG (2024) Significance of landfill microbial communities in biodegradation of polyethylene and nylon 6,6 microplastics. J Hazard Mater 462:132786. https://doi.org/10.1016/j.jhazmat.2023.132786
Tsakona M, Baker E, Rucevska I, Maes T, Rosendahl Appelquist L, Macmillan-Lawler M, et al. (2021) Drowning in plastics – marine litter and plastic waste vital graphics. Kenya: United Nations Environments Programme. DEP/2386/NA. ISBN:978-92-807-3888-9
Wang L, Li H, Cai J (2023) Degradation of microplastics during wet air oxidation treatment. Chem Eng J 474:145653. https://doi.org/10.1016/j.cej.2023.145653
Wang S, Yu H, Li W, Song E, Zhao Z, Xu J, Gao S, Wang D, Xie Z (2024) Biodegradation of four polyolefin plastics in superworms (Larvae of Zophobas atratus) and effects on the gut microbiome. J Hazard Mater 477:135381. https://doi.org/10.1016/j.jhazmat.2024.135381
Wu Z, Shi W, Valencak TG, Zhang Y, Liu G, Ren D (2023) Biodegradation of conventional plastics: candidate organisms and potential mechanisms. Sci Total Environ 885:163908. https://doi.org/10.1016/j.scitotenv.2023.163908
Xi B, Wang B, Chen M, Lee X, Zhang X, Wang S, Yu Z, Wu P (2022) Environmental behaviors and degradation methods of microplastics in different environmental media. Chemosphere 299:134354. https://doi.org/10.1016/j.chemosphere.2022.134354
Zeenat, Elahi A, Bukhari DA, Shamim S, Rehman A (2021) Plastics degradation by microbes: a sustainable approach. J King Saud Univ – Sci 33(6):101538. https://doi.org/10.1016/j.jksus.2021.101538
Zhang K, Hamidian AH, Tubic A, Zhang Y, Fang JKH, Wu C, Lam PKS (2021) Understanding plastic degradation and microplastic formation in the environment: a review. Environ Pollut 274:116554. https://doi.org/10.1016/j.envpol.2021.116554

3 Micro and Nano Plastics: Identification and Quantification Methods

Amit Samanta, Chayanika Goswami, Jashabir Chakraborty, Saumendu Deb Roy, Dibyendu Shil, Manab Deb Adhikari, and Ahed J. Alkhatib

3.1 Introduction

Millions of tonnes of plastic have contaminated the environment over time. Plastics are produced frequently for single use and come in a wide variety. The primary materials used in the plastic products industry are high-density polyethylene (hdpe), polypropylene (pp), polystyrene, polyvinyl chloride (pvc), nylon, and PET (polyethylene terephthalate). Ocean plastic products—especially those in the form of packaging, fishing son-tags, and even vehicle's tires—are often subject to the sun's degrading rays, they also transverse by wave action, within marine life, sand rubs, and underwater as well. Plastics are very durable. As a result of the breakdown of these fragments, the quantity of "secondary micro plastics" and "secondary nano plastics" is continuously rising (Prüst et al. 2020). The techniques of sampling and analysis are often employed to investigate the dimensions of micro/nano plastics. Micro plastics mostly range from the common size of 1–5 mm, whereas nano plastics are only approximately 100 nm in length (Murray and Örmeci 2020). The composition of microplastics (MPs) and nano plastics (NPs) changes with their shapes, processes of weathering or wearing away, and periods of stay in the environment (Murray and Örmeci 2020). It has been estimated that more than half, or 52.7%, of both sewage and urban atmospheric deposits predominantly consist offibers and fragments (Annenkov et al. 2021). This review aims to evaluate the various analytical techniques that have been employed including their detection, characterization, and identification of microplastics and nano plastics of various environmental sample matrices. When it comes to the various analytical techniques that may be employed in chemical characterization, apart from FTIR, thermal analysis coupled with

A. Samanta (✉)
Department of Pharmaceutical Analysis, Mata Gujri College of Pharmacy, Mata Gujri University, Kishanganj, Bihar, India

C. Goswami · J. Chakraborty
Department of Pharmacology, Mata Gujri College of Pharmacy, Mata Gujri University, Kishanganj, Bihar, India

S. D. Roy · D. Shil
Department of Pharmacognosy, Mata Gujri College of Pharmacy, Mata Gujri University, Kishanganj, Bihar, India

M. D. Adhikari
Department of Biotechnology, University of North Bengal, Darjeeling, West Bengal, India

A. J. Alkhatib
Department of Forensic Science and Toxicology, Faculty of Medicine, Jordan University of Science and Technology, Ar-Ramtha, Jordan

S. Goswami et al. (eds.), *Micro-Nano Plastics Exposure, Environmental Degradation and Public Health Crisis*, Sustainable Development Goals Series, https://doi.org/10.1007/978-981-96-5047-7_3

Raman and GC-MS spectroscopy techniques, and some other microscopy techniques can also be adapted for studying and defining the various physical and chemical properties of micro and nano plastics. These include transmission electron microscopy, scanning electron microscopy, atomic force microscopy, polarizing, and fluorescence microscopy. In particular, we are introducing some new methods to tackle the problems associated with the identification of micro and nano plastics. MALDI-TOF MS and Py-GC/MS are among those methods.

3.2 Sampling Techniques

A smaller sample volume can be used because smaller particles (1–10 m) have been found to be more common than larger ones in drinking water samples. Given that it is unlikely that plastic levels will change quickly, the current measures should be adequate for water treatment facilities that draw their water from lakes or the ground. Depending on the seasonality and variability of the water quality for those who receive river water, in some cases, especially when surprises arise, it may be deemed necessary to obtain composite samples. Grab samples are also known as instantaneous samples, meaning that they capture the state of any particular matrix (for example, water or sediment) only once and usually from the top layer. Composite Samples: A sample that is created through the amalgamation of a minimum of two samples that are collected at different times or places for the central location of the research (Prata et al. 2019).

Sampling of micro plastic uses different techniques like water sampling, sediment sampling, and biota sampling. The sampling of microplastics within water bodies may encompass their vertical dispersal in the water column due to various inherent characteristics such as density, configuration, dimensions, chemical adsorption, biological fouling characteristics, and external environmental factors such as immature water density, water pressure, or wind and tidal waves. Hence, the collection and quality of the retrieved MNPs are greatly determined by the altitude and depth of sampling. In-haul sampling of such waterbodies is frequently carried out using sieves, nets, or pumps, the mesh size of which has a salient bearing on the material caught (Vermaire et al. 2017). The determination of MPs found in benthic sediments located on the ocean floor, within estuaries, and in the water column is carried out through the efficient process of sediment sampling (Claessens et al. 2011). It is important to emphasize the notion of sampling depth when attempting to estimate the concentrations of MPs present in sediments. This is because the first 1–5 cm layers are more concentrated than in the 10 cm layer (Besley and Vijver 2017). The methods that scientists use for bird sampling focus on retrieving the carcasses of birds from marine or coastal breeding grounds while methods for biota sampling differentiate on how the animals would be characterized in terms of their movements, for instance, lazy, active, slow runners, or fast runners (GESAMP 2019). For the purpose of sampling, NPs (160 nm) from filtered and unfiltered waters, as well as from ultrapure water, can be kept using continuous flow centrifugation (Robertson et al. 2016).

3.2.1 Sediment Sampling

Synthetic polymers, known for their buoyancy and durability, accumulate in sediments worldwide. Microplastics, resistant to degradation, are easily accessible to aquatic organisms and can contaminate both freshwater bodies and benthic habitats. These particles can affect ecosystem life quality, leading to fewer species and juvenile isopods in certain species (Bagheri et al. 2020).

3.2.1.1 Manual Grab Samplers

Manual grab samplers facilitate sediment sampling in littoral and beach environments, allowing for comprehensive analysis of nonbuoyant particles and providing a better understanding of microplastic pollution.

3.2.1.2 Deep Sediment Samplers

Deep sediment sampling using grabs and corers can study benthic sediment abundance and microplastic debris. Glass bottles or plastic bags are recommended for specimen transition, with blank control for bias prevention (Merga et al. 2020).

3.2.2 Water Sampling

Water samples have a lower microplastic burden, requiring larger sample volumes for analysis. On-site techniques like net filtration and sieving reduce volume, resulting in smaller, concentrated samples.

3.2.2.1 Trawls and Nets

Samplers such as trawls and specialized nets—which include the manta, neuston, plankton, and bongo nets—are employed for sampling water. Trawls are submerged and towed at low speed. To determine the area, a person needs only to make use of the below-mentioned formula: the area equals the towing distance multiplied by the trawl width. Water volume passing through the net is, however, quite challenging to gauge (Campanale et al. 2020). Factors like time, location, and trawling distance should be considered to ensure representative sampling. The kind of microplastics is also dependent on the mesh size, and the employment of tandem nets with different mesh sizes is effective in describing the smaller particles. Single net sampling provides a reliable approximation in evaluating the conditions of microplastic particles; however, this might not reflect the general trend of microplastic pollution (Michida 2019).

3.2.2.2 Pump Samplers and Grab Samples

For studies focused on microplastics, researchers employ both volumetric pump samplers and grab samples. With respect to the characteristics enabling the collection of large quantities of water, it is the pumps that are suitable, but these may not be exact as the regions consist of areas with different characteristics (Mao et al. 2020). Merging volume-reduced net methods with bulk sampling allows for a better estimation of missing fractions and enhances spatial specificity (Han et al. 2020).

3.2.3 Biota Sampling

Microplastics have been detected in many organisms both in the field and in the lab. The majority of these studies, however, are laboratory based as only about 40% deal with the species from their natural habitat. Organisms can be collected through various means such as grasps, traps, creels, bottom crawling, using manta or bongo nets, trawls, hand collection, electrofishing, or freshwater gill netting. A clear description of representative sample collection, however, is lacking in the studies. After collection, living organisms need to be incorporated through freezing, desiccation, or fixing them in fixatives. It is therefore suggested that microplastics be incorporated in other routine samplings of 50 different specimens taking into account different sizes and types of specimens, time, and space distribution (Stock et al. 2019).

3.3 Separating Techniques

Micro and nano plastics (MNPs) separation employs several techniques such as density separation, filtration, and sieving among others. The specific gravity of plastic materials varies a lot with the type of polymer in use and the manufacturing methods em ployed. This is used to reduce the content of the sediment by separating lighter plastic particles from the denser sedimentary grains while the mixture is shaken for a set period of time after combining test sediment with a saturated solution of some kind. Once mixing occurs, it is anticipated that the lighter particles will either remain suspended within or rise to the liquid's surface, with the sediment rapidly settling to the base. Following that, the plastic-laden clear liquid is retrieved for the next steps. The mix containing a lot of plastic is often filtered through a mesh after the suspension created by density separation of the plastic particles has been produced, with a vacuum to assist (Ng and Obbard 2006). Paper filters with pore sizes 1, 1.6, and 2 μm were used (Norén 2007). Micro plastics separation can be done through the use of muscles with different mesh sizes of sieves. The sieved-out materials (and sorted) are materials that remain inside the sieve after being pressed. All retained in the sieve, with the exception of those squeezed out, are commonly discarded. Sieves with different mesh sizes can be used to create subdivisions in the micro plastic size categories (Claessens et al. 2011).

The aspects of plastic that facilitate adhesion have made oil an appropriate first-stage separation process. Still, recovery rates remain low (55.0%). This is especially true for high-density plastics, where the combination of pine oil and a froth conditioner improves wetting and reduces surface tension thereby enhancing the separation of plastics from silt under de-ionized water. With recovery rates of 96.1%, a shorter sampling period (2 h) as opposed to salt-saturated solutions with low retention of organic matter has also been used for canola oil. Olive oil has also been helpful in the process of recovering plastics from the supernatant of salt-saturated solutions where the number of plastic debris recovered increased from 64% to 82%. In spite of what seems to be a constraint in the recovery of plastic using oils, the oils can be mixed with saturated solutions to enhance recovery rates though washing off the oils requires a detergent (Woo et al. 2021).

3.3.1 Sieving

The elemental technique of removing the micro plastics from water samples implements the use of various types of sieves with different mesh sizes. The particles trapped on the uppermost covering are gathered and thrown away. For instance, some works employ a sieve with a specific mesh size, while others join 3 up to 4 meshes for the purpose of sorting out different sizes. The sieves used predominantly ranged from 5 mm to roughly about 0.2 μm in scope (Tiwari et al. 2019; Wang et al. 2017).

3.3.2 Filtration

Filtration is one of the common techniques that is applied to separate solids from a liquid phase with the aid of a device that allows the flow of liquid only. The process of filtration includes the usage of funnel, filter, and vacuum pump systems with appropriate pore size membranes. Instead, the cloth contains ragged bits of cellulose and synthetic fiberglass, which could cause contamination. Transfer of the filter to petri plates may lead to loss of microplastics, and the use of a nylon filter plate can also contaminate the sample (Lares et al. 2019; Cai et al. 2019).

3.3.3 Density Separation

Microplastic separation refers to the process of segregating microplastics from sediments or any other inorganic materials. This is because plastics generally have a low density which is even lower than that of most sediments; thus, it is most convenient to work with the two. In this case, the suspension containing the sample is prepared in a certain proportion with the concentrated solution of sodium chloride or another salt and acid a photography is used for a specific amount of time while the separation occurs because of differences in densities within the sample. It has been established that the NaCl solution is both inexpensive and least costly, hence the best alternative. However, this alternative has limitations in that it does not separate high molecular weight substances like polyvinyl chloride and polyethylene terephthalate. Some other sodium iodide (NaI), lithium meta tungstate solution, and zinc chloride (ZnCl2), as well as canola oil, have also been analyzed on their performance in retrieval rates for higher density salts. A new separation technique called electrostatic separation, along with hydrophobic iron nanoparticles, has also been introduced to enhance the processes of separation in a different type of microplastics. The presented methods are designed to facilitate the separation of microplastics from the sediment and other inorganic materials (Stock et al. 2019; Scopetani et al. 2020a).

3.4 Characterization and Identification Techniques

3.4.1 Fourier Transform Infrared Spectroscopy (FTIR)

FTIR can identify plastic materials made of carbon and differentiate them from other organic and inorganic substances based on the nature of their bonding (Löder et al. 2015). First, FTIR was used to pinpoint MPs in sediment and water samples (Vianello et al. 2013). As per the sources, the association of FTIR with micro-FTIR made it possible to carry out in situ infrared absorption studies on various plastics without any destruction or being overly intrusive (Mecozzi et al. 2016). For the purposes of their identification, noninfrared transparent strata, for instance, filters, and focal plane array (FPA) detectors are usually characterized using their reflectance spectra. FPA detectors aided in the acquisition of spatially resolved spectra over other numerous pixel-edged frames, each corresponding to a unique IR spectrum. This technique expeditiously eases the separation of small portions and microfibers within bigger surfaces (Tagg et al. 2015). For the FTIR imaging identification examination, a sample quantity of 1 mg is sufficient to establish the particle size distribution of polyethylene, polypropylene, polyvinyl chloride, and polystyrene. However, this technique was not successful for quantification (Tagg et al. 2015).

3.4.2 Microscopy

As a faster and simpler alternative to FTIR, the microscope method can identify micro plastics. Micro plastics or nonplastics are mistaken for plastics and are more likely to be overlooked or undercounted if they are smaller than 1 mm. Therefore, micro plastics larger than one millimetre can be identified using only a microscope (Song et al. 2015).

3.4.2.1 Fluorescence Microscopy

Fluorescence microscopy is a useful technique to identify MPs based on their inherent ability to emit fluorescence, especially for white and transparent plastics (Norén 2007). When used in conjunction with imaging, this method lowers the size limit of the detected MPs and lowers the failure rate for MPs detection. The process of determining the presence of microplastics can also be carried out by quantifying fluorescent beads which are then counted using microscopes (Catarino et al. 2019). Cell culture assays employing labeled nanoparticles (NPs) typically result in the leaching of excess dye into the media due to pH, salts, and biological interactions. Additionally, NPs exposed cells can exhibit total fluorescence, which is sometimes misconstrued as fluorescence caused by NPs when in fact it is just a free dye present in the system. For these reasons, the knowledge of the evidence for MP biological membrane crossing has not been satisfactory (Andreozzi et al. 2013). The introduction of various chemical agents during the polymerization process, such agents are likely to have an effect on the fluorescent character, limits the imaging of the MP's (Piruska et al. 2005). These contaminants must be removed to the maximum possible and proper prior treatment (Elert et al. 2017). This is very often the case for enzyme treatment and washing the surfaces with acid or oxidizing agents (such as hydrogen peroxide) (Löder et al. 2015). This stain could on the other hand help overcome these challenges. Also, Nile Red (NR) or 9-diethylamino-5-benzofuran-3--carboxylic acid is a dye that stains self-polymerizing substrates more than organic materials encouraging quick enumeration of microplastics (Maes et al. 2017).

3.4.2.2 Transmission Electron Microscopy

If we turn our attention to the chronology of modulated transmission imagery, it is apparent that this concept is by no means recent. However, the first papers developed on the authors' approach appeared only recently, at the beginning of this millennium (Sims and Hardin 2007). There is no reason why TMP should be regarded as one of the techniques available for MP characterization. In most cases, TEM is used for studying the influence of MPs on the model systems. For example, to study the effects of microplastics (PP, PE, PET, and PVC) on algae, specifically growth inhibition and changes in cellular structure, transmission electron microscopy was employed (Song et al. 2020).

3.4.2.3 Scanning Electron Microscopy

Scanning electron microscopy (SEM) is a microscopical technique that aids in the study of the surface topology of MPs by enabling the acquisition of high-definition images of the specific surface without alteration. Additionally, its potential of integrating energy-dispersive X-ray spectroscopy (EDS) detectors tends to take it a notch higher, and it evaluates the samples more thoroughly, and qualitatively, on the composition of elements present in them. Bogner et al. (2007) asserts that despite the widespread belief that there is a skyrocketing cost associated with the use of SEM-EDS, the technique is less labor intensive and economical. Imaging of the microsphere particles is also achievable without the application of nitrogen gas under thermal emission scanning electromicroscopy (Wang and Wang 2018). Another important distinction is the ability to incorporate a holder for sample grids since the design of the SEM makes the examination of nanoparticles possible as well. Small spherical objects which are the subject of studies do not require extremely high energies and voltages which are the characteristics of standard techniques of transferring electron microscopes (Mitrano et al. 2019).

3.4.2.4 Atomic Force Microscopy

Nanomaterials such as powdered or manufactured grains of soil, polymer membranes, and other similar structures have been examined to a greater extent in the AFM style (Fu and Zhang 2018). On the other hand, while in SEM-based analysis one cannot carry out other additional characterizations, in AFM analysis one can carry stiffens, hydrophobicity, conductivity, or magnetization characterization (Fu and Zhang 2018; Trotsenko et al. 2016). An advantage that is associated with the use of AFM is the analysis of the surfaces of nonconducting polymers, as well as very high-resolution imaging of the order of a few nanometers (Dussud et al. 2018). Some reasons that make this technique unavoidable include blocking of external influences, such as pollution, tip artifacts, namely, destruction caused by surfaces in contact with the tip, and inaccurate imaging in the case of sticky polymers as a result of debris dislodgment from contact with the tip (Ukraintsev et al. 2012). In conclusion, while AFM displays a lot of promise in imaging the details of microspheres or even nanoparticles, it is imperative that the technique be integrated with other methods.

3.4.2.5 Polarizing Microscopy

In toxicological research, PE particles were recognized by using a polarized light microscope. Any crystalline structure in the plastic will modify the measurable transmission of polarized light. The crystallinity and its degrees vary also within a single polymer due to the different processes of production and the nature of the polymer. Still, microplastic samples, on the other hand, should be appropriately thin to allow sufficient passage of polarized light. This approach is not applicable to opaque micro and nano plastic samples.

3.4.3 Raman Spectroscopy

In contrast to microscopic identification alone, Raman spectroscopy is a reliable and nonpresumptive method of identifying plastic polymers (Zhao et al. 2017). Raman offers a more specific fingerprint sensor and has more spectral coverage than FTIR; the capacity to visualize targets is what is drawing more and more public interest (Ivleva et al. 2017). Raman hyperspectral imaging is combined with curve resolution-alternating least squares Data Analysis with Multivariate Techniques—A New Tool for Mapping MNPs in Cluttered Environments. This method sought to directly image and detect radio labeled compounds either as single entities or in mixtures without sample pretreatment, washing, or removal of the organic matrix. MNPs have been successfully identified in situ in biological samples like fish feces and environmental sand samples using multivariate surface modeling of Raman images using an alternating least-squares approach. Also, through the use of multivariate analysis of resolution-enhanced curve imaging using alternate least squares, this technique showed that MNP particles could be observed even in more complex backgrounds, where fluorescence and the presence of interfering Raman bands are accounted for (Tian et al. 2022). For the traditional means of MNP detection using Raman spectra, a laser light serves as the incident ray, directly striking the MNP samples (Zada et al. 2018). The molecular analyte's bindings scatter the laser light, and an image of the Raman spectroscopic spectrum is created from the elastically scattered light (Liu et al. 2020). The simple Raman spectroscopy-assisted sensor technique has been successfully used to purposefully identify and analyze specific substances whose individual spectra can be observed in the background of the sample. This approach is widely applicable in the fields of biology, food, and the environment (Valsesia et al. 2021).

3.4.4 Surface Identification for MPs and NPs

3.4.4.1 Time of Flight Secondary Ion Mass Spectroscopy

Time-of-flight secondary ion mass spectrometry (ToF-SIMS) is an effective method that analyzes the surfaces of different materials because of its excellent mass resolution and spatial resolution. It can deliver tons of information for all types of materials (Sodhi 2004). Using the ToF-SIMS method, particles of polyethylene that range from 1 to 5 μm in size can also be conveniently located. Furthermore, ToF-SIMS examination can reveal the mass spectrum of polymeric substances in great detail, focusing on specific ions/protons that relate to the polymer surface (Jungnickel et al. 2016).

3.4.4.2 Mass Spectrometry of the Matrix-Supported Laser Desorption Ionization Time-of-Flight Type Spectroscopy

MALDI-TOF mass spectrometry is mostly preferred in the analysis of large high molecular weight biomolecules for their ionization and detection. The process consists of three basic stages: ionization, separation, and detection (Badía et al. 2011). In the process, a minimal amount of sample is placed on a stainless-steel target which is subjected to vacuum where laser energy removes the sample from the target and ionizes it, and the ions are then separated by a mass spectrometer. Among these techniques, matrix-assisted laser desorption ionization time of flight mass spectrometry has been gaining more importance, especially in the evaluation of these paradigms of environmental pollutants because of its analytical sensitivity, speed of analysis, and user friendliness (Yang et al. 2019). Studies demonstrated that, use of cationization compounds, such as sodium, potassium, and even silver salts, together with the drawn [Polymer] composite, created Metal+ ions, and that MALDI TOF MS could also be used to assess the clean polymers as well. Since polymers are the major components of the above-mentioned MNPs, it could be inferred that the MALDI-TOF MS technique would also be useful for tracing MNPs in the environment. Other polymers that are contained within the matrix vapor along with the TOF machine after they have been ionized by a laser are collected and sorted by their m/z values. Especially after aging or degradation, the most significant attributes for MNP assessment in this analysis are the ones offered by the MS with regard to the following: repeating unit weight, end groups or terminal components, and molecular composition (Wu et al. 2020).

3.4.4.3 Inductively Coupled Plasma Mass Spectroscopy

Single-event mode operation has made inductively coupled plasma mass spectrometry a powerful tool for quick trace metal measurements. This technique may be applied indeed for NPs and MP characterization (Mozhayeva and

Engelhard 2020; Laborda et al. 2014). This method, known as single-particle (SP) ICP-MS, allows for determining the size distribution, chemical composition, and mass concentration of the analyzed particles (Laborda et al. 2014). The inductively coupled plasma mass spectrometry technique was also beneficial for the identification of various micro and nanoplastics from samples collected from the environment.

3.4.5 Thermal Analysis

The detection of micro and nano plastics has been attempted recently with the help of thermal analysis techniques wherein the physical and chemical properties of thermally stable polymers are altered. These techniques use the principle of detecting polymers in the by-products formed upon degradation of the sample.

3.4.5.1 Differential Scanning Calorimetry

Differential scanning calorimetry or DSC is a technique that can study thermal properties. As much support is also advanced for the dimensional properties of polymers, it shows how the melting, crystallization, and transition temperatures, as well as changes in the enthalpy and entropy, vary. All the polymer resins have specific features making it difficult to group them due to variations in each plastic product. However, it is worthwhile pointing out some limitations regarding the interfacial application of DSC, especially when characterizing microplastics that have melting temperature patterns that resemble one another. With injecting or getting brands of sheets one cannot utilize it also for primary microplastic types.

3.4.5.2 Thermogravimetric Analysis

Another technique used in analyzing materials thermally, thermogravimetric analysis (TGA) to be specific is a fundamental thermal analysis approach that aims at understanding the effects that time and temperature have on any given specimen in order to make sure that the qualitative and quantitative data that are availed about the specimen is accurate through the observation of the mass change that the specimen undergoes while subjected to a heating process at a controlled rate and constant conditions. Composite materials of microplastic present in wastewater were earlier examined using DSC TGA. However, only certain results concerning microplastics were reported in regard to polyethylene and polypropylene. There was not a single mention of PVC or polyvinyl chloride, polyamide, or even any polyesters such as PA of SBS. The phase transition signals of materials like PET and PU are extremly bonded making it inseperable.

3.4.6 Pyrolysis Techniques

Py-GC/MS has various operational modes like "single-shot" analyses, so-called "double-shot" (or "multi-shot") analyses, evolved gas analysis (EGA-MS), and reactive as well as thermos chemolysis Py-GC/MS, to name a few. Pyrolysis, so-called "single-shot" mode, is performed at a specific and usually high temperature which is usually beyond 500 °C. The sample, immediately previous to the pyrolysis, is brought up to the heating pyrolysis temperature from room temperature. The heat inside the pyrolyzer is sufficient to break up the macromolecules within a few seconds, and then the volatile degradation products formed are passed through the GC column leading to the MS for polymer(s) and some of the additives detection (Primpke et al. 2020). The "double-shot" (or "multi-shot") mode, also known as thermo desorption (TD) Py-GC/MS, is more complicated because it consists of two or more steps so different types of species can be analyzed within the same run. For instance, low-boiling point volatiles produced at low temperatures during thermal desorption and high temperatures and the destruction of nonvolatile molecular cords at high/s during the thermal pyrolysis step (Herrera et al. 2003). In addition, thermos chemolysis Py-GC/MS inverts the function of the traditional pyrolysis procedure by combining it with chemical derivatization, wherein a tetrakismethylamine hydroxide (TMAH) solution is added causing an enzyme-assisted methylation that cleaves an ester and ether (Primpke et al. 2020).

3.4.7 Hyperspectral Imaging

Each camera captures a pixel in an image, which is then displayed in a matrix. These values can be either reflectance, transmittance, or fluorescence. Hence, the need arises for a cheap and easy method in order to cater to the high sample analytical luxury techniques. In recent years, hyperspectral imaging (HSI) has been used for imaging MNPs with some success. HSI was primarily intended for satellites used in earth observations, but it found applications in a wide range of contexts relating to materials identification. More specifically, it has found a place in facilities that deal with plastic recycling for sorting different variations of plastic according to the polymer. In the recent past, HSI has been put into use in the analysis of MP and has close similarities to FPA-FTIR. One of the advantages of HSI is that the whole filter can be analyzed, making it unnecessary to pick up single particles or to analyze only small areas of the filter. Like the FPA-FTIR and Raman imaging techniques, HSI also captures images containing a complete spectrum for every pixel of the image. An image captures a scene with various objects, and to capture the chemical signatures of the objects, image captures are subjected to a process known as classification of spectra using multivariate methods. Therefore, this did not only imply the identification of the type of polymers in the MNPs present in the image but it also included their distribution, number, and morphology. The applications of HSI are numerous. The remote sensing using HSI has been taken to outer space and even down to the ocean floor. HSI is employed even in waste management centers, especially in the recycling of plastic where it is useful for sorting plastics. When the segregation of plastic waste is carried out by its polymeric type, the polymers of superior grades are retrieved for reprocessing. In this case, the same approach can be used for the analysis of environmental samples containing MNPs (Fig. 3.1).

3.5 Future Perspectives and Recommendation

The most harmful anthropogenic debris in the environment is now plastic because its production and use by humans more frequently over the past few decades. Nevertheless, the data regarding the occurrence of MPs in various environmental matrices, their implications toward health, and rapid assessment are lacking. Among all techniques, including mass-based and particle-based ones, the first MP study already used IR spectroscopy. The issue of significant water absorption can be solved by thoroughly drying the samples before analysis. This technique might be used for quantification at the particle level (i.e., number, size/size distribution, and shape of particles) along with microscopic techniques at a spatial resolution of about 10–20 m (μ-FTIR). It is possible to quickly analyze large filter areas using μ-FTIR by using FPA-based detection, which uses multiple detectors arranged in a grid pattern. This technique is used to produce what is known as hyperspectral imaging.

For the purpose of characterizing and quantifying microplastics in the 1 μm and above range, Raman spectroscopy, which is another kind of vibrational spectroscopy, is a splendid method for work with NPLs as well (and even NPLs down to about 300 nm). The true advantage of the Raman technique is that particles smaller than 10–20 μm can be analyzed which is impossible to analyze with any conventional IR method. While this technique is able to examine the particles much smaller than what Eurofins μ-FTIR spectroscopy can analyze, it has however been stated that the method of size range is not only limited to that particle size. As they enhance each other's weaknesses, IR and Raman spectroscopies can be employed to study the MP samples in greater detail. In this scenario, a new approach using an intelligent imaging system allows for the dual

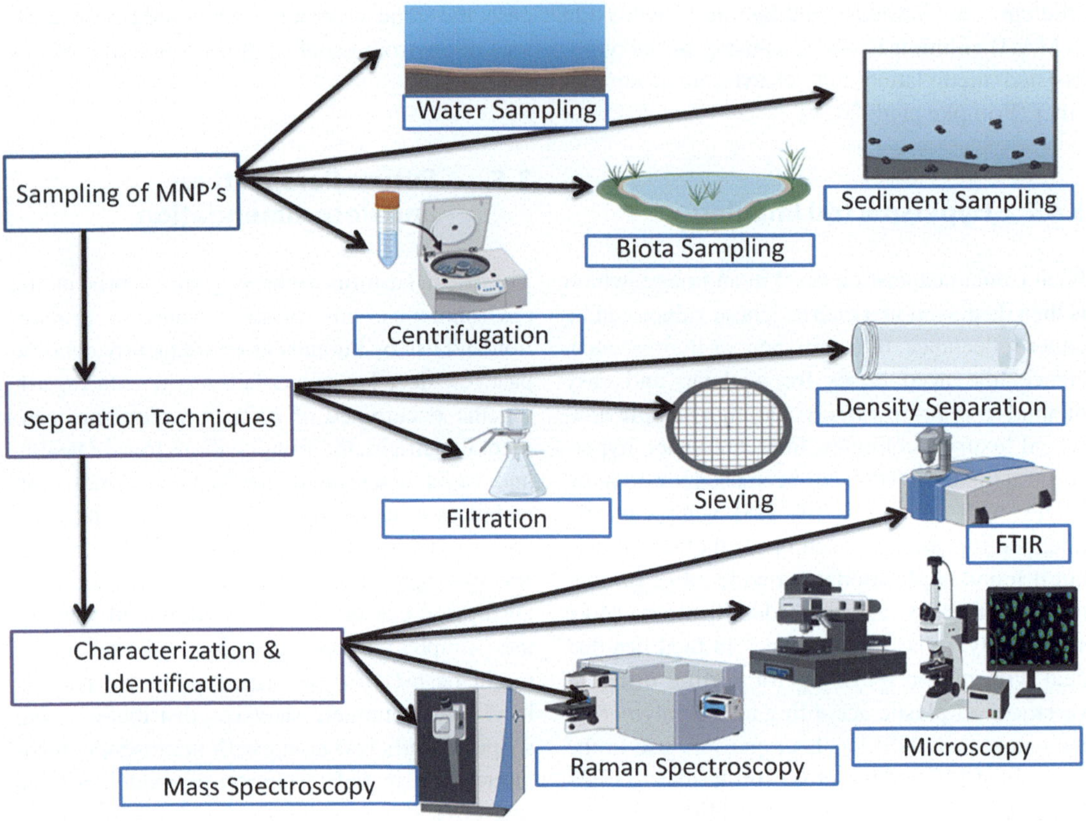

Fig. 3.1 Overview of sampling, separation, characterization, and identification techniques

measurement of Raman and IR signals at a particular spot without any variation in spatial resolution, creating unusual opportunities for the complementary MNP analysis down to about 1 m and even lower. With the help of these techniques, it is possible to recognize, characterize, chemically identify, and quantify (plastic) micro particles as well as automate data processing. This makes it possible to conduct an accurate and thorough MP analysis. Combining size fractionation and chemical identification can also be used to effectively characterize nano plastics. The techniques that have been developed present new avenues for implementation in areas that are not directly linked to the study of NMP, where information about the number of particles, the size/size distribution, and shapes is required alongside the knowledge of the chemical makeup of complexity including nonorganic and microbiological particles.

3.6 Conclusion

As of now, there is no universal system of identification and characterization that can be plausible in all cases presented here. In this chapter, we focused on the most commonly employed methods for responding to the issue of how microplastics should be classified and identified in environmental samples, putting stress on the advantages and drawbacks of these methods. Some of the issues raised might be solved by combining analytical methods with microscopic techniques. The determination of the extent of MP pollution and the efficiency of various measures taken to mitigate plastic contamination, for which and more readily and frequently MP analysis can be performed, requires sensitive chemical identification and mass-based quantitative methods, among others.

References

Andreozzi P, Martinelli C, Carney RP et al (2013) Erythrocyte incubation as a method for free-dye presence determination in fluorescently labelled nanoparticles. Mol Pharm 10(3):875–882. https://doi.org/10.1021/mp300530c

Annenkov VV, Danilovtseva EN, Zelinskiy SN et al (2021) Submicro-and nano plastics: how much can be expected in water bodies? Environ Pollut 278:116910. https://doi.org/10.1016/j.envpol.2021.116910

Badía JD, Strömberg E, Ribes-Greus A et al (2011) A statistical design of experiments for optimizing the MALDI-TOF-MS sample preparation of polymers. An application in the assessment of the thermo-mechanical degradation mechanisms of poly (ethylene terephthalate). Anal Chim Acta 692(1–2):85–95. https://doi.org/10.1016/j.aca.2011.02.063

Bagheri T, Gholizadeh M, Abarghouei S et al (2020) Microplastics distribution, abundance and composition in sediment, fishes and benthic organisms of the Gorgan Bay, Caspian Sea. Chemosphere 257:127201. https://doi.org/10.1016/j.chemosphere.2020.127201

Besley A, Vijver MG, Behrens P et al (2017) A standardized method for sampling and extraction methods for quantifying micro plastics in beach sand. Mar Pollut Bull 114(1):77–83. https://doi.org/10.1016/j.marpolbul.2016.08.055

Bogner A, Jouneau PH, Thollet G et al (2007) A history of scanning electron microscopy developments: towards "wet-STEM" imaging. Micron 38(4):390–401. https://doi.org/10.1016/j.micron.2006.06.008

Cai H, Du F, Li L et al (2019) A practical approach based on FT-IR spectroscopy for identification of semi-synthetic and natural celluloses in microplastic investigation. Sci Total Environ 669:692–701. https://doi.org/10.1016/j.scitotenv.2019.03.124

Campanale C, Stock F, Massarelli C et al (2020) Microplastics and their possible sources: the example of Ofanto river in Southeast Italy. Environ Pollut 258:113284. https://doi.org/10.1016/j.envpol.2019.113284

Catarino AI, Frutos A, Henry TB (2019) Use of fluorescent-labelled Nano plastics (NPs) to demonstrate NP absorption is inconclusive without adequate controls. Sci Total Environ 670:915–920. https://doi.org/10.1016/j.scitotenv.2019.03.194

Claessens M, De Meester S, Van Landuyt L et al (2011) Occurrence and distribution of micro plastics in marine sediments along the Belgian coast. Mar Pollut Bull 62(10):2199–2204. https://doi.org/10.1016/j.marpolbul.2011.06.030

Dussud C, Meistertzheim AL, Conan P et al (2018) Evidence of niche partitioning among bacteria living on plastics, organic particles and surrounding seawaters. Environ Pollut 236:807–816. https://doi.org/10.1016/j.envpol.2017.12.027

Elert AM, Becker R, Duemichen E (2017) Comparison of different methods for MP detection: what can we learn from them, and why asking the right question before measurements matters? Environ Pollut 231:1256–1264. https://doi.org/10.1016/j.envpol.2017.08.074

Fu W, Zhang W (2018) Measurement of the surface hydrophobicity of engineered nanoparticles using an atomic force microscope. Phys Chem Chem Phys 20(37):24434–24443. https://doi.org/10.1039/C8CP04676J

GESAMP (2019) Guidelines for the monitoring and assessment of plastic litter and microplastics in the ocean. In: Kershaw PJ, Turra A, Galgani F (eds) GESAMP Joint Group of Experts on the Scientific Aspects of Marine Environmental Protection, London, 130pp. GESAMP Reports and Studies, No. 99. http://hdl.handle.net/11329/889. Accessed 2019

Han M, Niu X, Tang M et al (2020) Distribution of microplastics in surface water of the lower Yellow River near estuary. Sci Total Environ 707:135601. https://doi.org/10.1016/j.scitotenv.2019.135601

Herrera M, Matuschek G, Kettrup A (2003) Fast identification of polymer additives by pyrolysis-gas chromatography/mass spectrometry. J Anal Appl Pyrolysis 70(1):35–42. https://doi.org/10.1016/S0165-2370(02)00078-5

Ivleva NP, Wiesheu AC, Niessner R (2017) Microplastic in aquatic ecosystems. Angew Chem Int Ed 6(7):1720–1739. https://doi.org/10.1002/anie.201606957

Jungnickel H, Pund R, Tentschert J et al (2016) Time-of-flight secondary ion mass spectrometry (ToF-SIMS)-based analysis and imaging of polyethylene micro plastics formation during sea surf simulation. Sci Total Environ 563:261–266. https://doi.org/10.1016/j.scitotenv.2016.04.025

Laborda F, Bolea E, Jimenez-Lamana J (2014) Single particle inductively coupled plasma mass spectrometry: a powerful tool for nanoanalysis. Anal Chem 86(5):2270–2278. https://doi.org/10.1021/ac402980q

Lares M, Ncibi MC, Sillanpää M et al (2019) Intercomparison study on commonly used methods to determine microplastics in wastewater and sludge samples. Environ Sci Pollut Res 26(12):12109–12122. https://doi.org/10.1007/s11356-019-04584-6

Liu J, Zhang X, Du Z et al (2020) Application of confocal laser Raman spectroscopy on marine sediment micro plastics. J Oceanol Limnol 38(5):1502–1516. https://doi.org/10.1007/s00343-020-0129-z

Löder MGJ, Kuczera M, Mintenig et al (2015) Focal plane array detector-based micro-Fourier-transform infrared imaging for the analysis of micro plastics in environmental samples. Environ Chem 12(5):563–581. https://doi.org/10.1071/EN14205

Maes T, Jessop R, Wellner N et al (2017) A rapid-screening approach to detect and quantify micro plastics based on fluorescent tagging with Nile Red. Sci Rep 7. https://doi.org/10.1038/srep44501

Mao Y, Li H, Gu W et al (2020) Distribution and characteristics of microplastics in the Yulin River, China: role of environmental and spatial factors. Environ

Pollut 265:115033. https://doi.org/10.1016/j.envpol.2020.115033

Mecozzi M, Pietroletti M, Monakhova YB (2016) FTIR spectroscopy supported by statistical techniques for the structural characterization of plastic debris in the marine environment: application to monitoring studies. Mar Pollut Bull 106(1–2):155–161. https://doi.org/10.1016/j.marpolbul.2016.03.012

Merga LB, Redondo-Hasselerharm PE, Van den Brink PJ et al (2020) Distribution of microplastic and small macroplastic particles across four fish species and sediment in an African lake. Sci Total Environ 741:140527. https://doi.org/10.1016/j.scitotenv.2020.140527

Michida Y (2019) Guidelines for harmonizing ocean surface microplastic monitoring methods. Version 1.1. Chiyoda-ku Tokyo Japan Ministry of the Environment Japan, 74pp. http://hdl.handle.net/11329/1361. Accessed 2019

Mitrano DM, Beltzung A, Frehland S et al (2019) Synthesis of metal-doped nano plastics and their utility to investigate fate and behaviour in complex environmental systems. Nat Nanotechnol 14(4):362–368. https://doi.org/10.1038/s41565-018-0360-3

Mozhayeva D, Engelhard C (2020) A critical review of single particle inductively coupled plasma mass spectrometry–a step towards an ideal method for nanomaterial characterization. J Anal At Spectrom 35(9):1740–1783. https://doi.org/10.1039/C9JA00206E

Murray A, Örmeci B (2020) Removal effectiveness of nano plastics (< 400 nm) with separation processes used for water and wastewater treatment. Water 12(3):635. https://doi.org/10.3390/w12030635

Ng KL, Obbard JP (2006) Prevalence of micro plastics in Singapore's coastal marine environment. Mar Pollut Bull 52(7):761–767. https://doi.org/10.1016/j.marpolbul.2005.11.017

Norén F (2007) Small plastic particles in coastal Swedish waters. Kimo Sweden. Available from: https://www.researchgate.net/profile/Fredrik-Noren/publication/284312290_Small_plastic_particles_in_Coastal_Swedish_waters/links/571203c608ae4ef74525ec38/Small-plastic-particles-in-Coastal-Swedish-waters.pdf. Accessed 29 Aug 2007

Piruska A, Nikcevic I, Lee SH et al (2005) The autofluorescence of plastic materials and chips measured under laser irradiation. Lab Chip 5(12):1348–1354. https://doi.org/10.1039/B508288A

Prata JC, da Costa JP, Duarte AC et al (2019) Methods for sampling and detection of microplastics in water and sediment: a critical review. Trends Anal Chem 110: 150–159. https://doi.org/10.1016/j.trac.2018.10.029

Primpke S, Christiansen SH, Cowger W et al (2020) Critical assessment of analytical methods for the harmonized and cost-efficient analysis of micro plastics. Appl Spectrosc 74(9):1012–1047. https://doi.org/10.1177/00037028209214

Prüst M, Meijer J, Westerink RH (2020) The plastic brain: neurotoxicity of micro-and nano plastics. Part Fibre Toxicol 17(1):1–16. https://doi.org/10.1186/s12989-020-00358-y

Robertson JD, Rizzello L, Avila-Olias et al (2016) Purification of nanoparticles by size and shape. Sci Rep 6(1):1–9. https://doi.org/10.1038/srep27494

Scopetani C, Chelazzi D, Mikola J et al (2020a) Olive oil-based method for the extraction, quantification and identification of microplastics in soil and compost samples. Sci Total Environ 733:139338. https://doi.org/10.1016/j.scitotenv.2020.139338

Sims PA, Hardin JD (2007) Fluorescence-integrated transmission electron microscopy images. In: Kuo J (ed) Electron microscopy. Methods in molecular biology™, vol 369. Humana Press. https://doi.org/10.1007/978-1-59745-294-6_14

Sodhi RN (2004) Time-of-flight secondary ion mass spectrometry (TOF-SIMS): versatility in chemical and imaging surface analysis. Analyst 129(6):483–487. https://doi.org/10.1039/B402607C

Song YK, Hong SH, Jang M et al (2015) A comparison of microscopic and spectroscopic identification methods for analysis of micro plastics in environmental samples. Mar Pollut Bull 93(1–2):202–209. https://doi.org/10.1016/j.marpolbul.2015.01.015

Song C, Liu Z, Wang C et al (2020) Different interaction performance between micro plastics and microalgae: the bio-elimination potential of Chlorella sp. L38 and Phaeodactylum tricornutum MASCC-0025. Sci Total Environ 723:138146. https://doi.org/10.1016/j.scitotenv.2020.138146

Stock F, Kochleus C, Bänsch-Baltruschat B et al (2019) Sampling techniques and preparation methods for microplastic analyses in the aquatic environment–a review. Trends Anal Chem 113:84–92. https://doi.org/10.1016/j.trac.2019.01.014

Tagg AS, Sapp M, Harrison JP et al (2015) Identification and quantification of micro plastics in wastewater using focal plane array-based reflectance micro-FT-IR imaging. Anal Chem 87(12):6032–6040. https://doi.org/10.1021/acs.analchem.5b00495

Tian M, Morais CL, Shen H et al (2022) Direct identification and visualisation of real-world contaminating micro plastics using Raman spectral mapping with multivariate curve resolution-alternating least squares. J Hazard Mater 422:126892. https://doi.org/10.1016/j.jhazmat.2021.126892

Tiwari M, Rathod TD, Ajmal PY et al (2019) Distribution and characterization of microplastics in beach sand from three different Indian coastal environments. Mar Pollut Bull 140:262–273. https://doi.org/10.1016/j.marpolbul.2019.01.055

Trotsenko O, Koestner R, Roiter Y et al (2016) Probing rough composite surfaces with atomic force microscopy: nafion ionomer in fuel cell electrodes. Polymer 102:396–403. https://doi.org/10.1016/j.polymer.2015.11.021

Ukraintsev E, Kromka A, Kozak H, Remeš Z, Rezek B (2012) Artifacts in atomic force microscopy of biological samples. In: Christopher F (ed) Force microscopy investigations into biology–from cell to protein. InTech, Rijeka

Valsesia A, Quarato M, Ponti J et al (2021) Combining microcavity size selection with Raman microscopy for the characterization of nanoplastics in complex matrices. Sci Rep 11(1):1–12. https://doi.org/10.1038/s41598-020-79714-z

Vermaire JC, Pomeroy C, Herczegh SM et al (2017) Micro plastic abundance and distribution in the open water and sediment of the Ottawa River Canada and its tributaries. Facets (Ott) 2(1):301–314. https://doi.org/10.1139/facets-2016-0070

Vianello A, Boldrin A, Guerriero P et al (2013) Micro plastic particles in sediments of Lagoon of Venice, Italy: first observations on occurrence, spatial patterns and identification. Estuar Coast Shelf Sci 130:54–61. https://doi.org/10.1016/j.ecss.2013.03.022

Wang W, Wang J (2018) Investigation of micro plastics in aquatic environments: an overview of the methods used, from field sampling to laboratory analysis. Trends Anal Chem 108:195–202. https://doi.org/10.1016/j.trac.2018.08.026

Wang ZM, Wagner J, Ghosal S et al (2017) SEM/EDS and optical microscopy analyses of microplastics in ocean trawl and fish guts. Sci Total Environ 603:616–626. https://doi.org/10.1016/j.scitotenv.2017.06.047

Woo H, Seo K, Choi Y et al (2021) Methods of analyzing microsized plastics in the environment. Appl Sci 11(22):10640. https://doi.org/10.3390/app112210640

Wu P, Tang Y, Cao G et al (2020) Determination of environmental micro (nano) plastics by matrix-assisted laser desorption/ionization–time-of-flight mass spectrometry. Anal Chem 92(21):14346–14356. https://doi.org/10.1021/acs.analchem.0c01928

Yang C, Lee HK, Zhang Y et al (2019) In situ detection and imaging of PFOS in mouse kidney by matrix-assisted laser desorption/ionization imaging mass spectrometry. Anal Chem 91(14):8783–8788. https://doi.org/10.1021/acs.analchem.9b00711

Zada L, Leslie HA, Vethaak AD et al (2018) Fast micro plastics identification with stimulated Raman scattering microscopy. J Raman Spectrosc 49(7):1136–1144. https://doi.org/10.1002/jrs.5367

Zhao S, Danley M, Ward JE et al (2017) An approach for extraction, characterization and quantitation of micro plastic in natural marine snow using Raman microscopy. Anal Methods 9(9):1470–1478. https://doi.org/10.1039/C6AY02302A

Micro and Nanoplastics: Contamination of Air, Water, and Soil

4

Murad Muhmmad, Muhammad Majeed, Wen-Jun Li, Sani Begum, Nimra Bashir Qureshi, Dwaipayan Sinha, Muqadas Batool, Abdul Wahab, Yong-Hong Liu, and Li Li

4.1 Introduction

Today, there are over 5300 polymers; "plastics" refers to a wide range of these synthetic and semi-synthetic materials (Zimmermann 2021). A high-molecular-weight polymer is a significant component of plastics, which can flow moulded at some point in their production into final goods (ISO 472:2013). Specifically, the International Organization for Standardization provided this explanation (ISO). Vocabulary concerning plastics: tiny plastic particles can be categorized as microplastics or nanoplastics (1 micrometre to 5 millimetres), while the size boundary between the categories (1000 vs 100 nm) is still up for discussion (Babick et al. 2016) (less than 1000 nm). The smallest size deemed a microplastic has no official upper limit. However, in recent years, scientists have started referring to particles smaller than a few micrometres in size as "nanoplastic." It is challenging to remove such tiny particles from their surroundings using straightforward processes like filtering, which can be employed for microplastics; therefore, the distinction is more than just a formality; it is helpful. Thanks to Bakelite, the first synthetic polymer ever created, at least 15 different kinds of synthetic polymers were produced in large quantities. Modern manufactured polymers saw a dramatic surge in use worldwide starting in the late 1950s (Geyer 2020).

M. Muhmmad (✉)
State Key Laboratory of Desert and Oasis Ecology, Key Laboratory of Ecological Safety and Sustainable Development in Arid Lands, Xinjiang Institute of Ecology and Geography, Chinese Academy of Sciences, Urumqi, People's Republic of China

University of Chinese Academy of Sciences, Beijing, People's Republic of China
e-mail: muradbotany1@uop.edu.pk

M. Majeed
Department of Botany, University of Gujrat, Gujrat, Punjab, Pakistan
e-mail: m.majeed@uog.edu.pk

W.-J. Li (✉)
State Key Laboratory of Desert and Oasis Ecology, Key Laboratory of Ecological Safety and Sustainable Development in Arid Lands, Xinjiang Institute of Ecology and Geography, Chinese Academy of Sciences, Urumqi, People's Republic of China

State Key Laboratory of Biocontrol, Guangdong Provincial Key Laboratory of Plant Resources and Southern Marine Science and Engineering Guangdong Laboratory (Zhuhai), School of Life Sciences, Sun Yat-Sen University, Guangzhou, People's Republic of China
e-mail: liwenjun3@mail.sysu.edu.cn

S. Goswami et al. (eds.), *Micro-Nano Plastics Exposure, Environmental Degradation and Public Health Crisis*, Sustainable Development Goals Series, https://doi.org/10.1007/978-981-96-5047-7_4

In 1907, phenol-formaldehyde resin, popularly known as bakelite, debuted. Since then, it has developed into one of the most often-used materials. The amount of plastic produced globally in 1950 was only 1.5 million tonnes (Luan et al. 2022). The United Nations world population sensor report for 2019 confirmed this, estimating that there were 7.7 billion people on Earth in the middle of the year (Vankayala and Vankayala 2019).

There has been a rise in the number of studies looking at the environmental effects of nanoplastics since 2017, and that number is predicted to rise by another 60% over the next 2 years (Allen et al. 2022). Recent reviews show that the vast bulk of nanoplastics research is focused on polystyrene nanoparticles. Our analysis of the already available scientific literature further supported this. Polystyrene nanoparticles are frequently employed as test materials in (eco) toxicity investigations because they are easily synthesized into nanoparticles and are widely accessible, affordable, and cost-effective. The type of interactions between plastics and biota in the environment is greatly influenced by the amount of plastic waste, significantly affecting ecological dynamics. This interaction is exacerbated since larger debris poses a greater risk of damage. Plastic's physical and chemical qualities mean it may take on a broad range of shapes and sizes, affecting the plastic's mobility, biodistribution, environmental distribution, and animal interactions (Paul et al. 2020). The extent to which we understand the destiny, occurrence, quantity, and dispersion of nanoplastics in the environment is hampered by our inability to recognize and isolate them from various complicated environmental matrices. The primary reason for this is that more methods are needed. One of the main reasons there are so few publications on nanoplastics is the lack of approaches that are both practical and capable of analysing nanoplastics in the environment. Because of this, accurate forecasts of the harmful effects of nanoplastics on human health are difficult to achieve. Experiments show that plastic exposed to UV light for 3 weeks breaks down into particles measuring in the nanometre (nm) range. Compared to more extensive, macroscopic particles of the same substance, these nanoparticles' physical and chemical properties, which range in size from 1 to 100 nm, are pretty distinct (Khan et al. 2019). It is all the rage to use nanoscale polymers with dimensions of a thousand nanometres (nano) or less in width, thickness, or depth (Mellor and Uchegbu 2022).

This chapter explores the critical issue of micro and nanoplastics, defining their characteristics and sources while assessing their environmental impact on air, water, and soil ecosystems. It will evaluate the potential human health risks associated with these pollutants and review current detection methods used to quantify their presence in various environments. Furthermore, the chapter will discuss effective mitigation strategies to reduce contamination and emphasize the importance of public awareness and policy development. Finally, it will highlight knowledge

S. Begum
Department of Botany, Hazara University, Mansehra, Pakistan

N. B. Qureshi
University of Chinese Academy of Sciences, Beijing, People's Republic of China

State Key Laboratory of Microbial Resources, Institute of Microbiology, Chinese Academy of Sciences, Beijing, China

D. Sinha
Department of Botany, Government General Degree College, Mohanpur, Paschim Medinipur, West Bengal, India

M. Batool
Department of Botany, (Govt Graduate College Layyah) Bahauddin Zakariya University, Multan, Pakistan

A. Wahab
University of Chinese Academy of Sciences, Beijing, People's Republic of China

Shanghai Center for Plant Stress Biology, C AS. Center for Excellence in Molecular Plant Sciences, Chinese Academy of Sciences, Shanghai, China
e-mail: wahabcrop_science@mails.ucas.ac.cn

Y.-H. Liu · L. Li
State Key Laboratory of Desert and Oasis Ecology, Key Laboratory of Ecological Safety and Sustainable Development in Arid Lands, Xinjiang Institute of Ecology and Geography, Chinese Academy of Sciences, Urumqi, People's Republic of China

gaps and suggest future research directions to understand better and address the implications of micro and nanoplastics in our environment.

4.2 Sources of Micro and Nanoplastics

One of the leading causes of the destruction of the natural environment is the human practice of constantly discarding large quantities of plastic (Ali et al. 2021). In this context, "plastics" might mean any synthetic polymers produced when organic molecules are reacted. To obtain nanoplastics, you can do any of the following. Microplastics (MPs) and nanoplastics (NPs) originate from various sources, significantly impacting the environment. Primary microplastics are intentionally manufactured in small sizes, commonly found in personal care products like exfoliants and in industrial applications such as abrasives. These particles enter the environment through wastewater discharge, bypassing treatment processes due to their small size (Rezai and Allahkarami 2021). Secondary microplastics, on the other hand, result from the breakdown of more oversized plastic items, such as bottles, bags, and packaging materials. This degradation is accelerated by environmental factors like UV radiation, mechanical abrasion, and chemical processes, releasing smaller plastic fragments into ecosystems (Andrady et al. 2022). Agricultural practices contribute to soil contamination with microplastics, mainly through sewage sludge and plastic mulching (Muhammad et al. 2023). Additionally, tyre wear particles from vehicles are a significant source of MPs in both urban and rural environments, as these particles are dispersed through air and water runoff (Gieré and Dietze 2022).

MPs and NPs originate in aquatic environments from synthetic textiles, which shed fibres during washing, and from the breakdown of larger plastics in oceans and rivers, as shown in Fig. 4.1. These particles can also enter the

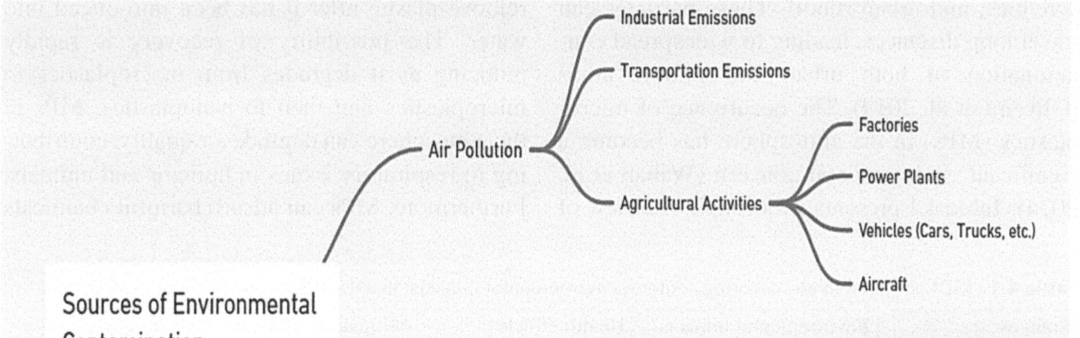

Fig. 4.1 The image is a mind map illustrating the sources of environmental contamination. It branches into three main categories: air pollution, water pollution, and soil contamination. Air pollution sources include industrial emissions, transportation emissions, and agricultural activities such as factories, power plants, vehicles, and aircraft. Water pollution sources are industrial discharges from chemical factories and manufacturing plants, agricultural runoff including pesticides and fertilizers, and sewage and wastewater disposal. Soil contamination sources are the dumping of toxic chemicals and industrial waste

atmosphere through airborne dispersion, where they are transported over long distances, contributing to air pollution and subsequent deposition into water and soil (Janhäll 2015). It highlights specific sources of contamination, the environmental impacts associated with MPs in each medium, potential health effects on humans and wildlife, and various strategies for mitigation. By examining these interconnected aspects, the table underscores the urgent need for comprehensive approaches to address microplastic pollution and its environmental and public health implications.

4.3 MPs and Air Pollution

The occurrence of microplastics (MPs) in the atmosphere has become a significant environmental concern. Recent studies indicate that MPs are prevalent in aquatic and terrestrial environments and the air we breathe. Microplastics in the air originate from various sources, including the degradation of larger plastic debris, industrial activities, and urban runoff. These particles can travel long distances, leading to widespread contamination of both urban and remote areas (Oliveira et al. 2019). The occurrence of microplastics (MPs) in the atmosphere has become a significant environmental concern (Wahab et al. 2024). Table 4.1 presents a thorough overview of microplastics (MPs) pollution in the air. Recent studies indicate that MPs are prevalent in aquatic and terrestrial environments and the air we breathe. Microplastics in the air originate from various sources, including the degradation of larger plastic debris, industrial activities, and urban runoff. These particles can travel long distances, leading to widespread contamination of both urban and remote areas (Dong et al. 2023). Coastal cities exhibit MP levels between 500 and 2500 particles per cubic metre, linked to marine activities and urban pollution. Jambeck et al. (2023) estimated that MP levels in the air could double by 2050, even if all sources of plastic pollution are eliminated.

Additionally, it was suggested that agricultural practices contribute to MP levels in the air, particularly in agricultural regions, with concentrations between 100 and 800 particles per cubic metre (Jambeck et al. 2023). Due to the difficulty of collecting plastic and the fact that plastic does not decompose in the natural environment in which it is located, it is highly challenging to remove plastic after it has been introduced into water. The possibility of recovery is rapidly reducing as it degrades from macroplastics to microplastics and then to nanoplastics. MPs in the atmosphere can degrade air quality, contributing to respiratory issues in humans and animals. Furthermore, MPs can adsorb harmful chemicals

Table 4.1 MPs pollution in air, covering sources, environmental impacts, health effects, and mitigation strategies

Sources	Environmental impacts	Health effects	Mitigation strategies	References
Tyre wear from vehicles	Deposition of airborne microplastics onto land and water	Inhalation leading to respiratory problems	Promote electric vehicles to reduce tyre wear	Arole et al. (2023) and Gehrke et al. (2023)
Industrial emissions	Transportation of microplastics over long distances	Long-term exposure linked to lung inflammation	Implement stricter air quality controls for industries	Munsif et al. (2021)
Textile fibres released during drying	Deposition on agricultural fields, worsening soil contamination	Inhalation leading to lung diseases	Improve filtration systems in industries and households	Bhat et al. (2022)
Fragmentation of plastics in urban areas	Dispersal over urban and rural regions	Chronic respiratory issues due to long-term exposure	Encourage the use of biodegradable materials in cities	Karbalaei et al. (2018)
Microplastics from construction materials	Contamination of indoor and outdoor air quality	Potential effects on human lungs and respiratory tract	Develop alternatives to microplastic-laden construction materials	Kacprzak and Tijing (2022)

from the atmosphere, which may then be inhaled by living organisms, leading to toxicological effects (Muhammad et al. 2024b). Airborne MPs also pose risks to atmospheric organisms, including insects and birds. These organisms may ingest MPs, resulting in physical and chemical stress that disrupts ecosystems (Muhammad et al. 2024b).

4.4 MPs and Water Pollution

Since the beginning of the Industrial Revolution, human activity has been one of the primary causes of trash in both marine and freshwater habitats (Goudie 2018). The widespread belief that rubbish floating away from shore is a relatively minor problem compared to the more significant problem of plastic pollution has led to the growing impact of plastic pollutants being overlooked in the past (Waheed et al. 2023). This is because of the widespread belief that plastic pollution is a more significant problem. Nevertheless, the poisoning of water is now generally recognized as a severe issue on a global scale. Since the 1940s, when mass manufacture of plastic first began, it has been, by a significant margin, the most critical contributor to pollution in marine environments (Brander and Betjemann 2023). Concerns on an ecotoxicological level have been raised about plastic in its macro form and the microparticles and nanoparticles generated due to its breakdown. Table 4.2 presents a thorough overview of microplastics (MPs) pollution in water.

- *MNP Concentration in Aquatic Ecosystems*
  The five most frequent methods for collecting samples of marine anthropogenic particles and plastic trash are beach cleanups, marine observational surveys, biological samples, trawling the ocean, and excavation through sediment (Mearns et al. 2015). Since nanoparticles are too small to be seen with the human eye, beachcombing and marine observational surveys cannot be utilized to monitor their potential environmental presence. Marine and avian species are especially vulnerable to swallowing rubbish; however, biological sampling can discover this issue (Muhammad et al. 2024b). Marine trawling is the practice of gathering rubbish from the ocean floor

Table 4.2 MPs pollution in water, covering sources, environmental impacts, health effects, and mitigation strategies

Sources	Environmental impacts	Health effects	Mitigation strategies	References
Textile fibres from washing clothes (e.g. synthetic fabrics)	Accumulation in oceans, rivers, and lakes	Ingestion of microplastics through contaminated seafood	Improve wastewater treatment	Mishra et al. (2020)
Plastic waste (littering, mismanagement)	Ingestion by marine organisms, disrupting food chains	Potential digestive and hormonal disturbances in humans	Reduce plastic use and waste and eco-friendly replacement for plastic must be addressed	Evode et al. (2021) and Kumar et al. (2021)
Microbeads in personal care products (e.g. cosmetics)	Altered water quality	Exposure to toxic chemicals adhering to microplastics	Ban microbeads in cosmetics	Mdlalose and Chimuka (2022)
Industrial effluents (plastic production)	Habitat destruction for aquatic life	Bioaccumulation of pollutants in the food chain	Regulate industrial effluent discharge	Ritter et al. (2002) and Baby et al. (2010)
Runoff from landfills	Contamination of freshwater sources	Long-term accumulation affecting human health	Implement proper landfill management practices	Mondal et al. (2023) Wijekoon et al. (2022) Paul et al. (2019)

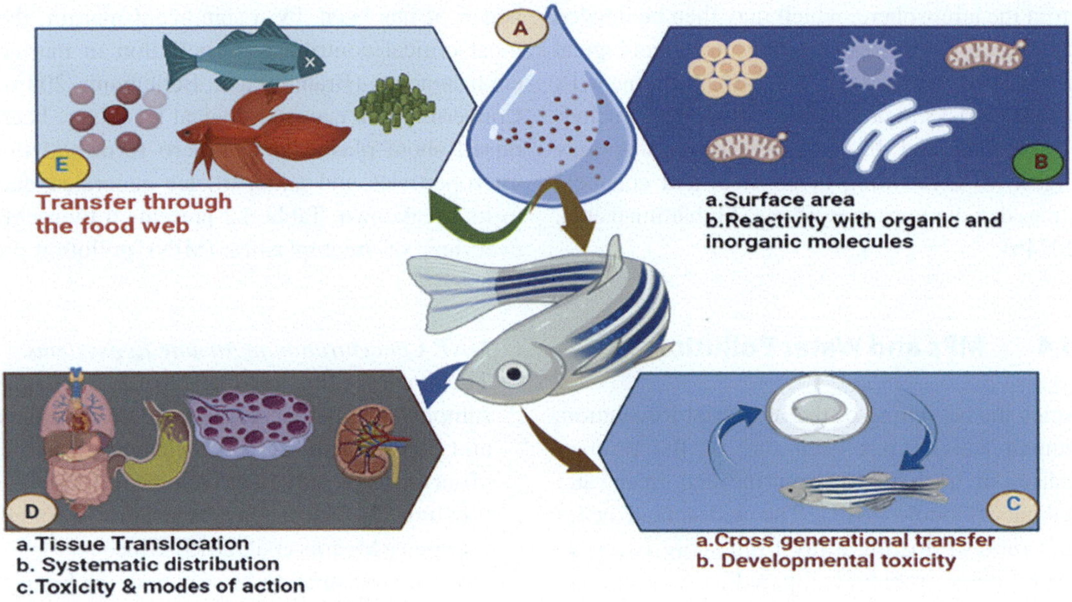

Fig. 4.2 Diagram illustrating the impact of nanoparticles on fish, showing five stages. Stage A: Nanoparticles with varying surface areas and reactivities. Stage B: Interaction with organic and inorganic molecules. Stage C: Cross-generational transfer and developmental toxicity in fish. Stage D: Tissue translocation, systematic distribution, and toxicity modes. Stage E: Transfer through the food web, depicting fish and other aquatic organisms. Central image of a fish connects all stages, indicating the flow and impact of nanoparticles

using a boat equipped with filters that are in motion. Most of these techniques involve flowing surface water through a mesh, whose size determines the smallest particle size that can pass through. The standard mesh size for particle collection is approximately 300 microns, making nanoparticle detection difficult (Montaño et al. 2016). In addition, sediment sampling is employed to collect benthic material from beaches, estuaries, and the ocean floor, as well as plastic litter samples from the same places. Plastic particles rise to the surface because they are less dense than the surrounding silt and must be separated using a process known as density separation. This approach loses some of its utility because it cannot be used to separate nanoparticles with low density. However, it is possible to segregate nanoparticles with a lower density than those that tend to sink. Hence, MNPs are becoming part of our food web, as shown in Fig. 4.2.

- *Aquatic Species: Cellular and Molecular Effects of MNPs*

  Since NPs are so much smaller than other nanoparticles, their surface-to-volume and mass-to-volume ratios are significantly higher than those of the latter (Fig. 4.3). Because of this, their interactions with organic biomolecules and inorganic contaminants may be more powerful. The study's findings can be found in the following: the buildup of NPs in the gills, brain, heart, liver, yolk sac, gonads, and digestive organs of vertebrate and invertebrate species demonstrates that this makes it easier for aquatic organisms to consume and absorb NPs when the size values are less than 100 nm (Turan et al. 2019). This section is for particles that are less than 100 nm in size. Nanoparticles are so small that they can be carried and distributed all over the body by endocytosis, transcytosis, paracellular diffusion between tight junctions, and absorption over enterocyte barriers (Barua and Mitragotri

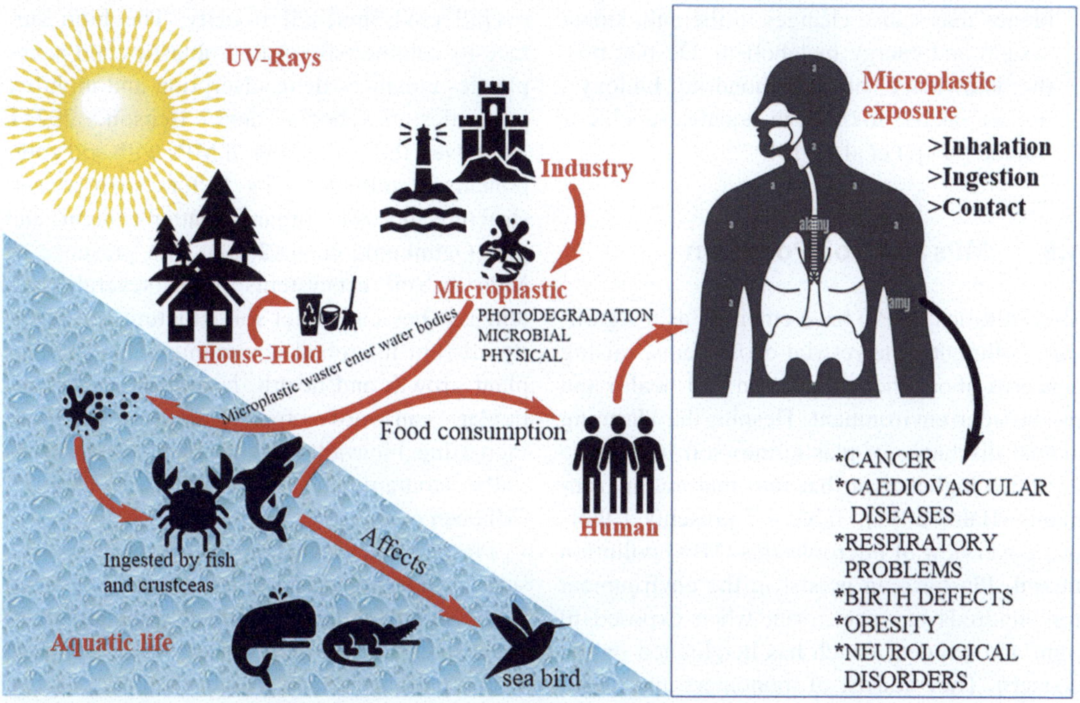

Fig. 4.3 Diagram illustrating the microplastic pollution cycle. UV rays and industries contribute to microplastic generation, which enters water bodies through household waste. Aquatic life, including fish and crustaceans, ingests microplastics, affecting the food chain. Humans consume these contaminated foods, leading to microplastic exposure through inhalation, ingestion, and contact. Potential health impacts include cancer, cardiovascular diseases, respiratory problems, birth defects, obesity, and neurological disorders

2014). Because of their diminutive size, nanoparticles are more susceptible to tissue translocation and systemic dispersion, which explains why they have been found in various animals' tissues exposed to them. Investigations have shown that nanoparticles (NPs) with sizes of less than 50 nm have a high potential to be internalized in endolysosomes, to be sustained for varied durations of time, and to bioaccumulate in a wide variety of organs (Foroozandeh and Aziz 2018). This information was gleaned from the findings of the studies (Muhammad et al. 2024a). This, in turn, may imply a high possibility of the trait being passed down from one generation to the next. This is, without a doubt, a highly significant finding. This means that the presence of plastic materials in aquatic ecosystems, the breakdown of bigger plastics into smaller plastics, and the small size of NPs are all significant indicators of the potential health concerns linked with being exposed to NPs by organisms (Bour et al. 2015). One, the NP mixes in water; two, how NPs interact with molecules in the environment; three, how NPs affect the growth of animals; four, how NPs are distributed internally and their toxicity; and five, how NPs can be biomagnified.

- *Toxicity of MNPs and Its Principal Biological Effects and Mechanisms*

  Toxic levels of NPs are most often seen in the reproductive organs, digestive system, liver, pancreas, heart, lungs, brain, and immune cells (Jiang et al. 2019). This allows them to harm cells and molecules, impacting many organs and biological functions. Damage to cells and tissues, inflammation, changes in metabolism and neurotransmitter transmission, and even disintegration of organelles are all possible side effects of NPs (Sawicki et al. 2019). There may be a connection between molecular, physiological, and organismal cascade effects and the overall population's odds of survival. NPs' interaction with lipid membranes may cause changes in the antioxidant system and energy metabolism. Despite this, the knowledge of mitochondrial biology's impact on NP toxicity in aquatic species is limited (Naqvi et al. 2018).

4.5 MPs and Soil Pollution

Microplastics (MPs) have emerged as a significant pollutant in terrestrial ecosystems, raising concerns about their impact on soil health and the broader environment. Despite the alarming annual discharge of plastic into soil being 4 to 23 times higher than that into marine environments (Bates 2019). Table 4.3 presents a thorough overview of microplastics (MPs) pollution in soil. Plastics can persist in the environment for hundreds of years, even when exposed to light and oxygen, which has heightened public concern. They consist of monomers and oligomers, such as styrene and bisphenol A, known carcinogens linked to cancer and endocrine disruption when used as polyvinyl chloride (PVC) monomers. Additionally, additives like phthalates and brominated flame retardants, which serve as plasticizers, pose significant human health risks (Bui et al. 2016). During the decomposition of plastics in soil, harmful monomers and additives can leach into waterways, threatening aquatic ecosystems (Waheed et al. 2023). Furthermore, plastics can carry hydrophobic organic and inorganic pollutants, increasing overall environmental toxicity. The high surface-to-volume ratio of microplastics and nanoplastics enhances their adsorption and mobility properties (Agboola and Benson 2021). Moreover, the interaction of MPs with soil components is influenced by factors such as soil structure, porosity, organic matter content, and pH (Ogunkunle et al. 2021). The presence of MPs in soil ecosystems poses several challenges. They can affect soil structure and fertility, disrupt microbial communities, and impair plant growth and health. Nanoplastics, in particular, can move freely through the soil, increasing bioavailability to plants and potentially contaminating groundwater supplies (Mubeen et al. 2023).

Plastic particles float on water but sink into the soil through pores and cracks formed by plant roots. Soil pH influences NP adhesion at the air–water interface and their efficacy, as altering pH affects particle surface charges (Liu et al. 2018). The functional groups of NPs reduce their transportability by impacting the stability of organic molecules in soil. Bioturbation, aided by organisms like earthworms, also influences NP distribution (Edwards and Arancon 2022). In conclusion, the growing presence of microplastics in soil highlights the urgent need for further research into their environmental impact and the development of strategies to mitigate soil pollution. Addressing the sources of plastic pollution is essential for protecting soil health and ensuring the sustainability of terrestrial ecosystems.

Table 4.3 MPs pollution in soil, covering sources, environmental impacts, health effects, and mitigation strategies

Sources	Environmental impacts	Health effects	Mitigation strategies	References
Agricultural plastic films	Soil degradation and reduced fertility	Potential uptake of microplastics by crops	Use biodegradable alternatives for agricultural films	Qi et al. (2020)
Biosolids (wastewater sludge)	Disruption of soil ecosystems, harming soil organisms	Ingestion through contaminated food	Limit microplastic content in biosolids	Mohapatra et al. (2016)
Plastic litter in agricultural fields	Reduced water retention and soil quality	Long-term exposure through food crops	Educate farmers about plastic waste management	Kumar et al. (2020)
Fertilizers with plastic coatings	Altered soil structure and reduced plant growth	Potential gastrointestinal problems from crop contamination	Promote the use of natural fertilizers	Stubenrauch and Ekardt (2020)
Runoff from plastic-coated fertilizers	Leaching of plastics into groundwater	Contaminated water supplies	Regulate plastic use in agricultural products	Stubenrauch and Ekardt (2020)

4.6 Strategies for Managing MNP Pollution

This situation is anticipated to not improve in the foreseeable future due to the dramatic rise in the use of plastic over the past few decades and the surplus of both MP and NP particles that has resulted as a direct consequence of this trend (Shen et al. 2020). We can identify the origins of these plastics and NPs. In that case, we can reduce the amount of microplastics and nanoparticles that make their way into the bodies of marine species and people. In addition, increasing education on microplastics and nanoplastics in both the public and private spheres will be beneficial to raising awareness of microplastics and nanoplastics (Allan et al. 2021). According to research that was carried out by the Australian National Centre for Ocean Resources and Security (ANCORS) at the University of Wollongong (UOW), marine debris can be managed in part by adhering to national policy and legal solutions (Warner 2015).

- Addressing marine garbage "hot areas"
- Implementing technical litter traps
- National rubbish disposal strategy and programmes
- Adopting Extended Producer Responsibility (EPR) schemes with consumer involvement
- Including financial incentives to get the private sector involved, regulations to reduce garbage production, and the use of economic methods to promote recycling

4.7 Future Perspectives and Recommendation

Scientists and the public are concerned about nanoplastics. Numerous studies have shown that nanoplastics pollute the entire human environment, threatening the survival of many species. Research also shows that nanoplastics in the environment are harmful to animals. The most plausible pathways of neoplastic exposure and their cumulative biological effects were examined. NPs affect plant growth and development, soil microbiota, mesofauna, and macrofauna. These changes may be produced by the chemical's toxicity to soil organisms or by its reactivity with other foreign organic or inorganic substances in the soil. In addition to phytotoxicity, NPs may affect soil microbiota, plant diversity, and community structure. Rhizosphere bacteria help plants grow and develop. To improve theoretical and statistical support for plastic waste control and treatment, microplastic discharge, and diffusion, household garbage classification education should be increased, and viable plastic waste treatment technologies should be improved. If nanoplastic is digested, it is uncertain if the stomach acid or cellular digestive organelles will further break it down (lysosomes). The intestinal epithelial barrier may prevent microplastic substances from entering the circulation. Despite fears that chemicals may seep from plastics into the food chain, it is unknown if nanoplastics boost the body's uptake of chemical hitchhikers. Plastics might have no ingredients or a wide variety of sizes and shapes. Integrating exposure and impact data into meaningful risk categories is complicated by the inconsistent use of units, exposure medium, and habitats. Because of this, building a plastic risk assessment paradigm is problematic. Polymer nanoparticles are ubiquitous, making it harder to generalize about their effects. Therefore, it is vital to collect data on nature's many plastics. Polystyrene is cheap, easy to synthesize, and widely utilized in polymer nanoparticle research. Polyolefins (polyethylenes and polypropylenes), polyesters, and polyurethanes are employed more in industry. We believe that precise prediction of long-term exposure and comprehensive study using different types of plastics at tolerable quantities can provide a sufficient picture of the dangers involved. Given the little about the Earth's surface, exploration should be a significant priority. Future studies should examine plastic additives' impacts on NPs' effects on DNA, RNA, and MPs and NPs' molecular interactions with contaminants and biomolecules.

4.8 Conclusion

Microplastics (MPs) and nanoplastics (NPs) represent a growing threat to environmental health, affecting air, water, and soil quality. This chapter has highlighted the multifaceted sources of MP and NP contamination, particularly emphasizing the role of agricultural practices, such as the application of sewage sludge, in exacerbating pollution. The interactions of these pollutants with soil ecosystems are particularly concerning, as they disrupt microbial communities crucial for soil health and plant vitality. Moreover, the release of MPs and NPs into aquatic systems through wastewater further complicates the pollution cycle, linking sediment, water, and air in a continuous cycle of contamination. Despite the critical implications of these pollutants, there remains a significant gap in comprehensive studies examining their life cycle, especially within wastewater treatment plants (WWTPs). Understanding these dynamics is essential for developing effective management strategies. The ecotoxicity of MPs and NPs and their interactions with various organisms underscores the urgent need for innovative solutions to combat this pervasive issue. Strategies such as developing biodegradable plastics, implementing biotechnology, and promoting public awareness are vital in mitigating the adverse effects of micro and nanoplastic pollution. Additionally, microbial remediation offers promising avenues for restoring contaminated environments. In conclusion, addressing the challenges posed by MPs and NPs requires a coordinated approach that combines scientific research, technological innovation, and community engagement to safeguard environmental health for future generations.

References

Agboola OD, Benson NU (2021) Physisorption and chemisorption mechanisms influencing micro (nano) plastics-organic chemical contaminants interactions: a review. Front Environ Sci 9:678574

Ali SS, Elsamahy T, Koutra E, Kornaros M, El-Sheekh M, Abdelkarim EA, Zhu D, Sun J (2021) Degradation of conventional plastic wastes in the environment: a review on current status of knowledge and future perspectives of disposal. Sci Total Environ 771:144719

Allan J, Belz S, Hoeveler A, Hugas M, Okuda H, Patri A, Rauscher H, Silva P, Slikker W, Sokull-Kluettgen B (2021) Regulatory landscape of nanotechnology and nanoplastics from a global perspective. Regul Toxicol Pharmacol 122:104885

Allen S, Allen D, Karbalaei S, Maselli V, Walker TR (2022) Micro (nano) plastics sources, fate, and effects: what we know after ten years of research. J Hazard Mater Adv 6:100057

Andrady A, Barnes P, Bornman J, Gouin T, Madronich S, White C, Zepp R, Jansen M (2022) Oxidation and fragmentation of plastics in a changing environment; from UV-radiation to biological degradation. Sci Total Environ 851:158022

Arole K, Velhal M, Tajedini M, Xavier PG, Bardasz E, Green MJ, Liang H (2023) Impacts of particles released from vehicles on environment and health. Tribol Int 184:108417

Babick F, Mielke J, Wohlleben W, Weigel S, Hodoroaba V-D (2016) How reliably can a material be classified as a nanomaterial? Available particle-sizing techniques at work. J Nanopart Res 18:1–40

Baby J, Raj JS, Biby ET, Sankarganesh P, Jeevitha MV, Ajisha SU, Rajan SS (2010) Toxic effect of heavy metals on aquatic environment. Int J Biol Chem Sci 4(4). https://doi.org/10.4314/ijbcs.v4i4.62976

Barua S, Mitragotri S (2014) Challenges associated with penetration of nanoparticles across cell and tissue barriers: a review of current status and future prospects. Nano Today 9(2):223–243

Bates A (2019) Transforming plastic: from pollution to evolution. GroundSwell Books, Summertown, Tennessee

Bhat MA, Eraslan FN, Gedik K, Gaga EO (2022) Impact of textile product emissions: toxicological considerations in assessing indoor air quality and human health. In: Ecological and health effects of building materials. Springer, Cham, pp 505–541

Bour A, Mouchet F, Silvestre J, Gauthier L, Pinelli E (2015) Environmentally relevant approaches to assess nanoparticles ecotoxicity: a review. J Hazard Mater 283:764–777

Brander S, Betjemann P (2023) Marine pollution: science, art, and the accumulations of time. In: Oceans and society. Routledge, London, pp 78–98

Bui TT, Giovanoulis G, Cousins AP, Magnér J, Cousins IT, de Wit CA (2016) Human exposure, hazard and risk of alternative plasticizers to phthalate esters. Sci Total Environ 541:451–467

Dong X, Liu X, Hou Q, Wang Z (2023) From natural environment to animal tissues: a review of microplastics (nanoplastics) translocation and hazards studies. Sci Total Environ 855:158686

Edwards CA, Arancon NQ (2022) The role of earthworms in organic matter and nutrient cycles. In: Biology and ecology of earthworms. Springer, New York, pp 233–274

Evode N, Qamar SA, Bilal M, Barceló D, Iqbal HM (2021) Plastic waste and its management strategies for environmental sustainability. Case Stud Chem Environ Eng 4:100142

Foroozandeh P, Aziz AA (2018) Insight into cellular uptake and intracellular trafficking of nanoparticles. Nanoscale Res Lett 13(1):339

Gehrke I, Schläfle S, Bertling R, Öz M, Gregory K (2023) Mitigation measures to reduce tire and road wear particles. Sci Total Environ 904:166537

Geyer R (2020) A brief history of plastics. In: Mare plasticum. Springer, Cham, pp 31–47

Gieré R, Dietze V (2022) Tire-abrasion particles in the environment. In: Degradation of elastomers in practice. Springer, Cham, pp 71–101

Goudie AS (2018) Human impact on the natural environment. John Wiley & Sons, Hoboken

Jambeck J, Moss E, Dubey B, Arifin Z, Godfrey L, Hardesty BD, Hendrawan IG, Hien TT, Junguo L, Matlock M (2023) Leveraging multi-target strategies to address plastic pollution in the context of an already stressed ocean. In: The blue compendium. Springer, Cham, pp 141–184

Janhäll S (2015) Review on urban vegetation and particle air pollution–deposition and dispersion. Atmos Environ 105:130–137

Jiang Z, Shan K, Song J, Liu J, Rajendran S, Pugazhendhi A, Jacob JA, Chen B (2019) Toxic effects of magnetic nanoparticles on normal cells and organs. Life Sci 220:156–161

Kacprzak S, Tijing LD (2022) Microplastics in indoor environment: sources, mitigation and fate. J Environ Chem Eng 10(2):107359

Karbalaei S, Hanachi P, Walker TR, Cole M (2018) Occurrence, sources, human health impacts and mitigation of microplastic pollution. Environ Sci Pollut Res 25:36046–36063

Khan I, Saeed K, Khan I (2019) Nanoparticles: properties, applications and toxicities. Arab J Chem 12(7):908–931

Kumar M, Xiong X, He M, Tsang DC, Gupta J, Khan E et al (2020) Microplastics as pollutants in agricultural soils. Environ Pollut 265:114980

Kumar R, Verma A, Shome A, Sinha R, Sinha S, Jha PK et al (2021) Impacts of plastic pollution on ecosystem services, sustainable development goals, and need to focus on circular economy and policy interventions. Sustain For 13(17):9963

Liu F, Xu B, He Y, Brookes PC, Tang C, Xu J (2018) Differences in transport behavior of natural soil colloids of contrasting sizes from nanometer to micron and the environmental implications. Sci Total Environ 634:802–810

Luan X, Kou X, Zhang L, Chen L, Liu W, Cui Z (2022) Estimation and prediction of plastic losses to the environment in China from 1950 to 2050. Resour Conserv Recycl 184:106386

Mdlalose L, Chimuka L (2022) Mitigation approaches to prevent microplastics effects in the aquatic environment: exploration of microbeads from personal care and cosmetic products. Int J Environ Res 16(5):84

Mearns AJ, Reish DJ, Oshida PS, Ginn T, Rempel-Hester MA, Arthur C, Rutherford N, Pryor R (2015) Effects of pollution on marine organisms. Water Environ Res 87(10):1718–1816

Mellor RD, Uchegbu IF (2022) Ultrasmall-in-nano: why size matters. Nanomaterials 12(14):2476

Mishra S, Singh RP, Rath CC, Das AP (2020) Synthetic microfibers: source, transport and their remediation. J Water Process Eng 38:101612

Mohapatra DP, Cledón M, Brar SK, Surampalli RY (2016) Application of wastewater and biosolids in soil: occurrence and fate of emerging contaminants. Water Air Soil Pollut 227:1–14

Mondal T, Choudhury M, Kundu D, Dutta D, Samanta P (2023) Landfill: an eclectic review on structure, reactions and remediation approach. Waste Manag 164:127–142. https://doi.org/10.1016/j.wasman.2023.03.034

Montaño MD, Olesik JW, Barber AG, Challis K, Ranville JF (2016) Single particle ICP-MS: advances toward routine analysis of nanomaterials. Anal Bioanal Chem 408:5053–5074

Mubeen B, Hasnain A, Wang J, Zheng H, Naqvi SAH, Prasad R, Rehman AU, Sohail MA, Hassan MZ, Farhan M (2023) Current progress and open challenges for combined toxic effects of manufactured nano-sized objects (MNO's) on soil biota and microbial community. Coatings 13(1):212

Muhammad M, Waheed A, Wahab A, Majeed M, Nazim M, Liu Y-H, Li L, Li W-J (2023) Soil salinity and drought tolerance: an evaluation of plant growth, productivity, microbial diversity, and amelioration strategies. Plant. Stress 11:100319

Muhammad M, Basit A, Ali K, Li W-J, Li L, Mohamed HI (2024a) Endophytic fungi as potential biocontrol agents of soil-borne pathogen. J Crop Health 76:617–636

Muhammad M, Wahab A, Waheed A, Mohamed HI, Hakeem KR, Li L, Li W-J (2024b) Harnessing bacterial endophytes for environmental resilience and agricultural sustainability. J Environ Manag 368:122201. https://doi.org/10.1016/j.jenvman.2024.122201

Munsif R, Zubair M, Aziz A, Zafar MN (2021) Industrial air emission pollution: potential sources and sustainable mitigation. In: Environmental emissions. IntechOpen, London

Naqvi S, Kumar V, Gopinath P (2018) Nanomaterial toxicity: a challenge to end users. In: Applications of nanomaterials. Elsevier, Duxford, pp 315–343

Ogunkunle CO, Oyedeji S, Okoro HK, Adimula V (2021) Interaction of nanoparticles with soil. In: Nanomaterials for soil remediation. Elsevier, Amsterdam, pp 101–132

Oliveira M, Almeida M, Miguel I (2019) A micro (nano) plastic boomerang tale: a never ending story? TrAC Trends Anal Chem 112:196–200

Paul S, Choudhury M, Deb U, Pegu R, Das S, Bhattacharya SS (2019) Assessing the ecological impacts of ageing on hazard potential of solid waste landfills: a green approach through vermitechnology. J Clean Prod 236:117643. https://doi.org/10.1016/j.jclepro.2019.117643

Paul MB, Stock V, Cara-Carmona J, Lisicki E, Shopova S, Fessard V, Braeuning A, Sieg H, Böhmert L (2020) Micro-and nanoplastics–current state of knowledge with the focus on oral uptake and toxicity. Nanoscale Adv 2(10):4350–4367

Qi R, Jones DL, Li Z, Liu Q, Yan C (2020) Behavior of microplastics and plastic film residues in the soil environment: a critical review. Sci Total Environ 703:134722

Rezai B, Allahkarami E (2021) Wastewater treatment processes—techniques, technologies, challenges faced, and alternative solutions. In: Soft computing techniques for solid waste and wastewater management. Elsevier, Amsterdam, pp 35–53

Ritter KS, Sibley P, Hall K, Keen P, Mattu G, Linton B (2002) Sources, pathways, and relative risks of contaminants in surface water and groundwater: a perspective prepared for the Walkerton inquiry. J Toxicol Environ Health A 65(1):1–142

Sawicki K, Czajka M, Matysiak-Kucharek M, Fal B, Drop B, Męczyńska-Wielgosz S, Sikorska K, Kruszewski M, Kapka-Skrzypczak L (2019) Toxicity of metallic nanoparticles in the central nervous system. Nanotechnol Rev 8(1):175–200

Shen M, Huang W, Chen M, Song B, Zeng G, Zhang Y (2020) (Micro) plastic crisis: un-ignorable contribution to global greenhouse gas emissions and climate change. J Clean Prod 254:120138

Stubenrauch J, Ekardt F (2020) Plastic pollution in soils: governance approaches to foster soil health and closed nutrient cycles. Environments 7(5):38

Turan NB, Erkan HS, Engin GO, Bilgili MS (2019) Nanoparticles in the aquatic environment: usage, properties, transformation and toxicity—a review. Process Saf Environ Prot 130:238–249

Vankayala P, Vankayala M (2019) Use of modern communication technologies IoT for Natural resource conservation-Water. In: 2019 4th international conference on internet of things: smart innovations and usages (IoT-SIU). IEEE. https://doi.org/10.1109/IoTSIU.2019.8777484

Wahab A, Muhammad M, Ullah S, Abdi G, Shah GM, Zaman W, Ayaz A (2024) Agriculture and environmental management through nanotechnology: eco-friendly nanomaterial synthesis for soil-plant systems, food safety, and sustainability. Sci Total Environ 926:171862

Waheed A, Haxim Y, Islam W, Ahmad M, Muhammad M, Alqahtani FM, Hashem Y, Alshahrani S (2023) Climate change reshaping plant-fungal interaction. Environ Res 238:117282

Warner R (2015) Stemming the Black tide: cooperation on oil pollution preparedness and response in the South China Sea and East Asian seas. J Ocean Policy 18(2):184–197

Wijekoon P, Koliyabandara PA, Cooray AT, Lam SS, Athapattu BC, Vithanage M (2022) Progress and prospects in mitigation of landfill leachate pollution: risk, pollution potential, treatment and challenges. J Hazard Mater 421:126627

Zimmermann L (2021) Toxicity of plastic consumer products: a biological, chemical and social-ecological analysis. Dissertation, Frankfurt am Main, Johann Wolfgang Goethe-Universität, 2021

Nano- and Micro-plastics Pollution Effects on Physical/Abiotic and Biotic Environment and United Nations' Sustainable Development Goals

5

Md Niamul Haque, Most Shirina Begum, and Aswathy Vijayakumar

Abstract

Nanoplastics (NPs) and micro-plastics (MPs) are persistent pollutants present in multiple spheres of the environment. Terrestrial and aquatic ecosystems are significant sinks of these contaminants; however, studies on the interaction of NPs and MPs with ecosystem health (*e.g.*, soil) and inhabitants are limited to date. NPs and MPs can modify soil physico-chemical attributes (*e.g.*, bulk density and aggregation) and biogeochemistry (*e.g.*, nutrient- and C-cycling) depending on their biodegradable or non-biodegradable characteristics. NPs and some MPs can be absorbed through plant roots and transported to different plant structures, hindering plant growth and crop production. The detrimental effect of NPs and MPs on soil and plants can further affect crop yield and food safety, impeding the United Nations' Sustainable Development Goals. Therefore, studying associations between the environment (soil and plant) and NPs and MPs is crucial in policymaking for sustainable agricultural production under global change.

Keywords

Plastics · Ecosystem · Biotic environment · Sustainable development goals

M. N. Haque (✉)
Department of Marine Science, College of Natural Sciences & Research Institute of Basic Sciences, Incheon National University, Incheon, Republic of Korea

Department of Biology, Trent University, Peterborough, ON, Canada

M. S. Begum
Department of Biology, Trent University, Peterborough, ON, Canada
e-mail: mshirinabegum@trentu.ca

A. Vijayakumar
Department of Nutritional Science and Food Management, Ewha Womans University, Seoul, Republic of Korea

5.1 Introduction

Plastics are extensively utilized in every aspect of daily life because of their diverse properties, such as low cost, versatility, biological dullness, and oxygen resistance (Arif et al. 2024). Generally, plastics derived from petrochemicals/crude oil exhibit high molecular mass and plasticity and are generally presumed chemically inert. During manufacturing, various harmful ingredients (e.g.,

S. Goswami et al. (eds.), *Micro-Nano Plastics Exposure, Environmental Degradation and Public Health Crisis*, Sustainable Development Goals Series, https://doi.org/10.1007/978-981-96-5047-7_5

bisphenol A) are added to improve their quality and properties, such as fillers and stabilizers (Horton et al. 2017). Due to mismanagement and poor natural degradation, enormous amounts of plastic have been inevitably thrown into the environment (Rajpal et al. 2024). Global plastics generation has dramatically expanded in the past years (Choudhury et al. 2022a; Statista 2017), and annual plastic production is anticipated to rise to 33 billion tons by 2050 (Rochman et al. 2013). Prominent plastics such as polyethylene, polyvinyl chloride, polystyrene, polyurethane, and polycarbonate formulate roughly 30% of total plastic generation (Lithner et al. 2011). Additionally, the COVID-19 pandemic has intensified plastic contamination through the utilization of gloves and masks. The land is a significant sink for anthropogenic/artificial materials, including plastic. However, plastic can also be transported and accumulated in the aquatic environment and the atmosphere (Mondal et al. 2023; Paul et al. 2019). Under different environmental conditions, larger plastics are subjected to progressive fragmentation and degradation to NPs and MPs (Gewert et al. 2015).

Nanoplastics (NPs; ≤1 μm) and microplastics (MPs; ≤5 mm) are emerging contaminants (Arif et al. 2024), and they have been detected across various environmental subsystems (hydrosphere, lithosphere, and atmosphere) and even in humans (Ragusa et al. 2021). The breakdown of large plastics produces secondary MPs and NPs, while primary MPs and NPs are purposely produced in minuscule sizes to serve as elements in paints, cosmetics, or fabrics (Alimba and Faggio 2019). In the ecosystem, NPs and MPs can affect the properties, function, and movement of co-existing chemicals (e.g., heavy metals and natural organic matter) because of their unique physicochemical properties including strong hydrophobicity, high surface charge, and larger surface area. Consequently, the occurrence of NPs and MPs in municipal wastewater (Ziajahromi et al. 2017), bottled water (Mason et al. 2018), oceans (Law and Thompson 2014), and soils (Zhang and Liu 2018) poses serious risks for living organisms, underscoring the urgent need for further research and public awareness.

Soil and water are essential natural resources for animal- and plant-based food production systems, including agriculture and aquaculture. Therefore, the sustainability of agricultural food production largely depends on the health of terrestrial and aquatic ecosystems (Kumar et al. 2020). NPs and MPs might enter the food web via diverse pathways, one of them being diet; thus, the negative consequences of NPs and MPs on soil and ocean productivity are forthcoming. It is vital to understand the interaction of NPs and MPs with abiotic and biotic environments, which will be highlighted in this chapter. Furthermore, potential approaches to attain the United Nations' Sustainable Development Goals (UN SDGs) correlated to food security impacted by plastic waste management will be reviewed.

5.2 Migration and transformation of NPs and MPs in Soil and Water

About 20% of the plastic waste is recycled (Yu et al. 2022), and the remaining 80% ends up in the rivers, oceans, and soil (Trevor 2020). The primary origin of MPs and NPs, including greenhouse materials, plastic film mulching, wastewater irrigation, atmospheric deposition, pesticide and fertilizer packaging wastes, runoff, and tire wear, play a significant role in their wide distribution in soil by wind, surface runoff, and animal movements (Yu et al. 2022) (Fig. 5.1). Mulched soils (mulch acts against water loss and preserves soil moisture) contain more plastic films compared to unmulched soils (Zhou et al. 2020). Residues of plastic film are cracked into small pieces through farming, photodegradation, exposure to temperature, and ultraviolet radiation (Yu et al. 2022). Indirect sources involve littering, compost, biosolids, and treated wastewater (Horton et al. 2017). Additionally, plastic powders, industrial scrubbers, micro-beads in household goods such as face wash and toothpaste, cosmetics, lip balm, and fabric softeners are also sources of NPs and MPs in agroecosystems (Guo

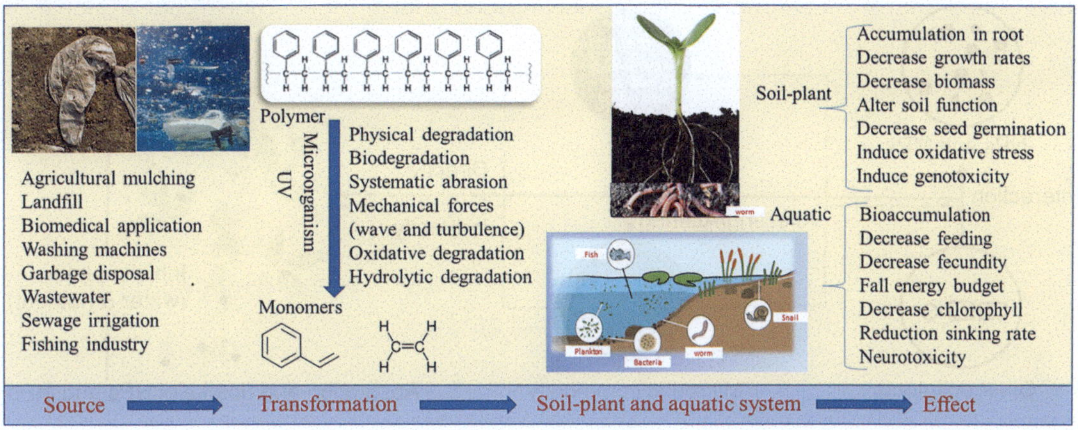

Fig. 5.1 Some of the major ways of nano- and micro-plastics pollution in soil and aquatic environment. (Modified from Sarkar et al. (2022))

et al. 2020). The occurrence and prevalence of NPs and MPs decrease with soil depth and vary based on soil texture (Liu et al. 2018). Yu et al. (2021) reported a greater abundance of MPs in sandy loam soil compared to loam soils. The degradation of plastic and the prevalence of NPs and MPs are dependent on soil texture and composition (e.g., soil organic matter). Subsequently, the future pollution of agroecosystems by secondary NPs and MPs could be extensive.

In the aquatic ecosystem, NPs and MPs enter via runoff (rain and irrigation) and through chemical, physical, and biological weathering of macroplastics and mesoplastics. Rivers, as major transport routes, play a crucial role in transferring NPs and MPs into the ocean. In rivers and lakes with large surface areas, strong waves can break down macroplastics into MPs (Imhof et al. 2013). Degradation of plastics in water generally occurs via an autocatalytic pathway introduced by ultraviolet solar radiation followed by thermo-oxidative degradation (Andrady 2011). There are various types of degradation such as physical, biological, and chemical degradation. Several factors including crystallinity and biofilm formation regulate the MPs degradation rate (Yuan et al. 2020). Microorganisms are also involved in biological degradation through enzyme catalysis and hydrolysis. Chemical degradation mainly comprises degradation through thermal oxidation and hydrolysis, whereas physical degradation includes thermal and photodegradation. Among all the processes, photodegradation is the predominant method of plastic degradation. The mechanisms of NPs production in nature are similar to those for MPs; however, NPs are deliberately synthesized in the lab for their utilization in nanotechnology including electronic devices, medicine, cosmetics, and the airplane industry (Roex et al. 2013). Active degradation and transportation of plastic ultimately contribute to the transfer of NPs and MPs in water and soil and eventually into the food web (Fig. 5.1). A number of studies in the last decade reported the occurrence of MPs in marine salt (Selvam et al. 2020), bottled water (Mason et al. 2018), sugar, honey, and beer (Karbalaei et al. 2018), and placenta (Ragusa et al. 2021) and stool (Zhang et al. 2021).

5.3 Interactions of NPs and MPs Pollution with Plants and Soil Characteristics

NPs and MPs can deposit in soil and change soil's chemical and physical properties, affecting soil structure, biodiversity, and function (Rillig 2012). Biodegradable and nonbiodegradable elements of NPs and MPs can perturb nutrient and carbon cycles in soil, impacting agricultural crop production yields. Cycling nitrogen and carbon in the soil is essential for soil sustainability, maintaining soil health for soil-dwelling organisms, and enabling food production (Dey et al.

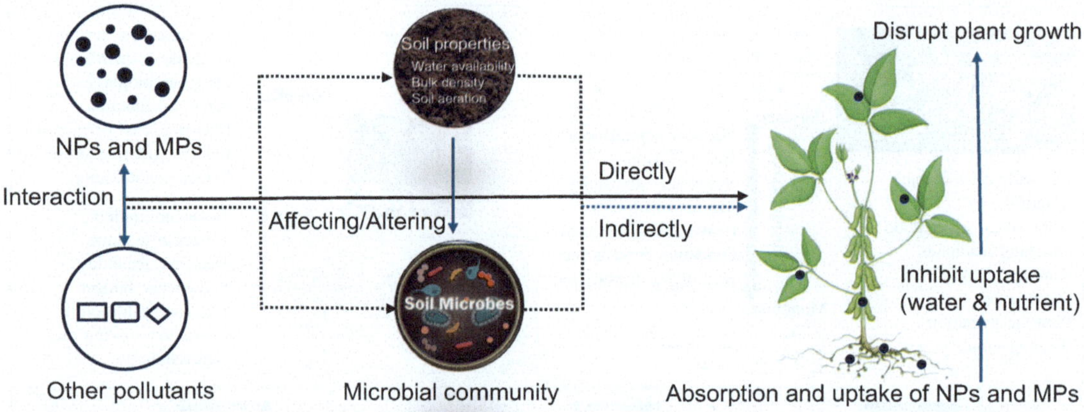

Fig. 5.2 Interaction and transportation of nano- and micro-plastics with soil properties, soil microbes, and plant growth. (Modified from Chen et al. (2022))

2024). The three most crucial soil health indicators are physical, chemical, and microbiological properties (Kibblewhite et al. 2008). Alteration of soil physical attributes including soil moisture, porosity, and soil structure by NPs and MPs affects plants and soil fauna and microbes (Wang et al. 2022; Chen et al. 2022) (Fig. 5.2).

5.3.1 Effects of NPs and MPs on Soil Physical Properties

Soil structure is correlated to soil hydraulic attributes (infiltration and water retention), thermal attributes (i.e., heat capacity and thermal conductivity), chemical properties (ion exchange and transfer, soil fertility), and soil aeration (Wang et al. 2022). Soil structure regulates soil porousness (water availability and soil aeration), degree of soil tightness (development and growth and plant roots), transformation of energy, and movement of soil nutrients and salts. Soil aggregation is a significant factor in regulating soil structure including soil pores. The pore space and soil aggregates allow air and water exchange, availability, and retention. Diverse studies documented that NPs and MPs adversely alter soil aggregation (Zhang and Liu 2018; Rillig and Lehmann 2020; Lehmann et al. 2021). A recent study reports that polystyrene microfiber significantly decreased soil aggregation (de Souza Machado et al. 2019). Shape (e.g., pellets), type of NPs and MPs, soil biota, and organic matter are prominent factors for their effect on soil aggregation and physical structure (Wang et al. 2022).

Soil pores provide a home for specific soil fauna, including earthworms, mites, and plant rhizosphere microorganisms such as viruses and fungi. Brownian diffusion in soil particles allows NPs and MPs to enter soil pores (Wang et al. 2010). The diameter of NPs and MPs blocks the pore, and the hydrophobicity of NPs and MPs causes soil water repellency. The soil pore occupancy by NPs and MPs affects the movement and distribution of soil fauna; for example, springtail (*Lobella sokamensis*) movement was disrupted by polystyrene MPs influx at a concentration of 8 mg/kg (Kim and An 2019). Furthermore, soil porousness primarily regulates water flow and soil aeration; consequently, NPs and MPs indirectly alter the relative richness of aerobic and/or anaerobic microbial populations and the uptake of water and nutrients by the plant root system.

Soil bulk density is associated with plant rooting, and it is an essential sign of soil fecundity (Dexter 2004). The soil bulk density is reduced as NPs and MPs accumulate in the soil as the density of NPs and MPs is generally lower compared to the density of soil particles (Wang et al. 2022). This mainly increases aeration and soil porosity, which can support and accelerate aerobic processes and facilitate root penetration.

While soil water regulates nutrients and pollutants availability, reproduction, and survival of soil plants and organisms, the impact of NPs and MPs through their hydrophobic surfaces on water-holding capacity can indirectly alter soil physical structure, aggregation, and porousness (de Souza Machado et al. 2018, 2019). MPs can interrupt vertical soil water flow and change soil water transport. The altered water-holding capacity by NPs and MPs may affect evapotranspiration, soil moisture content, and water availability, thus exacerbating drought (Wan et al. 2019; de Souza Machado et al. 2019). Nevertheless, the impacts of NPs and MPs on water-holding capacity are variable depending on the content, dimension, type, and shape. An increase in water-holding availability and capacity can attenuate drought and promote plant growth; on the contrary, a decrease in water-holding capacity can exacerbate soil drought if it enhances water infiltration and evapotranspiration (Lozano et al. 2021). Soil water is vital in ecosystem services and functions, especially agricultural production.

5.3.2 Effects of NPs and MPs on Soil Chemical Properties

One of the key abiotic factors that regulates soil attributes including soil organic carbon, adsorption of nutrients, binding potentiality of minerals, and microbial performances is soil pH (Rousk et al. 2010). NPs and MPs, depending on polymeric compositions, shape and sizes, doses, and incubating time (Dong et al. 2021), can increase (Qi et al. 2020; Yang et al. 2021), decrease, or not affect soil pH (Boots et al. 2019). Small-sized NPs and MPs appeared to have detrimental consequences on soil pH in comparison to larger ones. The implications of microplastics on soil pH are further influenced by agricultural practices, drought, buried depth (deep or superficial) in soil, soil moisture, and soil factors (fluvo-aquic, paddy, or fertilized red soil) (Bandow et al. 2017; Lozano et al. 2021).

MPs might modify soil cation exchange, enabling proton exchange in soil water and potentially affecting soil pH levels (Boots et al. 2019). Notably, the pH change can affect the microbial community by regulating the community structure of microbes (Kim et al. 2020; Feng et al. 2022); however, the exact mechanism by which NPs and MPs affect soil pH remains unclear. Undoubtedly, soil pH alterations are a reason for huge concern as they affect soil organic matter (SOM), and nutrient availability, eventually influencing plant growth and crop productivity.

SOM is linked to microbial activity and soil fertility. Compared to acidic environments, MPs altered SOM more strongly under alkaline situations. The SOM is reduced with the increase of NPs and MPs concentration impacting the soil microbial populations (Liu et al. 2021). The consequences of NPs and MPs on SOM currently focus on dissolved organic matter, which contributes to the relocation of contaminants in soil and the transportation of soil nutrients, namely, P and N (Kalbitz et al. 2000; Feng et al. 2022). Though existing studies have found contradicting effects (Li et al. 2021), adding NPs and MPs can significantly elevate the soil's dissolved organic carbon level (Feng et al. 2022).

5.3.3 Effects of NPs and MPs on Soil Organisms and Microbial Processes

Owing to their greater surface-to-volume ratio, NPs are more detrimental than MPs to soil microorganisms. The interaction of microorganisms with NPs and MPs can influence the P and N cycle and the amount of greenhouse gas production in soil (Wang et al. 2022). Nonetheless, we expect that changes in soil physical and chemical structure by NPs and MPs could influence the diversity of microbial communities (e.g., N-fixers, rhizosphere microbes, and pathogens) (Rillig et al. 2019) (Fig. 5.2). The presence of NPs and MPs was reported to modify the activity and performances of various soil microbial enzymes (e.g., alkaline-phosphatase, dehydrogenase) (Kim et al. 2021). The described negative consequences of NPs and MPs exposition include hindering plant growth (Liu et al. 2014), declin-

Fig. 5.3 Schematic diagrams on pathways to achieve the Sustainable Development Goals by managing plastic waste. (Adopted from Sarkar et al. (2022))

ing diversity, and the efficacy of soil microbes (Besseling et al. 2013; Mbachu et al. 2021). Moreover, the bioaccumulation and trophic transfer of NPs and MPs via the food chain could negatively disturb the tertiary predator (e.g., humans) (Guo et al. 2020).

5.4 Plastic Pollution and Sustainable Development Goals

The *Sustainable Development Goals (SDGs)* were set forth by the United Nations (UN) in 2015 to address societal, economic, and environmental challenges and to promote a sustainable future by 2030. The UN SDGs aim to conserve our planet, eradicate poverty, and ensure that everyone lives in prosperity and security by 2030. Healthy seas and oceans are vital for human health and existence. Among the UN SDGs, SDG 14 ("conserve and sustainably use the oceans, seas, and marine resources for sustainable development") mainly concentrates on influences of NPs and MPs pollution from anthropogenic activities (e.g., marine debris and land-based sources). At least 12 of the UN SDGs, such as SDG 2 ("end hunger, achieve food security and improved nutrition and promote sustainable agriculture"), SDG 3 ("good health and well-being"), and SDG 6 ("clean water and sanitation") face an indirect threat from NPs and MPs (Walker 2021; Huck 2022) (Fig. 5.3).

Maintaining proper soil and aquatic ecosystem health is important to ensure the sustainability of our food systems. Recent research suggests NPs and MPs accumulation in agricultural soils (Kumar et al. 2020), fruits and vegetables (Conti et al. 2020), and seafood (Barboza et al. 2018; Karbalaei et al. 2018). This accumulation is a threat to human beings by the consumption of food items like salt, bottled water, and seafood

(Danopoulos et al. 2020; Choudhury et al. 2022b). Plastic food packaging has also been identified as a source of NPs and MPs in human food (Sobhani et al. 2020). Cox et al. (2019) documented that every year the amount of MPs intake via air, water, and food consumption, including poultry, seafood, and sugar, is 39–52 thousand particles per person. There is overwhelming evidence that shows the contamination of seafood and soil by NPs and MPs potentially affects overall human health, food safety, and sustainability.

5.5 Conclusions

NPs and MPs enter terrestrial and aquatic environments from primary and secondary sources. However, according to the available data, their interactions with soil and aquatic organisms have not been fully understood. This has a tremendous detrimental effect on human well-being and food production. Consequently, with adequate plastic reduction or proper recycling, achieving the UN SDGs is much easier for governments and organizations (Fig. 5.3). Hence, a growing number of studies suggested various ways to mitigate the negative consequences of NPs and MPs on soil and plants, including soil remediation, switching to biodegradable plastic, and raising awareness about the harmful impacts of plastic pollution. Future research needs to incorporate empirical evidence of plastic contamination and experiment with potential mitigation strategies to minimize the effects on soil and plants.

References

Alimba CG, Faggio C (2019) Microplastics in the marine environment: current trends in environmental pollution and mechanisms of toxicological profile. Environ Toxicol Pharmacol 68:61–74

Andrady AL (2011) Microplastics in the marine environment. Mar Pollut Bull 62(8):1596–1605

Arif Y, Mir AR, Zieliński P et al (2024) Microplastics and nanoplastics: source, behavior, remediation, and multi-level environmental impact. Environ Manag 356:120618

Bandow N, Will V, Wachtendorf V et al (2017) Contaminant release from aged microplastic. Environ Chem 14:394–405

Barboza LGA, Vethaak AD, Lavorante BR et al (2018) Marine microplastic debris: an emerging issue for food security, food safety and human health. Mar Pollut Bull 133:336–348

Besseling E, Wegner A, Foekema EM et al (2013) Effects of microplastic on fitness and PCB bioaccumulation by the lugworm *Arenicola marina* (L.). Environ Sci Technol 47(1):593–600

Boots B, Russell CW, Green DS (2019) Effects of microplastics in soil ecosystems: above and below ground. Environ Sci Technol 53:11496–11506

Chen G, Li Y, Liu S (2022) Effects of micro (nano) plastics on higher plants and the rhizosphere environment. Sci Total Environ 807:150841

Choudhury M, Sahoo S, Samanta P et al (2022a) COVID-19: an accelerator for global plastic consumption and its implications. Int J Environ Res Public Health 2022(1):1066350. https://doi.org/10.1155/2022/1066350

Choudhury M, Sharma A, Pervez A et al (2022b) Growing menace of microplastics in and around the coastal ecosystem. In: Coastal ecosystems: environmental importance, current challenges and conservation measures, pp 117–137. https://doi.org/10.1007/978-3-030-84255-0_6

Conti GO, Ferrante M, Banni M et al (2020) Micro- and nano-plastics in edible fruit and vegetables. The first diet risks assessment for the general population. Environ Res 187:109677

Cox KD, Covernton GA, Davies HL et al (2019) Human consumption of microplastics. Environ Sci Technol 53(12):7068–7074

Danopoulos E, Jenner LC, Twiddy M et al (2020) Microplastic contamination of seafood intended for human consumption: a systematic review and meta-analysis. Environ Health Perspect 128(12):126002

de Souza Machado AA, Lau CW, Till J et al (2018) Impacts of microplastics on the soil biophysical environment. Environ Sci Technol 52(17): 9656–9665

de Souza Machado AA, Lau CW, Kloas W et al (2019) Microplastics can change soil properties and affect plant performance. Environ Sci Technol 53(10):6044–6052

Dexter AR (2004) Soil physical quality: part I. Theory, effects of soil texture, density, and organic matter, and effects on root growth. Geoderma 120:201–214

Dey S, Santra M, Choudhury M et al (2024) Food waste generation and its industrial utilization: an overview. Environ Sci Pollut Res:1–20. https://doi.org/10.1007/s11356-024-34252-3

Dong Y, Gao M, Qiu W et al (2021) Effect of microplastics and arsenic on nutrients and microorganisms in rice rhizosphere soil. Ecotoxicol Environ Saf 211:111899

Feng X, Wang Q, Sun Y et al (2022) Microplastics change soil properties, heavy metal availability and bacterial

community in a Pb-Zn-contaminated soil. J Hazard Mater 424:127364
Gewert B, Plassmann MM, MacLeod M (2015) Pathways for degradation of plastic polymers floating in the marine environment. Environ Sci Process Impacts 17(9):1513–1521
Guo JJ, Huang XP, Xiang L et al (2020) Source, migration and toxicology of microplastics in soil. Environ Int 137:105263
Horton AA, Walton A, Spurgeon DJ et al (2017) Microplastics in freshwater and terrestrial environments: evaluating the current understanding to identify the knowledge gaps and future research priorities. Sci Total Environ 586:127–141
Huck W (2022) Goal 14 conserve and sustainably use the oceans, seas and marine resources for sustainable development. Sustainable development goals: article-by-article commentary. Nomos, CH Beck, Hart Publishing
Imhof HK, Ivleva NP, Schmid J et al (2013) Contamination of beach sediments of a subalpine lake with microplastic particles. Curr Biol 23(19):867–868
Kalbitz K, Solinger S, Park JH et al (2000) Controls on the dynamics of dissolved organic matter in soils: a review. Soil Sci 165:277–304
Karbalaei S, Hanachi P, Walker TR et al (2018) Occurrence, sources, human health impacts and mitigation of microplastic pollution. Environ Sci Pollut Res 25:36046–36063
Kibblewhite MG, Ritz K et al (2008) Soil health in agricultural systems. Philos Trans R Soc B Biol Sci 363(1492):685–701
Kim SW, An YJ (2019) Soil microplastics inhibit the movement of springtail species. Environ Int 126:699–706
Kim SW, Waldman WR, Kim TY et al (2020) Effects of different microplastics on nematodes in the soil environment: tracking the extractable additives using an ecotoxicological approach. Environ Sci Technol 54(21):13868–13878
Kim YN, Yoon JH, Kim KHJ (2021) Microplastic contamination in soil environment–a review. Soil Sci Annu 71(4):300–308
Kumar M, Xiong X, He M et al (2020) Microplastics as pollutants in agricultural soils. Environ Pollut 9:114980
Law KL, Thompson RC (2014) Microplastics in the seas. Science 345:144–145
Lehmann A, Leifheit EF, Gerdawischke M et al (2021) Microplastics have shape-and polymer-dependent effects on soil aggregation and organic matter loss-an experimental and meta-analytical approach. Micro Nanoplast 1:1–14
Li HZ, Zhu D, Lindhardt JH et al (2021) Long-term fertilization history alters effects of microplastics on soil properties, microbial communities, and functions in diverse farmland ecosystem. Environ Sci Technol 55:4658–4668
Lithner D, Larsson Å, Dave G (2011) Environmental and health hazard ranking and assessment of plastic polymers based on chemical composition. Sci Total Environ 409(18):3309–3324
Liu J, Bu L, Zhu L et al (2014) Optimizing plant density and plastic film mulch to increase maize productivity and water-use efficiency in semiarid areas. Agron J 106(4):1138–1146
Liu M, Lu S, Song Y et al (2018) Microplastic and mesoplastic pollution in farmland soils in suburbs of Shanghai, China. Environ Pollut 242(Pt A):855–862
Liu Y, Huang Q, Hu W et al (2021) Effects of plastic mulch film residues on soil-microbe-plant systems under different soil pH conditions. Chemosphere 267:128901
Lozano YM, Aguilar-Trigueros CA, Onandia G et al (2021) Effects of microplastics and drought on ecosystem functions and multifunctionality. J Appl Ecol 58:988–996
Mason SA, Welch VG, Neratko J (2018) Synthetic polymer contamination in bottled water. Front Chem 6:407
Mbachu O, Jenkins G, Kaparaju P et al (2021) The rise of artificial soil carbon inputs: reviewing microplastic pollution effects in the soil environment. Sci Total Environ 780:146569
Mondal T, Choudhury M, Kundu D et al (2023) Landfill: an eclectic review on structure, reactions and remediation approach. Waste Manag 164:127–142
Paul S, Choudhury M, Deb U et al (2019) Assessing the ecological impacts of ageing on hazard potential of solid waste landfills: a green approach through vermitechnology. J Clean Prod 236:117643
Qi Y, Ossowicki A, Yang X et al (2020) Effects of plastic mulch film residues on wheat rhizosphere and soil properties. J Hazard Mater 387:121711
Ragusa A, Svelato A, Santacroce C et al (2021) Plasticenta: first evidence of microplastics in human placenta. Environ Int 146:106274
Rajpal A, Choudhury M, Goswami S et al (eds) (2024) Waste management and treatment: advances and innovations, 1st edn. CRC Press. https://doi.org/10.1201/9781003258377
Rillig MC (2012) Microplastic in terrestrial ecosystems and the soil? Environ Sci Technol 46:6453–6454
Rillig MC, Lehmann A (2020) Microplastic in terrestrial ecosystems. Science 368:1430–1431
Rillig MC, Ryo M, Lehmann A et al (2019) The role of multiple global change factors in driving soil functions and microbial biodiversity. Sci 366:886–890
Rochman CM, Browne MA, Halpern BS et al (2013) Policy: classify plastic waste as hazardous. Nature 494:169–171
Roex E, Vethaak D, Leslie H et al (2013) Potential risk of microplastics in the fresh water environment. STOWA, Amersfoort
Rousk J, Bååth E, Brookes PC et al (2010) Soil bacterial and fungal communities across a pH gradient in an arable soil. ISME J 4:1340–1351
Sarkar B, Dissanayake PD, Bolan NS et al (2022) Challenges and opportunities in sustainable management of microplastics and nanoplastics in the environment. Environ Res 207:112179

Selvam S, Manisha A, Venkatramanan S et al (2020) Microplastic presence in commercial marine sea salts: a baseline study along Tuticorin Coastal salt pan stations, Gulf of Mannar, South India. Mar Pollut Bull 150:110675

Sobhani Z, Lei Y, Tang Y et al (2020) Microplastics generated when opening plastic packaging. Sci Rep 10(1):4841

Statista (2017) Global plastic production from 1950 to 2015. https://www.statista.com/statistics/282732/global-production-of-plastics-since-1950/. Accessed Aug 2017

Trevor ML (2020) Chapter 1 – introduction to plastic waste and recycling. In: Plastic waste and recycling. Academic Press, New York, pp 3–12

Walker TR (2021) (Micro) plastics and the UN sustainable development goals. Curr Opin Green Sustain Chem 30:100497

Wan Y, Wu C, Xue Q et al (2019) Effects of plastic contamination on water evaporation and desiccation cracking in soil. Sci Total Environ 654:576–582

Wang Y, Li Y, Kim H et al (2010) Transport and retention of fullerene nanoparticles in natural soils. J Environ Qual 39(6):1925–1933

Wang F, Wang Q, Adams CA et al (2022) Effects of microplastics on soil properties: current knowledge and future perspectives. J Hazard Mater 424:127531

Yang W, Cheng P, Adams CA et al (2021) Effects of microplastics on plant growth and arbuscular mycorrhizal fungal communities in a soil spiked with ZnO nanoparticles. Soil Biol Biochem 155:108179

Yu L, Zhang J, Liu Y et al (2021) Distribution characteristics of microplastics in agricultural soils from the largest vegetable production base in China. Sci Total Environ 756:143860

Yu H, Zhang Y, Tan W et al (2022) Microplastics as an emerging environmental pollutant in agricultural soils: effects on ecosystems and human health. Front Environ Sci 10:855292

Yuan J, Ma J, Sun Y et al (2020) Microbial degradation and other environmental aspects of microplastics/plastics. Sci Total Environ 715:136968

Zhang GS, Liu YF (2018) The distribution of microplastics in soil aggregate fractions in southwestern China. Sci Total Environ 642:12–20

Zhang N, Bin LY, He HR et al (2021) You are what you eat: microplastics in the feces of young men living in Beijing. Sci Total Environ 767:144345

Zhou B, Wang J, Zhang H et al (2020) Microplastics in agricultural soils on the coastal plain of Hangzhou Bay, East China: multiple sources other than plastic mulching film. J Hazard Mater 388:121814

Ziajahromi S, Neale PA, Rintoul L et al (2017) Wastewater treatment plants as a pathway for microplastics: development of a new approach to sample wastewater-based microplastics. Water Res 112:93–99

Micro and Nanoplastics: Contamination of Air

6

Prerna Singh, Sarva Mangala Praveena, and Manoj Kumar

6.1 Introduction

Plastic has become integral to our lives because of its properties: low manufacturing cost, easy accessibility, light weight, and versatility. Its extensive use has put tremendous pressure on the environment, i.e. air, water, and soil. Increased production and usage of plastic among the urban populace have resulted in a massive generation of micro and nanoplastics (MPs/NPs), which today have emerged as new sources of environmental pollutants (UNEP 2018; Choudhury et al. 2022). Research on the existence of MPs/NPs in the ambient atmosphere has gained momentum over the last few years (Dris et al. 2015; Zhang et al. 2020a); societal and ecological concerns due to atmospheric MPs/NPs have resulted in more studies. Since the studies on atmospheric MPs/NPs are still in their initial stage, more active research is needed to understand their atmospheric origin or sources, distribution, life cycle, impact on different living organisms, etc. This chapter addressed the issue of MPs and NPs in the atmosphere, including information on their sources, distribution, and impacts on many living beings and air quality. At first, it focused on the importance of clean air for life support and the need for air quality monitoring policies for MPs and NPs. Afterwards, it examined the pathways by which MPs and NPs entered the atmosphere, highlighting their ubiquitousness in various environmental compartments. Furthermore, it discussed the harmful impacts of MPs and NPs on the environment and human health, focusing on how they affect various plant, animal, and human species. Then, it included an overview of atmospheric MPs and NP studies, focusing primarily on the limited research carried out in developing nations like India. Lastly, it described the approaches for detecting and analysing MPs/NPs in the air, highlighting the significance of standardized methods and additional research to understand this environmental problem's scope thoroughly.

P. Singh · M. Kumar (✉)
Environmental Sciences & Biomedical Metrology Division, CSIR—National Physical Laboratory, New Delhi, India

Academy of Scientific and Innovative Research, Ghaziabad, Uttar Pradesh, India
e-mail: prerna.npl22a@acsir.res.in; manojkumar.nplindia@csir.res.in

S. M. Praveena
Department of Environmental & Occupational Health, Faculty of Medicine & Health Sciences, University Putra Malaysia, Serdang, Selangor, Malaysia
e-mail: smpraveena@upm.edu.my

S. Goswami et al. (eds.), *Micro-Nano Plastics Exposure, Environmental Degradation and Public Health Crisis*, Sustainable Development Goals Series, https://doi.org/10.1007/978-981-96-5047-7_6

6.2 Importance of Air in Sustaining Life

A clean and healthy atmosphere is one of the fundamental requirements for the existence and survival of all living things. The Earth's atmosphere is composed of nitrogen (78%), oxygen (20.9%), argon (0.9%), and many trace gases or aerosols coming from different sources (Schlesinger and Bernhardt 2020). Many gases and aerosols present in trace amounts act as pollutants that have the potential to impact the abiotic and biotic components negatively. Aerosols are considered pollutants suspended in the atmosphere in multiple phases, such as solid or liquid. Aerosols are divided into two categories—primary aerosols (such as black carbon released directly into the atmosphere) and secondary aerosols (such as sulphate and nitrate) (Bellouin et al. 2020). The air quality index (AQI) is one of the methods used to monitor the changes in air quality and the overall pollution of a region. Depending on the harmful effects on human beings, the AQI scale is divided into categories: good, satisfactory, moderately polluted, poor, very poor, and severe. At present, this index is provided only for eight pollutants, including PM_{10}, $PM_{2.5}$, carbon monoxide (CO), oxides of sulphur (SOx), oxides of nitrogen (NOx), tropospheric ozone (O_3), ammonia (NH_3), and lead (Pb) (CPCB 2014). A separate parameter in AQI needs to be developed for other emerging pollutants, for example, MPs and NPs, which today have emerged as new sources of contaminants in the atmospheric region (Sridharan et al. 2021).

6.3 Micro and Nanoplastics: Route of Entry to Atmosphere

Plastics, widely used in industries such as storage, clothing, and packaging, contribute significantly to environmental pollution because of their broad use and slow disintegration rate (UNEP 2018). These plastics degrade into MPs (1 μm to 5 mm) and NPs (<1 μm). MPs/NPs come from a variety of sources, including primary sources such as microbeads used in personal care items and microfibres from textile industries, as well as secondary sources resulting from the breakdown of larger plastic particles due to environmental factors such as UV radiation and wind abrasion (Henry et al. 2019). Synthetic fibres, tyre abrasion, sewage effluent, urban runoff, and atmospheric deposition are some familiar sources (Dris et al. 2015; Sommer et al. 2018; Gatidou et al. 2019; De Falco et al. 2019; Galvão et al. 2020) (Fig. 6.1). The slow disintegration of MPs/NPs allows them to exist in the atmosphere in various forms, including persistent fibres and fragments. They can absorb a wide range of contaminants, such as pesticides and heavy metals, and their transportation is influenced by various meteorological conditions such as temperature and wind patterns (Naik et al. 2019; Narmadha et al. 2020; Wright et al. 2020). Although atmospheric MPs/NP pollution has recently received attention, it has been an environmental issue for many years but is not widely recognized (Bhat et al. 2022).

6.4 Micro and Nanoplastics: Air Contamination Scenario Worldwide

Globally, settled and suspended MPs and NPs have become prominent air pollutants in atmospheric dust. Generally speaking, settled dust is less dangerous than suspended dust because the latter is more likely to be inhaled and stay in the air for extended periods. Most prior research on MPs in suspended particulate matter (PM) has used air samplers, which might only partially capture the diversity of MPs/NPs. Because MPs and NPs come in various sizes, shapes, and compositions, only a tiny percentage of them can be detected in the current air quality evaluations that mainly focus on particulate matter sizes like $PM_{2.5}$ and PM_{10} (European

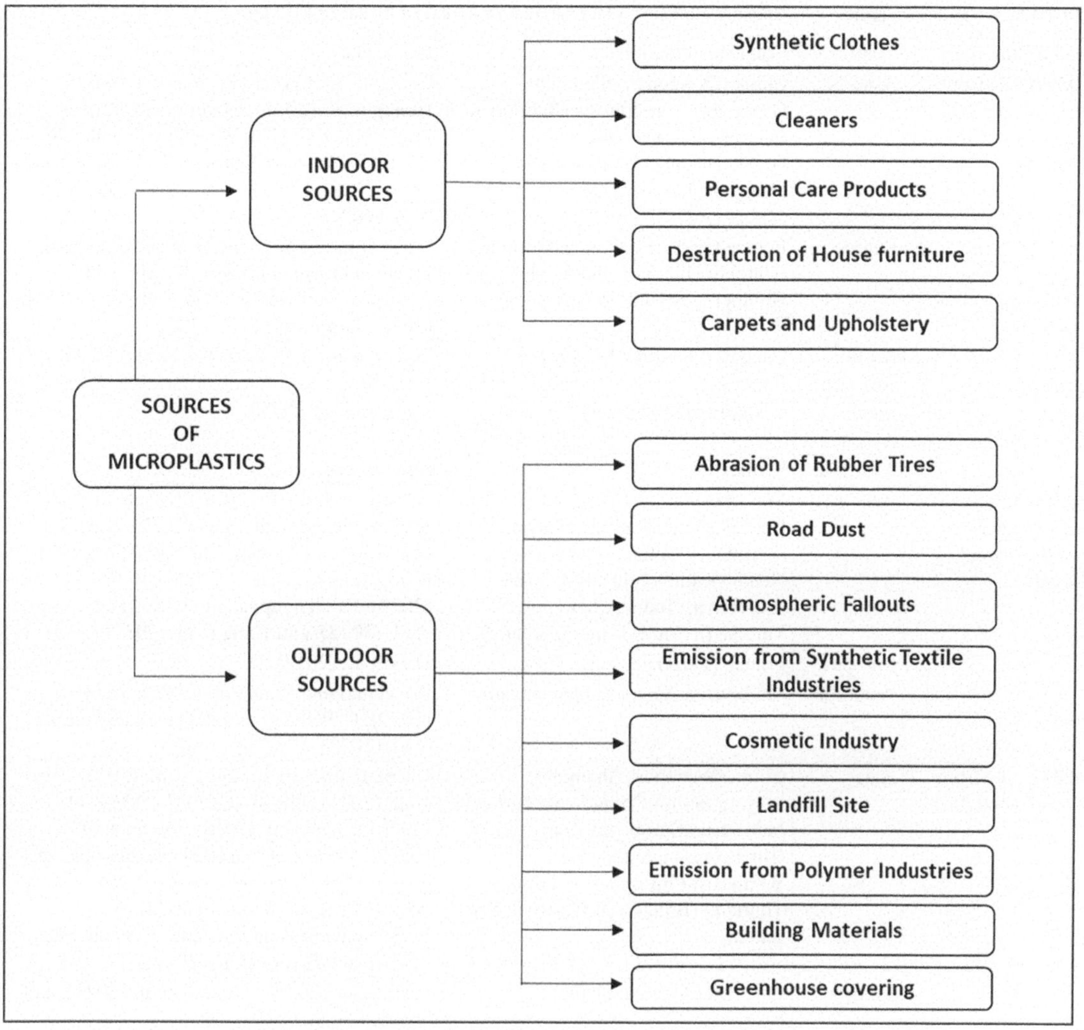

Fig. 6.1 Various indoor and outdoor sources of atmospheric micro and nanoplastics (MPs/NPs)

Commission 2019). This limitation makes it more challenging to comprehend the impact of MPs and NPs on air quality. Previous studies have revealed complex, non-linear relationships between MPs and PMs, emphasizing the need for additional in-depth analyses of MPs/NPs in PM concentrations (Zhu et al. 2021). Despite the growing number of studies on MPs and NPs in the atmosphere using various sampling methods (Table 6.1), there is a pressing need for improved monitoring and more comprehensive air quality standards to mitigate their significant impact on human health and ecosystems (Bhat et al. 2022).

6.5 Consequences of Atmospheric Micro and Nanoplastics

6.5.1 Effects on Air Quality

MPs and NPs significantly degrade air quality by raising particulate matter (PM_{10} and $PM_{2.5}$) levels, especially in urban areas. These particles can infiltrate the respiratory system, causing inflammation and exacerbating asthma (Blackburn and Green 2022). MPs absorb and transport harmful chemicals, such as heavy metals, releasing them into the environment (Naik et al. 2019). They act

Table 6.1 Global overview of atmospheric MPs/NPs research categorized by sampling type

Sampling type	Continent	Countries/locations	References
MPs/NPs from atmospheric dust deposition/ fallout	Asia	China (Dongguan, Shanghai, Guangzhou); Indonesia (Surabaya, Jakarta); Iran (Shiraz, Mount Derak, Sistan); Vietnam (Ho Chi Minh); India (Mumbai)	Cai et al. (2017), Zhang et al. (2020b), Huang et al. (2021), Cui et al. (2022), Pratiwi et al. (2020), Purwiyanto et al. (2022), Abbasi and Turner (2021), Shahraki et al. (2022), Truong et al. (2021), and Yadav et al. (2022)
	Europe	France (Paris, Pyrenees); Germany (Hamburg); UK (London, Hull and Humber); Ireland; Poland (Krakow)	Dris et al. (2015, 2016), Allen et al. (2019), Klein and Fischer (2019), Wright et al. (2020), Jenner et al. (2021), Roblin et al. (2020), and Jarosz et al. (2022)
	North America	Canada (Ontario); USA (New Jersey)	Welsh et al. (2022) and Yao et al. (2022)
	South America	Colombia (Cartagena)	Joaquin et al. (2022)
	Australia	Sydney	Soltani et al. (2021)
MPs/NPs from atmospheric settled street dust	Asia	Iran (Bushehr, Tehran, Shiraz); China (39 major cities); Japan (Kusatsu); Vietnam (Da Nang); Nepal (Kathmandu); Indonesia (Surabaya); India (Chennai, Varanasi); South Korea (Goyang)	Abbasi et al. (2017, 2022), Dehghani et al. (2017), Nematollahi et al. (2022), Kashfi et al. (2022), Liu et al. (2019a), Zhang et al. (2020a), Yukioka et al. (2020), Bahrina et al. (2020), Patchaiyappan et al. (2021), Pandey et al. (2022), and Kang et al. (2022)
	Europe	France (Paris)	Dris et al. (2017)
	Australia	Melbourne, Victoria, Queensland	Roychand and Pramanik (2020), Su et al. (2020), O'Brien et al. (2021), and Monira et al. (2022)
MPs/NPs from suspended dust	Asia	China (Wenzhou, Shanghai, Beijing, South China Sea); Turkey; Indonesia (Surabaya); Iran (Bushehr); India (Nagpur, Varanasi); Philippines (Manila); Thailand (Bangkok); Pakistan; Sri Lanka	Liu et al. (2019b, c), Ding et al. (2021), Liao et al. (2021), Zhu et al. (2021), Xie et al. (2022), Kaya et al. (2018), Syafei et al. (2019), Syafina et al. (2022), Akhbarizadeh et al. (2021), Narmadha et al. (2020), Pandey et al. (2022), Romarate et al. (2023), Sarathana and Winijkul (2022), Aslam et al. (2022), and Perera et al. (2022)
	Europe	Portugal (Aveiro); Denmark (Aarhus)	Prata et al. (2020), Xumiao et al. (2021), and Vianello et al. (2019)
	North America	USA (California, New Jersey); Mexico	Gaston et al. (2020), Shruti et al. (2022), and Yao et al. (2022)

as cloud condensation nuclei, influencing weather patterns, and interact with other pollutants to promote the generation of secondary pollutants in the air. MPs also affect atmospheric optical properties by scattering light, which impairs visibility and disrupts the radiative balance (Parvez et al. 2024). Given their widespread distribution and ability to travel long distances (Allen et al. 2019), MPs/NPs should be included in air quality monitoring and regulated under the National Ambient Air Quality Standards (NAAQS) (Sridharan et al. 2021). Primary studies on atmospheric MPs highlighted the severity of this issue; MPs concentrations in Paris ranged from 2 to 355 particles per square metre per day, according to Dris et al. (2016), whereas Allen et al. (2019) reported an average of 365 particles per square metre per day in the isolated French Pyrenees, with polystyrene and polyethene being the predominant polymers.

Similarly, in Nagpur, India, the average number of MP particles in the air is between 50 and 120 per day. Polymers detected include low-density polyethene, rubber fibre, polystyrene, polyolefin, rayon, polyaniline, and chlorinated polyvinyl chloride (Narmadha et al. 2020). Numerous studies have documented these minuscule plastic particles in the atmosphere across a wide range, highlighting the urgent need for further investigations.

6.5.2 Effects on Plants and Animals

MPs and NPs in the atmosphere significantly affect soil health by disrupting organic carbon levels, biodiversity, nutrient exchange, and microbial activity, which reduces crop productivity and plant growth (Rillig 2018; Rillig and Lehmann 2020). These contaminants alter soil properties such as composition, aeration, water retention, and permeability (Zhang et al. 2018; Boots et al. 2019). Different polymers have distinct impacts; for example, polystyrene (PS) impairs seed germination, early growth and hinders nutrient uptake in rice, lettuce, and wheat; polyethylene (PE) affects maize by impairing nitrogen and water uptake; polyvinyl chloride (PVC) reduces plant height and biomass in sweet potatoes and also inhibits leaf multiplication in *Spirodela polyrhiza*; polyethylene terephthalate (PET) and polyethylene furanoate (PEF) inhibit lettuce development by producing reactive oxygen species (ROS) and decreasing nutrient accumulation; and high-density polyethylene (HDPE) negatively impacts ryegrass by reducing seed germination and root–shoot length (Jia et al. 2023). Also, MPs/NPs with different shapes affect the soil biota, plant growth, and microbial activity, e.g. fibres enhance the porosity and aeration of soils, and films reduce soil bulk density; however, fragments and foams improve porosity and soil aeration (Lozano et al. 2021). MPs/NPs, along with heavy metals, disrupt root symbiosis, soil organisms, and nutrient cycling and can be ingested by animals such as earthworms and birds, impairing health and perpetuating a cycle of plastic pollution through the food chain to humans (Saeedi 2024; Boots et al. 2019; Wang et al. 2020). Investigating the effects on forest canopies and emergent layers is essential for understanding their overall impact on forest ecosystems (Sridharan et al. 2021). A comprehensive examination of these effects will help develop solutions to reduce the adverse effects of MPs on plant health and ecosystem stability.

6.5.3 Effects on Human Health

Atmospheric MPs and NPs pose significant health concerns because they can enter the human body by inhaling airborne MPs particles, consuming contaminated food and water, and through dermal contact with polluted air and surfaces (López et al. 2023). Studies have identified MPs in various body fluids and tissues. The human placenta included 12 MPs fragments varying in size from 5 to 10 μm, with both spherical and irregular forms (Rasgua et al. 2021). In contrast, particles ≥700 nm are present in the blood at a mean concentration of 1.6 μg/ml (Leslie et al. 2022). MPs in human breastmilk range from 2 to 12 μm and the most common MPs observed were polyvinyl chloride, polypropylene, and polyethylene (Rasgua et al. 2022). Human faeces include 15 different varieties of MPs, with polyethylene terephthalate ranging from 22.3% to 34.0% and polyamide ranging from 8.9% to 12.4% dominating, primarily in the form of sheets and fibres (Yan et al. 2021). Synthetic fibres are present in 87% of human lungs, with plastic workers facing increased exposure concerns (Blackburn and Green 2022; Ojha et al. 2021). MPs can transport hazardous substances that disrupt endocrine functions and may cause cancer (Rajmohan et al. 2019; Mohamed Nor et al. 2021). MPs' exposure has been associated with increased inflammation, DNA damage, and oxidative stress (Vethaak and Legler 2021). Children and older adults with diseases are particularly vulnerable (Mohamed Nor et al. 2021). Despite these findings, the extent of MPs/NPs' influence on human health is still being investigated, highlighting the need for more studies and public health actions to reduce exposure and address the environmental impacts of these tiny particles.

6.6 Identifying Gaps in Atmospheric MPs/NPs Research: An Indian Context

Research on atmospheric MPs and NPs has received worldwide attention, with pioneering studies such as Dris et al. (2015) first detecting airborne MPs in Paris. Developed countries have made significant progress in understanding the MPs' concentration and their potential health consequences, including respiratory and cardiovascular disorders. For example, Paris had an average fallout of 118 particles/m^2/day, while investigations in Germany, China, and Iran revealed elevated airborne MPs levels (Habibi et al. 2022). MPs can enter into the bloodstream and breast milk, with over 76% of samples containing small plastic fragments (2–12 μm) (Leslie et al. 2022; Rasgua et al. 2022). In contrast, India, being a developing country, has conducted little research in this field. Preliminary studies in locations such as Chennai, Nagpur, Patna, Varanasi, Mumbai, Kerala, and along the Ganga River have shown varied concentrations of airborne MPs, resulting mainly from human activities (Table 6.2). However, extensive studies are required due to a lack of research and inconsistent findings. Future studies should expand the geographical scope, standardize techniques, and focus on source apportionment techniques to completely understand and mitigate the risks associated with airborne MPs.

6.7 Determination of Micro and Nanoplastics in Air

A standard procedure for atmospheric MPs/NPs detection in the environment includes sampling, extraction, density-based separation, identification, and categorization (Shim et al. 2017). The MPs/NPs detection methodology from the atmospheric compartment is represented in Fig. 6.2.

6.7.1 Determination of Micro and Nanoplastics in Air: Sampling Methods

Sampling methods, categorized as active and passive (Fig. 6.3), vary depending on sample type and practical considerations. However, both types of sampling methods present unique challenges, outlined in Table 6.3. Passive sampling methods are preferred over active methods because they are simple, easy to use, have no electricity requirements, and are of low cost. Currently, the existence of MPs/NPs in atmospheric compartments is analysed by various methods, targeting deposited or suspended particles (Beaurepaire et al. 2021). Passive sampling methods are suitable for collecting deposited particles. They are commonly used to assess long-term data lasting for weeks or months; thus, they give complete information about the MPs/NPs' deposition over a long time interval. In earlier studies, deposited dust was collected with the help of passive samplers, e.g. atmospheric fallout (in dry or wet form) collected with the help of a metallic funnel leading to a glass bottle. It was first used by Dris et al. (2015) in their founding study and is also considered the most common passive sampling method. Street or road dust was collected with the help of a wooden brush (Pandey et al. 2022), vacuum cleaner (Dris et al. 2017), and hog bristle or paintbrush (Patchaiyappan et al. 2021) and then transferred to paper bags with the help of metal pans. Passive sampling methods result in collecting coarser fractions of plastic particles than finer ones because lighter or smaller particles remain suspended in the air for longer. Active sampling methods are suitable for collecting suspended dust, and it is based on pumping a fixed amount of air with the help of an active sampler over a certain period, leading to more accurate and precise results (Beaurepaire et al. 2021; Dris et al. 2017). The

Table 6.2 Studies on atmospheric microplastics (MPs) conducted at different locations in India

S. No.	Study area	References	Sampling time	Sampling type	Dominant shape of MPs	Dominant polymer of MPs
1	Nagpur	Narmadha et al. (2020)	2019–2020 (November–February)	Active: Roadside suspended dust (Outdoor)	Fibres	Polyethylene (PE)
2	Patna	Zhang et al. (2020a)	2010–2014 (Not specified)	Passive: Settled dust samples (Indoor)		Polyethylene terephthalate (PET)
3	Chennai	Patchaiyappan et al. (2021)	2019 (May)	Passive: Street dust (Outdoor)	Fragments	Polyvinyl chloride (PVC)
4	Varanasi	Pandey et al. (2022)	2019 (June–September)	Both, Passive: Street dust (Outdoor) Active: Suspended dust (Outdoor)	Fragments	Polyethylene (PE)
5	Mumbai	Yadav et al. (2022)	2019–2021 (June and March)	Passive: Atmospheric deposition/dust fallout (Indoor and outdoor)	Fragments	–
6	Patna	Parashar and Hait (2023)	2022 (mid-June–mid-October)	Passive: Wet atmospheric deposition (Outdoor)	Fibres	Polyethylene terephthalate (PET) and polypropylene (PP)
7	Seven sites along the Ganga River in India (Patna, Kannauj, Varanasi, Anupshahar, Sahibganj, Harsil, and Rishikesh)	Napper et al. (2023)	2019 (May–June)	Passive: Atmospheric dust deposition (Outdoor)	Fibres	Rayon (synthetically altered cellulose)
8	Kerala	Kannankai and Devipriya (2024)	October 2021–September 2022	Passive: Atmospheric dust deposition (Outdoor)	Fibres	Polyethylene (PE)
9	Ranchi	Nandi et al. (2024)	November 2022–February 2023	Active: Suspended dust (Indoor and Outdoor)	Fibres	Polyethylene (PE)

The major highlights of these studies are listed

quantification of MPs/NPs present in the atmospheric compartments is mainly performed in various units based on the origin of particles. Deposited street dust is quantified in MP or NP number/gram, while atmospheric dust deposition or dust fallout is measured in MP or NP number/m^2/day. Suspended dust, on the other hand, is expressed as MP or NP number/m^3. These distinct representations help assess MPs concentrations in various dust samples, which is helpful in comprehensive analyses of air quality dynamics.

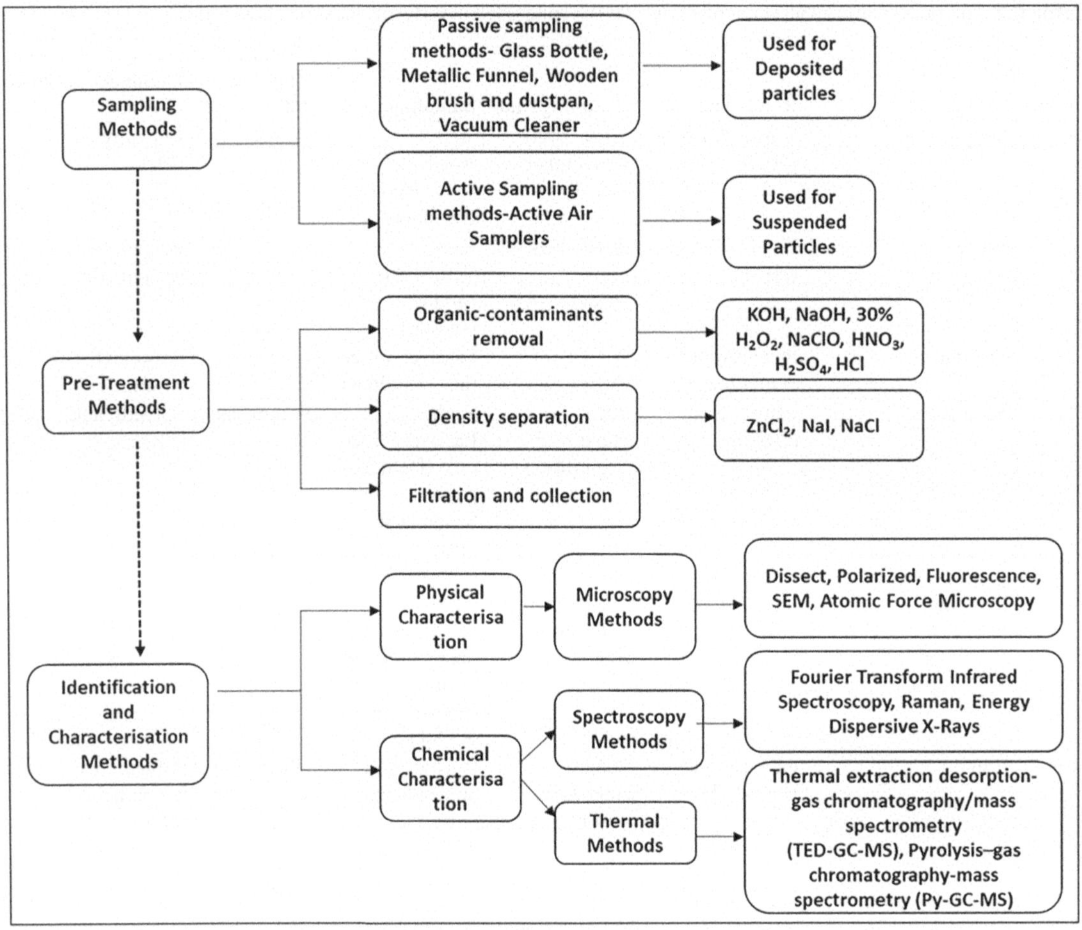

Fig. 6.2 Methodology for micro and nanoplastics (MPs/NPs) detection from the atmospheric compartment

6.7.2 Determination of Micro and Nanoplastics in Air: Pre-treatments and Extraction Methods

The isolation of MPs/NPs from the sample involves different pre-treatment and extraction methods, such as the destruction of organic contaminants or separation based on density. Beaurepaire et al. (2021) reviewed that the majority of the studies related to atmospheric compartments (i.e. 17 out of 26) did not apply any pre-treatment process due to various practical reasons like the absence of dense particles and less concentration of organic matter in air samples as compared to water or sediments. Organic digestion is performed by using potassium hydroxide (KOH), sodium hydroxide (NaOH), hydrogen peroxide (H_2O_2), enzymatic methods, or sodium hypochlorite (NaClO). Some studies have also performed acid digestion using nitric acid (HNO_3), hydrochloric acid (HCl), and sulphuric acid (H_2SO_4) (Allen et al. 2019; Dris et al. 2016; Uddin et al. 2020). The most widely used method among all the studies is digestion with a 30% H_2O_2 solution because it is more effective in digesting organic matter; thus, separation and identification of MPs/NPs can be easily carried out. Density separation is performed using chemicals such as zinc chloride ($ZnCl_2$), sodium iodide (NaI), and sodium chloride (NaCl) solution (Pandey et al. 2022; Narmadha et al. 2020; Patchaiyappan et al. 2021).

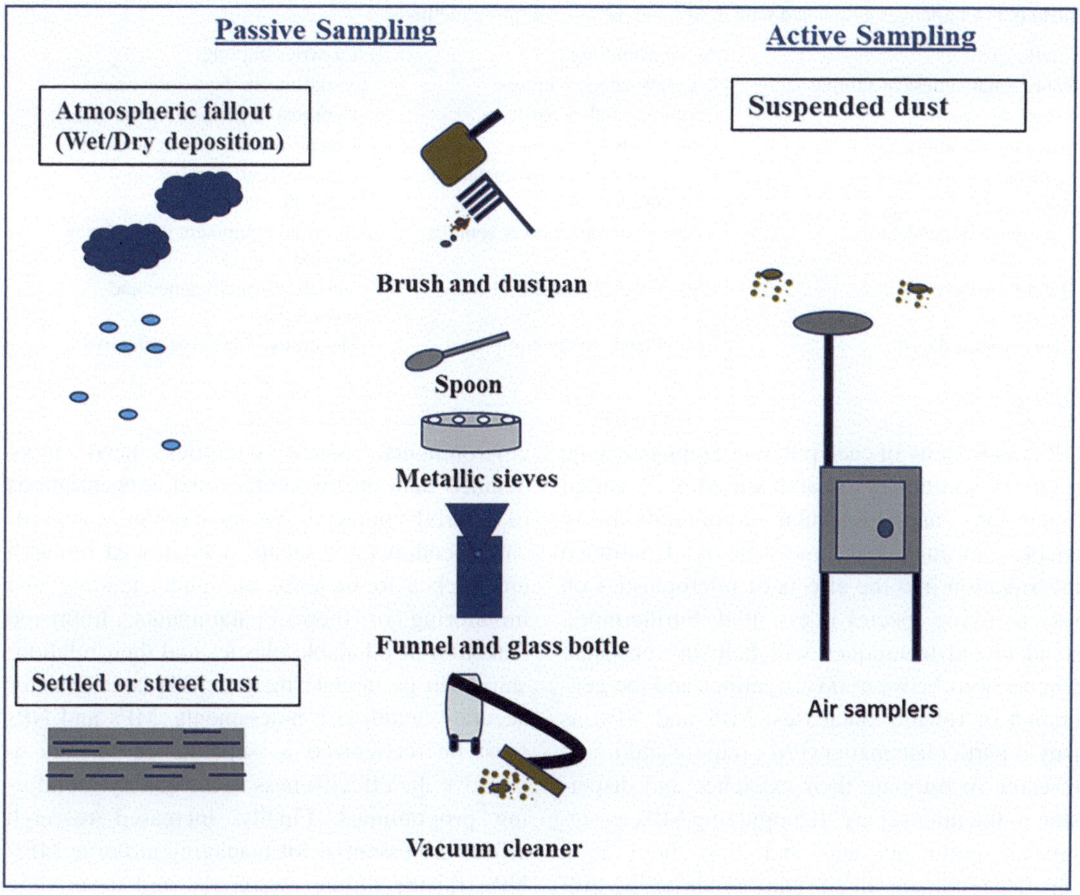

Fig. 6.3 Active and passive samplers for collecting micro and nanoplastics (MPs/NPs)

6.7.3 Determination of Micro and Nanoplastics in Air: Identification and Characterization Methods

MPs/NP analysis can be performed in two ways: (i) physical characterization (e.g. microscopy methods) and (ii) chemical characterization (e.g. spectroscopy methods) (Shim et al. 2017). Physical characteristics such as shape, size, colour, and number of airborne MPs/NPs are identified and counted under microscopes. Moreover, chemical or compositional details and polymer identification are performed with the help of spectroscopy methods. Before adopting a specific methodology, the factors considered in earlier studies were the detection limits, identification time, the need for samples pre-treatment, approximate costs, and the availability of instruments. Since there is no standardized method, both the characterization techniques (physical or chemical) would be better to quantify or analyse MPs/NPs in the sample, leading to more accurate results, e.g. Fourier transform infrared spectroscopy (FTIR) in combination with the microscopy technique adopted in most of the studies (Beaurepaire et al. 2021; Habibi et al. 2022).

6.8 Future Perspective and Recommendations

Future studies on MPs and NPs must address significant research gaps, particularly in developing countries with widespread pollution and plastic waste generation (Rajpal et al. 2024). The

Table 6.3 Challenges associated with active and passive sampling methods

Challenges	Active sampling	Passive sampling
Representativeness of sample	Location-specific biases	Variations in deposition rates
Sampling bias	Captures smaller particles	Captures larger settling particles
Sampling duration	Continuous operation	Longer durations for collection
Control over sampling	High control over parameters (flow rate, duration)	Limited control, relies on environmental factors
Equipment dependency	Requires pumps, power sources	Minimal equipment dependency needed
Quantification accuracy	Requires precise calibration	Variations in efficiency and techniques
Equipment and cost	Specialized, costly equipment	Generally, more cost-effective

different toxicity of each polymer emphasizes the need for additional research on MPs of varied dimensions and molecular configurations to enable thorough risk assessments. Continued investigation into the effects of microplastics on various living species is essential. Furthermore, standardized techniques will help in consistent comparisons between investigations and the generation of reliable databases. MPs and NPs, as part of particulate matter (PM), require additional research to estimate their existence and dispersion in the atmosphere. Recognizing MPs as significant pollutants and including them as a separate parameter in air quality monitoring programmes will improve regulatory plans.

Furthermore, investigating the consequences of MPs on climate change, developing source apportionment methodologies, and identifying tracers to detect the origins of atmospheric MPs could be the potential scope of future research. Also, studying meteorological conditions' effects on atmospheric MPs is critical for understanding their behaviour and dispersion in the atmosphere. Since MPs are considered critical environmental pollutants, addressing research gaps will improve our understanding and help manage MPs pollution, leading to an improved environment and more effective global regulations.

6.9 Conclusion

The impact of MPs and NPs on the atmospheric compartment has received comparatively little attention compared to aquatic and terrestrial environments. Many countries need more detailed data on the sources and concentrations of MPs/NPs in the air. We must optimize analytical procedures and create standardized research approaches to increase our understanding and monitoring of these contaminants. Improved detection of inhalable plastics and their relationship with particulate matter (PM) is critical for accurate health risk assessments. MPs and NPs must be recognized as separate parameters to improve the effectiveness of air quality monitoring programmes. Finally, increased research efforts are essential for managing airborne MPs/NPs, raising public awareness, and developing practical mitigation strategies.

References

Abbasi S, Turner A (2021) Dry and wet deposition of microplastics in a semi-arid region (Shiraz, Iran). Sci Total Environ 786:147358. https://doi.org/10.1016/j.scitotenv.2021.147358

Abbasi S, Keshavarzi B, Moore F, Delshab H, Soltani N, Sorooshian A (2017) Investigation of microrubbers, microplastics and heavy metals in street dust: a study in Bushehr city, Iran. Environ Earth Sci 76:1–19. https://doi.org/10.1007/s12665-017-7137-0

Abbasi S, Turner A, Sharifi R, Nematollahi Mohammad J, Keshavarzifard M, Moghtaderi T (2022) Microplastics in the school classrooms of Shiraz, Iran. Build Environ 207:108562. https://doi.org/10.1016/j.buildenv.2021.108562

Akhbarizadeh R, Dobaradaran S, Torkmahalleh MA, Saeedi R, Aibaghi R, Ghasemi FF (2021) Suspended fine particulate matter (PM2.5), microplastics (MPs), and polycyclic aromatic hydrocarbons (PAHs) in air: their possible relationships and health implications.

Environ Res 192:110339. https://doi.org/10.1016/j.envres.2020.110339

Allen S, Allen D, Phoenix VR, Le Roux G, Durántez Jiménez P, Simonneau A, Binet S, Galop D (2019) Atmospheric transport and deposition of microplastics in a remote mountain catchment. Nat Geosci 12(5):339–344. https://doi.org/10.1038/s41561-019-0335-5

Aslam I, Qadir A, Ahmad SR (2022) A preliminary assessment of microplastics in indoor dust of a developing country in South Asia. Environ Monit Assess 194:340. https://doi.org/10.1007/s10661-022-09928-3

Bahrina I, Syafei AD, Satoto R, Jiang JJ, Nurasrin NR, Assomadi AF, Nasir M (2020) An occupant-based overview of microplastics in indoor environments in the city of Surabaya, Indonesia. J Ecol Eng 21(8):236–242. https://doi.org/10.12911/22998993/126876

Beaurepaire M, Dris R, Gasperi J, Tassin B (2021) Microplastics in the atmospheric compartment: a comprehensive review on methods, results on their occurrence and determining factors. Curr Opin Food Sci 41:159–168. https://doi.org/10.1016/j.cofs.2021.04.010

Bellouin N, Quaas J, Gryspeerdt E, Kinne S, Stier P, Watson-Parris D, Stevens B (2020) Bounding global aerosol radiative forcing of climate change. Rev Geophys 58(1):e2019RG000660. https://doi.org/10.1029/2019RG000660

Bhat MA, Gedik K, Gaga EO (2022) Atmospheric micro (nano) plastics: future growing concerns for human health. Air Qual Atmos Health 16:233–262. https://doi.org/10.1007/s11869-022-01272-2

Blackburn K, Green D (2022) The potential effects of microplastics on human health: what is known and what is unknown. Ambio 51(3):518–530. https://doi.org/10.1007/s13280-021-01589-9

Boots B, Russell CW, Green DS (2019) Effects of microplastics in soil ecosystems: above and below ground. Environ Sci Technol 53(19):11496–11506. https://doi.org/10.1021/acs.est.9b03304

Cai L, Wang J, Peng J, Tan Z, Zhan Z, Tan X, Chen Q (2017) Characteristic of microplastics in the atmospheric fallout from Dongguan city, China: preliminary research and first evidence. Environ Sci Pollut Res 24:24928–24935. https://doi.org/10.1007/s11356-017-0116-x

Central Pollution Control Board (2014) National air quality index. https://cpcb.nic.in/National-Air-Quality-Index/. https://cpcb.nic.in/National-Air-Quality-Index/. Accessed 25 June 2024

Choudhury M, Sahoo S, Samanta P, Tiwari A, Tiwari A, Chadha U, Chakravorty A (2022) COVID-19: an accelerator for global plastic consumption and its implications. J Environ Public Health 2022(1):1066350. https://doi.org/10.1155/2022/1066350

Cui J, Chen C, Gan Q, Wang T, Li W, Zeng W, Jin B (2022) Indoor microplastics and bacteria in the atmospheric fallout in urban homes. Sci Total Environ 852:158233. https://doi.org/10.1016/j.scitotenv.2022.158233

De Falco F, Di Pace E, Cocca M, Avella M (2019) The contribution of washing processes of synthetic clothes to microplastic pollution. Sci Rep 9(1):6633. https://doi.org/10.1038/s41598-019-43023-x

Dehghani S, Moore F, Akhbarizadeh R (2017) Microplastic pollution in deposited urban dust, Tehran metropolis, Iran. Environ Sci Pollut Res 24:20360–20371. https://doi.org/10.1007/s11356-017-9674-1

Ding Y, Zou X, Wang C, Feng Z, Wang Y, Fan Q, Chen H (2021) The abundance and characteristics of atmospheric microplastic deposition in the northwestern South China Sea in the fall. Atmos Environ 253:118389. https://doi.org/10.1016/j.atmosenv.2021.118389

Dris R, Gasperi J, Rocher V, Saad M, Renault N, Tassin B (2015) Microplastic contamination in an urban area: a case study in Greater Paris. Environ Chem 12(5):592–599. https://doi.org/10.1071/EN1416

Dris R, Gasperi J, Saad M, Mirande C, Tassin B (2016) Synthetic fibers in atmospheric fallout: a source of microplastics in the environment? Mar Pollut Bull 104(1–2):290–293. https://doi.org/10.1016/j.marpolbul.2016.01.006

Dris R, Gasperi J, Mirande C, Mandin C, Guerrouache M, Langlois V, Tassin B (2017) The first overview of textile fibers, including microplastics, in indoor and outdoor environments. Environ Pollut 221:453–458. https://doi.org/10.1016/j.envpol.2016.12.013

European Commission (2019) Environmental and health risks of microplastic pollution. https://op.europa.eu/en/publication-detail/-/publication/f235d1e3-7c4d-11e9-9f05-01aa75ed71a1/language-en/format-PDF/source-108645429. Accessed 15 June 2024

Galvão A, Aleixo M, De Pablo H, Lopes C, Raimundo J (2020) Microplastics in wastewater: microfiber emissions from common household laundry. Environ Sci Pollut Res 27(21):26643–26649. https://doi.org/10.1007/s11356-020-08765-6

Gaston E, Woo M, Steele C, Sukumaran S, Anderson S (2020) Microplastics differ between indoor and outdoor air masses: insights from multiple microscopy methodologies. Appl Spectrosc 74(9):1079–1098. https://doi.org/10.1177/0003702820920652

Gatidou G, Arvaniti OS, Stasinakis AS (2019) Review on the occurrence and fate of microplastics in sewage treatment plants. J Hazard Mater 367:504–512. https://doi.org/10.1016/j.jhazmat.2018.12.081

Habibi N, Uddin S, Fowler SW, Behbehani M (2022) Microplastics in the atmosphere: a review. J Environ Expo Assess 1(1):6. https://doi.org/10.20517/jeea.2021.07

Henry B, Laitala K, Klepp IG (2019) Microfibres from apparel and home textiles: prospects for including microplastics in environmental sustainability assessment. Sci Total Environ 652:483–494. https://doi.org/10.1016/j.scitotenv.2018.10.166

Huang Y, He T, Yan M, Yang L, Gong H, Wang W, Qing X, Wang J (2021) Atmospheric transport and deposition of microplastics in a subtropical urban environment. J Hazard Mater 416:126168. https://doi.org/10.1016/j.jhazmat.2021.126168

Jarosz K, Janus R, Wądrzyk M, Wilczyńska-Michalik W, Natkański P, Michalik M (2022) Airborne microplastic in the atmospheric deposition and how to identify and quantify the threat: semi-quantitative approach based on Kraków case study. Int J Environ Res Public Health 19(19):12252. https://doi.org/10.3390/ijerph191912252

Jenner LC, Sadofsky LR, Danopoulos E, Rotchell JM (2021) Household indoor microplastics within the Humber region (United Kingdom): quantification and chemical characterisation of particles present. Atmos Environ 259:118512. https://doi.org/10.1016/j.atmosenv.2021.118512

Jia L, Liu L, Zhang Y, Fu W, Liu X, Wang Q, Huang L (2023) Microplastic stress in plants: effects on plant growth and their remediations. Front Plant Sci 14:1226484. https://doi.org/10.3389/fpls.2023.1226484

Joaquin HF, Puello-Polo E, Trilleras J (2022) Characterization of microplastics in total atmospheric deposition sampling from areas surrounding industrial complexes in Northwestern Colombia. Sustainability 14(20):13613. https://doi.org/10.3390/su142013613

Kang H, Park S, Lee B, Kim I, Kim S (2022) Concentration of microplastics in road dust as a function of the drying period—a case study in G City, Korea. Sustainability 14(5):3006. https://doi.org/10.3390/su14053006

Kannankai MP, Devipriya SP (2024) Atmospheric microplastic deposition in a coastal city of India: the influence of a landfill source on monsoon winds. Sci Total Environ 908:168235. https://doi.org/10.1016/j.scitotenv.2023.168235

Kashfi FS, Ramavandi B, Arfaeinia H, Mohammadi A, Saeedi R, De-la-Torre GE, Dobaradaran S (2022) Occurrence and exposure assessment of microplastics in indoor dusts of buildings with different applications in Bushehr and Shiraz cities, Iran. Sci Total Environ 829:154651. https://doi.org/10.1016/j.scitotenv.2022.154651

Kaya AT, Yurtsever M, Bayraktar SÇ (2018) Ubiquitous exposure to microfiber pollution in the air. Eur Phys J Plus 133:488. https://doi.org/10.1140/epjp/i2018-12372-7

Klein M, Fischer EK (2019) Microplastic abundance in atmospheric deposition within the Metropolitan area of Hamburg, Germany. Sci Total Environ 685:96–103. https://doi.org/10.1016/j.scitotenv.2019.05.405

Leslie HA, Van Velzen MJ, Brandsma SH, Vethaak AD, Garcia-Vallejo JJ, Lamoree MH (2022) Discovery and quantification of plastic particle pollution in human blood. Environ Int 163:107199. https://doi.org/10.1016/j.envint.2022.107199

Liao Z, Ji X, Ma Y, Lv B, Huang W, Zhu X, Fang M, Wang Q, Wang X, Dahlgren R, Shang X (2021) Airborne microplastics in indoor and outdoor environments of a coastal city in Eastern China. J Hazard Mater 417:126007. https://doi.org/10.1016/j.jhazmat.2021.126007

Liu C, Li J, Zhang Y, Wang L, Deng J, Gao Y, Yu L, Zhang J, Sun H (2019a) Widespread distribution of PET and PC microplastics in dust in urban China and their estimated human exposure. Environ Int 128:116–124. https://doi.org/10.1016/j.envint.2019.04.024

Liu K, Wang X, Fang T, Xu P, Zhu L, Li D (2019b) Source and potential risk assessment of suspended atmospheric microplastics in Shanghai. Sci Total Environ 675:462–471. https://doi.org/10.1016/j.scitotenv.2019.04.110

Liu K, Wang X, Wei N, Song Z, Li D (2019c) Accurate quantification and transport estimation of suspended atmospheric microplastics in megacities: implications for human health. Environ Int 132:105127. https://doi.org/10.1016/j.envint.2019.105127

López APA, Trilleras J, Arana VA, Garcia-Alzate LS, Grande-Tovar CD (2023) Atmospheric microplastics: exposure, toxicity, and detrimental health effects. RSC Adv 13(11):7468–7489. https://doi.org/10.1039/D2RA07098G

Lozano YM, Lehnert T, Linck LT, Lehmann A, Rillig MC (2021) Microplastic shape, polymer type, and concentration affect soil properties and plant biomass. Front Plant Sci 12:616645. https://doi.org/10.3389/fpls.2021.616645

Mohamed Nor NH, Kooi M, Diepens NJ, Koelmans AA (2021) Lifetime accumulation of microplastic in children and adults. Environ Sci Technol 55(8):5084–5096. https://doi.org/10.1021/acs.est.0c07384

Monira S, Roychand R, Bhuiyan MA, Hai FI, Pramanik BK (2022) Identification, classification and quantification of microplastics in road dust and stormwater. Chemosphere 299:134389. https://doi.org/10.1016/j.chemosphere.2022.134389

Naik RK, Naik MM, D'Costa PM, Shaikh F (2019) Microplastics in ballast water as an emerging source and vector for harmful chemicals, antibiotics, metals, bacterial pathogens and HAB species: a potential risk to the marine environment and human health. Mar Pollut Bull 149:110525. https://doi.org/10.1016/j.marpolbul.2019.110525

Nandi S, Kumar RN, Dhandapani A, Iqbal J (2024) Characterization of microplastics in outdoor and indoor air in Ranchi, Jharkhand, India: first insights from the region. Environ Pollut 346:123543. https://doi.org/10.1016/j.envpol.2024.123543

Napper IE, Baroth A, Barrett AC, Bhola S, Chowdhury GW, Davies BF, Koldewey H (2023) The distribution and characterisation of microplastics in air, surface water and sediment within a major river system. Sci Total Environ 901:166640. https://doi.org/10.1016/j.scitotenv.2023.166640

Narmadha VV, Jose J, Patil S, Farooqui MO, Srimuruganandam B, Saravanadevi S, Krishnamurthi K (2020) Assessment of microplastics in roadside suspended dust from urban and rural environment of Nagpur, India. Int J Environ Res 14(6):629–640. https://doi.org/10.1007/s41742-020-00283-0

Nematollahi MJ, Zarei F, Keshavarzi B, Zarei M, Moore F, Busquets R, Kelly F (2022) Microplastic occurrence in settled indoor dust in schools. Sci Total Environ 807:150984. https://doi.org/10.1016/j.scitotenv.2021.150984

O'Brien S, Okoffo ED, Rauert C, O'Brien JW, Ribeiro F, Burrows SD, Thomas KV (2021) Quantification of selected microplastics in Australian urban road dust. J Hazard Mater 416:125811. https://doi.org/10.1016/j.jhazmat.2021.125811

Ojha UC, Kumar A, Spalgais S, Choudhari OK, Anbhule A (2021) Chronic lung diseases in plastic fumes exposed workers: a case series. Indian J Chest Dis Allied Sci 63(3):193–197. https://doi.org/10.5005/ijcdas-63-3-193

Pandey D, Banerjee T, Badola N, Chauhan JS (2022) Evidences of microplastics in aerosols and street dust: a case study of Varanasi City, India. Environ Sci Pollut Res 29(54):82006–82013. https://doi.org/10.1007/s11356-022-21514-1

Parashar N, Hait S (2023) Plastic rain—atmospheric microplastics deposition in urban and peri-urban areas of Patna City, Bihar, India: distribution, characteristics, transport, and source analysis. J Hazard Mater 458:131883. https://doi.org/10.1016/j.jhazmat.2023.131883

Parvez MS, Ullah H, Faruk O, Simon E, Czédli H (2024) Role of microplastics in global warming and climate change: a review. Water Air Soil Pollut 235(3):201. https://doi.org/10.1007/s11270-024-07003-w

Patchaiyappan A, Dowarah K, Zaki Ahmed S, Prabakaran M, Jayakumar S, Thirunavukkarasu C, Devipriya SP (2021) Prevalence and characteristics of microplastics present in the street dust collected from Chennai metropolitan city, India. Chemosphere 269:128757. https://doi.org/10.1016/j.chemosphere.2020.128757

Perera K, Ziajahromi S, Bengtson Nash S, Manage PM, Leusch FD (2022) Airborne microplastics in indoor and outdoor environments of a developing country in South Asia: abundance, distribution, morphology, and possible sources. Environ Sci Technol 56(23):16676–16685. https://doi.org/10.1021/acs.est.2c05885

Prata JC, Castro JL, da Costa JP, Duarte AC, Rocha-Santos T, Cerqueira M (2020) The importance of contamination control in airborne fibers and microplastic sampling: experiences from indoor and outdoor air sampling in Aveiro, Portugal. Mar Pollut Bull 159:111522. https://doi.org/10.1016/j.marpolbul.2020.111522

Pratiwi A, Syafei AD, Assomadi AF, Boedisantoso R, Hermana J (2020) Microplastic characterization based on the number of occupants. AIP Conf Proc 2296:020061. https://doi.org/10.1063/5.0030816

Purwiyanto AIS, Prartono T, Riani E, Naulita Y, Cordova MR, Koropitan AF (2022) The deposition of atmospheric microplastics in Jakarta-Indonesia: the coastal urban area. Mar Pollut Bull 174:113195. https://doi.org/10.1016/j.marpolbul.2021.113195

Rajmohan KVS, Ramya C, Raja Viswanathan M, Varjani S (2019) Plastic pollutants: effective waste management for pollution control and abatement. Curr Opin Environ Sci Health 12:72–84. https://doi.org/10.1016/j.coesh.2019.08.006

Rajpal A, Choudhury M, Goswami S, Chakravorty A, Raghavan V (eds) (2024) Waste management and treatment: advances and innovations, 1st edn. CRC Press. https://doi.org/10.1201/9781003258377

Rasgua A, Svelato A, Santacroce C, Catalano P, Notarstefano V, Carnevali O, Papa F, Rongioletti MCA, Baiocco F, Draghi S, D'Amore E, Rinaldo D, Matta M, Giorgini E (2021) Plasticenta: first evidence of microplastics in human placenta. Environ Int 146:106274. https://doi.org/10.1016/j.envint.2020.106274

Rasgua A, Notarstefano V, Svelato A, Belloni A, Gioacchini G, Blondeel C, Giorgini E (2022) Raman microspectroscopy detection and characterisation of microplastics in human breastmilk. Polymers (Basel) 14(13):2700. https://doi.org/10.3390/polym14132700

Rillig MC (2018) Microplastic disguising as soil carbon storage. Environ Sci Technol 52(11):6079–6080. https://doi.org/10.1021/acs.est.8b02338

Rillig MC, Lehmann A (2020) Microplastic in terrestrial ecosystems. Science (New York, NY) 368(6498): 1430–1431. https://doi.org/10.1126/science.abb5979

Roblin B, Ryan M, Vreugdenhil A, Aherne J (2020) Ambient atmospheric deposition of anthropogenic microfibers and microplastics on the western periphery of Europe (Ireland). Environ Sci Technol 54(18):11100–11108. https://doi.org/10.1021/acs.est.0c04000

Romarate RA, Ancla SMB, Patilan DMM, Inocente SAT, Pacilan CJM, Sinco AL, Guihawan JQ, Capangpangan RY, Lubguban AA, Bacosa HP (2023) Breathing plastics in Metro Manila, Philippines: presence of suspended atmospheric microplastics in ambient air. Environ Sci Pollut Res 30:53662–53673. https://doi.org/10.1007/s11356-023-26117-y

Roychand R, Pramanik BK (2020) Identification of micro-plastics in Australian road dust. J Environ Chem Eng 8(1):103647. https://doi.org/10.1016/j.jece.2019.103647

Saeedi M (2024) How microplastics interact with food chain: a short overview of fate and impacts. J Food Sci Technol 61(3):403–413. https://doi.org/10.1007/s13197-023-05720-4

Sarathana D, Winijkul E (2022) Concentrations of airborne microplastics during the dry season at five locations in Bangkok Metropolitan Region, Thailand. Atmosphere 14(1):28. https://doi.org/10.3390/atmos14010028

Schlesinger WH, Bernhardt ES (2020) The atmosphere. In: Biogeochemistry, pp 51–97. https://doi.org/10.1016/B978-0-12-814608-8.00003-7

Shahraki M, Rezaei Kahkha MR, Piri J, Sharafi A, Kaykhaii M (2022) Microplastics in atmospheric dust samples of Sistan: sources and distribution. J Environ Health Sci Eng 20:931–936. https://doi.org/10.1007/s40201-022-00833-y

Shim WJ, Hong SH, Eo SE (2017) Identification methods in microplastic analysis: a review. Anal Methods 9(9):1384–1391. https://doi.org/10.1039/C6AY02558G

Shruti VC, Kutralam-Muniasamy G, Pérez-Guevara F, Roy PD, Martínez IE (2022) Occurrence and characteristics of atmospheric microplastics in Mexico City. Sci Total Environ 847:157601. https://doi.org/10.1016/j.scitotenv.2022.157601

Soltani NS, Taylor MP, Wilson SP (2021) Quantification and exposure assessment of microplastics in Australian indoor house dust. Environ Pollut 283:117064. https://doi.org/10.1016/j.envpol.2021.117064

Sommer F, Dietze V, Baum A, Sauer J, Gilge S, Maschowski C, Gieré R (2018) Tire abrasion as a major source of microplastics in the environment. Aerosol Air Qual Res 18(8):2014–2028. https://doi.org/10.4209/aaqr.2018.03.0099

Sridharan S, Kumar M, Singh L, Bolan NS, Saha M (2021) Microplastics as an emerging source of particulate air pollution: a critical review. J Hazard Mater 418:126245. https://doi.org/10.1016/j.jhazmat.2021.126245

Su L, Nan B, Craig NJ, Pettigrove V (2020) Temporal and spatial variations of microplastics in roadside dust from rural and urban Victoria, Australia: implications for diffuse pollution. Chemosphere 252:126567. https://doi.org/10.1016/j.chemosphere.2020.126567

Syafei A, Nurasrin N, Assomadi A, Boedisantoso R (2019) Microplastic pollution in the ambient air of Surabaya, Indonesia. Curr World Environ 14:290–298. https://doi.org/10.12944/CWE.14.2.13

Syafina PR, Yudison AP, Sembiring E, Irsyad M, Tomo HS (2022) Identification of fibrous suspended atmospheric microplastics in Bandung Metropolitan Area, Indonesia. Chemosphere 308:136194. https://doi.org/10.1016/j.chemosphere.2022.136194

Truong TNS, Strady E, Kieu-Le TC, Tran QV, Le TMT, Thuong QT (2021) Microplastic in atmospheric fallouts of a developing Southeast Asian megacity under tropical climate. Chemosphere 272:129874. https://doi.org/10.1016/j.chemosphere.2021.129874

Uddin S, Fowler SW, Saeed T (2020) Microplastic particles in the Persian/Arabian Gulf – a review on sampling and identification. Mar Pollut Bull 154:111100. https://doi.org/10.1016/j.marpolbul.2020.111100

United Nations Environment Programme (2018) Single-use plastics: a roadmap for sustainability. https://www.unep.org/resources/report/single-use-plastics-roadmap-sustainability. Accessed 15 June 2024

Vethaak AD, Legler J (2021) Microplastics and human health. Science 371(6530):672–674. https://doi.org/10.1126/science.abe504

Vianello A, Jensen RL, Liu L, Vollertsen J (2019) Simulating human exposure to indoor airborne microplastics using a Breathing Thermal Manikin. Sci Rep 9:8670. https://doi.org/10.1038/s41598-019-45054-w

Wang F, Zhang X, Zhang S, Zhang S, Sun Y (2020) Interactions of microplastics and cadmium on plant growth and arbuscular mycorrhizal fungal communities in an agricultural soil. Chemosphere 254:126791. https://doi.org/10.1016/j.chemosphere.2020.126791

Welsh B, Aherne J, Paterson AM, Yao H, McConnell C (2022) Atmospheric deposition of anthropogenic particles and microplastics in south-central Ontario, Canada. Sci Total Environ 835:155426. https://doi.org/10.1016/j.scitotenv.2022.155426

Wright SL, Ulke J, Font A, Chan KLA, Kelly FJ (2020) Atmospheric microplastic deposition in an urban environment and an evaluation of transport. Environ Int 136:105411. https://doi.org/10.1016/j.envint.2019.105411

Xie Y, Li Y, Feng Y, Cheng W, Wang Y (2022) Inhalable microplastics prevails in air: exploring the size detection limit. Environ Int 162:107151. https://doi.org/10.1016/j.envint.2022.107151

Xumiao L, Prata JC, Alves JR, Duarte AC, Rocha-Santos T, Cerqueira M (2021) Airborne microplastics and fibers in indoor residential environments in Aveiro, Portugal. Environ Adv 6:100134. https://doi.org/10.1016/j.envadv.2021.100134

Yadav H, Sethulekshmi S, Shriwastav A (2022) Estimation of microplastic exposure via the composite sampling of drinking water, respirable air, and cooked food from Mumbai, India. Environ Res 214:113735. https://doi.org/10.1016/j.envres.2022.113735

Yan Z, Liu Y, Zhang T, Zhang F, Ren H, Zhang Y (2021) Analysis of microplastics in human feces reveals a correlation between fecal microplastics and inflammatory bowel disease status. Environ Sci Technol 56(1):414–421. https://doi.org/10.1021/acs.est.1c03924

Yao Y, Glamoclija M, Murphy A, Gao Y (2022) Characterization of microplastics in indoor and ambient air in northern New Jersey. Environ Res 207:112142. https://doi.org/10.1016/j.envres.2021.112142

Yukioka S, Tanaka S, Nabetani Y, Suzuki Y, Ushijima T, Fujii S, Takada H, Van Tran Q, Singh S (2020) Occurrence and characteristics of microplastics in surface road dust in Kusatsu (Japan), Da Nang (Vietnam), and Kathmandu (Nepal). Environ Pollut 256:113447. https://doi.org/10.1016/j.envpol.2019.113447

Zhang M, Dong B, Qiao Y, Yang H, Wang Y, Liu M (2018) Effects of sub-soil plastic film mulch on soil water and salt content and water utilization by winter wheat under different soil salinities. Field Crop Res 225:130–140. https://doi.org/10.1016/j.fcr.2018.06.010

Zhang J, Wang L, Kannan K (2020a) Microplastics in house dust from 12 countries and associated human exposure. Environ Int 134:105314. https://doi.org/10.1016/j.envint.2019.105314

Zhang Q, Zhao Y, Du F, Cai H, Wang G, Shi H (2020b) Microplastic fallout in different indoor environments. Environ Sci Technol 54:6530–6539. https://doi.org/10.1021/acs.est.0c00087

Zhu X, Huang W, Fang M, Liao Z, Wang Y, Xu L, Mu Q, Shi C, Lu C, Deng H, Dahlgren R, Shang X (2021) Airborne microplastic concentrations in five megacities of Northern and Southeast China. Environ Sci Technol 55(19):12871–12881. https://doi.org/10.1021/acs.est.1c03618

Micro and Nanoplastics: Effect on Lentic Ecosystem

7

Amit Kumar Mandal, Atanu Kumar Das, Surjendu Bhattacharyya, Rajdeep Chowdhury, and Tridib Mondal

7.1 Introduction

Since their invention in the early 1900s, synthetic polymers have experienced significant growth due to their lightweight nature, affordability, resistance to corrosion, durability, and chemical stability against certain reactions (Sharma et al. 2017). As a result, plastic polymers have replaced various traditional materials like wood, ceramic, glass, and metal. A 2018 report indicated that global plastic production reached 348 million tons (Plastics Europe 2018), with projections suggesting it could reach 33 billion tons by 2050 (Rochman et al. 2013). Due to inadequate waste management, a substantial amount of plastic ends up in terrestrial and aquatic environments, where it persists for long periods. Each year, over eight million tons of plastic debris pollutes both standing and flowing water systems (Choudhury et al. 2022; Liu et al. 2022).

Plastic dispersed in the environment is broken down into smaller particles through the actions of biological organisms and physical processes such as photo-oxidation, mechanical forces (like wind and waves), and thermal degradation. These plastic fragments can range in size from meters to nanometers. Microplastics (MPs) are defined as particles smaller than 5 nm, while nanoplastics are less than 100 nm in size (Pan et al. 2019). MPs are classified based on their origin as either primary microplastics (MPs) or secondary microplastics (MPs). PMPs are intentionally manufactured in products like raw plastic resin, microbeads, microfibers, medicines, textiles, and personal care items (e.g., cosmetics, body scrubs, toothpaste) (Roy and Choudhury 2022; Hernandez et al. 2017). In contrast, sMPs result from the breakdown of larger plastic items, such as containers, tires, nets, and films, through biological, chemical, and physical processes (Wang et al. 2020a).

Plastic pollution, particularly in aquatic environments, is a significant global issue. MPs and

A. K. Mandal
Department of Chemistry, Bankura University, Bankura, West Bengal, India
e-mail: amitkumarmandal@bankurauniv.ac.in

A. K. Das
Department of Physics, Kandi Raj College, Kandi, Murshidabad, West Bengal, India

S. Bhattacharyya
SLAC National Accelerator Laboratory, Menlo Park, CA, USA

R. Chowdhury
Institute for Neuro- and Sensory Physiology, University of Göttingen Medical Center, Göttingen, Germany

Department of Chemistry, GITAM School of Science, GITAM, Hyderabad, Telangana, India

T. Mondal (✉)
Department of Chemistry, Sukanta Mahavidyalaya, Dhupguri, Jalpaiguri, West Bengal, India

S. Goswami et al. (eds.), *Micro-Nano Plastics Exposure, Environmental Degradation and Public Health Crisis*, Sustainable Development Goals Series, https://doi.org/10.1007/978-981-96-5047-7_7

nanoplastics can absorb organic, inorganic, and heavy metal contaminants due to their high surface-area-to-volume ratio and hydrophobic nature, facilitating their transport in water systems. Aquatic species, including fish, turtles, seabirds, amphipods, mussels, barnacles, and zooplankton, frequently ingest these particles, leading to harmful effects such as stunted growth, reduced reproduction, altered behavior, mechanical injury, neurotoxicity, and even death (Bhattacharya and Roychoudhury 2023; Chang et al. 2020). Through the food chain, these contaminants can make their way into the human body, potentially causing various health problems (Bhuyan 2022; Chang et al. 2020).

Most research on microplastics (MPs) and nanoplastics (NPs) has focused on marine environments, with relatively fewer studies addressing plastic contamination in freshwater ecosystems, particularly in lentic environments (Yin et al. 2020; Slootmaekers et al. 2019). In general, the concentration of MPs/NPs is higher in lentic waters compared to lotic waters, and sediments tend to be more contaminated than the water due to the sinking behavior of plastics. Globally, rivers, lakes, ponds, and reservoirs serve as crucial sources of drinking water, fisheries, and food production. As a result, recent research on plastic contamination in semi-lentic and lentic ecosystems has garnered significant attention in both developing and developed nations. Accordingly, this chapter focuses on identifying and quantifying MP/NP contamination in lentic ecosystems and examines its effects on phytoplankton, microorganisms, higher-order organisms, and human health.

7.2 Importance of Water in Lentic Ecosystem for Sustaining Life

Standing water is called lentic (from the Latin "lentus," meaning slow or stationary). Ponds, marshes, bogs, lakes, and swamps comprise the still-water lentic habitats. Lentic eco-arrangement normally lacks a clear origin; yet, it is thought to have originated at the location where groundwater and the earth's surface converge. As a consequence of the saturation of the basic solid ground, water occasionally stays on the surface of the planet, constituting the lentic ecosystem (Fig. 7.1). Light-impacted ecological coating surfaces may exist in extensive water, such as lakes. Since ponds offer superior light penetration, a wider variety of water plants may be housed within them. These ecosystems range from tiny ponds or pools to large lakes. Lakes and ponds are various places of landward freshwater domain that have existed covering the world, and these ecosystems provide essential supplies and environment for both aquatic and terrestrial animals.

Water is a vital basis of life. Water's fundamental roles are to promote plant growth, provide an appropriate environment for species that reside there, act as a provincial resident or breeding ground for numerous insects, amphibians, and other water-born organisms, and supply the minerals and nutrients required to maintain physical existence (Cantonati et al. 2020). The vast majority of the lentic ecosystem consists of invertebrate insects. They are found near the shore of almost any aquatic habitat and they lay eggs in fresh water. So lentic ecosystems are necessary for insect existence. Water is the most vital element in nature and is necessary for humans to live. Water is essential for industry, agriculture, cattle, and drinking, as well as for maintaining the ecosystems that support all life. It is also for sanitation and industrial purposes.

The quantity of oxygen in lentic water bodies depends on the contact between the water and air, water circulation, and the amount of energy produced and consumed (Cantonati et al. 2020). Different consumer species, e.g., worms, fish, crustaceans, snails, insects, amphibians, reptiles, fish, and birds, are found in lentic habitats (Cantonati et al. 2020). Herbivorous groups, such as snails, crustaceans, and amphibian larvae, are essential in controlling primary productivity (Cantonati et al. 2020).

7.3 Micro and Nanoplastics: Contamination of Lentic Ecosystem

Microplastics are a heterogeneous pollutant ensemble originating from multiple distinct product categories. They are made from several chem-

Fig. 7.1 Representative images of different lentic ecosystems were captured using a digital camera by the contributing authors

ical additives and polymers and are distinguished by a wide variety of different colors and textures. Plastics are highly persistent, so their degradation occurs slower while their accumulation is fast. In the current outline, the worldwide spread of a microscopic/nanoscopic fragment of plastics, i.e., MPs and NPs, is capturing notable attention from investigators extensively owing to their severe habitat consequences (Fig. 7.2).

MPs and NPs have the ability to infiltrate freshwater systems through three different channels: wastewater treatment output, agricultural runoff from sludge-treated land, and sewage water overflow from heavy precipitation (Rajpal et al. 2024; Zhou et al. 2023; Bigalke et al. 2022). The elimination process of MPs and NPs from the lentic ecosystem in the waste management system is also insufficient. Furthermore, there is evidence that storms and harsh weather play an important factor in the high proportion of MPs in the aqueous area (Wang et al. 2020a). Table 7.1 summarizes the diverse sources of MPs in lentic ecosystems globally.

Particularly in Western India, which comprises the states of Chhattisgarh, Daman and Diu, Gujarat, Madhya Pradesh, Maharashtra, and Dadra and Nagar Haveli, the plastics industry is growing quickly (Vaid et al. 2021). India produces 26 million metric tons of pollutant plastic waste roughly annually and consumes about 11 kg of plastics annually per person, making it a significant contributor to the global plastic waste generation industry. Numerous experts are looking into the amount of plastic waste that is present in different ecosystem components in India, which eventually will break down into fragments to form MPs (Shaikh and Shaikh 2021).

7.4 Consequences of Micro and Nanoplastics in Lentic Ecosystem

7.4.1 Effects on Lentic Water Quality

Ninety-seven percent (97%) of the water on the earth is found in the oceans, which cover two-thirds of the planet's surface. Consequently, the pollution of aquatic environments is widely recognized as an environmental issue. By accu-

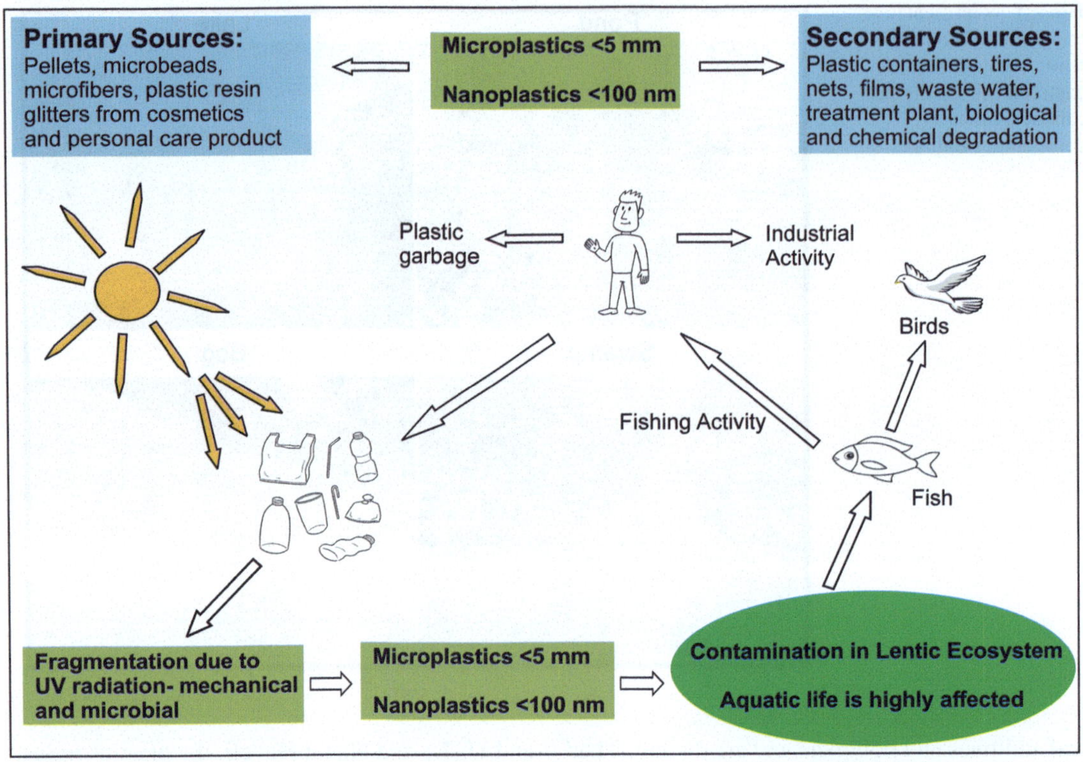

Fig. 7.2 A schematic showing contamination of the lentic ecosystem by MPs and NPs and how this affects humans and natural ecosystems

Table 7.1 Most common types of MPs/NPs are found in lentic ecosystems

MPs/NPs types in the lentic ecosystem	Source	References
Polyethylene (PE)	Food packaging materials, bottles, plastic bags, etc.	Miranda-Peña et al. (2023)
Polypropylene (PP)	Bottlecaps, food containers, automotive industry materials	Miranda-Peña et al. (2023)
Polystyrene (PS)	Disposable food and storage containers, insulation, foams and films, etc.	Miranda-Peña et al. (2023)
Poly amide (PA)	Silk, nylon, wool, fishing net, etc.	Yuan et al. (2022)
Polyvinyl chloride (PVC)	Pipes, cable insulation, films etc.	Yuan et al. (2022)
Polyethylene terephthalate (PET)	Beverage bottles, textile fibers, disposable food containers, etc.	Miranda-Peña et al. (2023)
Polyvinylidene fluoride (PVDF)	PVDF tubes, plastic, films, plate, wire, etc.	Miranda-Peña et al. (2023)
Polyester (PES) fibers	Various textile products, computer mouse materials, etc.	Yuan et al. (2022)

mulating dangerous pollutants from the environment and concurrently distributing chemicals (additives), MPs can serve as transport vectors. Plastics in water bodies break and shrink to produce micro- or nanoparticles because of the mechanical and photochemical processes that are enhanced by waves and sunshine, respectively (Wang et al. 2020a). The aquaculture industry, wastewater treatment plants, and agricultural activities all contribute to the deposition of microplastic debris in lentic water systems, including ponds, lakes, and rivers (Horton et al. 2017; Lyu et al. 2020). Various types of microplastics (MPs), including PET, PS, PE, PVC, and PE, are commonly found in large quantities in soil, lakes, ponds, and river water globally (Table 7.1). MP pollution levels are notably lower in rural lakes, ponds, and rivers compared to urban areas (Lyu et al. 2020). For instance, significant concentrations of PS, PE, and PP MPs have been detected in surface water and sediments near urban areas in the USA's Great Lakes, such as Lake Erie, Lake Superior, and Lake Huron (Eriksen et al. 2013).

According to a recent report, MP contamination is becoming a popular topic for limnological exploration, starting with lakes, followed by rivers and reservoirs (D'Avignon et al. 2022). When the MP's content is high, it is usually communicated as the number of particles per unit volume of water; when the MP's content is low, it is often communicated as the weight per unit volume. As reported by literature sources, surface waters have much fewer MPs than marine water systems. According to Horton et al. (2017), MP concentration is immensely changeable at the river's surface. Flow patterns, precipitation, season, and proximity to sites of entry, such as sludge discharge, road or agricultural runoff, sewage or stormwater effluent, and litter, impact it. Current environmental research indicates that surface inland waterways have lower mean particle levels (~1.9 particles/L) than estuaries (3.1 particles/L) or the ocean (16 particles/L) (D'Avignon et al. 2022). Alternatively, D'Avignon et al. (2022) report that the average concentrations of these compartments vary, with lotic systems detecting 0.007 particles/L and lentic systems detecting around (one) 1 particle/L. Although surface-water MPs are widely distributed, their concentrations vary significantly in areas with high people densities or streams undergoing significant alterations (D'Avignon et al. 2022).

7.4.2 Effects on Phytoplankton and Microorganisms

The phytoplankton is a crucial part of the freshwater and ocean biospheres. They typically obtain energy from photosynthesis, possess a large surface area, and respond promptly to environmental changes. They are an essential component of the global carbon cycle and create the substructure of freshwater food webs (Karlusich et al. 2020). In lakes and ponds, the chief photo-sensitizers are the phytoplankton. These organic species are important indicators within any ecosystem to assess freshwater pollution responsible for MPs and NPs. Aquatic MPs and NPs, particularly polystyrene NPs (1.6–40 mg/mL), have pathogenic effects on microorganisms and affect their development and reproduction (Bhattacharya and Roychoudhury 2023; Geremia et al. 2023). They are directly responsible for the rate and growth of phytoplankton photosynthesis. Preliminary evidence indicates that plastic particles with a nanoscale size can be more harmful to phytoplankton than those with a more significant size (Sharma et al. 2023).

Regarding MPs and NPs, the latter are more active and more likely to absorb additional pollutants from the water since they have a higher surface area-volume ratio of particles (Horton 2022). To freshwater organisms, MPs and NPs (like polystyrene) can positively induce chemical and physical harm through ingestion and biotransfer throughout the planktonic food chain. For instance, harmful effects at the cellular level have been reported in green algae exposed to MPs or NPs fragments (Li et al. 2021). The most frequent negative consequences are oxidative stress and changes in the cell structure. Specific subcellular effects, such as plasmolysis and vacuolation, are also noted (Zheng et al. 2021). Algal stress reactions are growth phase dependent, with the stationary phase's restoration to standard cell structure and the phases of algal homo- and hetero-aggregation being the most notable. A few freshwater algae are negatively affected by PVC MPs as well, including physical harm and reduced photosynthetic efficiency (Wang et al. 2020b).

7.4.3 Effects on Plants and Animals

The marine environment is more susceptible to the risks posed by nanoplastics (NPs) than microplastics (MPs) (Sharma et al. 2023). MPs like PE, PET, PVC, and PP are likely to be transferred up the planktonic food chain, moving through various secondary aquatic organisms at higher ecological levels (Geremia et al. 2023). Being nonbiodegradable, these plastics directly impact the growth and reproduction of plants and animals in aquatic ecosystems (Geremia et al. 2023). Numerous experimental studies have examined the ingestion of primary MPs and NPs, with reported toxicity thresholds of ~180 μg/L for PE, 0–0.86 mg/L for PET, ~0.1% [quality ratio] for PVC, and ~130 mg/L for PS, affecting organisms like zooplankton (Li et al. 2021) and fish (Rochman et al. 2013), along with other invertebrates. Recently, there has been growing interest in field studies on the interactions of ingested MPs/NPs in the aquatic food web, as new research shows the presence of MPs in the gastrointestinal and respiratory systems of aquatic organisms. The overall environmental impacts of MPs/NPs can be classified into physical, chemical, and biological pathways. MPs have a wide range of effects on aquatic ecosystems, including fish and the broader aquatic food chain (Li et al. 2021). Fish and other aquatic species may experience decreased food intake, stunted growth, oxidative stress, and abnormal behavioral changes. Furthermore, nano-scale MPs can penetrate biological barriers and accumulate in tissues, leading to the generation of reactive oxygen species (ROS) and disruptions in lipid metabolism. These impacts can also have detrimental effects on overall aquatic life (Geremia et al. 2023).

7.4.4 Impact on Human Health

While numerous studies have examined the environmental impacts of microplastics (MPs) and nanoplastics (NPs), there is limited research on their effects on human health. Humans may ingest particulate plastics, such as PS, PE, and PP, through contaminated seafood and water (Bhuyan 2022). Once ingested, these plastics can cause significant biological effects on the gastrointestinal system, including digestive issues, impaired immune function, lung cell inflammation, oxidative stress, and apoptosis. They may also increase the risk of metabolic disorders in children (Bhuyan 2022; Chang et al. 2020). A review by La Porta et al. (2023) highlighted a notable association between MPs and various health disorders (Table 7.2). Recent studies further support growing evidence of MPs/NPs prevalence, exposure, and transfer from lower trophic levels to humans across the food web. These MPs/NPs could have identical effects in human tissues as seen for other animals. It has already been reported that MPs/NPs are easily absorbed by human epithelial tissues (Magri et al. 2018). Adsorption of MPs/NPs-bound hydrophobic organic compounds and trace elements and bioaccumulation of polymeric particles inside the gastrointestinal and respiratory systems may negatively impact human health (Zazouli et al. 2022). The widespread occurrence of MPs/NPs in biota, including commercially important species, in our daily lives is a concern. It will be better for our society to restrict the levels of MPs/NPs in our food so that it does not harm us.

7.5 Determination of MPs/NPs in Lentic Ecosystems

A standardized sample collection, identification, and quantification strategy has yet to be available despite MPs/NPs being pollutants emerging in the environment for a long time. Different methods have been applied to collect the plastic sample from the aquatic environment such as the net and grab sampling method, pumps at different column methods, and the zooplankton sampling method (Sarijan et al. 2021). The grab sampling method is mainly used to examine MPs in a large water volume using a net with a mesh size ranging from 50 to 3000 μM; mostly 300 μM mesh size is used (Green et al. 2018). A small mesh size net can capture more minor MPs, but the filtered sample volume is reduced. Compared to the net sampling method, the grab sampling method

Table 7.2 Generated disorders in the human body by different kinds of MPs/NPs

Type of diseases/ Sample type	Total number of MPs/NPs detected in the whole sample	Number of MPs/NPs in normal sample	Number of microfibers in disease sample	Most abundance MPs/NPs in disease sample[a]
Pulmonary Ground Glass Nodules (GGNs) in human lungs	50	27	38	PET, phenoxy resin, acrylic microfibers
Liver of patients with cirrhosis	6	Not observed	6	PS, PVC, PET, PP, POM, and PMMA
Inflammatory bowel disease (IBD)	Various MPs	MPs ≤ than 50 μm significantly lower than the disease sample	MPs ≤ than 50 μm significantly higher than the normal sample	PET
Changes in Meconium Microbiota	Various MPs	Concentration of MPs lower	Concentration of MPs significantly high	PS and PE
Normal Breast milk	58	–	–	PE, PVC and PP
Colon Cancer	331	15.4 particles per g of colon tissue	28.1 particles per g of colon tissue	PC, PA and PP
Normal Blood Sample	–	77% of donors sample contain MPs	–	PET, PS, PE and PMMA

[a] La Porta et al. (2023)

captures a more significant amount of MPs and NPs (Green et al. 2018). Two different techniques collect samples from sediment in an aquatic ecosystem; in one method, stainless steel or a trowel is used to collect the sample from the top of the surface up to 2–5 cm depth (Mariano et al. 2021). The other method collects the sediment sample using a standardized sample-collecting instrument (Klein et al. 2018), thus providing the same result. Silva-Cavalcanti et al. (2017) proposed that fish's gastrointestinal (GI) tract can be considered an experimental tool for qualitatively measuring MPs/NP contamination in aquatic environments. However, some internal organs and fatty acid tissue are excluded from the GI tract by many authors to avoid interference at the time of digestion (Roch et al. 2019). Before removing the GI tract, the fish exterior is washed thoroughly with MilliQ water to avoid contamination.

A plastic polymer's detection and quantification process may interfere with the adsorbed organic, inorganic, and heavy metals in MPs/NPs. Thus, the analyzed samples are treated with different oxidizing agents before density separation to remove biogenic contamination (Sarijan et al. 2021). Sometimes, the samples are pretreated with HCl and NaOH to remove acidic and basic contaminations. Used chemicals may harm (degradation and discoloration) some MPs/NPs through degradation and discoloration. In density separation techniques, samples are treated with dense salt solutions (e.g., NaCl, KF, etc.), and MPs/NPs float over the solution (Klein et al. 2018). After this, the solutions need to be carefully filtered. Like water and sediment samples, the dissected GI tract of fish samples were treated with H_2O_2 and NaCl to isolate MPs/NPs. Recently, NaOH, KOH, and NaClO have also been used in fish and bivalves for digestion (Mariano et al. 2021).

MPs/NPs are qualitatively and quantitatively analyzed by microscopy, mass spectrometry, Raman spectrometry, infrared spectroscopy, and thermal analysis techniques. Stereo-microscopy can quickly analyze the sample size, shape, and color for samples with dimensions of ~100 μM. This technique cannot efficiently determine small, transparent samples and their composition. Fluorescence microscopy is used to visualize transparent and small-dimension samples, but the laser may induce photo-toxicity to the detecting sample. Scanning electron microscopy (SEM) offers knowledge about the surface morphology and chemical composition of MPs/NPs, but the technique is expensive and requires a long time to analyze the data (Neto et al. 2020). Like SEM, transmission electron microscopy (TEM) can visualize MPs/NPs with very high resolution and give sample elementary composition if coupled with energy-dispersive X-ray spectroscopy (EDS). Conventional Raman spectroscopy (detect MPs > 10 μM) and micro-Raman spectroscopy (detect ~1 μM and MPs and NPs < 1 μM) are widely used for the determination of plastic contamination in freshwater, sediments, different organisms, cosmetics, and drinks with high authenticity (Woo et al. 2021). This technique does not damage the analyzed sample and permits further investigation. Time-consuming data analysis, expensive methods, and noninterpretable Raman spectra of fluorescent plastic (due to contamination) are among a few disadvantages of this technique. FTIR and μ-FTIR quantitatively determine MPs/NPs composition, occurrence method, sample origin, and physio-chemical parameter (Woo et al. 2021). The high availability of energy in FTIR produces a high signal/noise ratio. Thus, the sample analysis time is significantly reduced. Gas chromatography-mass spectroscopy with the combination of the pyrolysis process (GC-MS-Py) can be employed to evaluate the chemical composition and structure of MPs/NPs (Fries et al. 2013).

Thermogravimetric analysis (TGA) and differential scanning calorimetry (DSC) are two dominant techniques used to assess the changes in the physio-chemical properties of plastics concerning heat or temperature change. TGA quantitatively measures the mass decrease when a sample warms up under certain temperatures and atmospheric conditions. On the other hand, DSC measures the enthalpy change of melting and the glass transition temperature of plastic polymer concerning a reference sample when heat flows between the sample and reference under pre-established programming conditions (Zainuddin and Syuhada 2020).

7.6 Future Perspectives and Recommendations

MPs/NPs contamination in lentic ecosystems increases exponentially because of increasing population, industrialization, and urbanization. Most of the plastic contamination reports were mainly focused on the marine ecosystem. However, the lentic freshwater ecosystem has a greater risk of contamination. Thus, more investigations are required all over the region, from the top water surface to bottom sediment in the lentic ecosystem, to understand the effect of MPs/NPs contamination on living organisms. This could address the behavioral change in their life cycle. The authors reported experimental results using different sorting, sampling, and analysis methods. Thus, it is essential to set up a standardized, reproducible protocol to compare results from different locations. A sustainable, cost-effective technological method is needed for biodegradation and cleaning MPs/NPs from different aquatic ecosystems utilizing microorganisms such as bacteria, fungi, and protozoans, especially before converting high-density MPs to low-density NPs, as the latter is more toxic. These pollutants adsorb harmful chemical substances and promote pathogenic microorganism growth. These materials and harmful chemicals may get transferred to humans through the food chain. However, the detrimental effect on the human body must be addressed rigorously. In summary, it is essential to establish an inclusive solution to control and deal with these pollutants.

7.7 Conclusion

In this chapter, we demonstrated several detrimental effects of MP/NPs on lentic ecosystems, which include still-water bodies like lakes and ponds. MP/NPs can absorb and transport harmful pollutants to aquatic organisms, leading to malnutrition, stunted growth, and even death. It can alter sediments' and water's physical and chemical properties, affecting light penetration and nutrient cycles. Lentic ecosystems often are sources of drinking water and food; hence, the presence of MP/NPs in these systems may pose potential risks to human health. The extent and nature of these risks are still under investigation. In summary, MP/NPs in lentic ecosystems are a growing environmental concern, and extensive research is essential to understand their impact and develop effective mitigation strategies fully.

References

Bhattacharya A, Roychoudhury A (2023) Microplastics in the aquatic environment – effects on ocean carbon sequestration and sustenance of marine life. In: Microplastics in the ecosphere: air, water, soil, and food. Wiley. https://doi.org/10.1002/9781119879534.ch12, Chapter 12

Bhuyan MS (2022) Effects of microplastics on fish and in human health. Front Environ Sci 10:827289. https://doi.org/10.3389/fenvs.2022.827289

Bigalke M, Fieber M, Foetisch A, Reyes J, Tollan P (2022) Microplastics in agricultural drainage water: a link between terrestrial and aquatic microplastic pollution. Sci Total Environ 806:150709. https://doi.org/10.1016/j.scitotenv.2021.150709

Cantonati M, Poikane S, Pringle CM et al (2020) Characteristics, main impacts, and stewardship of natural and artificial freshwater environments: consequences for biodiversity conservation. Water 12(1):260. https://doi.org/10.3390/w12010260

Chang X, Xue Y, Li J, Zou L, Tang M (2020) Potential health impact of environmental micro- and nanoplastics pollution. J Appl Toxicol 40:4–15. https://doi.org/10.1021/envhealth.3c00052

Choudhury M, Sharma A, Pervez A, Upadhyay P, Dutta J (2022) Growing menace of microplastics in and around the coastal ecosystem. Coastal ecosystems: environmental importance, current challenges and conservation measures. Springer International Publishing, Cham, pp 117–137. https://doi.org/10.1007/978-3-030-84255-0_6

D'Avignon G, Gregory-Eaves I, Ricciardi A (2022) Microplastics in lakes and rivers: an issue of emerging significance to limnology. Environ Rev 30(2):228–244. https://doi.org/10.1139/er-2021-0048

Eriksen M, Mason S, Wilson S, Box C, Zellers A, Edwards W, Farley H, Amato S (2013) Microplastic pollution in the surface waters of the Laurentian Great Lakes. Mar Pollut Bull 77:177–182. https://doi.org/10.1016/j.marpolbul.2013.10.007

Fries E, Dekiff JH, Willmeyer J, Nuelle M-T, Ebert M, Remy D (2013) Identification of polymer types and additives in marine microplastic particles using pyrolysis-GC/MS and scanning electron microscopy. Environ Sci Process Impacts 15:1949. https://doi.org/10.1039/c3em00214d

Geremia E, Tomajoli MTM, Murano C, Petito A, Fascilo G (2023) The impact of micro- and nanoplastics on aquatic organisms: mechanisms of oxidative stress and implications for human health-a review. Environments 10(9):161. https://doi.org/10.3390/environments10090161

Green DS, Kregting L, Boots B, Blockley DJ, Brickle P, Da Costa M, Crowley Q (2018) A comparison of sampling methods for sea water microplastics and a first report of the microplastic litter in coastal waters of Ascension and Falkland Islands. Mar Pollut Bull 137:695–701. https://doi.org/10.1016/j.marpolbul.2018.11.004

Hernandez E, Nowack B, Mitrano DM (2017) Polyester textiles as a source of microplastics from households: a mechanistic study to understand microfiber release during washing. Environ Sci Technol 51(12):7036–7046. https://doi.org/10.1021/acs.est.7b01750

Horton AA (2022) Plastic pollution: when do we know enough? J Hazard Mater 422:126885. https://doi.org/10.1016/j.jhazmat.2021.126885

Horton AA, Walton A, Spurgeon DJ, Lahive E, Svendsen C (2017) Microplastics in freshwater and terrestrial environments: evaluating the current understanding to identify the knowledge gaps and future research priorities. Sci Total Environ 586:127–141. https://doi.org/10.1016/j.scitotenv.2017.01.190

Karlusich JJP, Ibarbalz FM, Bowler C (2020) Exploration of marine phytoplankton: from their historical appreciation to the omics*era*. J Plankton Res 42:595–612. https://doi.org/10.1093/plankt/fbaa049

Klein S, Dimzon IK, Eubeler J, Knepper TP (2018) Analysis, occurrence, and degradation of microplastics in the aqueous environment. In: Wagner M, Lambert S (eds) Freshwater microplastics. The handbook of environmental chemistry, vol 58. Springer, pp 51–67. https://doi.org/10.1007/978-3-319-61615-5_3

La Porta E, Exacoustos O, Lugani F (2023) Microplastics and kidneys: an update on the evidence for deposition of plastic microparticles in human organs, tissues and fluids and renal toxicity concern. Int J Mol Sci 24:14391. https://doi.org/10.3390/ijms241814391

Li Y, Sun Y, Li J, Tang R, Miu Y, Ma X (2021) Research on the influence of microplastics on marine life. IOP Conf Ser: Earth Environ Sci 631:012006. https://doi.org/10.1088/1755-1315/631/1/012006

Liu S, Huang J, Zhang W, Shi L, Yi K, Yu H, Zhang C, Li S, Li J (2022) Microplastics as a vehicle of heavy metals in aquatic environments: a review of adsorption factors, mechanisms, and biological effects. J Environ Manag 302:113995. https://doi.org/10.1016/j.jenvman.2021.113995

Lyu W, Chen Q, Cheng L, Zhou W (2020) Microplastics in aquaculture systemsand their transfer in the food chain. In: He D, Luo Y (eds) Microplastics in terrestrial environments. The handbook of environmental chemistry. Springer, Cham, p 95. https://doi.org/10.1007/698_2020_455

Magri D, Sanchez-Moreno P, Caputo G, Gatto F, Veronesi M, Bardi G, Catelani T, Guarnieri D, Athanassiou A,

Pompa PP, Fragouli D (2018) Laser ablation as a versatile tool to mimic polyethylene terephthalate nanoplastic pollutants: characterization and toxicology assessment. ACS Nano 12:7690–7700. https://doi.org/10.1021/acsnano.8b01331

Mariano S, Tacconi S, Fidaleo M, Rossi M, Dini L (2021) Micro and nanoplastics identification: classic methods and innovative detection techniques. Front Toxicol 3:636640. https://doi.org/10.3389/ftox.2021.636640

Miranda-Peña L, Buitrago-Duque L, Rangel-Buitrago N, Gracia CA, Arana VA, Trilleras J (2023) Geographical heterogeneity and dominant polymer types in microplastic contamination of lentic ecosystems: implications for methodological standardization and future research. RSC Adv 13(39):27190–27202. https://doi.org/10.1039/D3RA04016J

Neto J, Rodrigues FL, Ortega I, Rodrigues L, Lacerda A, Coletto JL, Kessler F, Cardoso LG, Maira LM, Proietti MC (2020) Ingestion of plastic debris by commercially important marine fish in southeast-south Brazil. Environ Pollut 267:115508. https://doi.org/10.1016/j.envpol.2020.115508

Pan Z, Guo H, Chen H, Wang S, Sun X, Zou Q, Zhang Y, Lin H, Cai S, Huang J (2019) Microplastics in the north-western pacific: abundance, distribution, and characteristics. Sci Total Environ 650:1913–1922. https://doi.org/10.1016/j.scitotenv.2018.09.244

Plastics Europe (2018) Plastics-the facts 2018. An analysis of European plastics production, demand and waste data. https://plasticseurope.org/wp-content/uploads/2021/10/2018-Plastics-the-facts.pdf

Rajpal A, Choudhury M, Goswami S, Chakravorty A, Raghavan V (eds) (2024) Waste management and treatment: advances and innovations, 1st edn. CRC Press. https://doi.org/10.1201/9781003258377

Roch S, Walter T, Ittner LD, Friedrich C, Brinker A (2019) A systematic study of the microplastic burden in freshwater fishes of south-western Germany – are we searching at the right scale? Sci Total Environ 689:1001–1011. https://doi.org/10.1016/j.scitotenv.2019.06.404

Rochman CM, Manzano C, Hentschel BT, Simonich SLM, Hoh E (2013) Polystyrene plastic: a source and sink for polycyclic aromatic hydrocarbons in the marine environment. Environ Sci Technol 47(24):13976–13984. https://doi.org/10.1021/es403605f

Roy P, Choudhury M (2022) Bioeconomy and green plastic production. In: Baskar C, Ramakrishna S, Daniela La Rosa A (eds) Encyclopedia of green materials. Springer, Singapore. https://doi.org/10.1007/978-981-16-4921-9_189-1

Sarijan S, Azman S, Said MIM, Jamal MH (2021) Microplastics in freshwater ecosystems: a recent review of occurrence, analysis, potential impacts, and research needs. Environ Sci Pollut Res 28:1341–1356. https://doi.org/10.1007/s11356-020-11171-7

Shaikh IV, Shaikh VAE (2021) A comprehensive review on assessment of plastic debris in aquatic environment and its prevalence in fishes and other aquatic animals in India. Sci Total Environ 779:146421. https://doi.org/10.1016/j.scitotenv.2021.146421

Sharma VP, Singh RL, Singh RP (2017) Degradable polymers and plastics of the future: steps toward environmental sustainability, regulations, and safety aspects. In: Singh R (ed) Principles and applications of environmental biotechnology for a sustainable future. Springer, Singapore, pp 467–487. https://doi.org/10.1007/978-981-10-1866-4_15

Sharma VK, Ma X, Lichtfouse E, Robert D (2023) Nanoplastics are potentially more dangerous than microplastics. Environ Chem Lett 21:1933–1936. https://doi.org/10.1007/s10311-022-01539-1

Silva-Cavalcanti JS, Silva JDB, de-França EJ, Araújo MCB, Gusmão F (2017) Microplastics ingestion by a common tropical freshwater fishing resource. Environ Pollut 221:218–226. https://doi.org/10.1016/j.envpol.2016.11.068

Slootmaekers B, Catarci CC, Belpaire C, Saverwyns S, Fremout W, Blust R, Bervoets L (2019) Microplastic contamination in gudgeons (Gobiogobio) from Flemish rivers (Belgium). Environ Pollut 244:675–684. https://doi.org/10.1016/j.envpol.2018.09.136

Vaid M, Mehra K, Gupta A (2021) Microplastics as contaminants in Indian environment: a review. Environ Sci Pollut Res 28:68025–68052. https://doi.org/10.1007/s11356-021-16827-6

Wang S, Zhang C, Pan Z, Sun D, Zhou A, Xie S, Wang J, Zou J (2020a) Microplastics in wild freshwater fish of different feeding habits from Beijiang and Pearl River Delta regions, South China. Chemosphere 258:127345. https://doi.org/10.1016/j.chemosphere.2020.127345

Wang S, Wang Y, Liang Y, Cao W, Sun C, Ju P, Zheng L (2020b) The interactions between microplastic polyvinyl chloride and marine diatoms: physiological, morphological, and growth effects. Ecotoxicol Environ Saf 203:111000. https://doi.org/10.1016/j.ecoenv.2020.111000

Woo H, Seo K, Choi Y, Kim J, Tanaka M, Lee K, Choi J (2021) Methods of analyzing microsized plastics in the environment. Appl Sci 11(22):10640. https://doi.org/10.3390/app112210640

Yin L, Wen X, Du C, Jiang J et al (2020) Comparison of the abundance of microplastics between rural and urban areas: a case study from East Dongting Lake. Chemosphere 244:125486. https://doi.org/10.1016/j.chemosphere.2019.125486

Yuan Z, Nag R, Cummins E (2022) Human health concerns regarding microplastics in the aquatic environment – from marine to food systems. Sci Total Environ 823:153730. https://doi.org/10.1016/j.scitotenv.2022.153730

Zainuddin Z, Syuhada S (2020) Study of analysis method on microplastic identification in bottled drinking water. Macromol Symp 391(1):1900195. https://doi.org/10.1002/masy.201900195

Zazouli M, Nejati H, Hashempour Y, Dehbandi R, Nam VT, Fakhri Y (2022) Occurrence of microplastics (MPs) in the gastrointestinal tract of fishes: a global systematic review and meta-analysis and meta-regression. Sci Total Environ 815:152743. https://doi.org/10.1016/j.scitotenv.2021.152743

Zheng X, Zhang W, Yuan Y, Li Y, Liu X, Wang X, Fan Z (2021) Growth inhibition, toxin production and oxidative stress caused by three microplastics in microcystis aeruginosa. Ecotoxicol Environ Saf 208:111575. https://doi.org/10.1016/j.ecoenv.2020.111575

Zhou Y, Li Y, Yan Z, Wang H, Chen H, Zhao S, Zhong N, Cheng Y, Acharya K (2023) Microplastics discharged from urban drainage system: prominent contribution of sewer overflow pollution. Water Res 236:119976. https://doi.org/10.1016/j.watres.2023.119976

8 Micro and Nanoplastics: Contamination of Lotic Ecosystem

Shivani Garg, Pooja Arora, Deepak Kumar, Meenakshi Suhag, Zaira Zaman Chowdhury, and Nelson Pynadathu Rumjit

8.1 Introduction

The widespread use of materials, along with their incapacity to break down naturally, has led to an increase in trash, creating a huge environmental concern in modern times (Pandey et al. 2023). Scientists are now looking at the consequences of micro and nanoplastics, which are minute, and formed from bigger plastic goods such as bags and bottles. These minute plastic pollutants stem from the degradation of plastics and direct release from human-made sources (Ya et al. 2021). Lotic ecosystems, comprising rivers, streams, and other flowing water habitats, are assets impacted by micro and nanoplastic pollution. Due to the water flow and diverse organisms inhabiting these ecosystems, they are particularly vulnerable to contamination by micro and nanoplastics, with severe ecological repercussions (Dikareva and Simon 2019). Tiny plastic particles have been discovered in flowing water habitats globally, ranging from mountain streams to large river networks (Davidson et al. 2023; Blettler et al. 2018; Mani et al. 2015). Even nanoplastic fragments, which were previously understudied due to testing challenges, are now being identified in rivers and streams (Wang et al. 2021). This chapter focuses on micro and nanoplastic contamination in flowing water habitats, problems, and knowledge gaps in remediation strategies and research goals.

8.2 Sources of Micro and Nanoplastics

Micro and nanoplastics are small plastic particles that originate from various sources and can harm the environment. Weathering and worsening of large plastic items ultimately result in smaller fragments. Items like bottles, bags, and other plastic goods slowly grind down and break up into minuscule particles speckled throughout nature. These sources can be categorized as primary and secondary sources.

S. Garg (✉) · P. Arora · D. Kumar · M. Suhag
Institute of Environmental Studies, Kurukshetra University, Kurukshetra, Haryana, India
e-mail: shivanigarg2015@kuk.ac.in; 2329phd.deepak@kuk.ac.in; msuhag@kuk.ac.in

Z. Z. Chowdhury
Nanotechnology & Catalysis Research Centre (NANOCAT), Institute for Advanced Studies (IAS), University of Malaya (UM), Kuala Lumpur, Malaysia

N. P. Rumjit
Marian Centre for Advanced Research, St. Mary's College (Autonomus), Thrissur, Kerala, India

S. Goswami et al. (eds.), *Micro-Nano Plastics Exposure, Environmental Degradation and Public Health Crisis*, Sustainable Development Goals Series, https://doi.org/10.1007/978-981-96-5047-7_8

8.2.1 Primary Sources of Micro and Nanoplastics

Plastics with a size range of 1 µm to 5 mm are classified as primary microplastics (Mehra et al. 2020). Examples consist of car tyre wear, paint, sewage sludge, washing wastewater, plastic pellets, and micro-beads in personal care goods like makeup, toothpaste, and peeling scrubs (An et al. 2020). Synthetic fibres from clothing and textiles are a significant contributor to microplastic pollution. Figure 8.1 depicts various sources of microplastics, including personal care products, synthetic fibres, plastic degradation, industrial processes, agriculture, and marine debris, and illustrates how these microplastics can enter the environment through pathways such as wastewater, air, soil, and water bodies. During washing, these fibres can shed and enter wastewater systems (Napper et al. 2015; Browne et al. 2011). The examples in detail are discussed in the following subsections.

(a) *Personal Care Products*: Microplastics are commonly used as abrasives or exfoliants in personal cleanliness products, such as body washes, toothpaste, and face scrubs (Mehra et al. 2020; Leslie 2015). Nylon, polypropylene, and polyethene are ordinary materials used to make these microbeads. The microbeads are intentionally added to provide a scrubbing or exfoliating effect (Juliano and Magrini 2017), and these products are capable of damaging marine life and possibly making their way into the food chain when they sponge down the drain and make their way past water filtration systems and into rivers. A study by Napper et al. (2015) found that a single load of laundry could release up to 700,000 microplastic fibres into wastewater systems. Microplastics can slip past sewage treatment and into the environment, where they can interact with various pollutants and aid in spreading toxins in both terrestrial and aquatic environments (Bui et al. 2020). Microplastics can cause lower food intake, developmental problems, and behavioural abnormalities in almost 700 marine species, including sea turtles, penguins, and crabs (Zolotova et al. 2022).

(b) *Synthetic Textiles and Tyres*: Synthetic fibres created from petroleum-based compounds are used to make an array of modern fabrics, including polyester, nylon, and acrylic. These fabrics discharge microscopic plastic fibres, or microfibres, during washing. These

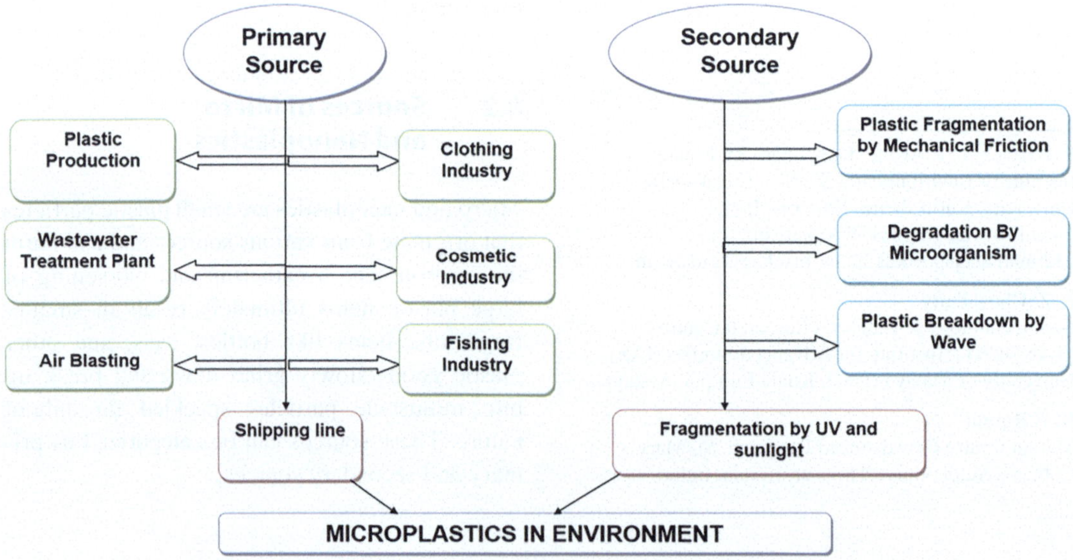

Fig. 8.1 Sources and pathways of microplastic pollution

microfibres are effortlessly able to get past wastewater treatment facilities and into bodies of water, where they put in danger marine life and may in the end find their way up the food chain (Mehra et al. 2020; Magnusson et al. 2016). Synthetic rubber and other materials found in tyres may tumble down over time as a result of rasping with the road. Tyre wear particles, often called tyre dust, are small pieces of rubber liberated into the environment due to wear and tear. Wind and precipitation can carry these particles into bodies of water, where they exaggerate the issue of microplastic contamination (Mehra et al. 2020; Sundt et al. 2015).

(c) *Paints and Wood Preservatives*: Microplastics are routinely used in paints and coatings to emphasize characteristics like texture, durability, or attractiveness (Karlsson et al. 2018). Polyethylene, polypropylene, and other synthetic polymers are examples of these microplastics. Some of these microplastics might not completely defend the paint during application, allowing them to run off into the surrounding area. Chemicals known as wood prophylactic are used on wood surfaces to ward off rot, fungi, insects, and other environmental invaders. Microplastics are a component in certain wood treatments (Havrlik and Ryparová 2015). The application process, weathering, or the deterioration of the wood could release these microplastics into the environment (Yang et al. 2021).

(d) *Plastic Pellet Manufacturers*: One of the main untreated materials used in the manufacture of plastic products is plastic pellets, sometimes referred to as nurdles and nibs (Khair and Matsana 2021). Plastic pellets may leak or break into the environment throughout manufacturing, shipping, and processing. These leaks frequently happen when the shipment is loaded and unloaded at ports, when goods are transported by trucks and railroads, or when manufacturing facilities mismanage materials (Karlsson et al. 2018). A few production procedures have the potential to produce microplastics openly. These pellets have the potential to decompose into smaller particles and turn into microplastics once they are in the ecosystem. Pellets kept outside in warehouses or manufacturing facilities may be carried away by wind or precipitation and end up in stormwater systems (Li et al. 2021). They can enter bodies of water, adding to the pollution from microplastics.

8.2.2 Secondary Sources of Micro and Nanoplastics

Secondary sources of micro and nanoplastics involve the fragmentation of larger plastic items. The degradation of plastic products, such as plastic bags, bottles, and other single-use plastics, can lead to the formation of smaller plastic particles (Andrady 2011). This process is accelerated by various environmental factors, including UV radiation, mechanical abrasion, and oxidation. Another significant source is the abrasion of synthetic rubber tyres, which releases tyre wear particles during driving (Kole et al. 2017). Photo-degradation (sunlight), mechanical scrape (waves, wind, etc.), and chemical degradation (oxidation) are some of the mechanisms that direct the disintegration of larger plastic products into even smaller pieces (Singh 2022). Figure 8.2 represents the fragmentation of larger plastic items like bags, bottles, and tyres, due to environmental factors like UV radiation, mechanical abrasion, and oxidation, leading to the formation of microplastics that can contaminate air, soil, and water bodies through various pathways.

(a) *Mechanical Breakdown*: Over time, unnecessary plastic bottles, bags, litter (toys, shoes, clothing, etc.) and packaging materials crash into smaller pieces because of chemical weathering (such as oxidation), and mechanical abrasion (such as wave action) (Raynaud 2014). This results in the formation of microplastic (size less than 5 mm) and nanoplastic (size less than 0.1 μm).

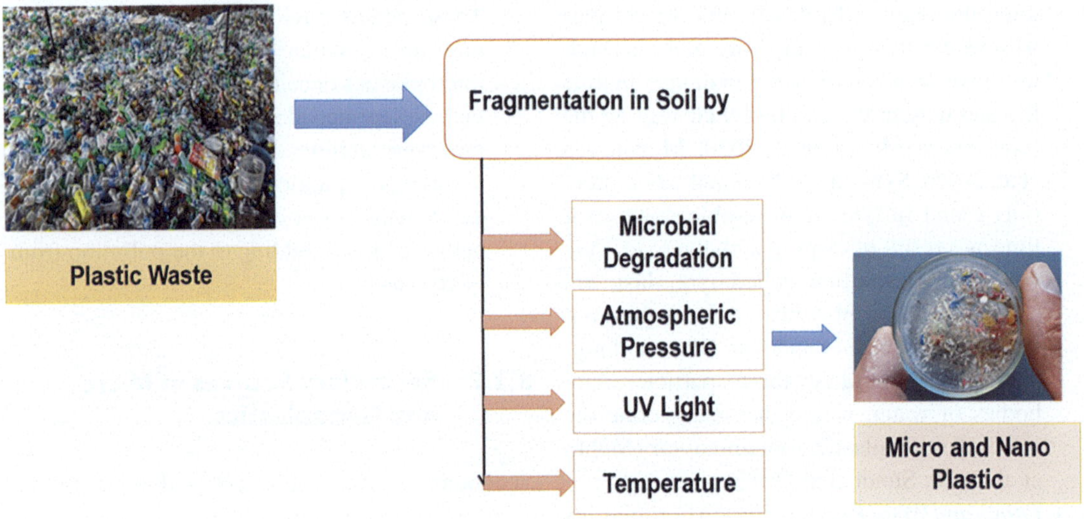

Fig. 8.2 Secondary sources and environmental pathways of microplastic pollution

(b) *Sunlight Degradation*: Plastics are phototainted by sunlight, which makes them brittle and breaks them into tiny pieces. In maritime areas, the strong UV radiation speeds up this process.

(c) *Landfill Leachate*: Plastics are among the numerous waste products found in landfills. Rainwater seeps through the waste, letting out dissolved organic and inorganic substances, such as microplastics, which can later on pollute nearby waterways and soil.

(d) *Domestic Items and Food Stuffs*: The United Nations Environment Programme (UNEP), 2014 (Raynaud 2014), states that domestic items are the principal reason for marine pollution. Oceans are full of stuff like straws, bottles, plastic cutlery, and glasses. Organizations construct plants, but they are more helpful in eliminating massive plastic waste than meso and microplastic rubbish. Micro and nanoplastic particles that infiltrate the food chain and have a range of effects on biotic ecosystems are present in wastewater treatment plant discharges, besides several unsafe substances (Mani et al. 2015; Carr et al. 2016; Prata 2018).

8.3 Transport and Fate of Micro and Nanoplastics in Lotic Ecosystems

The fate and transport of micro and nanoplastics in lotic systems are governed by the unique hydrodynamics of flowing water bodies. Once introduced, these tiny particles can undergo sustained downstream movement, vertical mixing, and potential accumulation hotspots influenced by factors like water velocity, turbulence, particle characteristics, and geomorphological features (Hurley et al. 2018). Dender microplastics may be preferentially transported along the benthic zone, while buoyant particles are more prone to surface dispersion (Allen et al. 2019). Nanoplastics, due to their diminutive size exhibit increased mobility and propensity for colloidal suspension compared to larger microplastics (Mattsson et al. 2018). Despite their tendency for downstream transport, micro and nanoplastics can deposit and accumulate within lotic sediments through sedimentation processes. This benthic sink is influenced by particle characteristics like density, shape, surface properties, and local hydrological conditions (Constant et al. 2019). Low-energy environments like quiescent zones, meanders, and floodplains facilitate sedimentation, acting as potential hotspots for micro/nanoplastic accumulation (Lu et al. 2023). Sediment-associated microplastics may undergo resuspension and redistribution during high flows or flooding events (Hurley et al. 2018). While sedimentation represents a temporary sink, the continuous downstream transport in lotic systems enables micro and nanoplastics to be distributed over considerable distances from their sources. Modelling studies indicate that microplastics can travel hundreds of kilometers along river networks, potentially traversing entire watersheds and facilitating far-reaching environmental dispersal (Nizzetto et al. 2016; van Wijnen et al. 2019). This long-range transport capacity raises concerns about the potential spread of micro/nanoplastic contamination to remote ecosystems, even those lacking direct emission sources. The environmental fate of micro and nanoplastics is further complexed by their physical transformation within lotic habitats. Weathering processes such as exposure to sunlight, oxidation reactions, and physical wear can modify the surface characteristics, size distribution, and chemical makeup of these plastic particles with time (Luo et al. 2022). Additionally, larger microplastics may undergo continuous fragmentation into smaller particles, potentially generating nanoplastic byproducts (Alimi et al. 2018). These ageing and fragmentation processes can modify the environmental behaviour, persistence and ecological impacts of micro/nanoplastics dynamically. Figure 8.3 shows the movement of the micro/nanoplastics in the flowing water bodies and the changes that occur during the transport from one place to another. Through atmospheric deposition, microplastics can also find their way into lotic ecosystems. When precipitation occurs, airborne microplastics can land on the surface of water bodies, adding to the contamination of rivers and streams (Seinfeld and Pandis 2016; Brahney et al. 2021).

8.4 Exposure and Fate of Micro and Nanoplastics in Lotic Ecosystems

8.4.1 Fate Due to Ingestion by Biota

The widespread occurrence of micro and nanoplastics in lotic environments raises concerns about their potential uptake and bioaccumulation by resident aquatic organisms. Ingestion represents a primary pathway for microplastic exposure, with numerous studies documenting the presence of these particles in the gastrointestinal tracts of various biota, including invertebrates, fish, amphibians, and birds (Phuong et al. 2016; Vaid et al. 2021). Nanoplastics, due to their minute size, can penetrate biological membranes, facilitating cellular internalization and systemic distribution (Isibor et al. 2024). The ingestion of micro/nanoplastics by aquatic organisms can occur through multiple routes, including indiscriminate feeding behaviours, respiratory intake, and adherence to external surfaces (Bhatt and Chauhan 2023). Polythene and polypropylene are the most prevalent types of microplastics ingested by freshwater fish in the Yangtze River (Jabeen et al. 2017). Factors such as the dimensions, physical form, and specific type of plastic polymer composing these particles play a role in determining their accessibility to organisms and the likelihood of being absorbed or ingested (Triebskorn et al. 2019). Moreover, microplastic ingestion can be exacerbated by the tendency of these particles to adsorb biomolecules, mimicking natural food sources (Huang et al. 2020).

8.4.2 Fate Associated with Trophic Transfer

Once ingested, micro and nanoplastics can undergo trophic transfer along aquatic food webs in lotic ecosystems. This process is driven by the consumption of contaminated prey by higher trophic levels, facilitating the potential biomagnification and food web accumulation of these persistent contaminants (Atugoda et al. 2022). Evidence of trophic transfer has been documented in various lotic environments, with microplastics 0.5 μm diameter in size detected in predatory invertebrates, fish, and waterbirds (Shi et al. 2021; Bisht and Negi 2020). The degree of trophic transfer is influenced by factors such as organism-feeding strategies, digestive physiology, and the polymer types and concentrations involved (Covernton et al. 2019). Additionally,

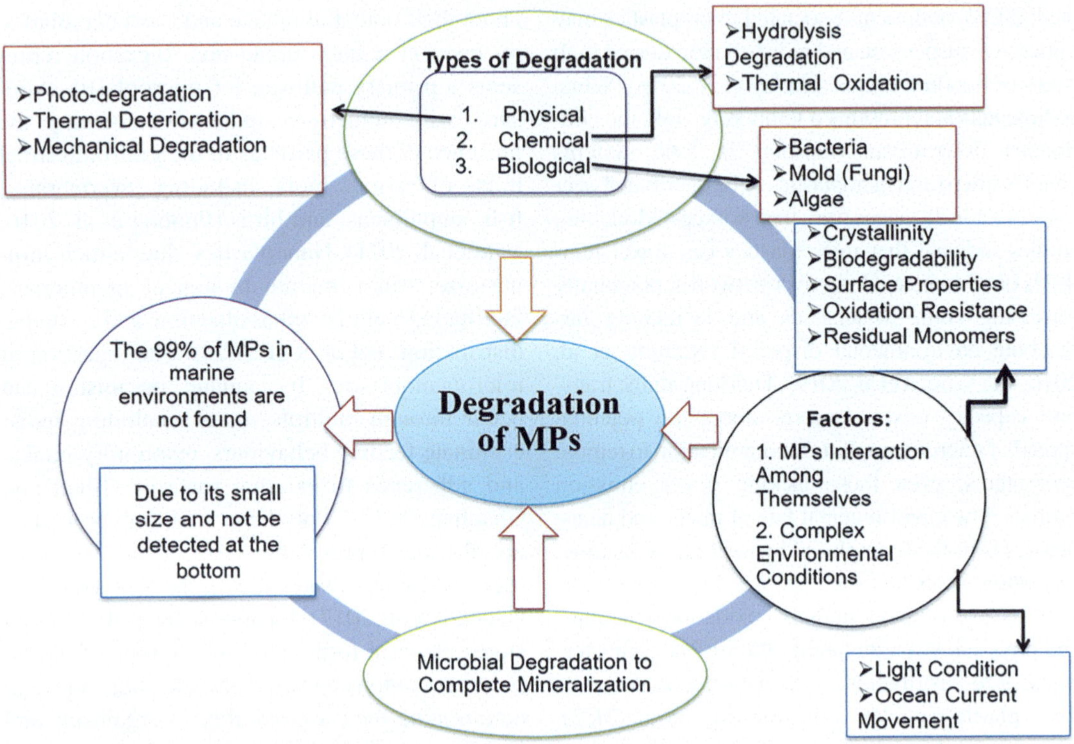

Fig. 8.3 Fate of micro and nanoplastics in lotic ecosystems

the potential for maternal transfer of microplastics has been demonstrated in fish, raising concerns about intergenerational impacts (Pitt et al. 2018).

8.4.3 Toxicological Impacts

Exposure to micro and nanoplastics can elicit a range of toxicological effects in aquatic biota inhabiting lotic ecosystems. These impacts are mediated by various mechanisms, including particle-induced physical damage, leaching of toxic additives, and the vectoring of harmful pollutants (Tang 2021). Physical toxicity can arise from the accumulation of indigestible microplastics, leading to intestinal blockages, abrasions, and inflammatory responses (Lei et al. 2018). Ding et al. (2018) reported that in freshwater fish, *Danio rerio* oxidative stress and intestinal damage were induced by exposure to polythene microplastics. Nanoplastics, due to their minute size, can translocate across biological barriers, potentially inducing oxidative stress, metabolic disruptions, and cytotoxicity (Li et al. 2023). Certain polymer types and additives like plasticizers and flame retardants can leach from micro/nanoplastics, exerting toxic effects on aquatic organisms (Sun et al. 2021).

8.4.4 Pollutant Interactions

Micro and nanoplastics can act as vectors for various harmful pollutants present in lotic environments, including persistent organic pollutants (POPs), heavy metals, and other contaminants (Tang 2021). These hydrophobic contaminants readily adsorb onto the surface of plastic particles, potentially enhancing their bioavailability and facilitating their uptake by aquatic organisms (Fauser et al. 2022). Furthermore, the ingestion of contaminated microplastics can enable the desorption and bioaccumulation of associated pollutants within organisms, a phenomenon known as the "Trojan horse effect" (Dai et al. 2022). These synergistic interactions between micro/nanoplastics and co-pollutants can amplify toxicological risks and contribute to adverse health effects in lotic biota (Pitt et al. 2018).

8.5 Mitigation Measures for Microplastic/Nanoplastic in Lotic System

Reducing the contamination caused by micro and nanoplastics requires a comprehensive approach that addresses the sources of the problem (Mandal and Mishra 2023). This involves improved waste handling, imposing restrictions on harmful products, and enforcing controls on industrial emissions. Upgrading sewage treatment plants with cutting-edge technologies such as membrane bioreactors and oxidation processes can also be beneficial (Deng et al. 2023). Additionally, implementing in situ remediation with coagulants/adsorbents or ex situ approaches like riverbank filtration can help to reduce plastic pollution (Poerio et al. 2019). Table 8.1 includes the removal mechanisms, advantages, and disadvantages of each treatment approach, providing a more comprehensive overview. It is also important to establish strong policy frameworks, regulations, and extended producer responsibility schemes that hold individuals and companies accountable and incentivize the reduction of microplastic pollution (Kumar et al. 2021). The following subsections explain the mitigation approaches in detail.

- *Source Control*
  Effective mitigation of micro and nanoplastic contamination in lotic environments necessitates a concerted effort to control their sources and reduce emissions. A critical strategy involves improved waste management practices, including the proper collection, containment, and environmentally sound recycling or disposal of plastic products (Ding et al. 2021). Banning or phasing out certain single-use plastics, microbeads in personal care products, and other avoidable sources can significantly curtail direct inputs (Paletta et al. 2019). Industrial emissions represent a major source of micro and nanoplastics, underscoring the need for stricter regulations, advanced filtration systems, and the adoption of circular economy principles in manufacturing processes (Napper and Thompson 2019). Additionally, addressing the shedding of synthetic textiles through improved washing machine filters and textile design can mitigate a significant source (De Falco et al. 2018). Recent studies have emphasized the importance of innovative solutions for mitigating micro and nanoplastic pollution. These include the development of biodegradable polymers and composites, which can significantly reduce the persistence of plastics in the environment. The mechanochemical treatments have been explored as a means to degrade microplastics into harmless substances (Sacco et al. 2023).
- *Wastewater Treatment*
  Wastewater treatment is a crucial step in mitigating the environmental impacts of micro and nanoplastics. Traditional wastewater treatment facilities are primarily engineered to eliminate organic waste, nutrient loads, and disease-causing microorganisms, but a substantial quantity of microscopic and nanoscopic plastic particles often pass through unfiltered (Enfrin et al. 2019). These toxins can then be released into water bodies, endangering aquatic ecosystems and perhaps entering the food chain (Bråte et al. 2016). To address the issue, several studies have been undertaken on enhanced treatment methods that can improve micro and nanoplastic removal from wastewater. Upgrading wastewater treatment plants with advanced treatment technologies during tertiary treatment, such as sand filtration, membrane bioreactors (MBRs), and advanced oxidation processes, can significantly enhance the removal of these contaminants (Ma et al. 2018; Mohana et al. 2021). One promising approach is the use of MBRs which combine a membrane filtration process with biological treatment (Zhao et al.

Table 8.1 Summary of treatment approaches for micro and nanoplastics in lotic systems

Treatment approach	Description	Removal mechanism	Advantages	Disadvantages	References
Source control	Reducing the release of micro and nanoplastics at the source	Prevention of pollution	Cost-effective, long-term solution	Requires widespread adoption, limited impact on existing pollution	Paletta et al. (2019) and Ding et al. (2021)
Wastewater treatment	Upgrading conventional WWTPs with advanced tertiary treatment technologies	Membrane filtration, physical straining, adsorption, advanced oxidation	High removal efficiency, integration with existing infrastructure	High capital and operational costs, potential byproduct formation	Enfrin et al. (2019), Lares et al. (2018), Michielssen et al. (2016), Mirza et al. (2020), and Talvitie et al. (2017),
In situ remediation	Application of coagulants, flocculants, or adsorbents directly in the water body	Coagulation, flocculation, sedimentation, adsorption	Potential for in situ treatment, relatively low-cost	Limited effectiveness in dynamic systems, potential for resuspension	Agarwal et al. (2024), Alimi et al. (2018), and Wang et al. (2020)
Ex situ remediation	Interception and removal before entry into lotic ecosystems	Physical straining, filtration, sedimentation, biological processes	Effective removal, integration with natural systems	Land availability, potential ecological impacts, long residence times	Bui et al. (2020), Fältström (2023), and Xu et al. (2022)

2024). MBRs have demonstrated high removal efficiencies for micro and nanoplastics ranging from 90% to 99.9% depending on the pore size and operating conditions (Talvitie et al. 2017). Another effective technology is sand filtration which can capture these tiny plastics through physical straining and adsorption mechanisms (Ziajahromi et al. 2016). According to a study, sand filtration can remove 98% of microplastics from treated wastewater (Hidayaturrahman and Lee 2019). Advanced oxidation processes (AOPs), such as ozonation and UV/H_2O_2 treatment, have also been explored for the removal of micro and nanoplastics (Enfrin et al. 2019). These processes generate highly reactive oxidizing agents that can degrade and mineralize organic contaminants, including micro and nanoplastics (Bracamontes-Ruelas et al. 2022). Despite the promising results of these advanced treatment technologies, their implementation in WWTPs remains a challenge due to economic and operational constraints. The additional costs associated with installation, operation, and maintenance can be substantial, particularly for smaller treatment facilities (Enfrin et al. 2019).

- *Remediation by Adsorbents*
  Remediation measures may be required to alleviate ecological concerns in lotic habitats that have already been polluted by micro and nanoplastics. In situ remediation strategies involve the direct application of remedies inside the polluted water body, such as coagulants or adsorbents, to aid in the removal of suspended micro and nanoplastics (Agarwal et al. 2024). Adsorbents, such as activated carbon or biochar, can be utilized to remove micro and nanoplastics by adsorption (Alimi et al. 2018). However, due to the constant flow of water and the likelihood of re-suspension of settled particles, in situ procedures may be limited in their effectiveness in dynamic flowing waters. Ex situ remediation strategies are

Table 8.2 List of some of the existing laws and regulations around the world

Laws/regulations	Country/region	References
Microbead-Free Waters Act (2015)	United States	The U.S. Congress (2015)
Plastics Waste Management Rules (2016, amended 2021)	India	Ministry of Environment, Forest and Climate Change (2021)
Microbeads in Toiletries Regulations (2017)	Canada	Government of Canada (2017)
Single-use Plastics Directive (2019)	European Union	Elliott et al. (2020)
Plastic Waste Management Plan (2020–2025)	China	Liu and Liu (2023)
National Plastics Plan (2021)	Australia	Bering and Karasik (2022)

methods used to remove micro and nanoplastics before they enter flowing bodies of water. For example, riverbank filtration systems or constructed wetlands can be installed to intercept and remove the plastics (Fältström 2023; Xu et al. 2022). These systems may successfully remove micro and nanoplastics, as well as other contaminants, by employing a combination of physical, chemical, and biological techniques. Constructed wetlands have shown promising results for removing micro and nanoplastics from wastewater and stormwater runoff (Liu et al. 2023a, b). However, the feasibility and cost-effectiveness of such therapies must be further evaluated.

- *Policies and Regulations*
  Strong policy frameworks and regulations are essential for motivating meaningful action and encouraging a coordinated worldwide response to micro and nanoplastic contamination. International agreements, such as the UN Environment Programme's resolution on marine trash and microplastics, lay the groundwork for coordinated initiatives (UNEP 2022). At the national and regional levels, specific laws addressing micro and nanoplastic sources, monitoring requirements, and environmental quality standards can help with successful management and mitigation (Munhoz et al. 2022). Table 8.2 provides an overview of key regulations in various countries and regions. Many countries and nations have implemented or are in the process of making and implementing acts and regulations on micro and nanoplastics issues. Furthermore, expanded producer responsibility programmes and economic incentives can promote sustainable practices in the plastics sector (Beaumont et al. 2019).

8.6 Future Directions and Recommendation

The ecological impacts across diverse lotic biota, particularly at environmentally relevant concentrations and under chronic exposure scenarios, warrant deeper investigation. To address these knowledge gaps, future research should prioritize developing advanced analytical techniques for nanoplastic detection and quantification in complex lotic matrices, by conducting long-term monitoring studies to elucidate the spatial and temporal trends of micro/nanoplastic contamination in lotic ecosystems and elucidating the fate, transport, and accumulation dynamics through integrated modelling and field-based approaches. From a sustainability perspective, the long-term persistence of micro and nanoplastics in aquatic environments challenges the notion of plastics as disposable materials. Addressing their sources, environmental accumulation, and ecological impacts is crucial for safeguarding the health of lotic ecosystems and promoting a more circular and sustainable plastics economy.

8.7 Conclusions

This chapter offers an in-depth study of the emerging environmental issue posed by contaminating river and stream ecosystems with micro and nanoplastic particles. These tiny plastic fragments arise from various sources, such as the breakdown of larger plastic waste materials, industrial discharges, products used for personal grooming and hygiene, and the shedding of synthetic fibres from clothing and textiles. Micro and nanoplastics can undergo long-range dispersal across the entire watershed and are subject to ageing processes that can alter their characteristics with time. Many toxicological impacts have been documented, including physical effects, chemical toxicity from leached additives, and interactions with other environmental pollutants. While mitigation strategies show promise, challenges remain as such. Despite growing research efforts, critical knowledge gaps persist regarding micro and nanoplastic pollution in lotic ecosystems.

References

Agarwal P, Prakash S, Saini G (2024) Natural coagulants (*Moringa oleifera* and *Benincasa hispida*) based removal of microplastics. Clean Water 1:100010

Alimi OS, Farner Budarz J, Hernandez LM, Tufenkji N (2018) Microplastics and nanoplastics in aquatic environments: aggregation, deposition, and enhanced contaminant transport. Environ Sci Technol 52(4):1704–1724

Allen S, Allen D, Phoenix VR, Le Roux G, Durántez Jiménez P, Simonneau A et al (2019) Atmospheric transport and deposition of microplastics in a remote mountain catchment. Nat Geosci 12(5):339–344

An L, Liu Q, Deng Y, Wu W, Gao Y, Ling W (2020) Sources of microplastic in the environment. In: Microplastics in terrestrial environments: emerging contaminants and major challenges. Springer, Cham, pp 143–159

Andrady AL (2011) Microplastics in the marine environment. Mar Pollut Bull 62(8):1596–1605

Atugoda T, Piyumali H, Liyanage S, Mahatantila K, Vithanage M (2022) Fate and behavior of microplastics in freshwater systems. In: Handbook of microplastics in the environment. Springer International Publishing, Cham, pp 781–811

Beaumont NJ, Aanesen M, Austen MC, Börger T, Clark JR, Cole M et al (2019) Global ecological, social and economic impacts of marine plastic. Mar Pollut Bull 142:189–195

Bering J, Karasik R (2022) Plastic pollution policy country profile: Australia. NI PB, 22-02

Bhatt V, Chauhan JS (2023) Microplastic in freshwater ecosystem: bioaccumulation, trophic transfer, and biomagnification. Environ Sci Pollut Res 30(4):9389–9400

Bisht VS, Negi D (2020) Microplastics in aquatic ecosystem: sources, trophic transfer and implications. Int J Fish Aquat Stud 8:227–234

Blettler MC, Abrial E, Khan FR, Sivri N, Espinola LA (2018) Freshwater plastic pollution: recognizing research biases and identifying knowledge gaps. Water Res 143:416–424

Bracamontes-Ruelas AR, Ordaz-Díaz LA, Bailón-Salas AM, Ríos-Saucedo JC, Reyes-Vidal Y, Reynoso-Cuevas L (2022) Emerging pollutants in wastewater, advanced oxidation processes as an alternative treatment and perspectives. Processes 10(5):1041

Brahney J, Mahowald N, Prank M, Cornwell G, Klimont Z, Matsui H, Prather KA (2021) Constraining the atmospheric limb of the plastic cycle. Proc Natl Acad Sci 118(16):e2020719118

Bråte ILN, Eidsvoll DP, Steindal CC, Thomas KV (2016) Plastic ingestion by Atlantic cod (Gadus morhua) from the Norwegian coast. Mar Pollut Bull 112(1–2):105–110

Browne MA, Crump P, Niven SJ, Teuten E, Tonkin A, Galloway T, Thompson R (2011) Accumulation of microplastic on shorelines worldwide: sources and sinks. Environ Sci Technol 45(21):9175–9179

Bui XT, Nguyen PT, Nguyen VT, Dao TS, Nguyen PD (2020) Microplastics pollution in wastewater: characteristics, occurrence and removal technologies. Environ Technol Innov 19:101013

Carr SA, Liu J, Tesoro AG (2016) Transport and fate of microplastic particles in wastewater treatment plants. Water Res 91:174–182

Constant M, Kerhervé P, Mino-Vercellio-Verollet M, Dumontier M, Vidal AS, Canals M, Heussner S (2019) Beached microplastics in the northwestern Mediterranean Sea. Mar Pollut Bull 142:263–273

Covernton GA, Pearce CM, Gurney-Smith HJ, Chastain SG, Ross PS, Dower JF, Dudas SE (2019) Size and shape matter: a preliminary analysis of microplastic sampling technique in seawater studies with implications for ecological risk assessment. Sci Total Environ 667:124–132

Dai Y, Zhao J, Sun C, Li D, Liu X, Wang Z et al (2022) Interaction and combined toxicity of microplastics and per-and polyfluoroalkyl substances in aquatic environment. Front Environ Sci Eng 16(10):136

Davidson J, Arienzo MM, Harrold Z, West C, Bandala ER, Easler S, Senft K (2023) Polymer characterization of submerged plastic litter from Lake Tahoe, United States. Appl Spectrosc 77(11):1240–1252

De Falco F, Gullo MP, Gentile G, Di Pace E, Cocca M, Gelabert L et al (2018) Evaluation of microplastic release caused by textile washing processes of synthetic fabrics. Environ Pollut 236:916–925

Deng S, Hu J, Ong SL, Li Q, Han J (2023) Advanced technologies for industrial wastewater reclamation. Front Environ Sci 11:1187335

Dikareva N, Simon KS (2019) Microplastic pollution in streams spanning an urbanisation gradient. Environ Pollut 250:292–299

Ding J, Zhang S, Razanajatovo RM, Zou H, Zhu W (2018) Accumulation, tissue distribution, and biochemical effects of polystyrene microplastics in the freshwater fish red tilapia (*Oreochromis niloticus*). Environ Pollut 238:1–9

Ding R, Tong L, Zhang W (2021) Microplastics in freshwater environments: sources, fates and toxicity. Water Air Soil Pollut 232:1–19

Elliott T, Gillie H, Thomson A (2020) European Union's plastic strategy and an impact assessment of the proposed directive on tackling single-use plastics items. In: Plastic waste and recycling. Academic Press, London, pp 601–633

Enfrin M, Dumée LF, Lee J (2019) Nano/microplastics in water and wastewater treatment processes–origin, impact and potential solutions. Water Res 161:621–638

Fältström E (2023) Mitigating microplastics pollution in urban water systems: flow, actors, and control measures. Doctoral dissertation, Linköping University Electronic Press

Fauser P, Vorkamp K, Strand J (2022) Residual additives in marine microplastics and their risk assessment–a critical review. Mar Pollut Bull 177:113467

Government of Canada (2017) Microbeads in toiletries regulations. Canada Gazette Part II, 151(12)

Havrlik M, Ryparová P (2015) Protection of wooden materials against biological attack by using nanotechnology. Acta Polytech 55(2):101–108

Hidayaturrahman H, Lee TG (2019) A study on characteristics of microplastic in wastewater of South Korea: identification, quantification, and fate of microplastics during treatment process. Mar Pollut Bull 146:696–702

Huang JN, Wen B, Zhu JG, Zhang YS, Gao JZ, Chen ZZ (2020) Exposure to microplastics impairs digestive performance, stimulates immune response and induces microbiota dysbiosis in the gut of juvenile guppy (*Poecilia reticulata*). Sci Total Environ 733:138929

Hurley R, Woodward J, Rothwell JJ (2018) Microplastic contamination of river beds significantly reduced by catchment-wide flooding. Nat Geosci 11(4):251–257

Isibor PO, Adeniyi OS, Jerry OG, Wurie AW, Opeyem AE, Sani AR, Oyewole OA (2024) Human health implications of environmental nanoparticles. In: Environmental nanotoxicology: combatting the minute contaminants. Springer Nature Switzerland, Cham, pp 235–266

Jabeen K, Su L, Li J, Yang D, Tong C, Mu J, Shi H (2017) Microplastics and mesoplastics in fish from coastal and fresh waters of China. Environ Pollut 221:141–149

Juliano C, Magrini GA (2017) Cosmetic ingredients as emerging pollutants of environmental and health concern. A mini-review. Cosmetics 4(2):11

Karlsson TM, Arneborg L, Broström G, Almroth BC, Gipperth L, Hassellöv M (2018) The unaccountability case of plastic pellet pollution. Mar Pollut Bull 129(1):52–60

Khair F, Matsana J (2021) Sustainable plastic waste management strategy: optimization of plastic manufacturing plant waste (gas and product transition). IOP Conf Ser Mater Sci Eng 1041(1):012043. IOP Publishing

Kole PJ, Löhr AJ, Van Belleghem FG, Ragas AM (2017) Wear and tear of tyres: a stealthy source of microplastics in the environment. Int J Environ Res Public Health 14(10):1265

Kumar R, Sharma P, Manna C, Jain M (2021) Abundance, interaction, ingestion, ecological concerns, and mitigation policies of microplastic pollution in riverine ecosystem: a review. Sci Total Environ 782:146695

Lares M, Ncibi MC, Sillanpää M, Sillanpää M (2018) Occurrence, identification and removal of microplastic particles and fibers in conventional activated sludge process and advanced MBR technology. Water Res 133:236–246

Lei L, Wu S, Lu S, Liu M, Song Y, Fu Z et al (2018) Microplastic particles cause intestinal damage and other adverse effects in zebrafish Danio rerio and nematode *Caenorhabditis elegans*. Sci Total Environ 619:1–8

Leslie HA (2015) Plastic in cosmetics: are we polluting the environment through our personal care?: plastic ingredients that contribute to marine microplastic litter. UNEP. https://wedocs.unep.org/bitstream/handle/20.500.11822/9664/-Plastic_in_cosmetics_Are_we_polluting_the_environment_through_our_personal_care_-2015Plas.pdf?sequence=3&isAllowed=y

Li P, Wang X, Su M, Zou X, Duan L, Zhang H (2021) Characteristics of plastic pollution in the environment: a review. Bull Environ Contam Toxicol 107:577–584

Li Z, Yan L, Junaid M, Chen X, Liao H, Gao D et al (2023) Impacts of polystyrene nanoplastics at the environmentally relevant and sub-lethal concentrations on the oxidative stress, immune responses, and gut microbiota to grass carp (*Ctenopharyngodon idella*). J Hazard Mater 441:129995

Liu C, Liu C (2023) Exploring plastic-management policy in China: status, challenges and policy insights. Sustain For 15(11):9087

Liu M, Liu J, Xiong F, Xu K, Pu Y, Huang J et al (2023a) Research advances of microplastics and potential health risks of microplastics on terrestrial higher mammals: a bibliometric analysis and literature review. Environ Geochem Health 45(6):2803–2838

Liu S, Zhao Y, Li T, Hu T, Zheng K, Shen M, Long H (2023b) Removal of micro/nanoplastics in constructed wetland: efficiency, limitations and perspectives. Chem Eng J 475:146033

Lu X, Wang X, Liu X, Singh VP (2023) Dispersal and transport of microplastic particles under different flow conditions in riverine ecosystem. J Hazard Mater 442:130033

Luo H, Liu C, He D, Xu J, Sun J, Li J, Pan X (2022) Environmental behaviours of microplastics in aquatic systems: a systematic review on degradation, adsorption, toxicity and biofilm under ageing conditions. J Hazard Mater 423:126915

Ma J, Dai R, Chen M, Khan SJ, Wang Z (2018) Applications of membrane bioreactors for water reclamation: micropollutant removal, mechanisms and perspectives. Bioresour Technol 269:532–543

Magnusson K, Eliaeson K, Fråne A, Haikonen K, Olshammar M, Stadmark J, Hultén J (2016) Swedish sources and pathways for microplastics to the marine environment. Swedish Environmental Research Institute, Stockholm

Mandal G, Mishra S (2023) A review on emerging micro and nanoplastic pollutants, heavy metals and their remediation techniques. Nano 8. https://doi.org/10.37819/nanofab.008.302

Mani T, Hauk A, Walter U, Burkhardt-Holm P (2015) Microplastics profile along the Rhine River. Sci Rep 5(1):17988

Mattsson K, Jocic S, Doverbratt I, Hansson LA (2018) Nanoplastics in the aquatic environment. In: Microplastic contamination in aquatic environments, pp 379–399. https://doi.org/10.1016/B978-0-12-813747-5.00013-8

Mehra S, Sharma K, Sharma G, Singh M, Chadha P (2020) Sources, fate, and impact of microplastics in the aquatic environment. In: Emerging contaminants. https://doi.org/10.5772/intechopen.93805

Michielssen MR, Michielssen ER, Ni J, Duhaime MB (2016) Fate of microplastics and other small anthropogenic litter (SAL) in wastewater treatment plants depends on unit processes employed. Environ Sci Water Res Tech 2(6):1064–1073

Ministry of Environment, Forest and Climate Change (2021) Plastic waste management (amendment) rules, 2021. The Gazette of India

Mirza NR, Huang R, Du E, Peng M, Pan Z, Ding H, Shan G, Ling L, Xie Z (2020) A review of the textile wastewater treatment technologies with special focus on advanced oxidation processes (AOPs), membrane separation and integrated AOP-membrane processes. Desalin Water Treat 206:83–107

Mohana AA, Farhad SM, Haque N, Pramanik BK (2021) Understanding the fate of nano-plastics in wastewater treatment plants and their removal using membrane processes. Chemosphere 284:131430

Munhoz DR, Harkes P, Beriot N, Larreta J, Basurko OC (2022) Microplastics: a review of policies and responses. Microplastics 2(1):1–26

Napper IE, Bakir A, Rowland SJ, Thompson RC (2015) Characterisation, quantity and sorptive properties of microplastics extracted from cosmetics. Mar Pollut Bull 99(1-2):178–185

Napper IE, Thompson RC (2019) Environmental deterioration of biodegradable, oxo-biodegradable, compostable, and conventional plastic carrier bags in the sea, soil, and open air over a 3-year period. Environ Sci Technol 53(9):4775–4783

Nizzetto L, Bussi G, Futter MN, Butterfield D, Whitehead PG (2016) A theoretical assessment of microplastic transport in river catchments and their retention by soils and river sediments. Environ Sci: Processes Impacts 18(8):1050–1059

Paletta A, Leal Filho W, Balogun AL, Foschi E, Bonoli A (2019) Barriers and challenges to plastics valorisation in the context of a circular economy: case studies from Italy. J Clean Prod 241:118149

Pandey P, Dhiman M, Kansal A, Subudhi SP (2023) Plastic waste management for sustainable environment: techniques and approaches. Waste Dispos Sustain Energy 5(2):205–222

Phuong NN, Zalouk-Vergnoux A, Poirier L, Kamari A, Châtel A, Mouneyrac C, Lagarde F (2016) Is there any consistency between the microplastics found in the field and those used in laboratory experiments? Environ Pollut 211:111–123

Pitt JA, Trevisan R, Massarsky A, Kozal JS, Levin ED, Di Giulio RT (2018) Maternal transfer of nanoplastics to offspring in zebrafish (*Danio rerio*): a case study with nanopolystyrene. Sci Total Environ 643:324–334

Poerio T, Piacentini E, Mazzei R (2019) Membrane processes for microplastic removal. Molecules 24(22):4148

Prata JC (2018) Microplastics in wastewater: state of the knowledge on sources, fate and solutions. Mar Pollut Bull 129(1):262–265

Raynaud J (2014) Valuing plastics: the business case for measuring. Managing and disclosing plastic use in the consumer goods industry. UNEP, 116 pp

Sacco NA, Zoppas FM, Devard A, González Muñoz MDP, García G, Marchesini FA (2023) Recent advances in microplastic removal from water with special attention given to photocatalytic degradation: review of scientific research. Microplastics 2(3):278–303

Seinfeld JH, Pandis SN (2016) Atmospheric chemistry and physics: from air pollution to climate change. John Wiley & Sons, Newark

Shi W, Sun S, Han Y, Tang Y, Zhou W, Du X, Liu G (2021) Microplastics impair olfactory-mediated behaviors of goldfish Carassius auratus. J Hazard Mater 409:125016

Singh VK (2022) Removal of microplastic pollution through waste water treatment: a review. J Res Appl Sci Biotechnol 1(4):5–12

Sun T, Zhan J, Li F, Ji C, Wu H (2021) Effect of microplastics on aquatic biota: a hormetic perspective. Environ Pollut 285:117206

Sundt P, Schulze PE, Syversen F (2015) Sources of microplastic pollution to the marine environment; 2014. M-321. Mepex for the Norwegian Environment Agency, Miljodirektoratet

Talvitie J, Mikola A, Koistinen A, Setälä O (2017) Solutions to microplastic pollution–removal of microplastics from wastewater effluent with advanced wastewater treatment technologies. Water Res 123:401–407

Tang KHD (2021) Interactions of microplastics with persistent organic pollutants and the ecotoxicological effects: a review. Trop Aquat Soil Pollut 1(1):24–34

Triebskorn R, Braunbeck T, Grummt T, Hanslik L, Huppertsberg S, Jekel M et al (2019) Relevance of nano-and microplastics for freshwater ecosystems: a critical review. TrAC Trends Anal Chem 110:375–392

U.S. Congress (2015). Microbead-free waters act of 2015. Public Law 114-114

UNEP (2022) Combating plastic pollution through regional cooperation and national policies: challenges, opportunities and best practices. United Nations Environment Programme

Vaid M, Sarma K, Gupta A (2021) Microplastic pollution in aquatic environments with special emphasis on riverine systems: current understanding and way forward. J Environ Manag 293:112860

van Wijnen J, Ragas AM, Kroeze C (2019) Modelling global river export of microplastics to the marine environment: sources and future trends. Sci Total Environ 673:392–401

Wang Z, Lin T, Chen W (2020) Occurrence and removal of microplastics in an advanced drinking water treatment plant (ADWTP). Sci Total Env 700:134520

Wang J, Zhao X, Wu F, Niu L, Tang Z, Liang W et al (2021) Characterization, occurrence, environmental behaviors, and risks of nanoplastics in the aquatic environment: current status and future perspectives. Fundam Res 1(3):317–328

Xu D, Yin X, Zhou S, Jiang Y, Xi X, Sun H, Wang J (2022) A review on the remediation of microplastics using constructed wetlands: bibliometric, co-occurrence, current trends, and future directions. Chemosphere 303:134990

Ya H, Jiang B, Xing Y, Zhang T, Lv M, Wang X (2021) Recent advances on ecological effects of microplastics on soil environment. Sci Total Environ 798:149338

Yang L, Zhang Y, Kang S, Wang Z, Wu C (2021) Microplastics in freshwater sediment: a review on methods, occurrence, and sources. Sci Total Environ 754:141948

Zhao H, Helgason A, Leng R, Chowdhury S, Clermont N, Dinh J et al (2024) Removal of microplastics/microfibers and detergents from laundry wastewater by microbubble flotation. ACS ES&T Water 4:1819

Ziajahromi S, Neale PA, Leusch FD (2016) Wastewater treatment plant effluent as a source of microplastics: review of the fate, chemical interactions and potential risks to aquatic organisms. Water Sci Technol 74(10):2253–2269

Zolotova N, Kosyreva A, Dzhalilova D, Fokichev N, Makarova O (2022) Harmful effects of the microplastic pollution on animal health: a literature review. PeerJ 10:e13503

Micro and Nanoplastics: Contamination of Soil

9

Ratul Pegu, Sarmistha Paul, Elkana Hezron, Bishal Paul, and Ruth Moirana

9.1 Introduction

Plastic is a polymeric synthetic and semi-synthetic material. Plastic materials are widely utilized in various fields, such as agriculture, manufacturing industry, medicine, and other areas, due to their long-lasting potential, good ductility, and low cost (Yu et al. 2022). The production and consumption of plastics increase day by day. In 1950, the world generated only two million tons of plastics annually, but in 2019, it climbed to 460 million tons (Kiener 2022). From the African context, the second continent after Asia, the release of plastic waste is reaching its peak, considering that over 70% of the daily generated plastic waste is mismanaged, making its environment more prone to degradation (Okeke et al. 2022). A study conducted in Australia depicted that the soil of an industrial area in Sydney had microplastics and nanoplastics abundance ranging from 300 to 67,500 mg/kg (Amobonye et al. 2021).

On the other hand, 16% of global plastic waste is anticipated to be produced by European countries, accounting for the least quantity compared to that of African and Asian countries, including India (Federici et al. 2022). In India, 15,342 tons of plastic waste are generated daily, and from the total volume, only 60% of plastic waste is reprocessed (Bhattacharya et al. 2018). However, India's average per capita plastic waste use is 11 kg, which is considerably low compared to the global average of 28 kg (Bhattacharya et al. 2018). Worldwide, most plastic materials become part of landfills or are thrown away into the soil, which causes pollution and endangers the ecology of the environment (Geyer et al. 2017).

Plastics, known for their resistance to decomposition, undergo a slow and steady decay process. Large plastic litter eventually breaks down

R. Pegu (✉) · B. Paul
Department of Environmental Sciences, Tezpur University, Tezpur, India

S. Paul (✉)
Department of Environmental Sciences, Tezpur University, Tezpur, India

Department of Botany, Visva Bharati University, Santiniketan, Bolpur, Birbhum, West Bengal, India

E. Hezron
Department of Sustainable Agriculture, Biodiversity and Ecosystem Management, School of Life Sciences and Bioengineering, The Nelson Mandela African Institution of Science and Technology, Arusha, Tanzania
e-mail: elkana.hezron@nm-aist.ac.tz

R. Moirana
School of Materials, Energy, Water, and Environmental Sciences, The Nelson Mandela African Institution of Science and Technology, Arusha, Tanzania
e-mail: ruth.moirana@nm-aist.ac.tz

S. Goswami et al. (eds.), *Micro-Nano Plastics Exposure, Environmental Degradation and Public Health Crisis*, Sustainable Development Goals Series, https://doi.org/10.1007/978-981-96-5047-7_9

into mesoplastics (5–40 mm), microplastics (1–5000 μm), and nanoplastics (0.1 or less) before complete decomposition. It is the micro and nanoplastics that raise specific concerns due to their unique features, particularly their small size, which allows them to swiftly spread throughout the environment. This rapid dispersion means they can easily enter the food chain and be transported across various tissues in contaminated animals, posing a significant risk to the environment and public health (Choudhury et al. 2022a, b).

Deterioration of soil health due to the pollution caused by microplastics and nanoplastics is one of the significant severe scientific issues of the environment and ecology. The microplastic abundance on land to that of the ocean outstands by 4–23 times (Yu et al. 2022). Moreover, the generation of microplastics and nanoplastics in cultivated soil, as compared to the ocean, acts as a much larger sink (Horton et al. 2017). The impact of plastic contamination in the soil environment is severe because humans and other organisms may ingest micro and nanoplastics via food. Therefore, this chapter demonstrates how micro/nanoplastics accumulate in the soil, their effect on soil quality, soil microbes, soil plant, animal health, human health, and their determination process from the soil environment.

9.2 Importance of Soil for Our Existence

The importance of soil in our lives cannot be overstated. Soil is crucial not only in our lives but also in the lives of plants and animals. The earth is the fundamental atmosphere and environment for crop growth that feeds humans, domestic animals, and wildlife. Soil provides many essential services in our daily life. They serve as a medium where different types of plants grow. The plant can emit and absorb gases such as carbon dioxide, methane, water vapour, and dust. Plants will provide us with enough food and materials for future technical needs (Gogoi et al. 2024). Soil also provides a habitat for soil-dwelling organisms, soaking up, retaining, releasing, and altering the water in the ecosystem. Soil is a significant switchyard for overall cycles such as the carbon, water, and nutrient cycles. Carbon (C), nitrogen (N), phosphorous (P), and other nutrients are also stored, transformed, and cycled through the soil (Luo et al. 2020). Soil also serves as the medium for engineering for the construction of foundations, roadbeds, dams, and buildings, and it also plays a role in water purification before the water moves into groundwater.

9.3 Micro-nanoplastics: Contamination in Soil

Plastics can be incorporated into the soil environment through different pathways and stay in different forms. Soil microplastics emerge as a result of agricultural mulching film, landfills, tire wear, sewage sludge, effluent soaking (Maity et al. 2022; Zhang and Liu 2018), and atmospheric precipitation (Liu et al. 2019; Maity et al. 2022). Plastics in the soil environment can also result from street runoff, littering, and wind disposal (Brahney et al. 2020; Maity et al. 2022). According to various researchers, soil containing 0.1% plastic by soil weight is considered polluted (Wang et al. 2019a, b).

Various research findings prove the occurrence of micro/nanoplastics in soil environments. Most of the research is conducted in East Asia, Europe, the Middle East, Australia, and the Americas. Recently, India also focused on the occurrence of micro/nanoplastics in the soil environment. However, compared to other countries, India conducts little research on the distribution of microplastics in soil, the types of microplastics present in soil, and the distribution of microplastics across different land uses. The East Asian countries conducted the most research on soil microplastics and found 0–690,000 items kg^{-1} dry weight of soil (Büks and Kaupenjohann 2020). In European countries, a few studies conducted on soil microplastics found very heterogeneous loadings ranging from 0 to 26 × 106 items kg^{-1}. Similar to the European countries, only a few reports are available on the presence of microplastics in soil environments in the Americas, the Middle East, and Australia. The loadings are obtained from 0 to 10,200 items

Table 9.1 Distribution characteristics of microplastics in some parts of the world in the soil environment

Country or region	Abundance	Composition	Size range	References
Mexico	2770 pcs/kg	Polyethylene, polystyrene	5–150 mm	Huerta Lwanga et al. (2017)
Switzerland	593 pcs/kg	Polyethylene, polystyrene, polyvinyl chloride	12.5–500 μm	Scheurer and Bigalke (2018)
Australia	300–67,500 mg/kg	Polyethylene, polystyrene, polyvinyl chloride	<5 mm	Fuller and Gautam (2016)
Hebei China	317 pcs/500 g	Polypropylene, polyvinyl chloride	1.56 ± 0.63 mm	Zhao et al. (2011)
Shandong China	1.3–14.7125 pcs/kg	Polyethylene, polypropylene, polyvinyl chloride	1 mm (60%)	Zhou et al. (2018)
China's Loess Plateau	<0.54 mg/kg	Polyethene	>100 μm	Zhang et al. (2020)
Yun Nan China	7100–42,960 pcs/kg	Polyethylene, polypropylene	10–0.05 mm	Zhang and Liu (2018)
Shanghai, China	62.59 ± 12.97 pcs/kg	Polyethylene, polypropylene, polyvinyl chloride	0.03–16 mm	Liu et al. (2019)
Maharashtra, India	40.46 pcs/kg	Polyacrylamide, Polyethylene, polypropylene, cellulose acetate, polyethylene terephthalate	–	Singh et al. (2024)
Bhopal, India	180 pcs/kg	Polyethylene, polypropylene, polyvinyl chloride, polyethylene terephthalate	–	Singh et al. (2023)

kg^{-1}, 0.2 mg kg^{-1} to 186 mg kg^{-1}, and 300 mg kg^{-1} to 67,500 mg kg^{-1}, respectively (Büks and Kaupenjohann 2020). Table 9.1 shows the types of microplastics detected in the soil and their prevalence in some parts of the global soil environment.

9.4 Consequences of Micro/Nanoplastics on Soil

9.4.1 Effects on Soil Physicochemical Properties

Micro/nanoplastics can change soil structure, reduce water retention capacity, hinder microbial growth and activity, and reduce soil fertility and quality (Seo et al. 2024). When MPs/NPs invade the soil condition, they incorporate other organic pollutants such as organochlorine pesticides and polychlorinated biphenyls and participate in soil property degradation. Taking into account their tiny stature and large surface area, their adsorption capacity is significantly higher. This affects the soil's physicochemical properties, soil health, and ecosystem and increases the possibility of toxicity to soil life forms (Li et al. 2020a, b; Yu et al. 2022). The effect of MPs/NPs on soil physicochemical properties relies upon the types of plastics. The MPs/NPs that degrade the soil quality are classified into several types; each type affects soil differently (Fig. 9.1). Some are primary MPs/NPs, secondary MPs/NPs, fibrous MPs/NPs, fragmented MPs/NPs, rubber MPs/NPs, and bioplastics. Primary microplastics alter soil structure, reduce porosity, and affect water retention (Guo et al. 2022). They also release toxic additives such as phthalates, which harm soil biota and reduce soil fertility. Primary microplastics include polyethylene (PE), polypropylene (PP), polystyrene (PS), polyvinyl chloride (PVC), polyester, nylon, and polymethyl methacrylate (PMMA) (Ziani et al. 2023). Secondary microplastics can change soil structure, increase compaction, and reduce aeration. Due to their irregular shape and size, secondary microplastics also affect soil texture, which reduces root penetration and decreases plant growth (Ziani et al. 2023). Secondary microplastics include fragments, fibers, foams, films, pellets, and microbeads. Fibrous microplastics mix with soil particles, alter soil structure, and reduce water

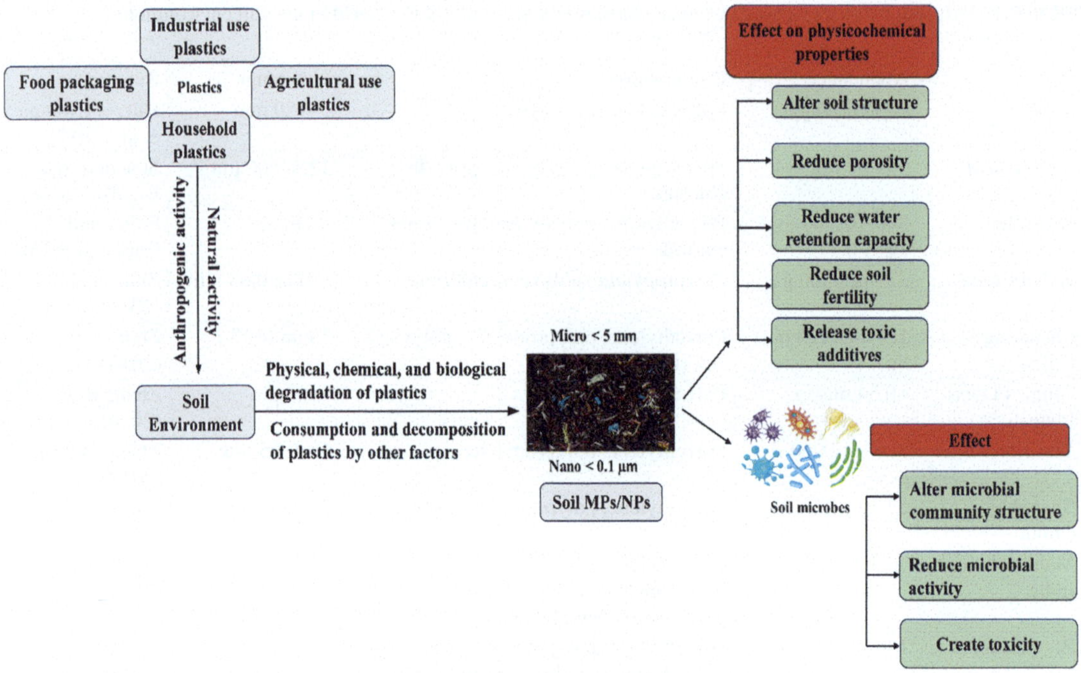

Fig. 9.1 Routes of micro/nanoplastics enter the soil and their effect on soil physicochemical properties and soil microbes

retention capacity (Zhang and Liu 2018). Fragmented microplastics alter soil surface area, influencing water retention and increasing the risk of soil erosion. They also carry pollutants such as heavy metals, which degrade soil properties (De Souza MacHado et al. 2018). Rubber microplastics can release toxic chemicals such as polycyclic aromatic hydrocarbons (PAHs), which harm soil organisms and reduce soil fertility. Rubber also blocks soil pores, reducing water infiltration and root growth (Sieber et al. 2020). Moreover, incomplete degradation of bioplastics can lead to the accumulation of microplastic residues, which disrupt microbial communities and nutrient cycles (Bandopadhyay et al. 2018).

9.4.2 Consequences of Micro/Nanoplastics on Soil Microorganisms

Soil microbes play a crucial role in maintaining soil physicochemical properties and health. Their enhanced performance leads to increased discharge of soil carbon (C), nitrogen (N), phosphorous (P), and nutrient cycles (Luo et al. 2020). However, the introduction of MPs/NPs into the soil can significantly disrupt the soil microbial community structure, reduce essential microbial activity, and increase the toxic effect on soil microorganisms (Roy et al. 2023). These changes in microbial activity have a lasting, negative impact on soil health, affecting ecosystem services and agricultural productivity (Fig. 9.1) (Judy et al. 2019).

Different types of micro/nanoplastics are found in the soil environment. Every plastic has a distinct impact on soil microorganisms. For example, Polyethylene (PE) microplastics can inhibit the activity of soil microorganisms such as bacteria and fungi, which decompose the organic matter. As a result, slow nutrient cycling is observed in that area (Sajjad et al. 2022). Polystyrene (PS) microplastics interfere with the activity of soil enzymes such as cellulase and ligninase, which play a significant role in the breakdown of lignin and cellulose (Jain et al. 2023). This disruption slows down the degradation process of plant residues in the soil, reducing nutrient availability for plant and soil microorganisms. Polyvinyl chloride (PVC) absorbs different types of pollutants such as metals, polycyclic aromatic hydrocarbons (PAHs), dioxins, pesticides, pharmaceuticals and personal care products (PPCPs), and hydrophobic organic compounds (HOCs) (Li et al. 2018). It makes them more toxic to soil microorganisms. PVC also alters the microbial community structure by reducing the abundance of decomposer-like saprophytic fungi, which play a significant role in the breakdown of dead organic matter (Zhou et al. 2021), and polypropylene (PP) inhibits the soil fungal decomposers, which plays a crucial role in the degradation of organic material-like plant residue. This can slow decomposition and alter soil health (Porter et al. 2023).

9.4.3 Soil Contamination and Its Effect on Plants and Animals

Micro/nanoplastic contamination in soil is one of the growing environmental concerns due to its potential impacts on soil animals, plants, and ecosystem functions. The effects of microplastics on soil animals and plants can be direct or indirect and may vary depending on the type, size, concentration, and composition of the MPs/NPs (Fig. 9.2). The overview of MPs/NPs impacts on soil plants and animals is discussed below.

- *Effect on Plants*
  Plants are a crucial part of the soil environs, and the soil regulates their maturation and advancements. Plants exposure to micro/nanoplastics is generally encountered in a natural environment which occurs in nature, including soil, water, and air (Wang et al. 2022). Micro/nanoplastic contamination in the soil environment alters plant growth and development by changing soil biomass content (Zang et al. 2020). Micro/nanoplastics have higher adsorption capacity, so they easily stick to the surface of any seeds, plant roots, and leaves, which affects plants' growth and developmental functions (Wang et al. 2022). The effect of micro/nanoplastics on plant growth depends on some other direct and indirect influences that are inclusive of the direct plastic toxicity on the plants, direct toxicity from adulterants present in plastics, and an indirect effect by changing soil physicochemical characteristics and microbial communities (Chang et al. 2022). Micro/nanoplastics impact the plant in various stages of its life span. Some of them are listed below.
  - *Effects on Seed Germination*
    Micro/nanoplastics negatively impact plant and seed germination by creating physical barriers and toxicity, altering soil structure, and interfering with the microbial community. However, their effect widely depends on the types, concentrations, sizes, shapes, and presence of additives or contaminants and the plant species involved. Different types of micro/nanoplastics are involved in the negative impact on seed germination. Some of them are as follows: polyethylene (PE), polypropylene (PP), polystyrene (PS), polyvinyl chloride (PVC), polyethylene terephthalate (PET), polyamide, microplastic fibers, and additives of plastics such as plasticizers and stabilizers.

    Microplastic fibers present a unique challenge. They entangle around seeds, roots, and soil particles, creating physical barriers that can impede seed germination and root growth. For instance, polyethylene (PE) creates a physical barrier around seeds, preventing water and nutrients from reaching them and reducing the seed germination rate (Bosker et al. 2019). Polypropylene (PP) obstructs seed germination by altering the soil's physical structure and absorbing toxic substances. Polystyrene (PS) interferes with seed germination by changing soil structure and reducing water retention capacity (Yuan et al. 2019). Polyvinyl chloride (PVC) releases harmful additives such as phthalates which reduce seed germination and early plant development. Polyethylene terephthalate (PET) reduces seed germination by altering soil physical properties and

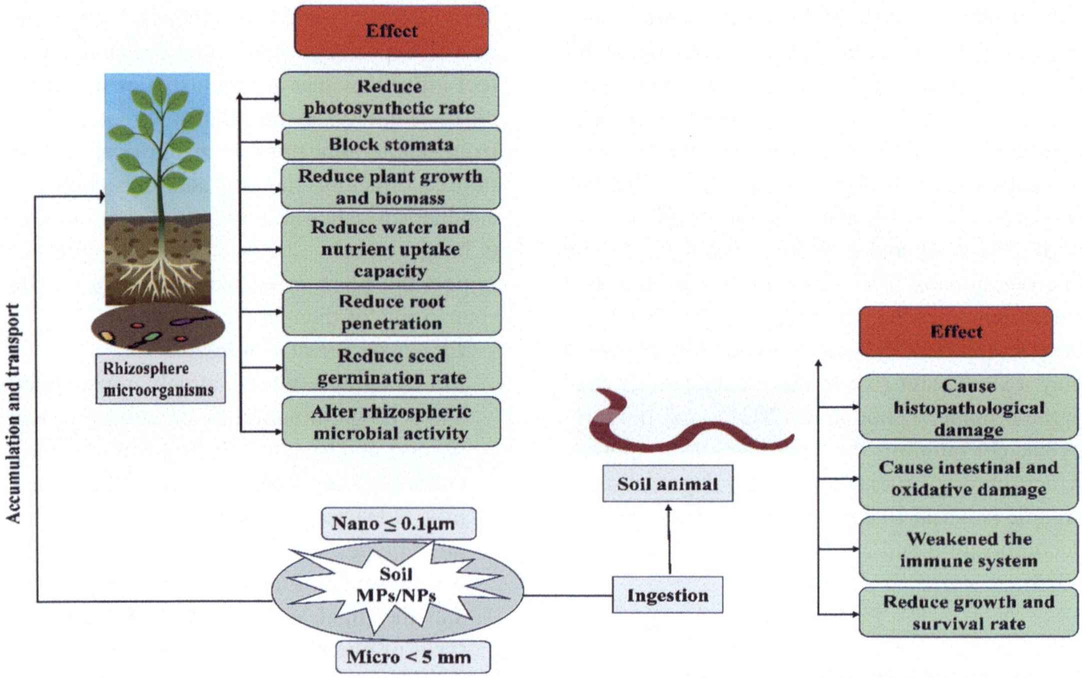

Fig. 9.2 Effect of micro/nanoplastics on plant and soil animals

leaching harmful chemicals. Polyamide creates a physical barrier in the soil that impedes seedling growth, and it also absorbs pollutants such as metals which hinder plant growth later. Additives such as phthalates and heavy metals introduce toxic elements which impact seed germination.

– *Effects on Plant Growth and Plant Biomass* Various research reports show the significant impact of micro/nanoplastics on plant growth and biomass. The effect of micro/nanoplastics plant growth and plant biomass depends on their types, sizes, concentrations, shapes, additives present on plastics, plant species, and soil conditions (Lozano et al. 2021; Wang et al. 2022). The effect is mediated through physical, chemical, and biological pathways such as altered soil physicochemical properties, reduced nutrient availability, increased toxicity, created oxidative stress, and disrupted microbial community structure. Different types of micro-nanoplastics are involved in the effect of plant growth and plant biomass. Some are polyethylene, polypropylene, polystyrene, polyvinyl chloride, nylon microfibers, polyethylene terephthalate, and polylactic acid. For example, the larger size of micro/nanoplastics such as polyethylene, polypropylene, polyethylene terephthalate, polyvinyl chloride, and polystyrene are not able to penetrate the plant's root system, so they stick to the root surface and clog stomata (Giorgetti et al. 2020; Li et al. 2022; Wang et al. 2022). This changes the root epidermal cells and reduces the capacity to uptake water and nutrients from the soil. As a result, plants' growth rate is inhibited and reduces plant biomass (Sun et al. 2020; Wang et al. 2022). Moreover, high-density polyethylene (HDPE) of 23–38 μm effortlessly sticks to the root surface of mung bean (*Vigna radiata*) and inhibits water uptake and plant biomass (Wang et al. 2021, 2022). Low-density polyethylene (LDPE) and biodegradable plastic significantly inhibit the seedlings of wheat (*Triticum aestivum*) (Qi et al. 2018).

- *Effects on Plant Nutrient Uptake*
  Macro- and micro-nutrient availability in soil significantly affects average plant growth. However, the presence of micro/nanoplastics in soil plays a crucial role in the availability of nutrients for plants. The exposure of micro/nanoplastics in soil reduces the plant nutrient availability and absorption of essential nutrients. For example, large-size micro/nanoplastics such as fibers, fragments, films, beads, and degraded nanoparticles can stick to the root surface area, which reduces the ability to absorb water and nutrients from the soil. Small-sized micro/nanoplastics such as polystyrene (PS), polyethylene (PE), polypropylene (PP), polyvinyl chloride (PVC), polyethylene terephthalate (PET), polylactic acid (PLA), etc., are accumulated by plants body and block the absorption and transportation of nutrients by obstructing the pores of the cell wall (Guo et al. 2020; Wang et al. 2022). Some other micro/nanoplastics such as polystyrene (PS), polyethylene (PE), polyvinyl chloride (PVC), and polypropylene (PP) generate oxidative stress in plant cells, which affects root morphology and function, impairing the plant's nutrient uptake (Li et al. 2020a; Ma et al. 2010; Qi et al. 2018).
- *Effects on Plant Photosynthesis*
  Photosynthesis is considered a good indicator of plant stress. Micro/nanoplastics cause "oxidative bursts" in plants, which destroy chlorophyll's structure. As a result, hamper chlorophyll formation and increase chlorophyll decomposition rate (Wang et al. 2022). For example, PS and HDPE plastics reduce the leaf chlorophyll content in Chinese cabbage (*Brassica rapa*) (Yang et al. 2021), and polyethylene (PE) reduces the chlorophyll percentage in lettuce leaves (Gao et al. 2019; Wang et al. 2022).
- *Effects on Plant Metabolism*
  Micro/nanoplastics can change primary metabolites—amino acids, proteins, organic acids, vitamins, carbohydrates, lipids, and secondary metabolites—phytohormone synthesis (Wang et al. 2022). For example, polyethylene (PE) microplastics inhibit the biosynthesis and bioaccumulation of amino acids, carbohydrates, organic acids, lipids, and polyols in rice leaves (Wang et al. 2022), whereas polyethylene (PE) nanoplastics interfere with the level of some vital amino acids such as lysine, tryptophan, threonine, isoleucine, leucine, and valine; some conditionally vital amino and some other non-essential amino acids such as proline, ornithine, and arginine (Lian et al. 2021; Wang et al. 2022); and polyethylene (PE) also decreases the concentration of indole-3-acetic acid (IAA and 3-IAA), indole-3-butyric acid, and dihydrozeatin concentration in leaves of barley plants (Li et al. 2021).

• *Effects on Soil Animals*
  Micro/nanoplastics pass to the soil animals through the feeding route (Yu et al. 2022). Micro/nanoplastics are introduced into the soil through various anthropogenic activities, and some of them are ingested by soil animals such as moles (Talpidae), hamsters (Cricetinae), earthworms (Lumbricina), and nematodes (Nematoda), which encourage the movement, alteration, and deterioration of micro/nanoplastics in soil (Rillig 2012). However, only a few reports are available on the molecular and biochemical impacts of micro/nanoplastics on soil animalia (Yu et al. 2022). In 2017, Cao and his team found that micro/nanoplastics significantly reduced earthworm growth (Cao et al. 2017). Rodriguez and his team found that polyethylene causes histopathological damage and earthworms' immune system increases their protein, lipid, and polysaccharide content (Rodriguez-Seijo et al. 2017). Micro/nanoplastics are introduced into the earthworm's body after ingestion, affecting the intestine, eating, and excretion, severely affecting growth and survival rate. In 2018, Lu et al. found that polystyrene micro/nanoplastics cause hepatic lipid metabolism disorder by

reducing mucin secretion in mice (Lu et al. 2018). Micro/nanoplastics also disrupt the microbe's community structure and significantly change the intestinal microbiota's abundance and biodiversity (Lu et al. 2018; Yu et al. 2022). Furthermore, micro/nanoplastics change the microbial assortment and community complex of collembola and nematodes (Zhu et al. 2018). In the case of nematodes, micro/nanoplastics cause intestinal and oxidative damage. As a result, the survival body length and reproductive capacity of nematodes are significantly reduced (Rillig 2012). However, there has been no investigation into the mode of action of micro/nanoplastics in soil animals. Therefore, these concerns must be eventually focused on in the forthcoming research.

9.4.4 Soil Contamination and Its Impact on Human Health

Many researchers already proved that micro/nanoplastics enter the human body in three ways—inhalation (Gasperi et al. 2018), dermal contact (Prata et al. 2020), and ingestion (Prata et al. 2020). Daily feeding habits, living habits, and a polluted environment are additional factors that lead to micro/nanoplastics entering the human body (Fig. 9.3). The overgrowing population and modern lifestyle have introduced large amounts of plastic waste into the environment. Considering the ubiquitous nature of micro/nanoplastics, they are unavoidable in almost every part of the human diet system.

Micro/nanoplastics from soil migrate to the surrounding environment through a medium like air and water by the action of anthropogenic or natural activity such as wind, dust, erosion, and surface runoff (Dris et al. 2017; Guo et al. 2020; O'Connor et al. 2019; Rezaei et al. 2019). Micro/nanoplastics from soil may also feasibly enter surface receptors such as aquifers. In 2020, Guo et al. showed that more than 60% of micro/nanoplastics in the soil intend to migrate to the river catchments, contaminating the water environment (Guo et al. 2020). Many researchers find that micro/nanoplastics enter aquatic organisms, including fish, mussels (Digka et al. 2018), turtles (Duncan et al. 2019), and various other organisms. Some of them are consumed by human beings, which may be the entrance path of micro/nanoplastic into the human digestive tract. Plastics are also found in other types of human diets and drinking water (Danopoulos et al. 2020; Koelmans et al. 2019). Approximately 74,000–121,000 plastic particles enter the human body through consumption and inhalation annually (Cox et al. 2019) (Fig. 9.3). Human exposure to microplastics has been hypothesized to create long-term effects on the human body system, such as respiratory disease, kidney disease, and nervous system, damage to digestive and excretory systems, and skin issues (Campanale et al. 2020). Many researchers proved the presence of micro/nanoplastics in the human body by detecting their concentration in human excreta (Schwabl et al. 2019; Wibowo et al. 2021; Zhang et al. 2021a, b), colectomy samples (Ibrahim et al. 2021), saliva, skins, head hair (Abbasi and Turner 2021), human placenta (Braun et al. 2021; Ragusa et al. 2021), and lung tissue (Amato-Lourenço et al. 2021). Although the intake of micro/nanoplastics in the human body is evident, there is still a scarcity of knowledge and information regarding the aftermath of these particles inside the human body.

9.5 Determination of MPs/NPs and Associated Components in Soil

Various numbers of methods are used to determine microplastic content from the soil environment, but no methods are still developed to determine nanoplastic content from the soil. Hence, here, we briefly introduce some methods that are used to determine microplastic content from soil, sediments, or sand.

- *Physical Separation (Sorting and Sieving)*
  This method recycles multiple thermoplastic polymers, such as polypropylene (PP), polyethylene (PE), polyethylene tere-

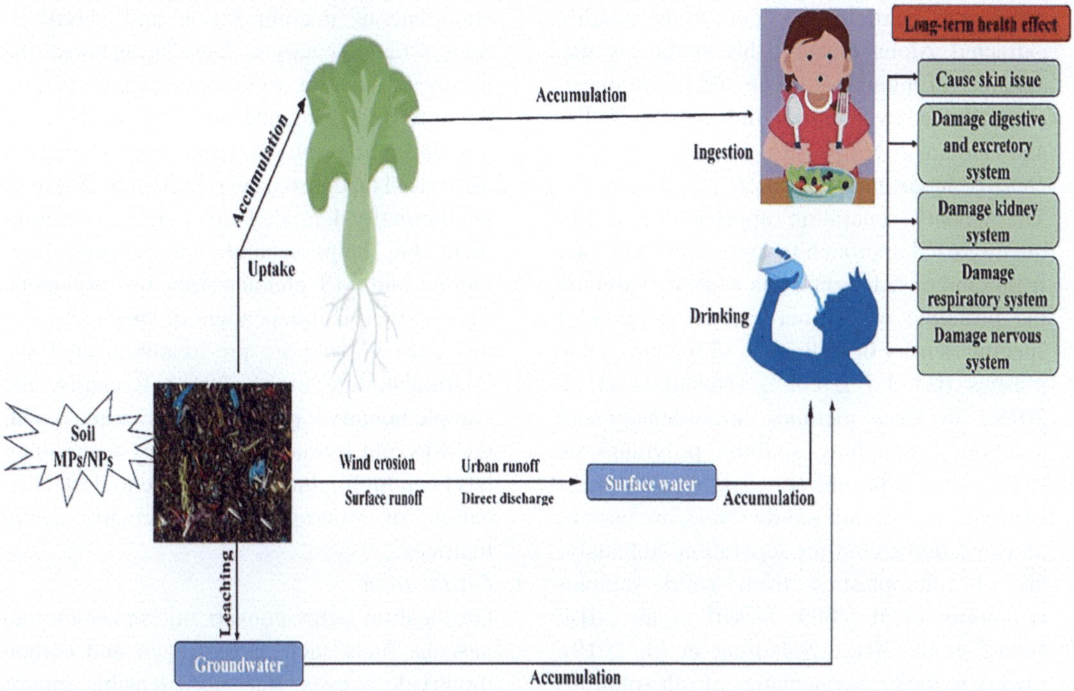

Fig. 9.3 Possible routes of soil micro/nanoplastics enter to the human health and their negative impact

phthalate (PET), and polyvinyl chloride (PVC). It may also help recycle waste materials into virgin raw materials. It is recommended to sort and sieve test soil samples based on their sizes to get fine-earth fractions. This method, in turn, helps remove and analyze fractions of microplastics and larger items.

- *Electrostatic Separation*
  An electrostatic separation unit (Felsing et al. 2018): This device was mainly used to recycle and isolate microplastics in industrial sectors (Silveira et al. 2018). This device allowed a higher throughput of samples, resulting in a bulk reduction of about 99% and recovery rates of approximately 95%. This was specifically for microplastics ranging from 63 μm to 5 mm. Moreover, to achieve such high recovery rates, soil fragments must undergo the process thrice at high temperatures. However, considering the unavoidable generation of aggregates, this method is unsuitable for moist samples (Felsing et al. 2018; He et al. 2021).
- *Magnetic Extraction*
  Iron nanoparticles were used in these methods to help bind to microplastic surfaces, allowing magnetic extraction (Grbic et al. 2019). Furthermore, it is desirable to sonicate the magnets in acid solutions, which would eliminate the iron nanoparticles from the surfaces of microplastics. This approach is claimed to have 49–90% recovery rates (Grbic et al. 2019). This method is effective after acid/alkali digestion and density separation since disadvantages may arise due to the lipophilic substances, resulting in non-specific binding.
- *Pressurized Fluid Extraction*
  This standard extraction technique can be exercised to compute and analyze microplastics in diverse environmental matrices with slight optimization and verification of functioning (PFE 2007). This technique helps extract plastics based on partial emulsification or solubilizing plastics. A solvent is used at subcritical pressure and temperature. This optimizing condition is feasible for quantitatively deriving and analyzing plastics from the soil environment. This process enables the

submicron particles to get more readily extracted. Alongside soil, this method is also suitable for municipal waste and liquid matrices, followed by an initial filtering step (Fuller and Gautam 2016).

- *Density Separation Approach*
  The density separation approach is a frequently used approach to extract microplastic from sand or sediments. It is working based on the principle of distinct characterization of specific density of sediments (2.6 g cm^{-3}) and plastics (0.1–1.7 g cm^{-3}) (Nakajima et al. 2019). In these methods, high-density salt solutions, including sodium polytungstate (SPT), zinc chloride ($ZnCl_2$), zinc bromide ($ZnBr_2$), and sodium iodide (NaI), are used as an extraction media for separation and analysis of microplastics from solid samples (Corcoran et al. 2009; Dekiff et al. 2014; Imhof et al. 2012; Nakajima et al. 2019). Owing to the higher densities of salt solutions in comparison to the microplastics, lesser dense microplastics float and get disassociated from the denser materials (Nakajima et al. 2019).
- *Acidic and Alkaline Digestion*
  Acid digestion using hydrochloric acid (HCl) and nitric acid (HNO3) can be a potential method for the removal of plastics from biota (Enders et al. 2017). These acids interfere with the microplastic particles, leading to deterioration and melting. Alkali, such as potassium hydroxide (KOH) and sodium hydroxide (NaOH), have also been reported to have degraded polycarbonate and polyethylene terephthalate. Acid and alkaline digestions are reported to discard microplastics from the soil matrix. Intense acid blends such as HNO3:HClO4 (4:1) can break down specific polymers, such as polystyrene and polyamide. Alkaline treatment with NaOH can also break down polyamide and polyethene (PE) fibers while leading to the melding or discoloration of other polymers (Gulizia et al. 2022).
- *Pyrolysis/Thermal Degradation*
  This method is subjected to thermal decomposition, maintaining an inert atmosphere. This method helps to focus on polymer decomposition at varying temperatures. This enables simultaneous decomposition and classifications, whereas catalytic degradation boosts the pyrolysis reaction while decreasing the temperature and residence time. Pyrolysis using gas chromatography and mass spectroscopy is also based on thermal decomposition. Samples get thermally degraded under an inert environment. GC helps separate the polymer fragments, and MS characterizes the polymers. This method is independent of shape and size and does not require pre-treatment analysis. (Hermabessiere et al. 2018) Recently, the complementary pyrolysis technique with GC-MS and Raman spectroscopy technology has been highlighted for the potential identification of microplastics in environmental matrices.
- *Gasification*
  Gasification helps convert microplastics into gaseous fuels such as hydrogen and carbon monoxide syngas. For more feasible microplastic removal, it is recommended to use a gasification process in fluidized bed technology (Chiemchaisri et al. 2010).
- *Thermal Analysis*
  This process requires cleaning and preconcentration procedures. The thermogravimetry procedure is dependent on time and temperature under an inert atmosphere. This method is applied to characterize microplastics in solid environmental matrices such as agricultural soils. David et al. recommended a process to identify and quantify the most common microplastics from the farm environment using the soil universal model method based on thermogravimetry analysis along with statistical calculations. These common microplastics include PE, PS, PVC, and PET (David et al. 2019).

9.6 Future Aspects and Recommendations

Man's introduction to synthetic plastics and their exposure to life and the environment have resulted in plastic pollution. Plastic waste has gradually become a menace and seriously threatens society. Plastic in soils may result from soil amendments and mulching containing many synthetic polymers. Flooding and littering are an addition to it. These have also resulted in soils being contaminated in remote areas. Thus, there is a need to come up with a standard protocol for soil-microplastic monitoring. Potential methods must be developed to identify nanoplastic particles in soil samples, and in situ microplastic detection methods must be promoted. The extent of microplastics in soil must be investigated. Exploring the pathways of sludge, runoff, etc., and measuring ways to lessen the microplastic entrance into the environmental matrices can be a precautionary step for potential environmental risks. Researchers must work together to study the impact and find new alternatives to quantify, control, and develop standards to determine microplastics more conveniently. Poor management of plastic waste is evident and has brought along some costs in the form of toxic plastic pollution and exposure to harmful chemicals. Government agencies and healthcare organizations must take practical steps to properly use and dispose of plastics. Due to a lack of awareness and a lack of standardized methods and devices, quantifying plastics/microplastics in the soil is still challenging in research. Social media can also play a necessary role in drawing the public's attention and policymakers to work actively to control the adverse effects on the environment. Due to a lack of awareness and a lack of standardized methods and devices, quantifying plastics/microplastics in the soil is still challenging in research. It is, therefore, high time to focus on the adverse human health and environmental conditions due to the potential plastic pollution threats.

9.7 Conclusion

Micro/nanoplastics in soil show significant and emerging environmental challenges with potential adverse effects on soil health, plant growth, and entire ecosystem and human health. Their persistence, ability to accumulate in soil ecosystems and potential to disrupt biological and chemical processes pose risks that need urgent attention. To mitigate or minimize such risk, we must reduce plastic pollution at source, enhance waste management practices, and advance research to fully understand and minimize their impacts on soil ecosystems. Cutting off their input into the soil and minimizing their effects is essential to protect soil health, boost agricultural products, and restore ecological balance.

References

Abbasi S, Turner A (2021) Human exposure to microplastics: a study in Iran. J Hazard Mater 403:123799. https://doi.org/10.1016/j.jhazmat.2020.123799

Amato-Lourenço LF, Carvalho-Oliveira R, Júnior GR, dos Santos Galvão L, Ando RA, Mauad T (2021) Presence of airborne microplastics in human lung tissue. J Hazard Mater 416:126124. https://doi.org/10.1016/j.jhazmat.2021.126124

Amobonye A, Bhagwat P, Raveendran S, Singh S, Pillai S (2021) Environmental impacts of microplastics and nanoplastics: a current overview. Front Microbiol 12. https://doi.org/10.3389/fmicb.2021.768297

Bandopadhyay S, Martin-Closas L, Pelacho AM, DeBruyn JM (2018) Biodegradable plastic mulch films: impacts on soil microbial communities and ecosystem functions. Front Microbiol 9:819. https://doi.org/10.3389/fmicb.2018.00819

Bhattacharya RRNS, Roy P, Khan A (2018) Challenges and opportunities: plastic waste management in India: Faculty Digital Archive: NYU Libraries. Energy Resour Inst 24(2):1–20.

Bosker T, Bouwman LJ, Brun NR, Behrens P, Vijver MG (2019) Microplastics accumulate on pores in seed capsule and delay germination and root growth of the terrestrial vascular plant Lepidium sativum. Chemosphere 226:774–781. https://doi.org/10.1016/j.chemosphere.2019.03.163

Brahney J, Hallerud M, Heim E, Hahnenberger M, Sukumaran S (2020) Plastic rain in protected areas of the United States. Science 368(6496):1257–1260. https://doi.org/10.1126/science.aaz5819

Braun T, Ehrlich L, Henrich W, Koeppel S, Lomako I, Schwabl P, Liebmann B (2021) Detection of microplastic in human placenta and meconium in a clinical setting. Pharmaceutics 13(7):921. https://doi.org/10.3390/pharmaceutics13070921

Büks F, Kaupenjohann M (2020) Global concentrations of microplastics in soils – a review. Soil 6(2):649–662. https://doi.org/10.5194/soil-6-649-2020

Campanale C, Massarelli C, Savino I, Locaputo V, Uricchio VF (2020) A detailed review study on potential effects of microplastics and additives of concern on human health. Int J Environ Res Public Health 17(4):1212. https://doi.org/10.3390/ijerph17041212

Cao D, Wang X, Luo X, Liu G, Zheng H (2017) Effects of polystyrene microplastics on the fitness of earthworms in an agricultural soil. IOP Conf Ser Earth Environ Sci 61:12148. https://doi.org/10.1088/1755-1315/61/1/012148. IOP Publishing

Chang X, Fang Y, Wang Y, Wang F, Shang L, Zhong R (2022) Microplastic pollution in soils, plants, and animals: a review of distributions, effects and potential mechanisms. Sci Total Environ 850:157857. https://doi.org/10.1016/j.scitotenv.2022.157857

Chiemchaisri C, Charnnok B, Visvanathan C (2010) Recovery of plastic wastes from dumpsite as refuse-derived fuel and its utilization in small gasification system. Bioresour Technol 101(5):1522–1527. https://doi.org/10.1016/j.biortech.2009.08.061

Choudhury M, Sahoo S, Samanta P, Tiwari A, Tiwari A, Chadha U et al (2022a) COVID-19: an accelerator for global plastic consumption and its implications. J Environ Public Health 2022(1):1066350. https://doi.org/10.1155/2022/1066350

Choudhury M, Sharma A, Pervez A, Upadhyay P, Dutta J (2022b) Growing menace of microplastics in and around the coastal ecosystem. In: Coastal ecosystems: environmental importance, current challenges and conservation measures, pp 117–137. https://doi.org/10.1007/978-3-030-84255-0_6

Corcoran PL, Biesinger MC, Grifi M (2009) Plastics and beaches: a degrading relationship. Mar Pollut Bull 58(1):80–84. https://doi.org/10.1016/j.marpolbul.2008.08.022

Cox KD, Covernton GA, Davies HL, Dower JF, Juanes F, Dudas SE (2019) Human consumption of microplastics. Environ Sci Technol 53(12):7068–7074. https://doi.org/10.1021/acs.est.9b01517

Danopoulos E, Twiddy M, Rotchell JM (2020) Microplastic contamination of drinking water: a systematic review. PLoS One 15:e0236838. https://doi.org/10.1371/journal.pone.0236838

David J, Weissmannová HD, Steinmetz Z, Kabelíková L, Demyan MS, Šimečková J et al (2019) Introducing a soil universal model method (SUMM) and its application for qualitative and quantitative determination of poly(ethylene), poly(styrene), poly(vinyl chloride) and poly(ethylene terephthalate) microplastics in a model soil. Chemosphere 225:810–819. https://doi.org/10.1016/j.chemosphere.2019.03.078

De Souza MacHado AA, Lau CW, Till J, Kloas W, Lehmann A, Becker R, Rillig MC (2018) Impacts of microplastics on the soil biophysical environment. Environ Sci Technol 52(17):9656–9665. https://doi.org/10.1021/acs.est.8b02212

Dekiff JH, Remy D, Klasmeier J, Fries E (2014) Occurrence and spatial distribution of microplastics in sediments from Norderney. Environ Pollut 186:248–256. https://doi.org/10.1016/j.envpol.2013.11.019

Digka N, Tsangaris C, Torre M, Anastasopoulou A, Zeri C (2018) Microplastics in mussels and fish from the Northern Ionian Sea. Mar Pollut Bull 135:30–40. https://doi.org/10.1016/j.marpolbul.2018.06.063

Dris R, Gasperi J, Mirande C, Mandin C, Guerrouache M, Langlois V, Tassin B (2017) A first overview of textile fibers, including microplastics, in indoor and outdoor environments. Environ Pollut 221:453–458. https://doi.org/10.1016/j.envpol.2016.12.013

Duncan EM, Broderick AC, Fuller WJ, Galloway TS, Godfrey MH, Hamann M et al (2019) Microplastic ingestion ubiquitous in marine turtles. Glob Chang Biol 25(2):744–752. https://doi.org/10.1111/gcb.14519

Enders K, Lenz R, Beer S, Stedmon CA (2017) Extraction of microplastic from biota: recommended acidic digestion destroys common plastic polymers. ICES J Mar Sci 74(1):326–331. https://doi.org/10.1093/icesjms/fsw173

Federici S, Ademovic Z, Amorim MJB, Bigalke M, Cocca M, Depero LE et al (2022) COST Action PRIORITY: an EU perspective on micro- and nanoplastics as global issues. Microplastics 1(2):282–290. https://doi.org/10.3390/microplastics1020020

Felsing S, Kochleus C, Buchinger S, Brennholt N, Stock F, Reifferscheid G (2018) A new approach in separating microplastics from environmental samples based on their electrostatic behavior. Environ Pollut 234:20–28. https://doi.org/10.1016/j.envpol.2017.11.013

Fuller S, Gautam A (2016) A procedure for measuring microplastics using pressurized fluid extraction. Environ Sci Technol 50(11):5774–5780. https://doi.org/10.1021/acs.est.6b00816

Gao M, Liu Y, Song Z (2019) Effects of polyethylene microplastic on the phytotoxicity of di-n-butyl phthalate in lettuce (Lactuca sativa L. var. ramosa Hort). Chemosphere 237:124482. https://doi.org/10.1016/j.chemosphere.2019.124482

Gasperi J, Wright SL, Dris R, Collard F, Mandin C, Guerrouache M et al (2018) Microplastics in air: are we breathing it in? Curr Opin Environ Sci Health 1:1–5. https://doi.org/10.1016/j.coesh.2017.10.002

Geyer R, Jambeck JR, Law KL (2017) Production, use, and fate of all plastics ever made. Sci Adv 3(7):e1700782. https://doi.org/10.1126/sciadv.1700782

Giorgetti L, Spanò C, Muccifora S, Bottega S, Barbieri F, Bellani L, Ruffini Castiglione M (2020) Exploring the interaction between polystyrene nanoplastics and Allium cepa during germination: internalization in root cells, induction of toxicity and oxidative stress. Plant Physiol Biochem 149:170–177. https://doi.org/10.1016/j.plaphy.2020.02.014

Gogoi N, Choudhury M, Ul Hasan MS, Changmai B, Baruah D, Samanta P (2024) Soil organic carbon pool in diverse land utilization patterns in North-East India: an implication for carbon sequestration. Environ Dev Sustain:1–27. https://doi.org/10.1007/s10668-024-05037-y(b)

Grbic J, Nguyen B, Guo E, You JB, Sinton D, Rochman CM (2019) Magnetic extraction of microplastics from environmental samples. Environ Sci Technol Lett 6(2):68–72. https://doi.org/10.1021/acs.estlett.8b00671

Gulizia AM, Brodie E, Daumuller R, Bloom SB, Corbett T, Santana MMF et al (2022) Evaluating the effect of chemical digestion treatments on polystyrene microplastics: recommended updates to chemical digestion protocols. Macromol Chem Phys 223(13):2100485. https://doi.org/10.1002/macp.202100485

Guo JJ, Huang XP, Xiang L, Wang YZ, Li YW, Li H et al (2020) Source, migration and toxicology of microplastics in soil. Environ Int 137:105263. https://doi.org/10.1016/j.envint.2019.105263

Guo ZQ, Li P, Yang XM, Wang ZH, Lu BB, Chen WJ et al (2022) Soil texture is an important factor determining how microplastics affect soil hydraulic characteristics. Environ Int 165:107293. https://doi.org/10.1016/j.envint.2022.107293

He D, Zhang X, Hu J (2021) Methods for separating microplastics from complex solid matrices: comparative analysis. J Hazard Mater 409:124640. https://doi.org/10.1016/j.jhazmat.2020.124640

Hermabessiere L, Himber C, Boricaud B, Kazour M, Amara R, Cassone AL et al (2018) Optimization, performance, and application of a pyrolysis-GC/MS method for the identification of microplastics. Anal Bioanal Chem 410(25):6663–6676. https://doi.org/10.1007/s00216-018-1279-0

Horton AA, Walton A, Spurgeon DJ, Lahive E, Svendsen C (2017) Microplastics in freshwater and terrestrial environments: evaluating the current understanding to identify the knowledge gaps and future research priorities. Sci Total Environ 586:127–141. https://doi.org/10.1016/j.scitotenv.2017.01.190

Huerta Lwanga E, Mendoza Vega J, Ku Quej V, de Los Angeles Chi J, Sanchez del Cid L, Chi C et al (2017) Field evidence for transfer of plastic debris along a terrestrial food chain. Sci Rep 7(1):1–7. https://doi.org/10.1038/s41598-017-14588-2

Ibrahim YS, Tuan Anuar S, Azmi AA, Wan Mohd Khalik WMA, Lehata S, Hamzah SR et al (2021) Detection of microplastics in human colectomy specimens. JGH Open 5(1):116–121. https://doi.org/10.1002/jgh3.12457

Imhof HK, Schmid J, Niessner R, Ivleva NP, Laforsch C (2012) A novel, highly efficient method for the separation and quantification of plastic particles in sediments of aquatic environments. Limnol Oceanogr Methods 10:524–537. https://doi.org/10.4319/lom.2012.10.524

Jain R, Gaur A, Suravajhala R, Chauhan U, Pant M, Tripathi V, Pant G (2023) Microplastic pollution: understanding microbial degradation and strategies for pollutant reduction. Sci Total Environ 905:167098. https://doi.org/10.1016/j.scitotenv.2023.167098

Judy JD, Williams M, Gregg A, Oliver D, Kumar A, Kookana R, Kirby JK (2019) Microplastics in municipal mixed-waste organic outputs induce minimal short to long-term toxicity in key terrestrial biota. Environ Pollut 252:522–531. https://doi.org/10.1016/j.envpol.2019.05.027

Kiener R (2022) Plastic pollution. CQ Press, Thousand Oaks, CA. https://doi.org/10.4135/cqresrre20220617

Koelmans AA, Mohamed Nor NH, Hermsen E, Kooi M, Mintenig SM, De France J (2019) Microplastics in freshwaters and drinking water: critical review and assessment of data quality. Water Res 155:410–422. https://doi.org/10.1016/j.watres.2019.02.054

Li J, Liu H, Paul Chen J (2018) Microplastics in freshwater systems: a review on occurrence, environmental effects, and methods for microplastics detection. Water Res 137:362–374. https://doi.org/10.1016/j.watres.2017.12.056

Li L, Luo Y, Li R, Zhou Q, Peijnenburg WJGM, Yin N et al (2020a) Effective uptake of submicrometre plastics by crop plants via a crack-entry mode. Nat Sustain 3(11):929–937. https://doi.org/10.1038/s41893-020-0567-9

Li M, Yu H, Wang Y, Li J, Ma G, Wei X (2020b) QSPR models for predicting the adsorption capacity for microplastics of polyethylene, polypropylene and polystyrene. Sci Rep 10(1):1–11. https://doi.org/10.1038/s41598-020-71390-3

Li S, Wang T, Guo J, Dong Y, Wang Z, Gong L, Li X (2021) Polystyrene microplastics disturb the redox homeostasis, carbohydrate metabolism and phytohormone regulatory network in barley. J Hazard Mater 415:125614. https://doi.org/10.1016/j.jhazmat.2021.125614

Li J, Yu S, Yu Y, Xu M (2022) Effects of microplastics on higher plants: a review. Bull Environ Contam Toxicol 109(2):241–265. https://doi.org/10.1007/s00128-022-03566-8

Lian J, Liu W, Meng L, Wu J, Zeb A, Cheng L et al (2021) Effects of microplastics derived from polymer-coated fertilizer on maize growth, rhizosphere, and soil properties. J Clean Prod 318:128571. https://doi.org/10.1016/j.jclepro.2021.128571

Liu K, Wang X, Fang T, Xu P, Zhu L, Li D (2019) Source and potential risk assessment of suspended atmospheric microplastics in Shanghai. Sci Total Environ 675:462–471. https://doi.org/10.1016/j.scitotenv.2019.04.110

Lozano YM, Lehnert T, Linck LT, Lehmann A, Rillig MC (2021) Microplastic shape, polymer type, and concentration affect soil properties and plant biomass. Front Plant Sci 12:616645. https://doi.org/10.3389/fpls.2021.616645

Lu L, Wan Z, Luo T, Fu Z, Jin Y (2018) Polystyrene microplastics induce gut microbiota dysbiosis and hepatic lipid metabolism disorder in mice. Sci Total Environ 631–632:449–458. https://doi.org/10.1016/j.scitotenv.2018.03.051

Luo G, Xue C, Jiang Q, Xiao Y, Zhang F, Guo S et al (2020) Soil carbon, nitrogen, and phosphorus cycling microbial populations and their resistance to global change depend on soil C:N:P stoichiometry. mSystems 5(3):10–1128. https://doi.org/10.1128/msystems.00162-20

Ma X, Geiser-Lee J, Deng Y, Kolmakov A (2010) Interactions between engineered nanoparticles (ENPs) and plants: phytotoxicity, uptake and accumulation. Sci Total Environ 408(16):3053–3061. https://doi.org/10.1016/j.scitotenv.2010.03.031

Maity S, Guchhait R, Sarkar MB, Pramanick K (2022) Occurrence and distribution of micro/nanoplastics in soils and their phytotoxic effects: a review. Plant Cell Environ 45(4):1011–1028. https://doi.org/10.1111/pce.14248

Nakajima R, Tsuchiya M, Lindsay DJ, Kitahashi T, Fujikura K, Fukushima T (2019) A new small device made of glass for separating microplastics from marine and freshwater sediments. PeerJ 2019(10):e7915. https://doi.org/10.7717/peerj.7915

O'Connor D, Pan S, Shen Z, Song Y, Jin Y, Wu WM, Hou D (2019) Microplastics undergo accelerated vertical migration in sand soil due to small size and wet-dry cycles. Environ Pollut 249:527–534. https://doi.org/10.1016/j.envpol.2019.03.092

Okeke ES, Olagbaju OA, Okoye CO, Addey CI, Chukwudozie KI, Okoro JO et al (2022) Microplastic burden in Africa: a review of occurrence, impacts, and sustainability potential of bioplastics. Chem Eng J Adv 12:100402. https://doi.org/10.1016/j.ceja.2022.100402

PFE, P. F. E (2007) Method 3545A pressurized fluid extraction (PFE)

Porter R, Černoša A, Fernández-Sanmartín P, Cortizas AM, Aranda E, Luo Y et al (2023) Degradation of polypropylene by fungi Coniochaeta hoffmannii and Pleurostoma richardsiae. Microbiol Res 277:127507. https://doi.org/10.1016/j.micres.2023.127507

Prata JC, da Costa JP, Lopes I, Duarte AC, Rocha-Santos T (2020) Environmental exposure to microplastics: an overview on possible human health effects. Sci Total Environ 702:134455. https://doi.org/10.1016/j.scitotenv.2019.134455

Qi Y, Yang X, Pelaez AM, Huerta Lwanga E, Beriot N, Gertsen H et al (2018) Macro- and micro- plastics in soil-plant system: effects of plastic mulch film residues on wheat (Triticum aestivum) growth. Sci Total Environ 645:1048–1056. https://doi.org/10.1016/j.scitotenv.2018.07.229

Ragusa A, Svelato A, Santacroce C, Catalano P, Notarstefano V, Carnevali O et al (2021) Plasticenta: first evidence of microplastics in human placenta. Environ Int 146:106274. https://doi.org/10.1016/j.envint.2020.106274

Rezaei M, Riksen MJPM, Sirjani E, Sameni A, Geissen V (2019) Wind erosion as a driver for transport of light density microplastics. Sci Total Environ 669:273–281. https://doi.org/10.1016/j.scitotenv.2019.02.382

Rillig MC (2012) Microplastic in terrestrial ecosystems and the soil? Environ Sci Technol. https://doi.org/10.1021/es302011r. ACS Publications

Rodriguez-Seijo A, Lourenço J, Rocha-Santos TAP, da Costa J, Duarte AC, Vala H, Pereira R (2017) Histopathological and molecular effects of microplastics in Eisenia andrei Bouché. Environ Pollut 220:495–503. https://doi.org/10.1016/j.envpol.2016.09.092

Roy T, Dey TK, Jamal M (2023) Microplastic/nanoplastic toxicity in plants: an imminent concern. Environ Monit Assess 195(1):27. https://doi.org/10.1007/s10661-022-10654-z

Sajjad M, Huang Q, Khan S, Khan MA, Liu Y, Wang J et al (2022) Microplastics in the soil environment: a critical review. Environ Technol Innov 27:102408. https://doi.org/10.1016/j.eti.2022.102408

Scheurer M, Bigalke M (2018) Microplastics in Swiss floodplain soils. Environ Sci Technol 52(6):3591–3598. https://doi.org/10.1021/acs.est.7b06003

Schwabl P, Koppel S, Konigshofer P, Bucsics T, Trauner M, Reiberger T, Liebmann B (2019) Detection of various microplastics in human stool: a prospective case series. Ann Intern Med 171(7):453–457. https://doi.org/10.7326/M19-0618

Seo Y, Zhou Z, Lai Y, Chen G, Pembleton K, Wang S et al (2024) Micro- and nanoplastics in agricultural soils: assessing impacts and navigating mitigation. Sci Total Environ 931:172951. https://doi.org/10.1016/j.scitotenv.2024.172951

Sieber R, Kawecki D, Nowack B (2020) Dynamic probabilistic material flow analysis of rubber release from tires into the environment. Environ Pollut 258:113573. https://doi.org/10.1016/j.envpol.2019.113573

Silveira AVM, Cella M, Tanabe EH, Bertuol DA (2018) Application of tribo-electrostatic separation in the recycling of plastic wastes. Process Saf Environ Prot 114:219–228. https://doi.org/10.1016/j.psep.2017.12.019

Singh S, Chakma S, Alawa B, Kalyanasundaram M, Diwan V (2023) Identification, characterization, and implications of microplastics in soil – a case study of Bhopal, central India. J Hazard Mater Adv 9:100225. https://doi.org/10.1016/j.hazadv.2022.100225

Singh S, Chakma S, Alawa B, Kalyanasundaram M, Diwan V (2024) Microplastic pollution in terrestrial environment: identification, characterization, and risk assessment in Indore, Central India. Soil Use Manag 40(2):e13053. https://doi.org/10.1111/sum.13053

Sun Q, Ren SY, Ni HG (2020) Incidence of microplastics in personal care products: an appreciable part of plastic pollution. Sci Total Environ 742:140218. https://doi.org/10.1016/j.scitotenv.2020.140218

Wang HT, Ding J, Xiong C, Zhu D, Li G, Jia XY et al (2019a) Exposure to microplastics lowers arsenic accumulation and alters gut bacterial communities of earthworm Metaphire californica. Environ Pollut 251:110–116. https://doi.org/10.1016/j.envpol.2019.04.054

Wang W, Gao H, Jin S, Li R, Na G (2019b) The ecotoxicological effects of microplastics on aquatic food web, from primary producer to human: a review. Ecotoxicol Environ Saf 173:110–117. https://doi.org/10.1016/j.ecoenv.2019.01.113

Wang L, Liu Y, Kaur M, Yao Z, Chen T, Xu M (2021) Phytotoxic effects of polyethylene microplastics on the growth of food crops soybean (Glycine max) and mung bean (vigna radiata). Int J Environ Res Public Health 18(20):10629. https://doi.org/10.3390/ijerph182010629

Wang F, Feng X, Liu Y, Adams CA, Sun Y, Zhang S (2022) Micro(nano)plastics and terrestrial plants: up-to-date knowledge on uptake, translocation, and phytotoxicity. Resour Conserv Recycl 185:106503. https://doi.org/10.1016/j.resconrec.2022.106503

Wibowo AT, Nugrahapraja H, Wahyuono RA, Islami I, Haekal MH, Fardiansyah Y et al (2021) Microplastic contamination in the human gastrointestinal tract and daily consumables associated with an indonesian farming community. Sustainability (Switzerland) 13(22):12840. https://doi.org/10.3390/su132212840

Yang M, Huang DY, Tian YB, Zhu QH, Zhang Q, Zhu HH, Xu C (2021) Influences of different source microplastics with different particle sizes and application rates on soil properties and growth of Chinese cabbage (Brassica chinensis L.). Ecotoxicol Environ Saf 222:112480. https://doi.org/10.1016/j.ecoenv.2021.112480

Yu J, Adingo S, Liu X, Li X, Sun J, Zhang X (2022) Micro plastics in soil ecosystem – a review of sources, fate, and ecological impact. Plant Soil Environ 68(1):1–17. https://doi.org/10.17221/242/2021-PSE

Yuan W, Zhou Y, Liu X, Wang J (2019) New perspective on the nanoplastics disrupting the reproduction of an endangered fern in artificial freshwater. Environ Sci Technol 53(21):12715–12724. https://doi.org/10.1021/acs.est.9b02882

Zang H, Zhou J, Marshall MR, Chadwick DR, Wen Y, Jones DL (2020) Microplastics in the agroecosystem: are they an emerging threat to the plant-soil system? Soil Biol Biochem 148:107926. https://doi.org/10.1016/j.soilbio.2020.107926

Zhang GS, Liu YF (2018) The distribution of microplastics in soil aggregate fractions in southwestern China. Sci Total Environ 642:12–20. https://doi.org/10.1016/j.scitotenv.2018.06.004

Zhang S, Liu X, Hao X, Wang J, Zhang Y (2020) Distribution of low-density microplastics in the mollisol farmlands of northeast China. Sci Total Environ 708:135091. https://doi.org/10.1016/j.scitotenv.2019.135091

Zhang J, Wang L, Trasande L, Kannan K (2021a) Occurrence of polyethylene terephthalate and polycarbonate microplastics in infant and adult feces. Environ Sci Technol Lett 8(11):989–994. https://doi.org/10.1021/acs.estlett.1c00559

Zhang N, Li YB, He HR, Zhang JF, Ma GS (2021b) You are what you eat: microplastics in the feces of young men living in Beijing. Sci Total Environ 767:144345. https://doi.org/10.1016/j.scitotenv.2020.144345

Zhao J, Wang Z, Liu X, Xie X, Zhang K, Xing B (2011) Distribution of CuO nanoparticles in juvenile carp (Cyprinus carpio) and their potential toxicity. J Hazard Mater 197:304–310. https://doi.org/10.1016/j.jhazmat.2011.09.094

Zhou Q, Zhang H, Fu C, Zhou Y, Dai Z, Li Y et al (2018) The distribution and morphology of microplastics in coastal soils adjacent to the Bohai Sea and the Yellow Sea. Geoderma 322:201–208. https://doi.org/10.1016/j.geoderma.2018.02.015

Zhou J, Gui H, Banfield CC, Wen Y, Zang H, Dippold MA et al (2021) The microplastisphere: biodegradable microplastics addition alters soil microbial community structure and function. Soil Biol Biochem 156:108211. https://doi.org/10.1016/j.soilbio.2021.108211

Zhu D, Chen QL, An XL, Yang XR, Christie P, Ke X et al (2018) Corrigendum to "Exposure of soil collembolans to microplastics perturbs their gut microbiota and alters their isotopic composition" [Soil Biology & Biochemistry 116, 302-310] (S0038071717306296) (10.1016/j.soilbio.2017.10.027). Soil Biol Biochem 124:277–278. https://doi.org/10.1016/j.soilbio.2018.05.031

Ziani K, Ioniță-Mîndrican CB, Mititelu M, Neacşu SM, Negrei C, Moroşan E et al (2023) Microplastics: a real global threat for environment and food safety: a state of the art review. Nutrients 15(3):617. https://doi.org/10.3390/nu15030617

10 Plastics in Groundwater of Floodplain Aquifers—a Review of Impacts

Md Moniruzzaman, Hazzaz Bin Hassan, Hafiz Al-Asad, and Maksym Gusyev

Abstract

Groundwater is the essential source of water and requires water quality protection to maintain ecosystems and socioeconomic growth globally. To date, aquifer systems are contaminated with heavy metals, metalloids, and microplastics (MPs) due to industry, urbanization, surface runoff from mulching garbage, and agriculture and food-related activities. In Bangladesh, floodplain aquifers receive precipitation and surface water recharge with MPs, heavy metals, and other contaminants due to horizontal and vertical transfer through soil. As a result, floodplain aquifers become the interim sinks for MPs, along with other pollution from anthropogenic activities posing a major risk due to the groundwater contamination. The book chapter reviews the existing knowledge of plastics in groundwater, examines the contamination mechanism of MPs and heavy metals in floodplain aquifers, and discusses MPs sources, distributions, and impacts on environment and human health due to groundwater pollution.

Keywords

Floodplain · Aquifers · Recharge · Plastics · Contamination

M. Moniruzzaman (✉)
Isotope Hydrology Division, Institute of Nuclear Science and Technology, Atomic Energy Research Establishment, Bangladesh Atomic Energy Commission, Ganakbari, Savar, Dhaka, Bangladesh

H. B. Hassan
Department of Environmental Science, Bangladesh University of Professionals (BUP), Dhaka, Bangladesh

H. Al-Asad
Department of Chemistry, Mawlana Bhashani Science & Technology University, Santosh, Tangail, Bangladesh

M. Gusyev
Institute of Environmental Radioactivity (IER), Fukushima University, Fukushima, Japan

10.1 Introduction

Given the vital role of water in the aquatic environment and the crucial need for groundwater in arid and semi-arid areas, numerous studies have underscored the urgent health hazards posed by poor groundwater quality (Hassan et al. 2023; Al-Asad et al. 2023; Khan et al. 2021; Choudhury et al. 2022; Gao et al. 2019; Moniruzzaman et al. 2018; Zhang et al. 2018; Rathi et al. 2021). Plastics and their derivatives, originating from a variety of sources such as rapidly expanding industrial areas, mine tailings, decommissioning of high metal wastes, livestock manure, fertiliza-

S. Goswami et al. (eds.), *Micro-Nano Plastics Exposure, Environmental Degradation and Public Health Crisis*, Sustainable Development Goals Series, https://doi.org/10.1007/978-981-96-5047-7_10

tion, pesticide residues, wastewater irrigation, sludges, leakage of hydrocarbons, combustion residues, atmospheric deposition, leaded gasoline, and paints (Mukherjee et al. 2021), can enter surface water and infiltrate to aquifers. Since the danger of plastics with heavy metals is still unknown (Liu et al. 2021), water contamination poses a major health risk to both humans and environment by ingestion and skin routes (Prasad et al. 2020; Choudhury et al. 2022; Baj et al. 2022).

Microplastics (MPs) and nanoplastics (NPs) are any particle of plastic debris that is between 1 μm and <5000 μm and <1 μm in size, respectively (Yang et al. 2021; Hartmann et al. 2019). When plastic waste is discarded into the environment, it diminishes into tiny chunks, causing secondary plastic pollution. The main sources of primary microplastics that are disposed directly into the environment are body and skin care products, industrial effluent, and textile applications (Bajt 2021). Secondary MPs are formed when larger plastic particles decompose and degrade due to chemical, physical, and mechanical weathering and exposure to UV light from the sun (Huber et al. 2022). Since soil is the most probable route for plastics to infiltrate floodplain aquifers, MP soil contamination has recently increased, making it an important topic to investigate plastic pollution in groundwater (Viaroli et al. 2022; Wei and Chen 2023). Major health concerns may arise when MP-contaminated groundwater is used for drinking water consumption or agricultural applications (Liu et al. 2021; Pelegrini et al. 2023).

The Ganges, Brahmaputra, and Meghna Rivers of the Ganges-Brahmaputra-Meghna (GBM) basin play a pivotal role in this study. The GBM transboundary basin spans a combined area of 1,712,700 km^2 across five countries, with Bangladesh alone comprising 120,400 km^2 of the GBM floodplain (BGS and DPHE 2001; FAO 2011). Bangladesh, with an estimated flood inundation area of about 50,000 km^2 (Gusyev et al. 2015; Kwak et al. 2015), recharges its floodplain aquifers with infiltrated flood water during high water levels over the flood-inundated area and stagnant water bodies, which can be identified with radioactive and stable isotopes (Jameel et al. 2022). In the GBM floodplain, shallow and intermediate-depth groundwater sources contain numerous solutes that influence the aggregation of plastics and heavy metals (Alam et al. 2023; Neelavannan et al. 2023). Studies of coastal and freshwater ecosystems have shown the presence of hazardous metals and plastics in the aquatic environment of Bangladesh (Kawsar et al. 2024; Vivekanand et al. 2021). Since the distribution of MPs in floodplain aquifers remains unclear, this study reviews the existing knowledge of plastics in groundwater, examines the contamination mechanism of MPs and heavy metals in floodplain aquifers, and discusses MPs sources, distributions, and impacts on environment and human health due to groundwater pollution.

10.2 Sources of Micro and Nano Plastics in the Floodplain Aquifer

Environmental health depends on groundwater quality and encompasses all elements, settings, and niches that can affect human health and wellbeing. While plastics are discovered almost throughout the environment, the most common primary sources are rubber, plastic pellets, synthetic textiles, lane markings, cosmetic products, and marine adhesives (Adetuyi et al. 2024; Chia et al. 2021). Plastic particles can pollute drinking groundwater sources via several pathways, which include surface runoff, sewage effluent, cumulative sewage overflows, industrial discharges, and deteriorated plastic garbage, including plastic bottles and lids (Chen et al. 2020). Wastewater effluent and surface runoff are the primary sources of plastics requiring more data to link the sources to plastics in waste streams.

10.2.1 Micro (Nano) Plastic Distribution Pattern in Groundwater

There are limited studies investigated plastics in groundwater at several countries (Table 10.1). Considering a large variety of solutes, the distribution pattern of MPs in groundwater remains unclear. In addition, the aquifer matrix of various

Table 10.1 Comparison of microplastics (MPs) in groundwater studies

#	Location	MP shapes	MP size	References
1	Mexico	Fibers, Fragments, Foams	<1002 μm	Alvarado-Zambrano et al. (2023)
2	USA	Fibers	<1500 μm	Panno et al. (2019)
3	China	Fibers, Flakes, Granular, Films	<3394 μm	Cheng et al. (2024)
4	India	Fibers, Foams, Pellets	<2500 μm	Selvam et al. (2021)
5	France	Plastic Debris	<2758 μm	Ledieu et al. (2023)
6	Italy	Fibers, Fragments, Beads	<1000 μm	Balestra et al. (2023)
7	Germany	Fragments, Fibers	<1008 μm	la Cecilia et al. (2024)
8	South Korea	Fragments, Fibers	<344 μm	Kim et al. (2023)
9	Australia	Fragments, Fibers	<491 μm	Samandra et al. (2022)

study areas influences how well heavy metals and plastics are adsorbed. Floodplain aquifers of Bangladesh have not been studied for MPs distribution in groundwater.

In Bangladesh, the redox state of the polluted area is changing due to the discharge of adsorbed heavy metals and plastics into groundwater deteriorating groundwater quality. The reduction process is enhanced by microbial decomposition of organic matter, extensive irrigation abstraction, and increased recharge to shallow aquifers. In productive aquifers, groundwater quality has deteriorated significantly due to overpumping, uneven groundwater resource exploitation, and low-intense recharge (Islam et al. 2018). Pollutants easily enter unconfined shallow and intermediate aquifers during recharge, whereas confined deep aquifers are more protected due to limited recharge with contaminants.

10.2.2 Microplastic Pollution and Surface-Groundwater Interactions in Floodplain Aquifers

Depending on hydrogeological conditions, MPs can enter the groundwater through various sources and polluting pathways in floodplain aquifers (Amrutha et al. 2021; Afrin et al. 2020). A major factor in spreading MPs into groundwater is the quantity and duration of the aquifer's recharge, and MPs may be transported by water seepage from the surface water features to groundwater (Islam et al. 2022, 2023; Parvin and Tareq 2021). Due to the limited research on MP contamination in groundwater, determining MP's primary origins is challenging (Singh and Bhagwat 2022). In floodplain aquifers of Bangladesh, the current findings suggest that soil migration, landfill leachates, sewage effluents, surface runoffs, and human activities, particularly industrial ones, might be the primary sources/causes of MP contamination in groundwater (Fig. 10.1). MPs in soils can be released into the vadose zone of floodplain aquifers and carried by the vertical water flow; both water movement and soil biota can trigger plastic migration (Moeck et al. 2023). For rivers, streams, and ponds, MPs have the potential to infiltrate with surface water into the hyporheic zone interacting with groundwater due to the surface-groundwater exchange (Viaroli et al. 2022; Wu et al. 2022).

10.3 Impact on Floodplain Aquifer and Associated Health Risk Quantification

Industrialization and other human-related processes have increased global exposure to heavy metals and plastics. The accumulation of plastic derivatives in tissues of animals and plants has been related to the adverse effects on human health due to contaminated liquids, foods, and airborne particles as potential sources for these pollutants. Since the accumulation of metals and plastics occurs in the human body, the prolonged exposure can disrupt metabolic processes and the immune system (Pelegrini et al. 2023; Burgos-Aceves et al. 2021). In human body, plastics are metabolized in liver and can obstruct the gastrointestinal tract by large fragments (Burgos-

Fig. 10.1 Contributors of micro (nano) plastics in the floodplain aquifer of Bangladesh

Aceves et al. 2021). These pollutants can also accumulate in the digestive system, cardiovascular system, kidneys, and brain, making difficult to perform their normal functioning, and even short-term exposure to MPs has been shown to cause serious health problems (Zhao et al. 2019).

10.3.1 Effects on Human Health

Contaminated drinking water and agricultural products pose health concerns, and MPs commonly enter the tissues of plants, animals, and humans in different routes such as respiration and consumption. Plastic derivatives are consumed by fish and livestock through consumption of contaminated food and water. After entering the gastrointestinal tract, MPs penetrate the liver to produce metallothionein, which is released into the blood (Hong et al. 2024). In addition, kidneys store these for long periods, which could also trigger tubular necrosis. NPs with heavy metals are frequently inhaled by individuals working in industries where they may be exposed to metal fumes for short or long time periods (Fu and Xi 2020) (Fig. 10.2).

10.3.2 Effects on Aquatic Organisms

Aquatic organisms can absorb MPs through direct feeding, filtering intake, and food chain transfer, and fish growth may be inhibited in contaminated waters (de Araújo and de Oliveira 2020; Covernton et al. 2021; Ferdous et al. 2023). MPs can significantly decrease freshwater organisms' eating rates and cause quasi-morphological changes. MPs are persistent and could be toxic due to accumulation in substantial quantities. For the long-term exposure, consumed MPs can cause chronic responses such as nerve damage, lowered immunoregulatory function, and reduced reproductivity and growth. Plastics enter aquatic habitats through wastewater effluents from metal-

Fig. 10.2 Schematic diagram of plastic contamination pathways shown by arrows in floodplain aquifers from land-use anthropogenic activities to drinking water intake

based materials, farming, mining, packaging materials, and personal care products (Kolarova and Napiórkowski 2021). a major environmental threat due to their persistence, bioaccumulation, and toxicity in aquatic environment as well as health risk they pose to people who rely on fish as a source of protein.

10.4 Fate of Plastics in the Environment

Plastic derivatives' persistence and toxicity in waterways have generated worldwide concern. Due to its inert nature, MPs can accumulate in the environment, becoming a new global risk. Alarmingly, MPs are found in large quantities in rivers with minimal population density and plastic consumption, demonstrating the scale of MPs distribution. Short- and long-term adverse health effects, including biological feeding, reproduction, impaired antioxidants, and innate immunity, have been linked to aquatic species exposed to persistent MPs (Vimalkumar et al. 2022). MPs can act as dipoles concentrating hazardous hydrophobic organic contaminants (Prajapati et al. 2022), and aquatic microorganisms' availability encourages MPs' interaction with heavy metals (Godoy et al. 2020). Numerous contaminants have also been discovered on the substratum of MPs, implying that MPs can serve as unavoidable transmitters of metal ions (Guan et al. 2020).

MP pollution is a recent example of human-caused environmental damages, with various MPs primarily traveling through fluvial systems. The fluvial sediments act as a sink of MPs for transporting these anthropogenic pollutants to floodplain aquifers. In floodplains, the presence of other anthropogenic pollutants like heavy metals and MPs in a broader and sedimentary context presents unique ecological challenges. The distinctiveness of floodplains as an ecological community type demands specific risk identification and quantification. The majority of MPs are found in floodplains rather than in the waterway itself. To date, the distribution of MPs from aquifers to estuaries remains a crucial area due to the lack of investigations requiring further examination, which may result in the overlooking of the persistent contamination in the GBM basin floodplain.

10.5 Concerns and Recommendations

Many studies indicate that bioaccumulation, persistence, and toxicity of plastic derivatives pose ecological and health threats. Plastic derivatives can enter the body via ingestion, inhalation, and skin contact routes, which leads to cardiovascular disease, neurotoxicity, renal damage, and infertility. Once MPs enter the body, MPs may build complex compounds with toxic metals as metal ion carriers, disrupting the immune system and triggering various chronic diseases. In aquatic habitats, plastics can bind in varying levels to sands, clay, metal oxides and hydroxides, humus, microorganisms, and other small-sized objects. When aquatic organisms are exposed to MPs and heavy metals, the adverse effects are amplified requiring major steps to protect the environment and people from the contamination.

10.5.1 Conclusion

Existing methods of releasing wastewater are potential contributors of plastic pollution, as plastic enters the water cycle in a wide range of sizes and via numerous pathways in floodplain aquifers. To date, the danger related to the contamination of plastics with heavy metals is still unknown requiring further studies. Determining the combined risk of plastics and hazardous metals to human health appears complex and challenging. As a result, MP and NP contamination is a significant threat to water quality in floodplain aquifers requiring removal for drinking water sources.

References

Adetuyi BO, Mathew JT, Inobeme A, Falana YO, Adetunji CO, Shahnawaz M, Oyewole OA, Kit E, Yerima MB (2024) Sources, uses and transport of secondary microplastics from land to marine environment. In: Microplastic pollution. Springer Nature Singapore, Singapore, pp 35–49

Afrin S, Uddin MK, Rahman MM (2020) Microplastics contamination in the soil from Urban Landfill site, Dhaka, Bangladesh. Heliyon 6(11):e05572

Alam MJ, Shammi M, Tareq SM (2023) Distribution of microplastics in shoreline water and sediment of the Ganges River Basin to Meghna Estuary in Bangladesh. Ecotoxicol Environ Saf 266:115537

Al-Asad H, Moniruzzaman M, Sarker AK, Bhuiyan MAQ, Ahsan MA (2023) Hydrogeochemical evaluation, groundwater contamination and associated health risk in southern Tangail, Bangladesh. Chemosphere 332:138806

Alvarado-Zambrano D, Rivera-Hernández JR, Green-Ruiz C (2023) First insight into microplastic groundwater pollution in Latin America: the case of a coastal aquifer in Northwest Mexico. Environ Sci Pollut Res 30:73600–73611. https://doi.org/10.1007/s11356-023-27461-9

Amrutha K, Unnikrishnan V, Shajikumar S, Warrier AK (2021) Current state of microplastics research in SAARC countries—a review. In: Microplastic pollution, pp 27–63

Baj J, Dring JC, Czeczelewski M, Kozyra P, Forma A, Flieger J, Kowalska B, Buszewicz G, Teresinski G (2022) Derivatives of plastics as potential carcinogenic factors: the current state of knowledge. Cancers 14:4637

Bajt O (2021) From plastics to microplastics and organisms. FEBS Open Bio 11(4):954–966

Balestra V, Vigna B, De Costanzo S, Bellopede R (2023) Preliminary investigations of microplastic pollution in karst systems, from surface watercourses to cave waters. J Contam Hydrol 252:104117

BGS and DPHE (2001) British Geological Survey (BGS) and Department of Public Health Engineering (DPHE), 2001. Arsenic contamination of groundwater in Bangladesh. Final report (February 2001). Dhaka: British Survey and Department of Public Health Engineering

Burgos-Aceves MA, Abo-Al-Ela HG, Faggio C (2021) Physiological and metabolic approach of plastic additive effects: immune cells responses. J Hazard Mater 404:124114

Chen X, Chen X, Zhao Y, Zhou H, Xiong X, Wu C (2020) Effects of microplastic biofilms on nutrient cycling in simulated freshwater systems. Sci Total Environ 719:137276

Cheng D, Liu H, Qian W, Yao R, Wang C (2024) Migration characteristics of microplastics in riparian soils and groundwater. Environ Monit Assess 196:796. https://doi.org/10.1007/s10661-024-12962-y

Chia RW, Lee JY, Kim H, Jang J (2021) Microplastic pollution in soil and groundwater: a review. Environ Chem Lett 19(6):4211–4224

Choudhury M, Jyethi DS, Dutta J, Purkayastha SP, Deb D, Das R et al (2022) Investigation of groundwater and soil quality near to a municipal waste disposal site in Silchar, Assam, India. Int J Energy Water Resour 6(1):37–47. https://doi.org/10.1007/s42108-021-00117-5

Covernton GA, Davies HL, Cox KD, El-Sabaawi R, Juanes F, Dudas SE, Dower JF (2021) A Bayesian analysis of the factors determining microplastics ingestion in fishes. J Hazard Mater 413:125405

de Araújo LCA, de Oliveira MBM (2020) Effect of heavy metals on the biofilm formed by microorganisms from impacted aquatic environments. In: Bacterial biofilms, vol 19. IntechOpen, London

FAO (2011) AQUASTAT Transboundary River Basins – Ganges-Brahmaputra-Meghna River Basin. Food and Agriculture Organization of the United Nations (FAO), Rome, Italy

Ferdous SR, Amin A, Hasan J, Alam MS, Shahjahan M (2023) Prevalence of microplastics in commonly consumed fish species of the river Old Brahmaputra, Bangladesh. Environ Sci Pollut Res 30(36):85639–85654

Fu Z, Xi S (2020) The effects of heavy metals on human metabolism. Toxicol Mech Methods 30(3):167–176

Gao B, Gao L, Gao J, Xu D, Wang Q, Sun K (2019) Simultaneous evaluations of occurrence and probabilistic human health risk associated with trace elements in typical drinking water sources from major river basins in China. Sci Total Environ 666:139–146

Godoy V, Martínez-Férez A, Martín-Lara MÁ, Vellido-Pérez JA, Calero M, Blázquez G (2020) Microplastics as vectors of chromium and lead during dynamic simulation of the human gastrointestinal tract. Sustain For 12(11):4792

Guan J, Qi K, Wang J, Wang W, Wang Z, Lu N, Qu J (2020) Microplastics as an emerging anthropogenic vector of trace metals in freshwater: significance of biofilms and comparison with natural substrates. Water Res 184:116205

Gusyev MA, Kwak Y, Khairul MI, Arifuzzaman MB, Magome J, Sawano H, Takeuchi K (2015) Effectiveness of water infrastructure for river flood management: part 1 - flood hazard assessment using hydrological models in Bangladesh. Proc IAHS 370:75–81. https://doi.org/10.5194/piahs-370-75-2015

Hartmann NB, Huffer T, Thompson RC, Hassellov M, Verschoor A, Daugaard AE, Rist S, Karlsson T, Brennholt N, Cole M, Herrling MP (2019) Are we speaking the same language? Recommendations for a definition and categorization framework for plastic debris. Environ Sci Technol 53(3):1039–1047

Hassan HB, Moniruzzaman M, Majumder RK, Ahmed F, Bhuiyan MAQ, Ahsan MA, Al-Asad H (2023) Impacts of seasonal variations and wastewater discharge on river quality and associated human health risks: a case of northwest Dhaka, Bangladesh. Heliyon 9(7):e18171

Hong SH, Shim WJ, Jang M (2024) Chemicals associated with marine plastic debris and microplastics: analyses and contaminant levels. In: Microplastic contamination in aquatic environments. Elsevier, pp 141–179

Huber M, Archodoulaki VM, Pomakhina E, Pukánszky B, Zinöcker E, Gahleitner M (2022) Environmental degradation and formation of secondary microplastics from packaging material: a polypropylene film case study. Polym Degrad Stab 195:109794

Islam ARM, Shen S, Haque MA, Bodrud-Doza M, Maw KW, Habib M (2018) Assessing groundwater quality and its sustainability in Joypurhat district of Bangladesh using GIS and multivariate statistical approaches. Environ Dev Sustain 20(5):1935–1959

Islam MS, Islam Z, Hasan MR (2022) Pervasiveness and characteristics of microplastics in surface water and sediment of the Buriganga River, Bangladesh. Chemosphere 307:135945

Islam MR, Ruponti SA, Rakib MA, Nguyen HQ, Mourshed M (2023) Current scenario and challenges of plastic pollution in Bangladesh: a focus on farmlands and terrestrial ecosystems. Front Environ Sci Eng 17(6):66

Jameel et al (2022) Shift in groundwater recharge of the Bengal Basin from rainfall to surface water. Nat Commun Earth Environ 4:1–16. https://doi.org/10.1038/s43247-022-00650-z

Kawsar MA, Munny FJ, Saif MU, Harun-Al-Rashid A, Rahman MA, Barman SK, Adikari D, Alam MT, Kunda M, Pandit D (2024) Unveiling the microplastic crisis: insights into Bangladesh's aquatic ecosystems - origins, impact, and solutions. J Hazard Mater Adv 14:100430. https://doi.org/10.1016/j.hazadv.2024.100430

Khan R, Saxena A, Shukla S, Sekar S, Senapathi V, Wu J (2021) Environmental contamination by heavy metals and associated human health risk assessment: a case study of surface water in Gomti River Basin, India. Environ Sci Pollut Res 28(40):56105–56116

Kim YI, Jeong E, Lee JY, Chia RW, Raza M (2023) Microplastic contamination in groundwater on a volcanic Jeju Island of Korea. Environ Res 226:115682

Kolarova N, Napiórkowski P (2021) Trace elements in aquatic environment. Origin, distribution, assessment and toxicity effect for the aquatic biota. Ecohydrol Hydrobiol 21(4):655–668

Kwak Y, Gusyev MA, Arifuzzaman MA, Khairul MI, Iwami Y, Takeuchi K (2015) Effectiveness of water infrastructure for river flood management: part 2 - flood risk assessment and changes in Bangladesh. Proc IAHS 2015(370):83–87. https://doi.org/10.5194/piahs-370-83-2015

la Cecilia D, Philipp M, Kaegi R, Schirmer M, Moeck C (2024) Microplastics attenuation from surface water to drinking water: impact of treatment and managed aquifer recharge – and identification uncertainties. Sci Total Environ 908:168378. https://doi.org/10.1016/j.scitotenv.2023.168378

Ledieu L, Phuong N-N, Flahaut B, Radigois P, Papin J, Guern C, Bechet B, Gasperi J (2023) May a former municipal landfill contaminate groundwater in microplastics? First investigations from the "Prairie de Mauves Site" (Nantes, France). Microplastics 2(1):93–106

Liu S, Shi J, Wang J, Dai Y, Li H, Li J, Liu X, Chen X, Wang Z, Zhang P (2021) Interactions between microplastics and heavy metals in aquatic environments: a review. Front Microbiol 12:652520

Moeck C, Davies G, Krause S, Schneidewind U (2023) Microplastics and nanoplastics in agriculture—a potential source of soil and groundwater contamination? Grundwasser 28:23–25. https://doi.org/10.1007/s00767-022-00533-2

Moniruzzaman M, Lee JH, Jung KM, Kwon JS, Kim KH, Yun ST (2018) Lithologic control of the hydrochemistry of a point-bar alluvial aquifer at the low reach of the Nakdong River, South Korea: implications for the evaluation of riverbank filtration potential. Water 10(12):1763

Mukherjee I, Singh UK, Singh RP (2021) An overview on heavy metal contamination of water system and sustainable approach for remediation. In: Water pollution and management practices. Springer, Singapore, pp 255–277

Neelavannan K, Sen IS, Sinha N, Thakur AK, Misra S (2023) Microplastics in the Ganga-Brahmaputra delta: sources and pathways to the sundarbans biosphere reserve-an UNESCO World Heritage Centre. Environ Adv 11:100350

Panno SV, Kelly WR, Scott J, Zheng W, McNeish RE, Holm N, Hoellein TJ, Baranski EL (2019) Microplastic contamination in karst groundwater systems. Groundwater 57(2):189–196

Parvin F, Tareq SM (2021) Impact of landfill leachate contamination on surface and groundwater of Bangladesh: a systematic review and possible public health risks assessment. Appl Water Sci 11(6):100

Pelegrini et al (2023) Micro- and nanoplastic toxicity: a review on size, type, source, and test-organism implications. Sci Total Environ 878:162954

Prajapati A, Narayan Vaidya A, Kumar AR (2022) Microplastic properties and their interaction with hydrophobic organic contaminants: a review. Environ Sci Pollut Res 29(33):49490–49512

Prasad S, Saluja R, Joshi V, Garg JK (2020) Heavy metal pollution in surface water of the Upper Ganga River, India: human health risk assessment. Environ Monit Assess 192(11):1–15

Rathi BS, Kumar PS, Vo DVN (2021) Critical review on hazardous pollutants in water environment: occurrence, monitoring, fate, removal technologies and risk assessment. Sci Total Environ 797:149134

Samandra S, Johnston JM, Jaeger JE, Symons B, Xie S, Currell M, Ellis AV, Clarke BO (2022) Microplastic contamination of an unconfined groundwater aquifer in Victoria, Australia. Sci Total Environ 802(1):149727. https://doi.org/10.1016/j.scitotenv.2021.149727

Selvam S, Jesuraja K, Venkatramanan S, Roy PD, Kumari VJ (2021) Hazardous microplastic characteristics and its role as a vector of heavy metal in groundwater and surface water of coastal south India. J Hazard Mater 402:123786

Singh S, Bhagwat A (2022) Microplastics: a potential threat to groundwater resources. Groundw Sustain Dev 19:100852

Viaroli S, Lancia M, Re V (2022) Microplastics contamination of groundwater: current evidence and future perspectives. A review. Sci Total Environ 824:153851

Vimalkumar K, Sangeetha S, Felix L, Kay P, Pugazhendhi A (2022) A systematic review on toxicity assessment of persistent emerging pollutants (EPs) and associated microplastics (MPs) in the environment using the Hydra animal model. Comp Biochem Physiol C: Toxicol Pharmacol 256:109320

Vivekanand AC, Mohapatra S, Tyagi VK (2021) Microplastics in aquatic environment: challenges and perspectives. Chemosphere 282:131151

Wei Y, Chen Y (2023) The urgent need to investigate microplastic contamination in groundwater: soil and groundwater interactions as key drivers. Environ Sci Technol 3(12):3736–3740. https://doi.org/10.1021/acsestwater.3c00645

Wu B, Li LW, Zu YX, Nan J, Chen XQ, Sun K, Li ZL (2022) Microplastics contamination in groundwater of a drinking-water source area, northern China. Environ Res 214:114048

Yang T, Luo J, Nowack B (2021) Characterization of nanoplastics, fibrils, and microplastics released during washing and abrasion of polyester textiles. Environ Sci Technol 55(23):15873–15881

Zhang Y, Wu J, Xu B (2018) Human health risk assessment of groundwater nitrogen pollution in Jinghui canal irrigation area of the loess region, northwest China. Environ Earth Sci 77:1–12

Zhao X, Li Z, Wang D, Li J, Zou B, Tao Y, Lei L, Qiao F, Huang J (2019) Assessment of residents' total environmental exposure to heavy metals in China. Sci Rep 9(1):1–12

Micro- and Nanoplastics: Its Impact on Antarctica and Similar Ecosystems

11

Laxmi Kant Bhardwaj, Moharana Choudhury, Linee Goswami, Farhana Rahman, and Jiajun Xu

11.1 Introduction

Antarctica is known as a pure and unspoiled continent. Now, this continent is becoming damaged day by day through plastic pollution. Current studies state that this untouched continent contains micro- and nanoplastics. Those studies' outcome concerns the region's isolated location and its well-known status of being unaltered (Bhardwaj 2023). The Antarctic Treaty System (ATS) has focused its research on the challenge of plastic pollution since the 1980s and is actively involved in the estimation of plastic contaminants. Plastics are divided into meso-plastics (5 to <25 mm), microplastics (1 to <5 mm), and macroplastics (≥25 mm) (Lee et al. 2013; Bhardwaj and Sharma 2021; Bhardwaj 2022a). These materials undergo degradation by chemical, physical, and biological processes (Alimi et al. 2018), resulting in the creation of smaller fragments that are now widely present in nature, including in the seas, soil, and sediments of Antarctica (Bhardwaj 2022b). Although plastic/microplastic pollution is widespread, extensive research on this issue is lacking in polar locations (Obbard 2018).

These particles persist long and have been reported worldwide from water, soil, sediments, etc. (Bhardwaj et al. 2025). They can be categorized as primary and secondary microplastics. Primary microplastics are formed from the direct ejection of small particles (e.g., toothpaste, skin cleansers, shampoos, shower gels, synthetic clothing, car tires, and cosmetics) (Rajpal et al. 2024). In contrast, secondary microplastics are produced from the obstruction of bigger plastics (Li et al. 2016). They enter Antarctica and its surrounding environment via wastewater, including fishing activities, tourism, scientific expeditions, and after the breakdown of microplastics (Bhardwaj 2023; Bhardwaj et al. 2024).

L. K. Bhardwaj (✉)
Amity Institute of Environmental Toxicology, Safety and Management, Amity University, Noida, Uttar Pradesh, India

M. Choudhury
Environmental Research and Management Division, Voice of Environment (VoE), Guwahati, Assam, India

L. Goswami
Department of Biology, School of Science & Technology, Örebro University, Örebro, Sweden

F. Rahman
University of Technology Sydney (UTS), Sydney, NSW, Australia

J. Xu
Institute for Advanced Studies, Universiti Malaya, Kuala Lumpur, Malaysia

S. Goswami et al. (eds.), *Micro-Nano Plastics Exposure, Environmental Degradation and Public Health Crisis*, Sustainable Development Goals Series, https://doi.org/10.1007/978-981-96-5047-7_11

Approximately 6.4 million tons of plastic garbage is dumped annually into the ocean (Ryan et al. 2016). Plastic production has risen quickly over the last 70 years, from 2 to 380 million tons per year from 1950 to 2015 (Geyer et al. 2017). 5.25 trillion micro-nano plastic pieces swim in the world's oceans (Eriksen et al. 2014). 15 to 51 trillion plastic particles were found moving on the surface water of the oceans (Van Sebille et al. 2015).

The existence of micro–nanoplastics presents substantial hazards to the Antarctic ecology. These particles are enduring and extremely difficult to eliminate because of their tiny size, which enables a wide variety of creatures to consume them, potentially having toxicological and physical effects (Waller et al. 2017). Scientists reported first-time ingestion of micro–nanoplastics by seabirds (Pachyptila species) (Harper and Fowler 1987) and in other species of seabird (Suaria et al. 2020) and Fur Seals of the Antarctic Ocean (Ryan et al. 2016). Because of their buoyancy, micro–nanoplastics can travel distances, primarily by floating over a long distance; these particles may also have been preserved on beaches or ice for several years. They can enter the ocean because of ice melting, falling or rising sea levels, wind transport, and other meteorological events (Lacerda et al. 2019). Several scientists have reported floating plastics, such as plastic bottles and polystyrene lumps, in the Antarctic Ocean (Lacerda et al. 2019). Other scientists have reported floating macroplastics from similar areas and stated that Polar Regions have not left the existence of micro–nanoplastics (Ryan et al. 2016).

Waller and Hughes (2018) reported that the Scientific Committee on Antarctic Research (SCAR) is working on the issue of plastic contamination in Antarctica. The occurrence of micro–nanoplastics in Antarctica is a dangerous issue and affects the marine ecosystem and humans. The scenario is concerning as it leads to environmental harm and necessitates additional research to preserve the untouched area near Antarctica. This chapter provides a comprehensive account of micro–nanoplastics' presence, their origins, and their detrimental impacts in Antarctica and its surrounding ecosystem. It will cover the origins of micro–nanoplastics, their presence in wastewater discharged from Antarctic research stations, their release from laundry and personal care products (PCPs), their formation from the deprivation of larger plastics, and the current state of micro–nanoplastics in the Antarctic region. In addition, we highlight the detrimental effects of these plastics and offer suggestions for future research and action. The detailed aspect of micro-nano plastic contamination in the Antarctic Region, which is to be covered in this chapter, is shown in Fig. 11.1.

11.2 Sources of Micro–nanoplastics

They may come from different sources, and the sources may be direct or indirect. The direct source may be the removal of waste created by research vessels and research stations (Waller et al. 2017), while transport by the currents of marine is the indirect source which can transport the micro–nanoplastics from distant parts located at lower latitudes (Fraser et al. 2018). The local sources of these tiny particles in the Antarctic region may be fewer due to the lack of human residents and the low shipping volume (Waller et al. 2017). Larger plastic particles are dumped into the ocean from terrestrial sources (Nerland et al. 2014). Human activities may contribute to micro-nano plastic pollution at the lowest level in Antarctica.

11.2.1 Wastewater Released from Antarctic Research Stations

Approximately 90% of micro-nano particles are retained by wastewater treatment plants (WWTPs), and non-retained particles are freed

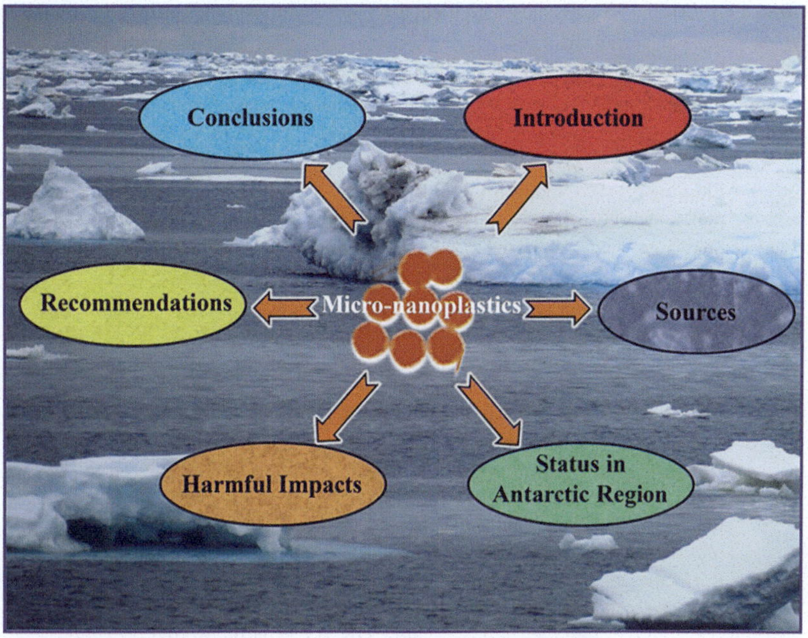

Fig. 11.1 The overall overview of micro-nano plastic and Antarctic Region

into the marine environment (Ziajahromi et al. 2016). Out of 71, only 39 research stations have WWTPs in Antarctica, and their efficiency for the removal of micro–nanoplastics is unknown (Waller et al. 2017). Very little research is available concerning the release of micro–nanoplastics through the wastewater that is released in Antarctica and its surrounding environment.

11.2.2 Laundry and Personal Care Products (PCPs)

PCPs and laundry are another cause of micro–nanoplastics in the Antarctica region. Micro–nanoplastics were used per person in PCPs, ranging between 2.4 and 27.5 mg/day (Napper and Thompson 2016). While washing synthetic clothes, micro–nanoplastics may be released into the wastewater of scientific research stations and tourist vessels. Synthetic fibers were released from shirts, fleece, and polyester blankets in the range of 680 to 1900 fibers per wash (Waller et al. 2017). About 7,28,000 microfibers may be freed into the sea through wastewater after washing 6 kg of clothes made of acrylic fabrics (Napper and Thompson 2016). Microfibers released from laundry to sewage may be a chief source of micro–nanoplastics compared to PCPs. Very few studies are available on the presence of microfiber in Antarctica's sediments, water column, and surrounding environment.

11.2.3 Breakdown of Macroplastic

Macroplastics are another contributor of micro-nano plastic pollution in the Antarctic region. In Antarctica, high UV radiation is the responsible for degradation of floating plastic debris through the photooxidation reaction. The low temperature decreases the rate of the breakdown of the plastics that are submerged (Hammer et al. 2012). Terrestrial macroplastics are the dominating source of micro–nanoplastics in the marine environment.

11.3 Status of Micro–nanoplastics in the Antarctic Region

The occurrence of micro–nanoplastics in different concentrations in different samples of the Antarctic region is presented in Table 11.1.

Table 11.1 Micro–nanoplastics with their detection methods, types, size, and color in the Antarctic region

S. no.	Sample Type	Location	Detection Procedure	Concentration of Micro–nanoplastics	Type/Color/Size of micro–nanoplastics	References
1	Sea surface water	Antarctic Ocean	Microscopy	0.55 to 56.58 gm/km^2	The dark color and small size	Eriksen et al. (2014)
2	Floating plastic debris	South of the Polar Front	Raman spectroscopy	0.100 to 0.514 gm/km^2	PE and industrial resin pellets	Cózar et al. (2014)
3	Seawater	Western Antarctic Peninsula	Microscopy	22 to 117 particles/L	Synthetic fiber	Waller et al. (2017)
4	Water	Ross Sea, Antarctica	FTIR	0.0032 to 1.18 particles/m^3	Large fragments and fiber	Cincinelli et al. (2017)
5	Sediment	Terra Nova Bay, Antarctica	Microscopy and FTIR	1 to 90 items/m^2	Black and yellow colors	Munari et al. (2017)
6	Sediment and macroalgal material	Admiralty Bay, South Shetland Islands	Visual detection	16 to 766 particles/m^2	Synthetic fiber	Waller et al. (2017)
7	Surface seawater	Antarctic Ocean	Stereoscopy, microscopy, and FTIR	46,000 to 99,000 particles/km^2	PS particles and fibers	Isobe et al. (2017)
8	Marine sediment	Rothera Research Station, Antarctica	FTIR	<5 particles/10 mL	White, vibrant reds and green	Reed et al. (2018)
9	Fecal material and Antarctic Krill	Antarctic Ocean	Enzyme digestion	6.0 ± 5.0 μm	PE micro-nano plastic beads and PE fragments	Dawson et al. (2018)
10	Gentoo Penguins	Bird Island, Antarctica	μ-FTIR	76 to 3980 μm	Green, transparent, red, blue, and black	Bessa et al. (2019)
		Signy Island, Antarctica		118 to 4945 μm	Blue and transparent	
11	Surface water	Antarctic Peninsula	FTIR	1794 items/km^2	PU, PA, and PE	Lacerda et al. (2019)
12	Surface water	Admiralty Bay, Antarctica	Scanning electron microscopy (SEM) and Raman spectroscopy	2.40 (±4.57) microfibers 100/m^3	Microfibers (blue, red, black, and clear), PEG, PU, PET, and PA	Absher et al. (2019)
13	Floating plastic debris	Antarctic Ocean	μ-FTIR	188 ± 589 particles/km^2	Floating micro–nanoplastics	Suaria et al. (2020)
14	Sea ice core	East Antarctica		11.71 particles/L	PE, PP, and PA	Kelly et al. (2020)
15	Sea surface water	Antarctic Ocean		188 ± 589 particles/km^2	PE, PP, PS, PVC and PA	Suaria et al. (2020)
16	Freshwater	Byers Peninsula, Livingston Island, Near Antarctic Peninsula		0.47 to 1.43 items/1000 m^3	Four PES fibers, two acrylic fibers and two transparent polytetrafluoroethylene films	González-Pleiter et al. (2020)

μ-FTIR micro-Fourier Transform Infrared Spectroscopy, *PE* Polyethylene, *PS* Polystyrene, *PP* Polypropylene, *PVC* Polyvinyl chloride, *PU* Polyurethane, *PA* Polyamide, *PEG* Polyethylene Glycol, *PET* Polyethylene terephthalate, *PES* Polyester, *PMMA* Poly methyl methacrylate

Table 11.2 Diverse impact of micro-nano plastic in the Antarctic region

S. no.	Impact area	Impact	References
1	Marine organisms	Ingestion causes internal cuts and digestive obstructions. Bioaccumulation and biomagnification in the food chain, threatening ecological balance	Wright et al. (2013); Weis (2014); Law and Thompson (2016); Bhardwaj (2023)
2	Toxicity	Leaching behavior potentially causes carcinogenic and endocrine-disrupting effects	Wright et al. (2013); Waller et al. (2017)
3	Contaminants adsorption	Adsorption of environmental contaminants posing risks to aquaculture products and food safety	Wright et al. (2013); Bhardwaj (2023)
4	Human health	Potential health effects through seafood consumption and exposure include immune system interaction and DNA alterations	Waller et al. (2017); Thushari and Senevirathna (2020); MacLeod et al. (2021); Xu et al. (2022)
5	Nutrient cycles	Disruption of nutrient cycles and alteration of microbial communities	Dawson (2017)
6	Soil invertebrates	Negative impacts on growth and survival, concentration-dependent effects, particularly for nematodes and earthworms	Ji et al. (2021)
7	Plant growth	Adverse effects on germination, root development, biomass, and chlorophyll levels	Zantis et al. (2023)
8	Waste management	Inefficient waste management leads to the accumulation of sediments and water	Jambeck et al. (2015); Paul et al. (2019);Caldwell et al. (2022); Mondal et al. (2023)
9	Research initiatives	Increased awareness and research initiatives focusing on mitigation and innovative solutions	Leistenschneider et al. (2024)

11.4 Harmful Impacts of Micro–nanoplastics in Antarctica

Micro- and nano-sized plastics significantly impact Antarctica's isolated and unaltered ecosystem (Table 11.2). The particles are derived from several sources, including marine trash generated by shipping and fishing activities and air deposition (Caldwell et al. 2022). These particles are putting a lot of stress on Antarctica's unique ecosystem. These micro–nanoplastics are consumed by marine species, resulting in the collection and progression of these compounds in the food chain. Moreover, micro- and nanoplastics interfere with nutrient cycles and alter microbial populations, crucial for preserving Antarctic ecosystems (Dawson 2017). Human health is endangered because plastic pollution can be transferred through seafood consumption, and those living in research stations may be exposed to these harmful particles (MacLeod et al. 2021).

Recent systematic evaluations emphasize the substantial ecological hazards presented by micro- and nanoplastics. In their study, Xu et al. (2022) emphasized the possible human health hazards caused by exposure to micro–nanoplastics. They observed that out of 133 studies, 117 revealed adverse effects in mammalian models, such as cytotoxicity and toxicity specific to certain organs. Additionally, Ji et al. (2021) presented data suggesting that micro–nanoplastics negatively affected soil

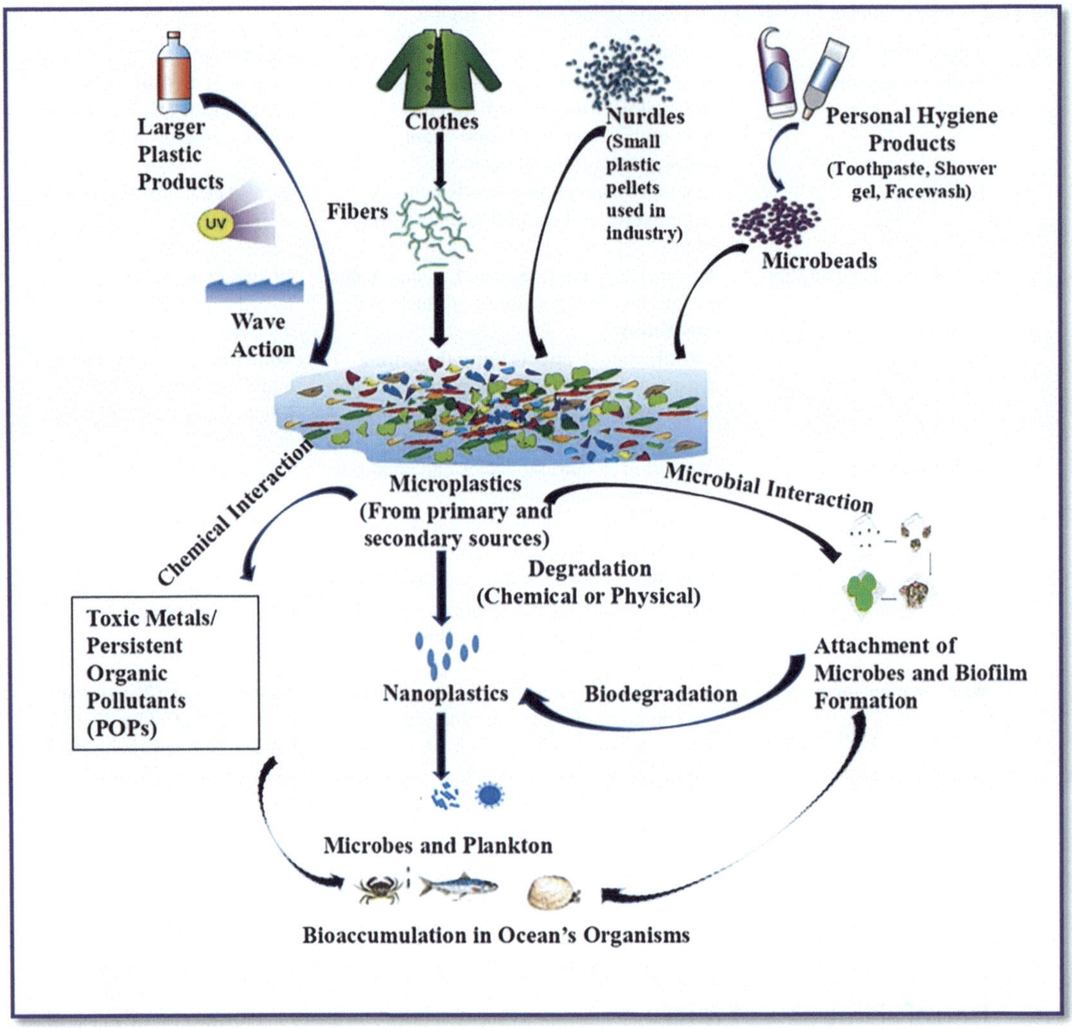

Fig. 11.2 Micro–nanoplastics and their degradation and bioaccumulation in different organisms

invertebrates. Laboratory experiments conducted under controlled conditions revealed that over half of the biological indicators were negatively impacted. Earthworms exhibited reduced growth and survival when exposed to higher concentrations of micro–nanoplastics, while nematodes experienced substantial impacts contingent upon the concentration. The studies show that we need to learn about the health and environmental effects of micro–nanoplastics right away, especially in places like the Antarctic that are easily damaged (MacLeod et al. 2021; Xu et al. 2022; Bhardwaj 2023; Zantis et al. 2023). The degradation and bioaccumulation pathway of micro–nanoplastics is shown in Fig. 11.2.

11.5 Future Aspects and Recommendations

Further research should be done regularly on micro-nano plastic pollution in Antarctica and its surrounding area. To solve the issue of micro-nano plastic, researchers should develop and standardize new analytical procedures for micro-nano plastic detection. Better-informed policy decisions and conservation efforts can be made using advanced methodologies for identifying and quantifying micro–nanoplastics, which can help us better understand their prevalence and consequences. So, to safeguard these delicate ecosystems from the ubiquitous threat of plastic contamination, combating plastic pollution in Polar Regions necessitates a multimodal strategy encompassing research, policy-making, public awareness, and establishing solid global collaboration.

11.6 Conclusions

Before 60 years, the production of plastics had started, and now, most valuable items are partially made of durable materials called plastics. The micro-nano plastic presence in Antarctica and its surrounding environment has been recognized as a main conservation problem and a primary concern for research. However, the most critical questions regarding plastic in the Antarctic region and the surrounding environment remain unanswered. In the Antarctic Ocean, subsurface currents bring micro–nanoplastics to the Polar Region. The bulks of micro-nano plastic particles in the aquatic environment come from the destruction of macroplastics and were reported in high amounts in the 1980s and 1990s. micro–nanoplastics in high quantities in the Polar Region mention that this region is not exempted from plastic pollution, as considered previously. Plastic concentrations on Antarctica's surface water have yet to be confirmed. This plastic pollution hurts the organisms of Antarctica and the Antarctic Ocean. However, it needs to be made clear which species are being affected more due to the scarcity of data.

References

Absher TM, Ferreira SL, Kern Y, Ferreira AL, Christo SW, Ando RA (2019) Incidence and identification of microfibers in ocean waters in Admiralty Bay, Antarctica. Environ Sci Pollut Res 26(1):292–298

Alimi OS, Farner Budarz J, Hernandez LM, Tufenkji N (2018) Microplastics and nanoplastics in aquatic environments: aggregation, deposition, and enhanced contaminant transport. Environ Sci Technol 52(4):1704–1724

Bessa F, Ratcliffe N, Otero V, Sobral P, Marques JC, Waluda CM et al (2019) Microplastics in gentoo penguins from the Antarctic region. Sci Rep 9(1):1–7

Bhardwaj LK (2022a) Evaluation of Bis (2-ethylhexyl) phthalate (DEHP) in the PET bottled mineral water of different brands and impact of heat by GC–MS/MS. Chem Afr 5(4):929–942

Bhardwaj LK (2022b) Microplastics in drinking water-uses, sources and transport. Futuristic trends in Agriculture Engineering & Food Sciences. IIP proceedings, volume 2, Book 9, Part 1, Chapter 28. ISBN:978-93-95632-65-2

Bhardwaj LK (2023) Occurrence of microplastics (MPs) in Antarctica and its impact on the health of organisms. Mar Technol Res 6(2):265418. https://doi.org/10.33175/mtr.2024.265418

Bhardwaj LK, Sharma A (2021) Estimation of physico-chemical, trace metals, microbiological and phthalate in PET bottled water. Chem Afr 4(4):981–991

Bhardwaj LK, Rath P, Yadav P, Gupta U (2024) Microplastic contamination, an emerging threat to the freshwater environment: a systematic review. Environ Syst Res 13(1):8

Bhardwaj LK, Rath P, Rehman F, Yadav P, Massahi T, Mittal R, Jain H (2025) Status of micro/Nano-plastic (MNP) pollution in the aquatic system. In: Societal and environmental ramifications of plastic pollution. IGI Global, pp 43–62

Caldwell J, Taladriz-Blanco P, Lehner R, Lubskyy A, Ortuso RD, Rothen-Rutishauser B, Petri-Fink A (2022) The micro-, submicron-, and nanoplastic hunt: a review of detection methods for plastic particles. Chemosphere 293:133514

Cincinelli A, Scopetani C, Chelazzi D, Lombardini E, Martellini T, Katsoyiannis A et al (2017) Microplastic in the surface waters of the Ross Sea (Antarctica): occurrence, distribution and characterization by FTIR. Chemosphere 175:391–400

Cózar A, Echevarría F, González-Gordillo JI, Irigoien X, Úbeda B, Hernández-León S et al (2014) Plastic debris in the open ocean. Proc Natl Acad Sci 111(28):10239–10244

Dawson AL (2017) Uptake dynamics and effects of traditional and emerging persistent environmental pollutants to the Antarctic keystone species, Antarctic krill

Dawson AL, Kawaguchi S, King CK, Townsend KA, King R, Huston WM, Bengtson Nash SM (2018) Turning microplastics into nanoplastics through digestive fragmentation by Antarctic krill. Nat Commun 9(1):1–8

Eriksen M, Lebreton LC, Carson HS, Thiel M, Moore CJ, Borerro JC et al (2014) Plastic pollution in the world's oceans: more than 5 trillion plastic pieces weighing over 250,000 tons afloat at sea. PLoS One 9(12):e111913

Fraser CI, Morrison AK, Hogg AM, Macaya EC, van Sebille E, Ryan PG et al (2018) Antarctica's ecological isolation will be broken by storm-driven dispersal and warming. Nat Clim Chang 8(8):704–708

Geyer R, Jambeck JR, Law KL (2017) Producción, uso y destino de todos los plásticos jamás fabricados. Sci Adv 3(7):1207–1221

González-Pleiter M, Edo C, Velázquez D, Casero-Chamorro MC, Leganés F, Quesada A et al (2020) First detection of microplastics in the freshwater of an Antarctic specially protected area. Mar Pollut Bull 161:111811

Hammer J, Kraak MH, Parsons JR (2012) Plastics in the marine environment: the dark side of a modern gift. Rev Environ Contam Toxicol:1–44

Harper PC, Fowler JC (1987) Plastic pellets in New Zealand storm-killed prions (Pachyptila spp.). Notornis 34(1):65–70

Isobe A, Uchiyama-Matsumoto K, Uchida K, Tokai T (2017) Microplastics in the southern ocean. Mar Pollut Bull 114(1):623–626

Jambeck JR, Geyer R, Wilcox C, Siegler TR, Perryman M, Andrady A et al (2015) Plastic waste inputs from land into the ocean. Science 347(6223):768–771

Ji Z, Huang Y, Feng Y, Johansen A, Xue J, Tremblay LA, Li Z (2021) Effects of pristine microplastics and nanoplastics on soil invertebrates: a systematic review and meta-analysis of available data. Sci Total Environ 788:147784

Kelly A, Lannuzel D, Rodemann T, Meiners KM, Auman HJ (2020) Microplastic contamination in East Antarctic Sea ice. Mar Pollut Bull 154:111130

Lacerda ALDF, Rodrigues LDS, Van Sebille E, Rodrigues FL, Ribeiro L, Secchi ER et al (2019) Plastics in sea surface waters around the Antarctic Peninsula. Sci Rep 9(1):1–12

Law K, Thompson RC (2016) Microplastics in the seas—concern is rising about widespread contamination of the marine environment by microplastics. Science 345:61–93

Lee J, Hong S, Song YK, Hong SH, Jang YC, Jang M et al (2013) Relationships among the abundances of plastic debris in different size classes on beaches in South Korea. Mar Pollut Bull 77(1–2):349–354

Leistenschneider C, Wu F, Primpke S, Gerdts G, Burkhardt-Holm P (2024) Unveiling high concentrations of small microplastics (11–500 μm) in surface water samples from the southern Weddell Sea off Antarctica. Sci Total Environ 927:172124

Li WC, Tse HF, Fok L (2016) Plastic waste in the marine environment: a review of sources, occurrence and effects. Sci Total Environ 566:333–349

MacLeod M, Arp HPH, Tekman MB, Jahnke A (2021) The global threat from plastic pollution. Science 373(6550):61–65

Mondal T, Choudhury M, Kundu D, Dutta D, Samanta P (2023) Landfill: an eclectic review on structure, reactions and remediation approach. Waste Manag 164:127–142. https://doi.org/10.1016/j.wasman.2023.03.034

Munari C, Infantini V, Scoponi M, Rastelli E, Corinaldesi C, Mistri M (2017) Microplastics in the sediments of Terra Nova Bay (Ross Sea, Antarctica). Mar Pollut Bull 122(1–2):161–165

Napper IE, Thompson RC (2016) Release of synthetic microplastic plastic fibres from domestic washing machines: effects of fabric type and washing conditions. Mar Pollut Bull 112(1–2):39–45

Nerland IL, Halsband C, Allan I, Thomas KV (2014) Microplastics in marine environments: occurrence, distribution and effects

Obbard RW (2018) Microplastics in polar regions: the role of long range transport. Curr Opin Environ Sci Health 1:24–29

Paul S, Choudhury M, Deb U, Pegu R, Das S, Bhattacharya SS (2019) Assessing the ecological impacts of ageing on hazard potential of solid waste landfills: a green approach through vermitechnology. J Clean Prod 236:117643. https://doi.org/10.1016/j.jclepro.2019.117643

Rajpal A, Choudhury M, Goswami S, Chakravorty A, Raghavan V (eds) (2024) Waste management and treatment: advances and innovations, 1st edn. CRC Press. https://doi.org/10.1201/9781003258377

Reed S, Clark M, Thompson R, Hughes KA (2018) Microplastics in marine sediments near Rothera research station, Antarctica. Mar Pollut Bull 133:460–463

Ryan PG, De Bruyn PN, Bester MN (2016) Regional differences in plastic ingestion among Southern Ocean fur seals and albatrosses. Mar Pollut Bull 104(1–2):207–210

Suaria G, Perold V, Lee JR, Lebouard F, Aliani S, Ryan PG (2020) Floating macro-and microplastics around the Southern Ocean: results from the Antarctic circumnavigation expedition. Environ Int 136:105494

Thushari GGN, Senevirathna JDM (2020) Plastic pollution in the marine environment. Heliyon 6(8)

Van Sebille E, Wilcox C, Lebreton L, Maximenko N, Hardesty BD, Van Franeker JA et al (2015) A global inventory of small floating plastic debris. Environ Res Lett 10(12):124006

Waller CL, Hughes KA (2018) Plastics in the Southern Ocean. Antarct Sci 30(5):269–269

Waller CL, Griffiths HJ, Waluda CM, Thorpe SE, Loaiza I, Moreno B et al (2017) Microplastics in the Antarctic marine system: an emerging area of research. Sci Total Environ 598:220–227

Weis JS (2014) Marine pollution: what everyone needs to know®. Oxford University Press

Wright SL, Thompson RC, Galloway TS (2013) The physical impacts of microplastics on marine organisms: a review. Environ Pollut 178:483–492

Xu JL, Lin X, Wang JJ, Gowen AA (2022) A review of potential human health impacts of micro-and nanoplastics exposure. Sci Total Environ 851:158111

Zantis LJ, Borchi C, Vijver MG, Peijnenburg W, Di Lonardo S, Bosker T (2023) Nano-and microplastics commonly cause adverse impacts on plants at environmentally relevant levels: a systematic review. Sci Total Environ 867:161211

Ziajahromi S, Neale PA, Leusch FD (2016) Wastewater treatment plant effluent as a source of microplastics: review of the fate, chemical interactions and potential risks to aquatic organisms. Water Sci Technol 74(10):2253–2269

12 Micro- and Nanoplastics: Impact on Wildlife and Biodiversity

Sumanta Bhattacharya

12.1 Introduction

Plastics pose a severe challenge to achieving sustainability due to their pernicious impacts on the environment, ecosystem, and biodiversity. The plastic particles having diameters within the range of 1 m to 1 nm exhibit a more severe threat, as these cannot be effectively removed from the system by the conventional plastic treatment processes. Moreover, due to their smaller size, these particles intervene in the system and effectively interact at the microscopic scale, resulting in severe internal damage. The micro- and nanoplastics (MNP) damage the physiological function of the microorganisms in the ecosystem and contaminate the food chain through undesirable interventions. The organisms responsible for the increase in soil fertility face a severe threat due to their interaction with the MNPs. When microplastics and nanoplastics are mixed with other contaminants, they become a complicated problem that has a fatalistic effect on phytotoxicity and ecotoxicity in nearby eco-habitats. The movement of microplastics and nanoplastics can change many properties of soil, including its structure, water-holding capacity, and rhizosphere pH. Multiple studies have also noted a negative effect on the richness and stability of the microbial communities in the rhizosphere. MNPs, on the other hand, interfere with macro animal physiological functions by consuming packaged food products, disrupting metabolism, blood transfusion, and other important biological activities. MNPs also intervene in the physiological activities of the plants through the translocation process. The photosynthetic quotient of the plant species declines due to the accumulation of MNPs with the photosynthetic pigments of the leaves, causing a decrease in the carbon dioxide absorption rate and deterioration of the photosynthetic efficiency of the plants. It directly impacts the healthy growth and development of plant species, as well as the ability of plants to utilize significant greenhouse gases such as carbon dioxide (Selley et al. 2020).

MNPs have been found to have detrimental effects on microbial communities and living organisms. In the rhizosphere, the region of soil surrounding plant roots, MNPs have been shown to reduce the richness and stability of microbial communities, which are essential for maintaining healthy soil ecosystems. This can have cascading effects on plant growth, nutrient cycling, and ecosystem functioning. Furthermore, when ingested through packaged food products, MNPs can disrupt various animal physiological functions, including metabolism, blood transfusion, and other vital processes, leading to potential harm. Additionally, MNPs can also interfere with plant

S. Bhattacharya (✉)
Asian International University, Manipur, India

S. Goswami et al. (eds.), *Micro-Nano Plastics Exposure, Environmental Degradation and Public Health Crisis*, Sustainable Development Goals Series, https://doi.org/10.1007/978-981-96-5047-7_12

physiological activities by accumulating in leaves and reducing photosynthetic efficiency. This can impede the healthy growth and development of plants and their ability to absorb carbon dioxide, a critical function for mitigating climate change.

Moreover, MNPs can also alter soil properties, affecting water retention, aeration, and fertility, further exacerbating their negative impacts. Overall, MNPs can have far-reaching and devastating consequences for the environment and living organisms, highlighting the need for careful consideration and regulation of their use (Jewett et al. 2022). This research thus concentrates on the in-depth examination of the effects of MNPs on wildlife biodiversity and animals. This chapter ends with a need for further investigation of the detrimental impacts of plastic fragments on people and wildlife, stressing the possible effects of MNPs on higher plants and the ecosystem.

12.2 Importance of Biodiversity in Ecological Balance

The suitable transfer of matter and energy in an ecosystem is necessary for the survivability of life. A suitable population of different biotic and abiotic components in the ecosystem is necessary for its sustenance population of different biotic and abiotic components in the ecosystem is necessary for its sustenance. The variety of living organisms, i.e., biodiversity, persists in the environment and is one of the most critical ecosystem factors that facilitate the development of a sustainable ecological relationship. The biological products generated by one living species become the energy source for another species. This food–consumer relationship between different components of the ecosystem, i.e., producer-primary consumer, primary consumer-secondary consumer, and secondary consumer-tertiary consumer, and their interaction with decomposers, can only be possible if the diversity of living organisms at every stage of the food chain is present in a proper quantum in the ecosystem. As all the living species in an ecosystem are intensely connected through the food–consumer relationship, the survival of one species depends on the existence of another (Selley et al. 2020). Therefore, maintaining biodiversity is essential for the sustenance of an ecosystem. The loss of biodiversity disrupts the circular flow of essential elements like carbon, sulfur, oxygen, and phosphorus resulting in the loss of ecological balance and environmental degradation. Some microorganisms and most plant species are critical in removing environmental toxicants. The uncontrolled exotic invasion of humans in an ecosystem causes biodiversity and ecological balance to be lost. Human activities also result in the accumulation of unwanted toxic materials, leading to biodiversity loss. Living organisms are not only sources of food but also an essential source of other resources. The variety of protozoa and plant species present in an ecosystem plays a vital role in the purification of the environment. The different types of flora present in terrestrial and aquatic ecosystems are the primary sources of oxygen and other food products essential for the survival of other consumers. The biodiversity of plant species facilitates the production of a variety of abiotic resources essential for the survival of human civilization. Biodiversity ensures the existence of all living and non-living components in the food chain and food web, leading to the sustenance of an ecosystem (Selley et al. 2020).

12.3 Micro- and Nanoplastics and Their Impact on Biodiversity

The accumulation of plastics in the environment poses a significant threat to ecosystems, as they create blockages in the flow of matter and energy and are non-biodegradable due to polymeric solid bonds. Micro- and nanoplastics, formed through the breakdown of microplastics, are hazardous due to their ease of access into physiological systems, high reactivity, and ability to be accepted by cells through endocytic machinery or phagocytosis, leading to damage to normal physiological activities. The accumulation of MNPs in various organs and systems disrupts their mechanisms, influences genomic sequences, and alters genotypic and phenotypic characteristics, signifi-

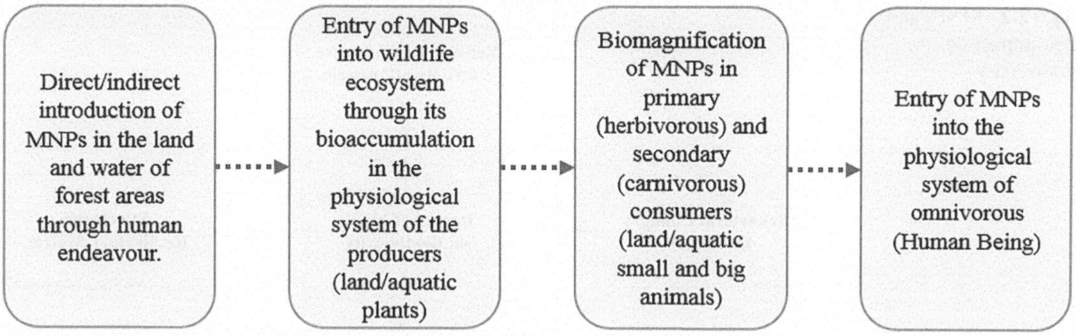

Fig. 12.1 Path of interaction of MNPs in wildlife ecosystem

cantly impacting biodiversity (Fig. 12.1). This has far-reaching consequences, threatening the survival of numerous species, contaminating terrestrial and aquatic ecosystems, and affecting the growth and reproduction of plants and animals. The large-scale accumulation of plastic particles in water bodies poses a significant threat to freshwater and marine biodiversity, challenging the growth and survival of phytoplankton, zooplankton, coral reefs, and various aquatic species, highlighting the urgent need for effective mitigation strategies. The ingestion of plastic particles by animals leads to their accumulation in the gut and stomach, followed by a gradual migration to vital organs such as the liver and brain, disrupting the animal's entire physiological process. These plastic particles release endocrine-disrupting chemicals, including bisphenol A, phthalates, and polychlorinated biphenyls, which increase oxidative stress and inflammation, impairing the metabolism system and causing chronic cellular damage, tissue effects, and organismal homeostasis in both animal and plant cells. Prolonged exposure to microplastics triggers the production of reactive oxygen species, leading to excessive cellular oxidation while also containing carcinogenic and mutagenic compounds that cause permanent genetic damage, resulting in cancer and other severe genetic diseases.

Furthermore, high concentrations of microplastics fragment DNA strands, effectively altering genotypic data in both animals and plants, disrupting the ecosystem's genetic biodiversity, and posing a significant threat to the health and survival of species. This underscores the urgent need to address the plastic pollution crisis and mitigate its far-reaching consequences (Masia et al. 2021). Figure 12.2 delineates the MNP's associated adverse impacts on wildlife biodiversity.

The cellular toxicity causes infertility in the female organisms, leading to a lower number of spawned eggs. The higher concentration of MNPs in animal cells also reduces sperm motility. The nanoplastics also cause a significant reduction in the number of germline cells and fertilized eggs, leading to a significant reduction in the reproductive capability of the animals. Increasing exposure of living organisms to MNPs causes a significant alteration of gonadal morphology in both sexes. It also leads to the qualitative and quantitative degradation of the gametes. The MNPs induce spermatogenic cell apoptosis by disrupting the cells' blood–testis barrier. It reduces the reproductive abilities of living organisms (Jewett et al. 2022). MNPs act as potential biological vectors of toxic pollutants in the ecosystem because they have a high surface area and strong hydrophobicity.

Different organic and inorganic pollutants, like phenanthrene, cadmium, lead, and mercury, combine with MNPs and contaminate the ecosystem. Plastics contain plasticizers and fillers to increase polymers' modulus and mechanical properties, i.e., impact strength and tensile strength. These plasticizers contain toxic inorganic and organic cadmium, barium, and lead salts that contaminate land and water when disposed of. Synthetic dyes and pigments are pri-

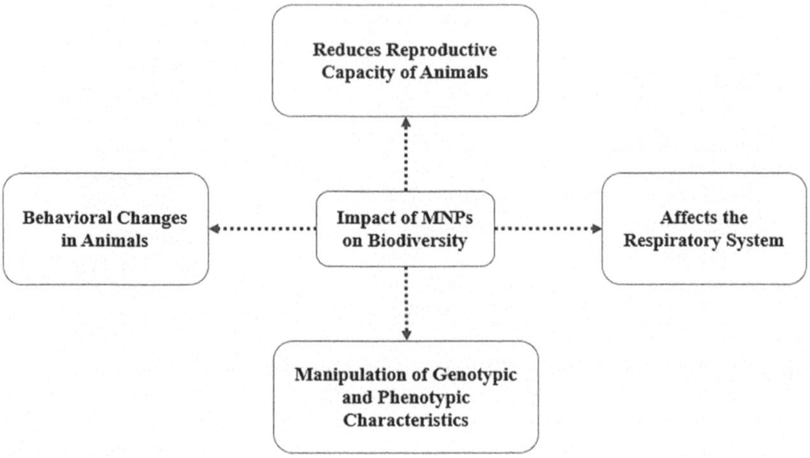

Fig. 12.2 MNPs and their impact on biodiversity

marily used to color plastics containing toxic chemicals (Gluge et al. 2022). Inorganic dyes contain heavy metals, whereas organic dyes contain chromophoric compounds, viz., azopigments, phthalocyanine pigments, anthraquinone chromophores, etc. Flame retardants used for raising the flashpoint of plastics contain antimony trioxide, aluminum hydroxide, tris (2,3-dibromo propyl) phosphate, chlorinated phosphates, polybrominated biphenyl, polybrominated ether, etc. (Campanale et al. 2020). These toxic contaminants also accelerate biotic and abiotic stress in the ecosystem by creating severe health issues in the animal community.

The physiological and behavioral activities of terrestrial and aquatic animals were also negatively affected by the accumulation of microplastics in their bodies. The significant population of microplastics in blood obstructs the hemoglobin molecules' ability to interact with oxygen atoms, resulting in a lack of oxygen supply in the blood. Microplastics also obstruct almost every physiological activity of the body, viz., blockage in the gastrointestinal tract, malfunction in the flow of neural responses, and interference in normal hormonal functions, resulting in metabolic disturbances, neurotoxicity, and endocrine disruption. The accumulation of microplastics in the lungs and other vital organs irritates, leading to inflammation that may further exhibit carcinogenic characteristics. The persistent organic pollutants (POPs) found in microplastics endanger the health of aquatic animals and biomagnify higher members of the food web. Various additives, viz., pigments, UV stabilizers, flame retardants, water repellants, stiffeners, and softeners, are extensively used in the production of synthetic plastics in order to improve the quality of plastic materials. However, these additives become an essential source of water, land, and soil pollution, which later creates serious health issues for the animals and interferes with the plants' physiological activities (Peters et al. 2022). MNP traces are also found in the lungs' pulmonary tissues, primarily administered during respiration. The microplastics enter the physiological structure of the animals through the dermal, inhalation, and ingestion pathways and accumulate in the respiratory organs of the animals. The excessive accumulation of MNPs in the lungs causes serious respiratory issues like asthma, pneumothorax, and bronchitis. Exposure to nanoplastics causes inflammatory responses in the alveolar basal epithelial cells and induces apoptosis, disrupting the respiratory system (Rodrigues et al. 2022).

12.4 Effect of MNPs in Terrestrial Ecosystem

Microplastics exhibit colloidal characteristics because they are the smaller particles of plastic, having a diameter of less than 5 mm. Because of their higher surface charge density and higher surface area to volume ratio, MNPs effectively react with soil abiotic components, such as perfluoroalkyls, and ionic components, such as mercury, lead, potassium, and sodium ions, increasing soil biotic and abiotic stresses. Microplastics are more polar and ionic than everyday plastics as the accumulation of charges on the surface increases, increasing their reactivity toward the ions in the salts. In this way, microplastics also alter the salinity and pH of the soil, leading to a paradigmatic alteration of the different parameters maintaining the quality of the soil (Azeem et al. 2021; Larue et al. 2021). The increase in the population of microplastics in soil reduces the bulk density notably. The water-holding capacity of the soil, i.e., water-stable aggregates, as well as the ability of the soil to resist degradation while exposed to external forces, also deteriorates due to the accumulation of microplastics, which consist of polymers like polyethylene terephthalate, polyethersulfone, and polypropylene. The muffling of soil pores due to the procurement of plastic molecules therein escalates the process of evapotranspiration, leading to the depletion of water content in the soil and the plants (Zhao et al. 2021). The lack of nutrients in the soil promotes the activity of fungi such as *Arbuscular mycorrhiza*, which interact with the root epidermis and cause root infection. The increasing population of microplastics in the soil enhances the possibility of microplastic–microplastic and microplastic–natural matter interactions in the soil that result in a paradigm shift in the soil texture and other biophysical properties of the soil (Yin et al. 2021). Figure 12.3 schematically represents the MNP's connection with the terrestrial ecosystem and its consequences.

The accumulation of microplastics in the soil affects the root traits of the plants, leading to a decrease in crop productivity. When the soil's water content decreases, the roots' length and surface area increase. However, the leaves' size decreases due to insufficient amounts of water and nutrients. It reduces the chlorophyll content of the leaves. The loss of water content reduces tissue density in plants' roots (Kumari et al. 2022). In this way, the contamination of micro- and nanoplastics in the environment causes the significant transformation of biotic and abiotic factors in the soil, causing the deprivation of resources for the producers of the ecosystem, i.e., plants. The microbial community plays a vital role in the ecosystem. It acts as a decomposer and is the food of many small animal species. The plants benefited from the microbes as they maintained the biotic and abiotic balance in the soil by facilitating the fixation of essential macro- and microelements. The microbes also increase the water and nutrient uptake capacity of the plants. This microbial community faces challenges due to the interaction with micro- and nanoplastics. In some studies, it is seen that the population of nitrifying and denitrifying bacteria experiences a declining trend due to the contamination with MNPs (Selley et al. 2020).

12.5 Effects of MNPs in Aquatic Ecosystems

The uncontrolled disposal of micro- and nanoplastics in bodies of water increases the likelihood of interaction between aquatic flora and fauna and MNPs. The aquatic fauna quickly consume these due to their smaller size, which further interacts with their physiological system. The accumulation of MNPs also causes animal behavioral changes, which largely hamper the ecosystem. Moreover, the MNPs facilitate the accumulation of more toxic chemicals in the water molecules, which causes severe repercussions in the aquatic ecosystem—almost 97% of the earth's water resources are occupied by oceans. Therefore, the marine ecosystem is mainly affected by plastic pollution (Choudhury et al. 2022). It has been found in several studies that almost 700 aquatic species in the world, including sea turtles, penguins, sea horses, and other crustaceans, are becoming endangered due to the accumulation of MNPs (Issac and Kandasubramanian 2021). The organic and inorganic toxicants present in plasticizers, flame retardants, and pigments of plastics, viz., DDT, polybrominated diphenyl ether, and other additives, contaminate the water, causing the degradation of the microbial population present therein. The additives present in the MNPs cause serious health issues in aquatic organisms

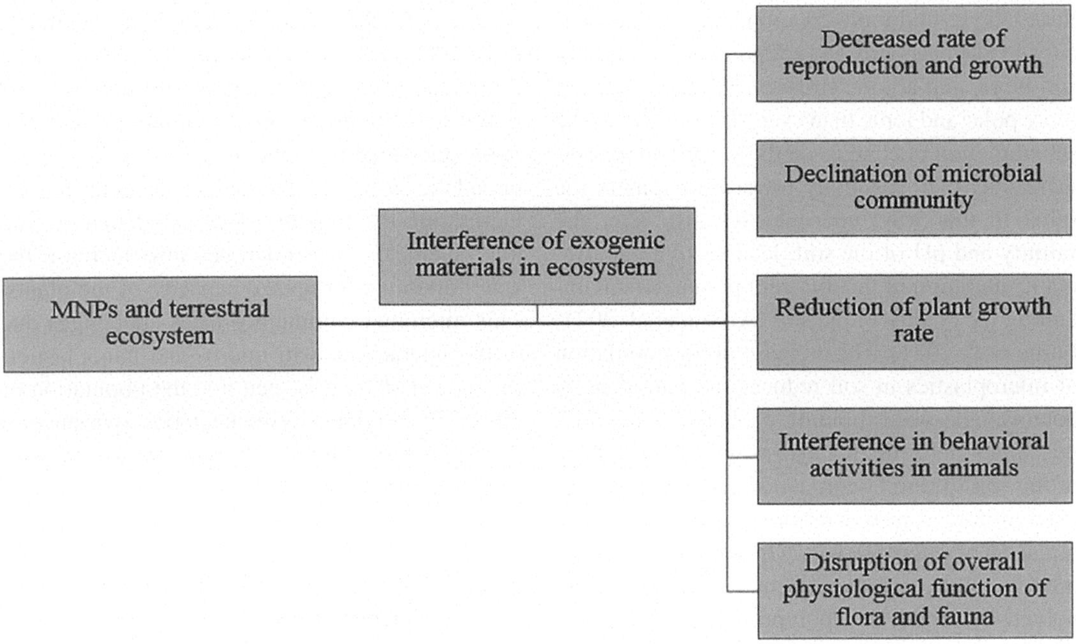

Fig. 12.3 MNPs and terrestrial ecosystem

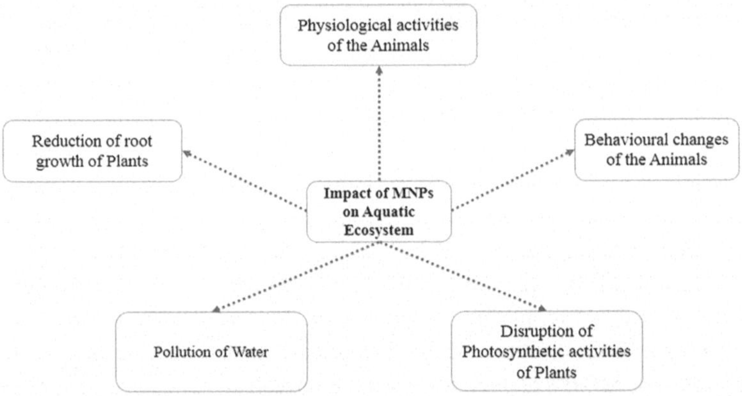

Fig. 12.4 MNPs and aquatic ecosystem

(Fig. 12.4). The bioaccumulation of ultraviolet chemicals present in the MNPs exhibits mutagenic effects in aquatic animals. The photodegradation of stabilizers present in the plastic additives affects the estrogenic activities of the animals, leading to their reluctance to reproduce. The components present in plasticizers cause reproductive, cardiac, renal, and neurotoxic effects in animals. The flame retardants disrupt the activities of the endocrine glands. The chemicals present in the pigments cause gut blockage. The surfactants destroy the mucus layer and cause damage to the gills. Once they enter the ecosystem, these components biologically magnify from one energy state to another and destroy the whole ecosystem through the food chain and food web. The ingestion of MNPs in the aquatic ecosystem causes the overall degradation of the animals' reproductive, metabolic, respiratory, nervous, and other physiological processes (Issac and Kandasubramanian 2021).

The higher concentration of MNPs also hampers the growth and development of aquatic plant species. The higher concentration of MNPs blocks the exposure of floating plants to sunlight, causing the quantum of photosynthetic pigments present in the plant leaves to degrade. It is seen that the amount of chlorophyll a and b pigments present in the floating freshwater plant duckweed has been reduced significantly due to the contamination with microplastics. MNPs also affect the root growth of coral plants. The rough, irregularly shaped microbeads cause damage to the membranes of root cells, leading to apoptosis. It obstructs the healthy growth of roots. Being the producers of aquatic ecosystems, the degradation of aquatic plant health becomes a serious concern for the existence of aquatic ecosystems.

12.6 Impact of MNPs in Food Chain

Micro- and nanoplastics act as exogenic materials in the ecosystem. It hampers the equilibrium between biotic and abiotic components, leading to the loss of ecological balance. MNPs easily enter the physiological systems of animals because of their smaller size, which further transforms other members of the food chain through the flow of matter and energy. Biomagnification causes MNPs to enter all stages of life, including human beings, the highest members of the food chain (Fig. 12.5). The aquatic and terrestrial food chains are related, as seafood becomes an important food source for terrestrial fauna. The accumulation of MNPs is mainly associated with water bodies due to the increasing frequency of disposal of plastic waste into the water (Masia et al. 2021). A significant proportion of MNPs enter the terrestrial ecosystem through contaminated seafood. New plastics are now replacing metallic household materials because of their cost-effectiveness, durability, and lightweight, among other essential characteristics. It simultaneously increases the amount of plastic waste. According to the data released by the United Nations Environment Programme, approximately 85% of the plastic ends up in landfills as unregulated waste, which affects the sewage disposal system effectively (Rodrigues et al. 2022). Weathering converts this plastic waste into smaller plastics; the stream energy of sewage water gradually dissolves them, and they enter water bodies. These micro- and nanosized plastic particles enter the forest ecosystem through the water bodies and affect the physiological functions of aquatic and terrestrial wildlife. The microplastics, after that, transform from a lower stage of the food web to a higher stage through the process of biomagnification and affect the whole forest ecosystem (Ramachandraiah et al. 2022). Wildlife refuges and national parks have become popular tourist destinations. The movement of tourists in the forests increases the number of plastics therein, which further enter the forest ecosystem through bioaccumulation and biomagnification.

These plastic particles contaminate soil quality directly, resulting in nutrient deficiency in wildlife plant species. The careless disposal of plastics in bodies of water results in a lack of water availability for wildlife animals. It directly affects aquatic wildlife, restricting the healthy growth of aquatic animals and flora.

12.7 Future Prospects and Recommendations

Replacing non-biodegradable plastics with biodegradable plastics is one of the most trending approaches to dealing with plastic solutions. To make plastic products biodegradable, synthesizing plastic products from bio-based natural products is becoming more common (Roy and Choudhury 2022). However, traces of bioplastics such as polylactic acetate, polystyrene, and others are found in significant amounts in terrestrial and aquatic biomasses. The use of toxic additives needs to be reduced in the synthesis of plastics. The application of advanced technology in the synthesis of plastics may be recommended to prepare eco-friendly and sustainable plastic products. The recent advancements in nanotechnology facilitate the successful mineralization of plastics

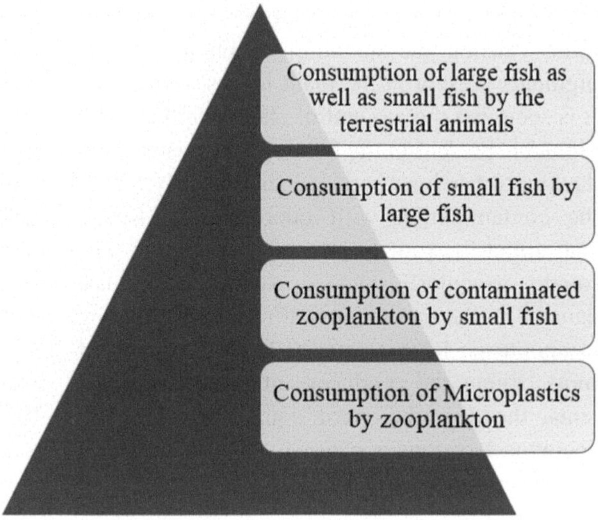

Fig. 12.5 Propagation of MNPs in terrestrial ecosystem from aquatic ecosystem via food web

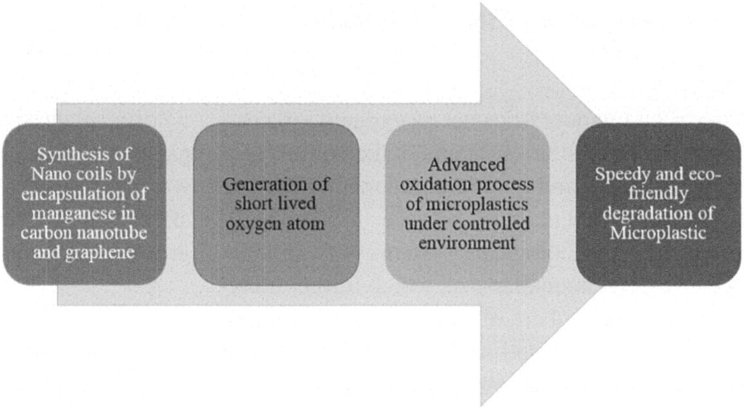

Fig. 12.6 Application of nanotechnology in degradation of microplastics

through oxidation in a controlled environment (Fig. 12.6). Nanocomposites are synthesized by encapsulating manganese carbide nanoparticles with nitrogen-doped carbon nanotubes for the advanced oxidation process of microplastics. A hollow nanostructure, nanocoils, is synthesized by introducing manganese molecules in carbon nanotubes. It facilitates the oxidation of microplastics in a controlled environment through the generation of short-lived oxygen molecules. Graphene and carbon nanotubes are also modified through advanced engineering for eco-friendly and faster degradation of microplastics (Issac and Kandasubramanian 2021). Figure 12.7 demonstrates how plastics research should be carried out shortly.

12.8 Conclusion

Microplastics can be directly synthesized from plastic molecules and used in different industries, known as primary microplastics. Microbeads are one type of primary microplastics, synthetically processed by polyethylene, that are used as exfoliants in several cosmetics and beauty products to remove dead skin cells. These microbeads can successfully escape wastewater treatment techniques and transfer to the marine food chain. However, most microplastics are derived from secondary sources, such as the shredding of large plastic wastes by weathering processes such as storms, wind, and rainfall, among others, and are called secondary microplastics. These secondary microplastics mainly affect the ocean and

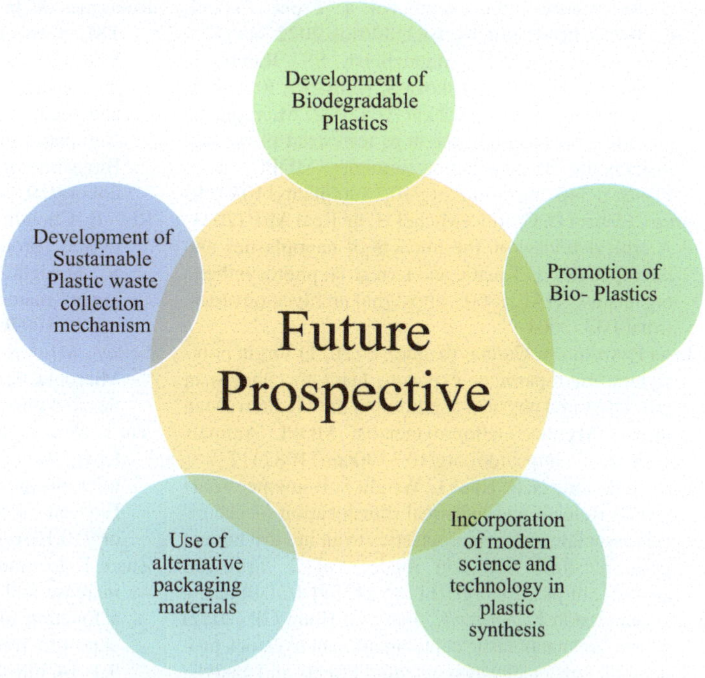

Fig. 12.7 Future prospective of the increasing contamination of ecosystem by MNPs

terrestrial ecosystems. About 4900 million metric tons of plastic waste end up in landfills, likely to be the most potential sources of secondary microplastics. If waste volume continues to rise at its current rate, this figure could reach 12,000 million metric tons by 2050. By 2018, microplastics were found in more than 114 aquatic species. Nowadays, plastic has become an integral part of our lives, and the demands are increasing continuously. Microplastic footprints formed from larger plastics, on the other hand, are found worldwide, including remote areas of Antarctica with no human population. The non-biodegradable, stable polymeric constituents of microplastics, including harmful additives, cause large-scale air, water, and soil contamination. It also hampers the normal biological functions of all the species of flora and fauna in the world.

References

Azeem I, Adeel M, Ahmad MA, Shakoor N, Jiangcuo GD, Azeem K, Ishfaq M, Shakoor A, Ayaz M, Xu M, Rui Y (2021) Uptake and accumulation of Nano/microplastics in plants: a Critical Review; MDPI. Nanomaterials 11(2935):1–22. https://doi.org/10.3390/nano11112935

Campanale C, Massarelli C, Savino I, Locaputo V, Uricchio VF (2020) A Detailed review study on potential effects of microplastics and additives of concern on human health; p. 1–26; MDPI. Int J Environ Res Public Health 17:1212. https://doi.org/10.3390/ijerph17041212

Choudhury M, Sharma A, Pervez A, Upadhyay P, Dutta J (2022) Growing menace of microplastics in and around the coastal ecosystem. In: Coastal ecosystems: environmental importance, current challenges and conservation measures, pp 117–137. https://doi.org/10.1007/978-3-030-84255-0_6

Gluge J, Ashta NM, Herzke D, Lebreton L, Scheringer M (2022) Correspondence regarding the perspective "addressing the importance of microplastic particles as vectors for long-range transport of chemical contaminants: perspective in relation to prioritizing research and regulatory actions", vol 2. Springer, pp 1–5. https://doi.org/10.1186/s43591-021-00021-z

Issac MN, Kandasubramanian B (2021) Effect of microplastics in water and aquatic systems; Nature Public Health Emergency Collection, National Library of Medicine, National Center for Biotechnology Information. Environ Sci Pollut Res Int 28(16):19544–19562. https://doi.org/10.1007/s11356-021-13184-2

Jewett E, Arnott G, Connolly L, Vasudevan N, Kevei E (2022) Microplastics and their impact on reproduction—can we learn from the C. elegans model; National Library of Medicine, National Center for

Biotechnology Information, vol 4. Front Toxicol, p 748912. https://doi.org/10.3389/ftox.2022.748912
Kumari A, Rajput VD, Mandzheiva SS, Rajput S, Minkina T, Kaur R, Sushkova S, Kumari P, Ranjan A, Kalintichenko VP, Glinushkin AP (2022) Microplastic pollution: an emerging threat to terrestrial plants and insights into its remediation strategies; MDPI, vol 11, Plants, p 340. https://doi.org/10.3390/plants11030340
Larue C, Sarret G, Castillo-Michel H, de Real AEP (2021) A critical review on the impacts of nanoplastics and microplastics on aquatic and terrestrial photosynthetic organisms. HAL:1–60. https://hal.archives-ouvertes.fr/hal-03373004
Masia P, Ardura A, García-Vázquez E (2021) Virgin polystyrene microparticles exposure leads to changes in gills DNA and physical condition in the Mediterranean mussel Mytilus Galloprovincialis, MDPI. Animals 11(8):2317. https://doi.org/10.3390/ani11082317
Peters R, de Jong N, de Haan L, Wright S, Bouwmeester H (2022) Release and intestinal translocation of chemicals associated with microplastics in an in vitro human gastrointestinal digestion mode, vol 2. Springer, pp 1–21. https://doi.org/10.1186/s43591-021-00022-y
Ramachandraiah K, Ameer K, Jiang G, Hong GP (2022) Micro- and nanoplastic contamination in livestock production: entry pathways, potential effects and analytical challenges. Sci Total Environ, Elsevier 844:157234
Rodrigues ACB, Jesus GP, Waked D, Gomes GL, Silva TM, Yariwake VY, Silva MP, Magaldi AJ, Veras MM (2022) Scientific evidence about the risks of micro and Nanoplastics (MNPLs) to human health and their exposure routes through the environment; National Library of medicine, National Center for Biotechnology Information. Toxics 10(6):308. https://doi.org/10.3390/toxics10060308
Roy P, Choudhury M (2022) Bioeconomy and green plastic production. In: Baskar C, Ramakrishna S, Daniela La Rosa A (eds) Encyclopedia of green materials. Springer, Singapore. https://doi.org/10.1007/978-981-16-4921-9_189-1
Selley ME, Song B, Passie R, Hale RC (2020) Microplastics affect sedimentary microbial communities and nitrogen cycling. Nat Commun 11:2372
Yin L, Wen X, Huang D, Du C, Deng R, Zhou Z, Tao J, Li R, Wang Z, Chen H (2021) Interactions between microplastics/ nanoplastics and vascular plants. Environ Pollut, Elsevier 290:1–11. https://doi.org/10.1016/j.envpol.2021.117999
Zhao T, Lozano YM, Rillig MC (2021) Microplastics increase soil pH and decrease microbial activities as a function of microplastic shape, polymer type, and exposure time; Frontiers in environmental. Science 9:1–14. https://doi.org/10.3389/fenvs.2021.675803

13 Micro-Nano Plastics—Impact on Coral Ecosystem

Sukhendu Dey, Priyajit Chatterjee, Palas Samanta, Apurba Ratan Ghosh, and Jinho Jung

13.1 Introduction

Despite accounting for a mere one percent of the planet's surface, coral reefs are among our planet's most enriched and diverse ecosystems. Coral reefs thrive primarily in tropical regions near the equator, with higher water temperatures (Munday et al. 2008). Just six nations—Indonesia, Australia, Papua New Guinea, the Philippines, Fiji, and the Maldives—are home to over half of the reefs found worldwide. They serve as the foundation of marine life, hosting nearly 4000 reef fish species, 840 corals, and over one million other aquatic organisms (Johnson et al. 2020). Beyond their ecological significance, coral reefs offer invaluable services to human populations. Due to their potential to alleviate wave energy before they reach the shore, they serve as natural barricades, safeguarding coasts from catastrophes and degradation (Norris et al. 2024). This protective function is crucial for many coastal communities, safeguarding lives and property from natural disasters (Kurniawan et al. 2024b). Additionally, coral reefs are a vital source of income for millions of people worldwide, supporting local economies through tourism, fishing, and other marine-related industries. More than 500 million individuals worldwide rely on underwater reefs for their daily sustenance, overall health, and ability to grow food (Golden et al. 2024).

Coral reefs are becoming more and more vulnerable and suffering significantly from regional disturbances like contamination, overfishing, and coastal development, as well as worldwide disturbances like climate change, seawater acidification, and plastic pollution, notwithstanding their enormous natural diversity (Wilson et al. 2006; Crabbe et al. 2024). In particular, plastic waste is now acknowledged as an emerging contaminant that coral reefs are increasingly concerned about as it is linked to a higher risk of illness (Pastorino 2024; Bian et al. 2024). Plastic pollution negatively impacts coral reefs in several ways, including the disturbance of ecological processes and the physiology of coral animals. Microplastics

S. Dey (✉) · A. R. Ghosh
Department of Environmental Science, The University of Burdwan, Burdwan, West Bengal, India

P. Chatterjee
University Science and Instrumentation Centre, The University of Burdwan, Burdwan, West Bengal, India

P. Samanta
Department of Environmental Science, Sukanta Mahavidyalaya, University of North Bengal, Dhupguri, West Bengal, India

J. Jung
Division of Environmental Science and Ecological Engineering, Korea University, Seoul, Republic of Korea

S. Goswami et al. (eds.), *Micro-Nano Plastics Exposure, Environmental Degradation and Public Health Crisis*, Sustainable Development Goals Series, https://doi.org/10.1007/978-981-96-5047-7_13

(MPs) also change how creatures living on coral reefs feed, endangering the ecosystems' overall health. According to (Lamb et al. 2018), there are an approximated 11.1 billion big plastic fragments, called MPs (diameter > 50 mm), "trapped" in coral reefs globally based on a study involving 124,000 reef-forming corals from 159 coral reefs around the Asia-Pacific region, and by 2025, that figure is anticipated to go up by 40%. These MPs more frequently impacted the structurally complicated reefs and boosted their sensitivity to reef-building coral illness from 4% to 89%, necessitating further investigation to comprehend their pathological pathways (Lamb et al. 2018). Furthermore, dangerous viruses and hazardous chemicals from marine waters are carried by microplastics and deposited onto coral reefs (Huang et al. 2021). Consequently, they have the potential to foster the dissemination of infections and ciliated protozoa, which can lead to coral-related illnesses such as skeletal erosion bands (Huang et al. 2021; Rahman et al. 2023). This makes these ecosystems even more vulnerable to disease and other environmental stresses, adding to their precarious state. For example, pathogens such as *Halofolliculina* and *Rhodobacterales* spp., whose presence is believed to cause occurrences of coral-related illnesses like skeletal corroding and white syndromes, could be transferred by floating plastic pieces (Rahman et al. 2023). Comparably, illnesses that trigger coral tissue impairment and discoloration, such as Rhodobacteraceae, Vibrionaceae, and Flavobacteraceae, could potentially increase MPs levels in coral reef locations (Peng et al. 2020).

It is yet unclear how commonplace and how MPs disrupt coral reef systems, even though MP contamination has received much of the spotlight. In the preceding decade, research on the distribution and MPs quantification in reef environments has progressively gained momentum both through field surveys and laboratory studies (Ding et al. 2019). Still, in-depth studies regarding MPs-associated environmental issues in the coral reef ecosystems are in their infancy. We must urgently require in-depth knowledge of MPs' prevalence, origins, consequences, and interaction mechanism in coral reef systems. Accordingly, the present chapter will address the following issues:

1. MPs distribution pattern and their sources in global reef regions.
2. The environmental consequences of MPs on reef systems.
3. The mechanism of MPs-associated consequences, and finally.
4. Policy recommendations for mitigating MPs contamination in the coral reef system.

13.2 Sources of MPs in Coral Reef

In 2011, the World Resources Institute hypothesized that the worldwide reef territories covered 249,713 km^2 of ocean flooring, broadly grouped into six reef zones: the Middle East (which accounts for 6% of worldwide reef territories), the Atlantic (10%), the Indian Ocean (13%), Australia (17%), the Pacific (26%), and Southeast Asia (28%) (https://www.wri.org/publication/reefs-risk-revisited) (Fig. 13.1). The origins of MPs to these coral ecosystems can be traced to various land-and sea-based origins, reflecting the pervasive nature of plastic pollution. Asia is the leading producer, accounting for 51% of the world's plastic generation, driven primarily by countries like China, South Korea, and Japan. Other significant contributors include NAFTA (North et al. countries—USA, Canada, and Mexico) and Europe, which significantly shape the global plastic landscape. The primary culprit causing MPs to be introduced into reef ecosystems is urbanization. Additionally, inadequate and ill waste management has significantly contributed to the introduction of these pollutants into coral ecosystems (Hernandez et al. 2019; Chen et al. 2021). They are called marine waste. As per the record, approximately 80% of marine waste originated from terrestrial sources such as ill-managed funerals in landfills, raw dirt, poor industrial management, amusing use of seaside areas, and the activities of the tourism industry (Bouwmeester et al. 2015; Torres et al. 2019; Paul et al. 2019;

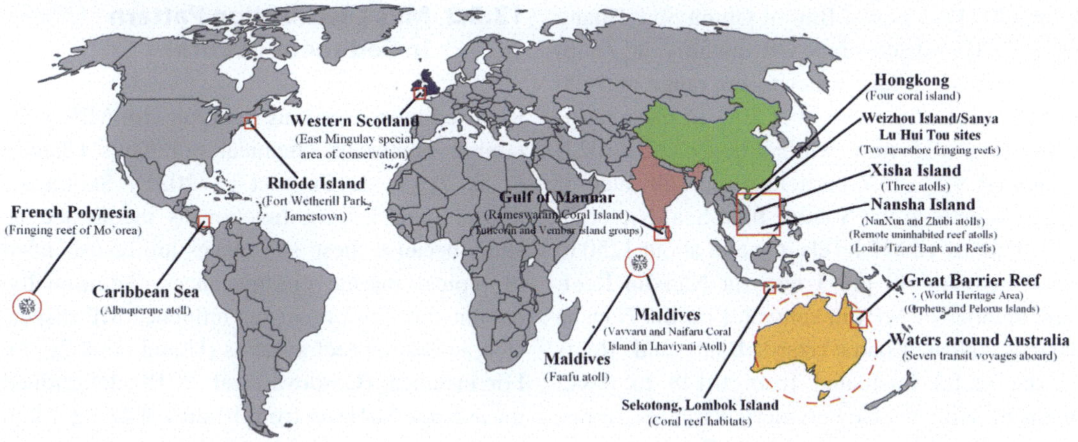

Fig. 13.1 The coral ecosystem distribution pattern across the world. (Adapted from Huang et al. 2021 with Elsevier license number 5877960906207)

Mondal et al. 2023; Rajpal et al. 2024). In addition, maritime trash is also linked with sea-based foundations such as port areas during shipment and fishing industries. Among the wide varieties of marine waste (as glass from different sources, such as the bottle and beverage industry and timber industry; aluminum containers being the most commonly found metal item, etc.), plastic materials, more specifically MPa, are the dominant type (Strafella et al. 2019; Peng et al. 2020). Among the various types of marine debris, plastic bottle caps and plastic bottles are among the most commonly gathered objects, followed by plastic grocery bags, food wrappers, plastic straws, glass beverage bottles, takeaway containers, and plastic carry bags (Ncube et al. 2021). Every year, the seabed contains a minimum 14 million tons of plastics (Kosior and Crescenzi 2020). According to data, plastic/MPs presumably will exceed every fish in seafloor by 2050. The seafloor contains between 50 and 75 trillion plastic pieces and MPs (Kurniawan et al. 2024a).

13.3 Occurrence and Distribution Pattern of MPs in Coral Reef

The MPs distribution and prevalence in coral environments, particularly the seawater, sediments, and corals that make up the reef system have been studied recently. Scientists have mentioned that MPs' prevalence in reef surface water varies between zero and tens of thousands of items/m^3 (Huang et al. 2021). However, the MP quantification in sediments and corals is a challenging issue. Further, the MPs discovered on reef systems are predominantly made up of fibers, flakes, pieces, films, and granules (like PE, PP-PE, PS, PP, PU, PET, UA, PA, and PVC) and appear as diverse dimensions and shadings. These components were mainly manufactured by garbage emissions from coastal communities and booming aquaculture practices (Rahman et al. 2023).

13.3.1 MPs Distribution Pattern in Surrounding Coral Surface Water

Generally, coastal reef water near densely populated regions exhibits higher MP levels than distant reef regions (Huang et al. 2021). For instance, Tan et al. (2020) acknowledged ~6×10^{-2} MP m^{-2} in nearby Nansha Islands reefs in South China Sea and Connors (2017) documented MPs quantities in Mo′orea's water 0.74 MP m^{-2}. The MP value in saltwater was determined to vary between 15.50 ± 1.50 and 22.14 ± 0.90 pieces/L, having a mean of 18.37 ± 2.60 pieces/L by Lei et al. (2021) in Sanya Bay of China Sea. Huang et al. (2019) discovered MP quantity in Zhubi Reef's ocean surface water in the range of 1400 to 8100 pieces/m^3, with over 80% of these particles being <0.5 mm, while Zhou et al. (2022) identified 9.5 ± 3.7 particles/L in saltwater of Xisha Islands in the South China Sea. Nie et al. (2019) assessed MP levels in water about 1250 to 3200 items/m^3(48 μm-5 mm) at Nanxun Reef. The number of MPs in saltwater on the Nansha Islands' Near Loaita/Tizard Bank and Reefs (seven reefs) fluctuated from 0.148 to 0.842 items/m^3 with diverse polymeric varieties, forms, and shades (Wang et al. 2019). Sabdono et al. (2022) documented MPs on reefs shoreline waters of Jepara, Java Sea, Central Java, Indonesia of four distinct kinds: sub-massive (*Galaxea fascicularis*: 9.50 ± 3.3 items kg^{-1}), enormous (*Porites lutea*: 9.75 ± 6.6), folious (*Montipora foliosa*: 11.50 ± 4.5) and grouping (*Acropora aspera*: 17.75 ± 2.3). The noteworthy majority of shades are blues and brown and originated from garbage from textiles, as well as fishing-related events. On the other hand, the Great Barrier Reef (GBR) water contained plastic fragments (0.4–82.6 mm, primarily MPs <5 mm) having a mean level of 4256.46 ± 757.79 pieces/km^2 (Reisser et al. 2013). Further, 547 intricate microfibers-trash (size 0.355–5 mm and 0.04–0.48 items/m^3) were obtained by (Jensen et al. 2019) during surface tows at 11 onshore (range 0–18 fsh$^{-1,}$ median = 4.5 fsh^{-1}) and 11 offshore (range 0–131 fsh^{-1}, median = 4.0 fsh^{-1}) corals of Australia's Central Great Barrier Reef (GBR), a World Heritage Area (WHA). The particles were composed of naturally derived (n = 25%), semi-synthetic (n = 39%), and synthetic (n = 36%) materials. Black, blue, white, and red colored polymers were identified as having the greatest incidence of PS. Moreover, in GBR surrounding the Pelorus and Orpheus Islands, Hall et al. (2015) observed up to two MPs (100–500 μm) per 11 m^3 of saltwater. Martí et al. (2017) considerably recorded smaller MP levels in water ($3.546 \times 10^{-3} \pm 8.154 \times 10^{-3}$MP m^{-2}) of Red Sea, presumably due to the minimal contribution of MP contamination from inland (Table 13.1).

13.3.2 MPs Distribution Pattern in Sediment of Coral

Reef sediments serve as reservoirs for MPs, containing numerous chemical pollutants (Sandre et al. 2019; de Smit et al. 2021). Sediments deeper than 3.5 cm function as inconspicuous sinks because these substances are resuspended in typical marine circumstances. Additionally, habitat-forming organisms influence MP dispersion across the reef systems (Utami et al. 2021). For instance, (Cordova et al. 2018) determined the average MP level from 10 sites of 48.3 ± 13.98 items/kg in Sekotong, Lombok Island, Indonesia coral sediments. According to their findings, PS was the predominant MP type, and the predominant morphologies were granulate, foaming, and fragments. (Cheang et al. 2018) Recorded mean MP value (0.3–5 mm) varied between 171.7 ± 57.6 and 223 ± 51.4 pieces/kg in reef sedimentary beds of Hong Kong, with PE and PET representing the most predominant polymer kinds, being followed by PP, PS, and PVC. The MP value in rock sediments was 280.9 ± 231.9 particles kg^{-1}by (Lei et al. 2021) in Sanya Bay of China Sea. Further, (Zhang et al. 2019) investigated MPs (0.3–5 mm) levels in two atolls located on Nansha and Xisha Islands, as well as two shoreline fringe reefs situated on Sanya Lu HuiTou and Weizhou Island sites in South China Sea, it was 40 ± 4 to 100 ± 2, 60 ± 3 to 610 ± 11, 50 ± 3 to 530 ± 7 and 60 ± 2 to 90 ± 5 pieces/kg, individually. (Portz et al. 2020) detected 0.5 items/m^2 (>80 μm) in the atoll inner lagoon saltwater and 90 MP items/m^2 (1–5 mm) on sandy beaches (primarily white in shade) of Albuquerque Atoll in the Caribbean Sea. Arreola-Alarcón et al. (2022) reported 680.25 items/100 and 321.75 pieces/100 g^{-1}d.w. in sandy sediments of CaboPulmo and Espiritu Santo Island, respectively. They recognized that, with almost every one of them being white, the most significantly common polymers are PP (70%), succeeded by PET (65%), HDPE (59%), and LDPE (50%).

(Imhof et al. 2017) The first identified MPs were in the sand sediment of an isolated Vavvaru Island of Maldives'sLhaviyani Atoll in the Indian Ocean. Subsequently, (Patti et al. 2020) assessed

Table 13.1 Levels and distribution of MPs in water of surrounding reef system in maritime ecosystem

Regions under investigation	Media	Methods of analysis	Color	Source	Reference
Coral Island, Naifaru, Lhaviyani atoll (Indian Ocean)	Beach (6 sites); reef flat (8 sites); and fore reef (8 sites) all contain sand or sediment	Microscope inspection, extraction, and in situ sampling	Mostly blue, with red, gray, orange, and black coming in order of preference	Coastal nations or surrounding islands; improper handling of local plastic garbage	Patti et al. (2020)
The Caribbean Sea's Albuquerque Atoll (Atlantic)	Sediment/sand; seawater	Trawl sampling, in situ sample, extraction, and microscope examination	Brown/orange, white, green, gray, blue	Fishing operations and marine resources	Portz et al. (2020)
The Indian Ocean's Gulf of Mannar (Tuticorin and Vembar)	Seawater (five locations)	FTIR, extraction, microscope inspection	Primarily transparent, blue, red, and black	Local water flow	Patterson et al. (2020)
17 reef atolls, South China Sea's Nansha Islands	24 samples of seawater	FTIR spectroscopy, extraction, microscope inspection	Clear, green, blue, yellow, black, red, and white	Residential islands nearby; fishing operations	Tan et al. (2020)
Nanxun Reef, South China Sea's	15 water samples	Raman spectroscopy, microscope examination	Blue, white, transparent, and other	Fishing operations; human waste from homes	Nie et al. (2019)
Zhubi Reef, South China Sea (Nansha Islands)	30 water samples	Microscope inspection	Green, transparent, pink, blue, and other colors	Fishing and pollution from coastal cities	Huang et al. (2019)
Central Great Barrier Reef (Australia)	22 samples of seawater	FTIR spectroscopy, microscope inspection	Mostly red, white, blue, and black	Land-based emissions of sewage effluent	Jensen et al. (2019)

MP levels in sand beds surrounding the NaifaruIsland of Maldives'sLhaviyani Atoll, with a mean level of 277.90 ± 24.98 items/kg. Further, (Saliu et al. 2021) recorded 0.32 ± 0.15 items/m^3 MPs in saltwater, whereas in beachfront sediments, the predominant polymeric forms were PP, PE, PS, PVC, PET, and PA, with an average of 22.8 ± 10.5 items/m^2. Many white, irregularly shaped plastic fragments were observed in sediments taken from Rameswaram Island in the Gulf of Mannar (GoM), India (Vidyasakar et al. 2018). A thorough investigation using saltwater and sediments in 98 sampling spots in GoMwas also undertaken by (Patterson et al. 2022). Researchers figured out that the water embodied 28.4–126.6 pieces L^{-1}, and the silt included 31.4–137.6 pieces kg^{-1}, with a predominance of PE and PP. An additional investigation of saltwater pollution from five different spots at Vembar and Tuticorin Island revealed MP affluences fluctuated between 60,000 to 126,600 items/m^3, with predominance of both translucent and blue-colored PE and PP (Patterson et al. 2022) (Table 13.2).

13.3.3 MPs Distribution Pattern in Coral

MPs ingesting has been described in broad spectrum reef biota, encompassing phylogenetic categories and trophic stages in this section. A fascinating investigation was conducted by (Ding et al. 2019; Samanta et al. 2022) on MPs prevalence in fish, coral, and saltwater specimens in 3 atolls of South China Sea's Xisha Islands. In the

Table 13.2 Summarizes quantity and distribution of MPs in coral sediment of reef system in maritime ecosystem

Regions	Biological or environmental samples	Media	Method of MP analysis	Color	Source of microplastics	References
Magoodhoo Island, Maldives, Faafu Atoll (Indian Ocean)	*Poriteslutea, Pocilloporaverrucosa,* and *Pavonavarians*	38 locations with seawater	μFT-IR analysis conducted in situ	Primarily pink and brown with traces of gray, white, black, and blue	On-site burning of waste	Raguso et al. (2022)
Maldives, Indian Ocean	*Haliclona* sp.	Coral from four locations	μFT-IR analysis conducted in situ	Brown (27%), black (5%), and gray (68%)	Biofouling, photo-oxidative deterioration, and nearby sewage pipe	Saliu et al. (2022)
The Gulf of Mannar (GoM)	Sediments	Sediment and seawater at 98 sample locations	ATR-FTIR is used, together with typical microscopy	The primary file has no indication of color	A high people density and severe problems with waste management on the mainland side	Patterson et al. (2022)
The Xisha Islands, South China Sea (SCS)	*Galaxeafascicularis, Plectroporadamicularis*, and *Porites lutea*	Five locations	μFT-IR	Transparency, blue, red, green, green, and other	Shipping, travel, residential sewage leaks, and both commercial and recreational fishing	Zhou et al. (2022)
Mexico's CaboPulmo and Espiritu Santo Island, Baja California Sur	Sediments and sand	Sand and sediment (eight locations)	FTIR spectroscopic principles	Black, red, green, pink, white, blue, and brown	Human-caused and tourism-related activities	Arreola-Alarcón et al. (2022)
Java Sea, Central Java, Jepara coastal waters, Indonesia	Massive, obedient, branching, foliated corals	In four sampling sites, sediment	Fluorescence microscopy analysis	Black dominant, and additional colors	Release of textiles or fishing activities	Sabdono et al. (2022)
China's Sanya Bay	*Galaxies fascicularis* and *Acroporamillepora*	32 water and coral samples (four sites)	Examined under stereomicroscope, and μ-Raman spectrometer	Primarily transparent and black, with traces of red, blue, and green	Extensive human activity, sewage discharge, and tourism growth	Lei et al. (2021)
Caribbean Sea, Albuquerque Atoll	Sediment and water	Sea water and sediment	Microscopic identification	Others are green, gray, and blue, along with brown and white	Seashores and aquatic resources	Portz et al. (2020)
Naifaru, Central Indian Ocean; Maldives	Sediments	Sediments from 22 locations of sample	Viewing under a stereomicroscope	Primarily blue, with red, gray, orange, and black	Poorly managed plastic trash, nearby inhabited islands, tourism	Patti et al. (2020)
Java and the Indonesian island of Lesser Sunda	*Rhincodon typus*	Fish and seawater in three locations	Identification through microscopic examination	No specific types	Regional waterways and human density	Germanov et al. (2019)

outermost reef slopes, reef flats, and lagoons, the MP proportions (7–4856 μm) appeared 200–11,200, 1000–12,200, and 1000–45,200 items/m^3, respectively. Researchers obtained MPs (24–4729 μm) in reefs with an average of 1.0–44.0 pieces/coral (0.02–1.3 pieces/g). Further, much investigation has already been done on MP distribution patterns in South China Sea. The MPs abundance (mean 189 ± 50 items/kg) in rocky coral species, including *P. lutea* and *Acropora* spp., was investigated by (Cheang et al. 2018) in sedimentary samples collected from 4 separate locations. The MPs spotted were constituted primarily of PE and PET, accounting for 51.9 and 29.3% of the overall MPs identified. Three scleractinian reef species (*P. lutea, G.fascicularis* and *Pocillopora damicornis*) from Xisha Islands were also studied by (Zhou et al. 2022) for investigating MPs distribution found MPs 2.5 ± 1.6, 1.2 ± 0.6 and 0.9 ± 0.5 particles/cm^2, respectively; representing CP (61.13%) and PET (33.49%), with colors ranging from green to red, transparent to black. (Tang et al. 2021), at the same time, the opulence of MPs in rocky coral organisms, notably *G. fascicularis* and *P. damicornis*, on Hainan Island was investigated. *P. damicornis* and *G. fascicularis* were reported to have 3.68 ± 3.94 and 5.89 ± 5.15 pieces/cm^3, respectively, with a higher prevalence of translucent and black MPs followed by CP and PET MPs. In another study, per polyp, *A. millepora*and *G. fascicularis* contained 0.27 ± 0.26 and 2.32 ± 0.86 MP fragments, respectively, while *P. damicornis* and *P. lutea* contained 0.9 ± 0.5 and 2.5 ± 1.6 items cm^{-2} in Sanya Bay of the China Sea, have been reported by Lei et al. (2021). Many MPs had been determined to be CP and PET, with clear blue, black, green, and red comprising the most popular hues.

In another intriguing study, Rotjan et al. (2019) recorded about 112 MP items/per polyp in four wild-captured coral populations (*Astrangia poculata*) in Fort Wetherill from Rhode Island, with synthetic cellulose-based fibers, PA and PEST being one of the most abundant fiber types. In corals, including *Pavonavarians*, *P. lutea,* and *P. verrucosa* in Faafu Atoll of Magoodhoo Island in Maldives (Raguso et al. 2022) acquired 8.9 pieces/g of brown and pink-colored MPs. In a different investigation, (Saliu et al. 2018) revealed that the MP (PS and PE polymers) quantity in Haliclona (Haplosclerida) tissues in the Indian Ocean was 1.2 particles/g. The principal sources of MPs (PE, PP, PA, PS, and PU polymers) arriving from ground-based maneuvers were detected in neuston and scleractinian coral reefs, where (Saliu et al. 2019) identified 0.12 ± 0.09 (0.03–0.65)pieces/m^3. These investigations disclosed that the manufacturing sectors that generate such microplastics in the maritime environment are the lucrative and recreational transportation, fishing, and tourism sectors. The discharging of human droppings via the riverine system, industrial wastewater, fishing, traveling, and leisurely pursuits are all designated the cradles of MP in coastal areas (Huang et al. 2021; Rahman et al. 2023) (Table 13.3).

13.4 Detrimental Effects of MPs Pollution on Coral Reefs Ecosystem

Recent studies have highlighted adverse impacts of marine plastic/MPs on coral reefs. By 2025, at least 155 million tons of MPs will have accumulated in the world's seafloor. This pollution severely threatens coral reefs, leading to physical damage, reduced resilience, and increased mortality rates among coral species (Mentzel et al. 2024).

13.4.1 Microplastic Ingestion and Its Consequences

Corals are non-selective feeders, meaning they ingest a wide variety of tiny organisms, including zooplankton. However, due to their similar size, corals can inadvertently ingest MPs from residing water. Once ingested, these MPs can cause severe tissue damage, impair the coral's immune system, diminish food intake, and hinder growth (Tang et al. 2018; Mouchi et al. 2019; Reichert et al. 2019). Additionally, MPs may permanently damage coral cells, rendering it more challenging

Table 13.3 Levels and distribution of MPs in corals of reef system in maritime ecosystem

Location	Coral species	MP concentration (particles/g)	Types of MPs found	Sources of MPs
Great Barrier Reef, Australia	*Acropora* spp.	0.2–0.6	Polyethylene, polypropylene	Urban runoff, marine debris
Caribbean Sea, Puerto Rico	*Montastraea cavernosa*	1.0–2.5	Polystyrene, nylon	Tourism, fishing activities
Red Sea, Egypt	*Pocillopora damicornis*	0.4–1.2	Polyethylene, polyester	Shipping lanes, coastal development
South China Sea, Malaysia	*Porites lutea*	0.5–1.5	Polyvinyl chloride, acrylic	River discharge, aquaculture
Hawaiian islands, USA	*Porites compressa*	0.3–0.8	Polypropylene, polyester	Marine debris, urban runoff

for the reefs to ingest and digest and making them susceptible to disease, thermal stress, behavioral changes, and even death (Santana et al. 2021). Over time, the MP accumulation can lead to necrosis (tissue death) and bleaching, a procedure whereby corals discharge the mutually beneficial algae that are housed in their muscle tissue, leading to discoloration and vitality (Hall et al. 2015; Reichert et al. 2018; Corona et al. 2020). Bleached corals are more vulnerable to disease and cannot recover from environmental stress (Mouchi et al. 2019; Reichert et al. 2019). Additionally, MP ingestion by corals affects the corals themselves and the broader reef ecosystem. However, MPs' consequences upon ingestion in coral species depend on factors like retention time, organism size, MP size, and their concentration in the surrounding water (Rahman et al. 2023). For example, smaller organisms may retain MPs longer, increasing potential adverse health effects (Rahman et al. 2023).

13.4.2 MPs-Associated Consequences of Toxic Chemicals

MPs in marine environments carry toxic chemicals that can further exacerbate the health risks to coral reefs. These contaminants include persistent organic pollutants (POPs), namely organochlorine pesticides, polychlorinated biphenyls (PCBs), and polycyclic aromatic hydrocarbons (PAHs), including heavy metals like cadmium, lead, and mercury. These chemicals can induce a range of toxic consequences, including diseases that weaken coral resilience and contribute to coral mortality (Wang et al. 2018; Syakti et al. 2019; Godoy et al. 2019; Lanctôt et al. 2020). The intricate equilibrium between MPs and nutrients in coral ecosystems could be upset by MPs, worsening coral development, thereby rendering them more susceptible to ecological stresses such as ocean acidification and climate instability (Patterson et al. 2020; Rahman et al. 2023). Additionally, biota that swallow corals or algae, like seafloor worms and certain fish species, are also vulnerable to these toxic effects, leading to bioaccumulation and biomagnification of harmful substances through food chain (Rakib et al. 2023).

13.4.3 Occurrence of Disease in Coral Reefs

Maritime plastic trash, acting as a vehicle for non-native species, rapidly introduces them to new environments where they can outcompete or disrupt native species, thereby escalating the health risks to reefs. This influx of invasive species stresses coral ecosystems, which are already grappling with pollution, climate change, and other human-induced pressures (Rahman et al. 2023). Promoting microbial growth and biofilm formation by plastic debris, creating a 'Plastisphere' for colonization of dangerous microbial vectors, is a challenge that requires collective action and collaboration (Goldstein et al. 2014; Amaral-Zettler et al. 2020). The chemical makeup and attributes of MP surface biofilms were also examined by (Feng et al. 2020) on nine different types of MPs (including PS, PP, PC, PA, LDPE, PVC, PET, EPS, and ABS) exposure at four reef-associated locations for two weeks. The 16S rRNA genome sequencing, obtained from colonizing testing, demonstrated that MPs types and reef locations substantially impacted makeup and biofilm arrangement on MPs surface. Remarkably, the most prevalent infectious agents in MPs biofilm are *Pseudomonas*, *Vibrionaceae*, *Bbellvibriocholerae*, *Flavobacteraceae*, Vibrio, and *Rhodobacteraceae*. These diseases can cause tissue injury and coral albinism, respectively. Further, Rotjan et al. (2019) explored PE beads ingestion (170.5–230.8 µm) containing *Escherichia coli* biofilm by a temperate coral species, *A. poculata*, and they discovered the presence of larger quantities of *E. coli* inside a fortnight. Conversely, research revealed that the scleractinian coral *P. damicornis's* immunity and detoxing mechanism were diminished by brief exposition to 1.0 µm PS MPs (9 × 1010 piecess/L) (Tang et al. 2018), which might boost coral vulnerability to infection. These few investigations revealed the non-negligible risk posed by pathogenic organisms that cause infection and handed over a fresh line of inquiry into the effects of "plastisphere" populations on various coral species.

13.4.4 MPs Impact on Coral-Zooxanthellae Symbiosis

Coral-zooxanthellae coexistence is the backbone of productive reef ecosystems because reefs-building genera of coral acquired their nutrients and energy from this endosymbiotic association with photosynthetic Symbiodiniacea (Rahman et al. 2023). The earliest proof showed that ingesting carboxylate microspheres (3, 6, 11 µm) and daily-use facewash-originated beads (3–60 µm) could significantly prevent the harmonious Symbiodiniacea from attacking their host (like sea anemone, *Aiptasia* sp. or coral, *Faviteschinensis*) and disrupt the harmonious interaction's starting (Okubo et al. 2018). Subsequently, they discovered that PS MPs (3 and 6 µm) could aggregate in the phagocytosis zone and be endocytosed by mesenterial filament cells (Okubo et al. 2020). Additionally, the endosymbiotic Symbiodiniaceae species' development was inhibited by exposition to 1 µm PS MPs (9.0 × 109 pieces/L), which impaired photosynthesis and detoxifying activities, exacerbated oxidative stresses, and disrupted apoptosis and metabolic processes (Su et al. 2020). In another investigation by Syakti et al. (2019), staghorn coral *A. formosa* ingesting LDPE MPs (<100, 100–200, and 200–500 µm) triggered the discharge of zooxanthellae and concomitant discoloration and death. Substantial effect in the scleractinian coral, *Stylophorapistillata,* upon their photosynthesis efficiency and a substantial transformation within their metabolism profiles following exposure to PE MPs (106–125 µm, 50,000 pieces/L) for 4 weeks has been described by Lanctôt et al. (2020). Comparatively, Reichert et al. (2019) demonstrated that four distinct types of scleractinian coral displayed species-specific consequences (i.e., improvement or absence of variation) on photosynthetic efficiency after six months of HDPE MPs exposure (65–410 µm, 200 pieces/L). However, symbiont and chlorophyll levels remained unchanged. Tang et al. (2018) reported that *P. damicornis* experienced adverse cytological, physiologic, and molecular reactions as well as an upsurge in symbiont concentration and levels of chlorophyll under 24-hour intoxication to 1 µm PS MPs (9 × 1010 pieces/L). Furthermore, Rocha et al. (2020) discovered that 96 h 10 mg/L PVC MPs (63–125 µm) exposure enhanced the photosynthetic rate without affecting energetic expenditures in the Hexacorallia coral, *Zoanthussociatus*. Considering the preceding, additional study must be conducted to determine whether MP exposure over time exerts a substantial adverse effect on the relationship between coral and zooxanthellae.

13.4.5 Impacts on Coral Reef Ecosystem

The adaptability and lifespan of reef environments are vulnerable to an extensive spectrum of possible stressors. For example, according to (Yang et al. 2020), the development of primary producer microalgae and the harmful effect of MPs are potentially affected by the combined impact of PS MPs (100 nm) and environmental determinants (namely CO_2, temperature, and light irradiance) that are altered by climate change. Higher water temperatures (30°C) did not influence significantly MPs ingested by scleractinian corals (*M. capitate* and *P.damicornis*). However, according to (Axworthy and Padilla-Gamiño 2019), they did minimize Artemia nauplii's heterotrophic eating. For MP to be fully acknowledged as a growing global contaminant, scientists need to look into the potential interface between ecological pressures linked to climate change and coral environment's MPs.

13.4.6 Impacts on Ecosystem Health and Human Health Risks

Plastic pollution's degradation of coral reefs has far-reaching consequences for ecosystem services and human well-being. As corals decline, the diversity they support is also endangered, leading to diversity loss and impaired ecosystem services they provide. This loss can impact fisheries, reduce coastal protection, and diminish the tourism appeal of coral-rich regions. Moreover, the health risks associated with MPs-induced coral degradation are another critical issue (Choudhury et al. 2022). Humans who consume seafood intoxicated with MPs are potentially exposed to toxic chemicals and pathogens, leading to public health concerns. Additionally, the bioaccumulation of harmful substances in the food chain can result in long-term health effects, including endocrine disruption, cancer, and other severe conditions. Accordingly, addressing plastic pollution is essential to preserving the integrity of coral reefs and ensuring the sustainability of the marine ecosystems they support.

13.5 Mechanism of Coral Toxicity

Till now, a conclusive mechanism of MP toxicity to reef systems has yet to be established. It is generally presumed that a confluence of chemical, biological, and physical processes generally causes MP toxicity/consequences in coral reefs. MPs can cause digestive obstructions, starvation, and physiological harm to tissue in corals at the physical level at the initial level. This is because MPs block light, which ultimately results in the reduction of symbiotic algae's (zooxanthellae) photosynthesis capability. This weakens coral and bleaches coral (Islam et al. 2022; Rahman et al. 2023).

At the chemical and biological level, MPs can leach harmful additives and absorb pollutants, such as POPs, heavy metals, and pathogenic organisms. These substances can accumulate in coral tissues, disrupting normal cellular processes and causing excessive oxidative damage and inflammatory responses. Moreover, MPs disrupt the delicate microbial ecosystems of corals, making them more susceptible to disease. This disruption of the coral's natural balance threatens reef ecological sustainability, hampers reproduction, and decreases coral health. The detailed mechanism of MP toxicity to the coral ecosystem is illustrated in (Fig. 13.2).

13.6 Mitigation Strategies of MPs Pollution in Coral Reefs

Plastic pollution is a pressing environmental issue that significantly threatens coral reefs, vital ecosystems supporting many marine lives. Plastic, especially MPs and NPs, has detrimental effects on coral health, leading to physical damage, disease, and disruption of ecological balance. To safeguard coral reefs for future generations, it is essential to adopt strategies that reduce or mitigate the entry and impact of plastic particles in these fragile ecosystems. This section outlines potential measures to address plastic pollution in coral reefs, emphasizing the need for coordinated global efforts.

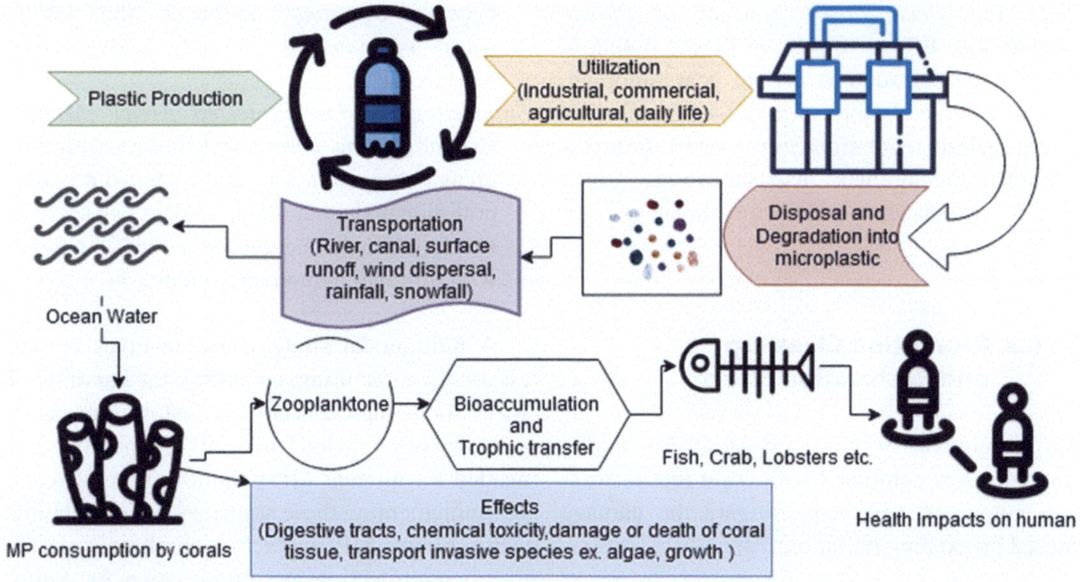

Fig. 13.2 Mechanism of MPs toxicity to coral ecosystem. (Adapted from Rahman et al. 2023 with Elsevier license number 5866351111644)

13.6.1 Reducing Plastic Production and Consumption

The most effective way to mitigate plastic pollution, particularly MPs pollution, is addressing the root cause, i.e., the excessive production and plastics utilization. Strategies to achieve this include:

Promoting Sustainable Alternatives: Encouraging the use of biodegradable or reusable materials instead of single-use plastics can reduce significantly the amount of plastic waste entering the environment. For example, replacing plastic packaging with materials like paper, glass, or bioplastics can help decrease plastic pollution.

Implementing Extended Producer Responsibility (EPR): EPR initiatives hold producers responsible for the complete lifecycle of their goods, including disposal after the customer. This may encourage manufacturers to create goods with reduced trash production or easier recycling.

Encouraging Consumer Behavior Change: Public awareness campaigns and education programs can help shift consumer behavior toward reducing plastic use, recycling, and proper waste disposal. Simple actions like carrying reusable bags, bottles, and containers can collectively make a significant impact.

13.6.2 Improving Waste Management and Recycling

Effective waste management is crucial in preventing plastic, particularly MP pollution, from entering marine environments. Key measures include:

Enhancing Waste Collection and Recycling Infrastructure: Improving waste collection systems, particularly in coastal areas, can prevent plastics from reaching the ocean. Expanding recycling facilities and making recycling more accessible can also help reduce plastic waste.

Implementing Waste-to-Energy Technologies: Converting non-recyclable plastics into energy through pyrolysis or gasification can minimize waste volume in landfills or oceans.

Preventing Ocean-Based Sources of Plastic Pollution: Efforts to prevent plastic pollution from sources such as fishing gear, shipping, and offshore platforms are also essential. This can include regulations to manage lost or discarded fishing nets, also known as "ghost nets," which are a significant threat to coral reefs.

13.6.3 Promoting Clean-up and Restoration Efforts

Active removal of plastic waste from marine environments, combined with coral reef restoration initiatives, can help mitigate the damage caused by existing pollution:

Organizing Coastal and Underwater Clean-Up Campaigns: Regular clean-up efforts in coastal areas and coral reefs can help to remove accumulated plastic waste. Engaging local communities, divers, and conservation organizations in these efforts can foster a sense of stewardship for marine environments.

Supporting Coral Reef Restoration Projects: Restoring damaged reefs through coral transplantation, artificial reefs, or other restoration techniques can help to rebuild resilient ecosystems that are better equipped to withstand environmental stressors, including plastic pollution.

13.6.4 Strengthening Policies and International Cooperation

Addressing plastic pollution, in particularly MP pollution on a global scale, requires coordinated efforts and strong policies:

Implementing International Agreements and Regulations: International treaties, such as the Basel Convention and the upcoming global treaty on plastic pollution, can set legally binding standards for managing plastic waste. These agreements can facilitate cooperation between countries and ensure that plastic waste is managed correctly and reduced worldwide.

Supporting Marine Protected Areas (MPAs): Expanding and enforcing MPAs can provide greater protection for coral reefs from plastic pollution and other threats. MPAs can serve as safe havens where coral ecosystems can thrive with minimal human interference.

A multimodal strategy that involves cutting plastic manufacturing, enhancing the handling of waste, boosting clean-up and restoring projects, and bolstering international cooperation is needed to mitigate MP pollution in coral reefs. By implementing these strategies, we can significantly reduce MPs impact on coral reefs, ensuring their protection and preservation for future generations. Additionally, global efforts are essential to address this environmental challenge and to safeguard the health of coral reefs, which are crucial to the biodiversity and sustainability of our oceans.

13.7 Future Concerns and Recommendations

Currently, MP/NP pollution in marine coral ecosystems is an urgent and very serious issue, but there is a lack of effective monitoring and proper treatment procedures. Given the significance of coral reefs to marine environments, it is pertinent to stress human health can be adversely impacted by MPs and NPs in marine environments. It is evident from the evidence at hand that lab-to-land approaches are necessary to disseminate scientific knowledge to society. Apart from that, many issues remain unsolved, which necessitates further studies and are now addressed as future recommendations as follows: (1) long-term surveillance measures should be adopted to investigate MPs level in the reef system as it is outlooked; (2) investigations should be carried out to delineate the details adverse outcome pathways (AOPs) of MPs on reef system; (3) size-dependent MPs effects on reef system is still unexplored one which should be carried out at utmost concern; (4) further investigation must be conducted on MPs consequences on coral reef, more specifically health consequences; (5) development of standard estimation/identification procedure should be improved for accurate estimation of MPs; and finally (6) public awareness and regulatory policy framework and their subsequent implementation should be improved to minimize the MPs contamination in the reef system, more specifically maritime ecosystem.

13.8 Conclusion

The contamination caused by MPs and the mechanism by which they become worse and harm coral reefs were highlighted in the present paper. The prevalent MP sources are also outlined in the current study. The distribution and prevalence nature of MPs in the surrounding coral environment's niche, namely water, sand/sediment, and the coral itself, has been demonstrated. Additionally, this paper outlined the MP toxicity mechanism in reef systems. It highlights MPs' dispersion, destiny, ecological implications, and health issues in important reef ecosystems. This paper also offers suggestions for reducing the MPs' contamination in terms of mitigation techniques and identifies areas of unmet research needs.

References

Amaral-Zettler LA, Zettler ER, Mincer TJ (2020) Ecology of the plastisphere. Nat Rev Microbiol 18:139–151. https://doi.org/10.1038/s41579-019-0308-0

Arreola-Alarcón IM, Reyes-Bonilla H, Sakthi JS et al (2022) Seasonal tendencies of microplastics around coral reefs in selected marine protected National Parks of Gulf of California, Mexico. Marine Pollut Bull 175:113333. https://doi.org/10.1016/j.marpolbul.2022.113333

Axworthy JB, Padilla-Gamiño JL (2019) Microplastics ingestion and heterotrophy in thermally stressed corals. Sci Rep 9:18193. https://doi.org/10.1038/s41598-019-54698-7

Bian W, Zeng Y, Li Y et al (2024) Microplastic pollution in tropical coral reef ecosystems from the coastal South China Sea and their impacts on corals in situ. J Hazard Mater 480:135898. https://doi.org/10.1016/j.jhazmat.2024.135898

Bouwmeester H, Hollman PCH, Peters RJB (2015) Potential health impact of environmentally released micro- and nanoplastics in the human food production chain: experiences from Nanotoxicology. Environ Sci Technol 49:8932–8947. https://doi.org/10.1021/acs.est.5b01090

Cheang CC, Ma Y, Fok L (2018) Occurrence and composition of microplastics in the seabed sediments of the coral communities in proximity of a metropolitan area. IJERPH 15:2270. https://doi.org/10.3390/ijerph15102270

Chen X, Chen X, Liu Q et al (2021) Used disposable face masks are significant sources of microplastics to environment. Environ Pollut 285:117485. https://doi.org/10.1016/j.envpol.2021.117485

Choudhury M, Sharma A, Pervez A, Upadhyay P, Dutta J (2022) Growing menace of microplastics in and around the coastal ecosystem. In: Coastal ecosystems: environmental importance, current challenges and conservation measures, pp 117–137. https://doi.org/10.1007/978-3-030-84255-0_6

Connors EJ (2017) Distribution and biological implications of plastic pollution on the fringing reef of Mo'orea, French Polynesia. PeerJ 5:e3733. https://doi.org/10.7717/peerj.3733

Cordova MR, Hadi TA, Prayudha B (2018) Occurrence and abundance of microplastics in coral reef sediment: a case study in Sekotong, Lombok-Indonesia. https://doi.org/10.5281/ZENODO.1297719

Corona E, Martin C, Marasco R, Duarte CM (2020) Passive and active removal of marine microplastics by a mushroom coral (Danafungia scruposa). Front Mar Sci 7:128. https://doi.org/10.3389/fmars.2020.00128

Crabbe MJC, Barker GR, Wong L, Zhou J (2024) The fragility of the ocean: from coral reef protection to deep-sea mining. In: Monaco E, Abe M (eds) Sustainable development across Pacific Islands. Springer Nature Singapore, Singapore, pp 187–212

de Smit JC, Anton A, Martin C et al (2021) Habitat-forming species trap microplastics into coastal sediment sinks. Sci Total Environ 772:145520. https://doi.org/10.1016/j.scitotenv.2021.145520

Ding J, Jiang F, Li J et al (2019) Microplastics in the coral reef systems from Xisha Islands of South China Sea. Environ Sci Technol 53:8036–8046. https://doi.org/10.1021/acs.est.9b01452

Feng L, He L, Jiang S et al (2020) Investigating the composition and distribution of microplastics surface biofilms in coral areas. Chemosphere 252:126565. https://doi.org/10.1016/j.chemosphere.2020.126565

Germanov ES, Marshall AD, Hendrawan IG et al (2019) Microplastics on the menu: plastics pollute Indonesian manta ray and Whale shark feeding grounds. Front Mar Sci 6:679. https://doi.org/10.3389/fmars.2019.00679

Godoy V, Blázquez G, Calero M et al (2019) The potential of microplastics as carriers of metals. Environ Pollut 255:113363. https://doi.org/10.1016/j.envpol.2019.113363

Golden CD, Hartmann AC, Gibbons E et al (2024) HIARA study protocol: impacts of artificial coral reef development on fisheries, human livelihoods and health in southwestern Madagascar. Front Public Health 12:1366110. https://doi.org/10.3389/fpubh.2024.1366110

Goldstein MC, Carson HS, Eriksen M (2014) Relationship of diversity and habitat area in North Pacific plastic-associated rafting communities. Mar Biol 161:1441–1453. https://doi.org/10.1007/s00227-014-2432-8

Hall NM, Berry KLE, Rintoul L, Hoogenboom MO (2015) Microplastic ingestion by scleractinian corals. Mar Biol 162:725–732. https://doi.org/10.1007/s00227-015-2619-7

Hernandez LM, Xu EG, Larsson HCE et al (2019) Plastic teabags release billions of microparticles and nanoparticles into tea. Environ Sci Technol 53:12300–12310. https://doi.org/10.1021/acs.est.9b02540

Huang Y, Yan M, Xu K et al (2019) Distribution characteristics of microplastics in Zhubi Reef from South China Sea. Environ Pollut 255:113133. https://doi.org/10.1016/j.envpol.2019.113133

Huang W, Chen M, Song B et al (2021) Microplastics in the coral reefs and their potential impacts on corals: a mini-review. Sci Total Environ 762:143112. https://doi.org/10.1016/j.scitotenv.2020.143112

Imhof HK, Sigl R, Brauer E et al (2017) Spatial and temporal variation of macro-, meso- and microplastic abundance on a remote coral Island of the Maldives, Indian Ocean. Mar Pollut Bull 116:340–347. https://doi.org/10.1016/j.marpolbul.2017.01.010

Islam MS, Phoungthong K, Islam ARMT et al (2022) Sources and management of marine litter pollution along the Bay of Bengal coast of Bangladesh. Mar Pollut Bull 185:114362. https://doi.org/10.1016/j.marpolbul.2022.114362

Jensen LH, Motti CA, Garm AL et al (2019) Sources, distribution and fate of microfibres on the Great Barrier Reef, Australia. Sci Rep 9:9021. https://doi.org/10.1038/s41598-019-45340-7

Johnson JE, Allain V, Basel B et al (2020) Impacts of climate change on marine resources in the Pacific Island region. In: Kumar L (ed) Climate change and impacts in the Pacific. Springer International Publishing, Cham, pp 359–402

Kosior E, Crescenzi I (2020) Solutions to the plastic waste problem on land and in the oceans. In: Plastic waste and recycling. Elsevier, pp 415–446

Kurniawan TA, Mohyuddin A, Othman MHD et al (2024a) Beyond surface: unveiling ecological and economic ramifications of microplastic pollution in the oceans. Water Environ Res 96:e11070. https://doi.org/10.1002/wer.11070

Kurniawan TA, Pasaribu B, Kusworo TD et al (2024b) Building disaster resilience in Thousand Islands (Indonesia): unlocking climate adaptation strategies to Navigate Sea level rise in coastal regions while safeguarding crop productivity and local biodiversity. ACS EST Water 4:3213–3224. https://doi.org/10.1021/acsestwater.4c00099

Lamb JB, Willis BL, Fiorenza EA et al (2018) Plastic waste associated with disease on coral reefs. Science 359:460–462. https://doi.org/10.1126/science.aar3320

Lanctôt CM, Bednarz VN, Melvin S et al (2020) Physiological stress response of the scleractinian coral Stylophora pistillata exposed to polyethylene microplastics. Environ Pollut 263:114559. https://doi.org/10.1016/j.envpol.2020.114559

Lei X, Cheng H, Luo Y et al (2021) Abundance and characteristics of microplastics in seawater and corals from reef region of Sanya Bay, China. Front Mar Sci 8:728745. https://doi.org/10.3389/fmars.2021.728745

Martí E, Martin C, Cózar A, Duarte CM (2017) Low abundance of plastic fragments in the surface waters of the Red Sea. Front Mar Sci 4:333. https://doi.org/10.3389/fmars.2017.00333

Mentzel S, Nathan R, Noyes P et al (2024) Evaluating the effects of climate change and chemical, physical, and biological stressors on nearshore coral reefs: a case study in the Great Barrier Reef, Australia. Integr Envir Assess Manag 20:401–418. https://doi.org/10.1002/ieam.4871

Mondal T, Choudhury M, Kundu D, Dutta D, Samanta P (2023) Landfill: an eclectic review on structure, reactions and remediation approach. Waste Manag 164:127–142. https://doi.org/10.1016/j.wasman.2023.03.034

Mouchi V, Chapron L, Peru E et al (2019) Long-term aquaria study suggests species-specific responses of two cold-water corals to macro-and microplastics exposure. Environ Pollut 253:322–329. https://doi.org/10.1016/j.envpol.2019.07.024

Munday PL, Jones GP, Pratchett MS, Williams AJ (2008) Climate change and the future for coral reef fishes. Fish Fish 9:261–285. https://doi.org/10.1111/j.1467-2979.2008.00281.x

Ncube LK, Ude AU, Ogunmuyiwa EN et al (2021) An overview of plastic waste generation and management in food packaging industries. Recycling 6:12. https://doi.org/10.3390/recycling6010012

Nie H, Wang J, Xu K et al (2019) Microplastic pollution in water and fish samples around Nanxun Reef in Nansha Islands, South China Sea. Sci Total Environ 696:134022. https://doi.org/10.1016/j.scitotenv.2019.134022

Norris BK, Storlazzi CD, Pomeroy AWM, Reguero BG (2024) Optimizing infragravity wave attenuation to improve coral reef restoration design for coastal defense. JMSE 12:768. https://doi.org/10.3390/jmse12050768

Okubo N, Takahashi S, Nakano Y (2018) Microplastics disturb the anthozoan-algae symbiotic relationship. Mar Pollut Bull 135:83–89. https://doi.org/10.1016/j.marpolbul.2018.07.016

Okubo N, Tamura-Nakano M, Watanabe T (2020) Experimental observation of microplastics invading the endoderm of anthozoan polyps. Mar Environ Res 162:105125. https://doi.org/10.1016/j.marenvres.2020.105125

Pastorino P (2024) Sunscreens and micro(nano)plastics: are we aware of these threats to the Egyptian coral reefs? Sci Total Environ 910:168587. https://doi.org/10.1016/j.scitotenv.2023.168587

Patterson J, Jeyasanta KI, Sathish N et al (2020) Microplastic and heavy metal distributions in an Indian coral reef ecosystem. Sci Total Environ 744:140706. https://doi.org/10.1016/j.scitotenv.2020.140706

Patterson J, Jeyasanta KI, Laju RL et al (2022) Microplastic in the coral reef environments of the Gulf of Mannar, India – characteristics, distributions, sources and ecological risks. Environ Pollut 298:118848. https://doi.org/10.1016/j.envpol.2022.118848

Patti TB, Fobert EK, Reeves SE, Burke da Silva K (2020) Spatial distribution of microplastics around an inhabited coral Island in the Maldives, Indian Ocean. Sci Total Environ 748:141263. https://doi.org/10.1016/j.scitotenv.2020.141263

Paul S, Choudhury M, Deb U, Pegu R, Das S, Bhattacharya SS (2019) Assessing the ecological impacts of ageing on hazard potential of solid waste landfills: a green approach through vermitechnology. J Clean Prod 236:117643. https://doi.org/10.1016/j.jclepro.2019.117643

Peng L, Fu D, Qi H et al (2020) Micro- and nano-plastics in marine environment: source, distribution and threats — a review. Sci Total Environ 698:134254. https://doi.org/10.1016/j.scitotenv.2019.134254

Portz L, Manzolli RP, Herrera GV et al (2020) Marine litter arrived: distribution and potential sources on an unpopulated atoll in the Seaflower Biosphere Reserve, Caribbean Sea. Marine Pollut Bull 157:111323. https://doi.org/10.1016/j.marpolbul.2020.111323

Raguso C, Saliu F, Lasagni M et al (2022) First detection of microplastics in reef-building corals from a Maldivian atoll. Mar Pollut Bull 180:113773. https://doi.org/10.1016/j.marpolbul.2022.113773

Rahman MN, Shozib SH, Akter MY et al (2023) Microplastic as an invisible threat to the coral reefs: sources, toxicity mechanisms, policy intervention, and the way forward. J Hazard Mater 454:131522. https://doi.org/10.1016/j.jhazmat.2023.131522

Rajpal A, Choudhury M, Goswami S, Chakravorty A, Raghavan V (eds) (2024) Waste management and treatment: advances and innovations, 1st edn. CRC Press. https://doi.org/10.1201/9781003258377

Rakib MRJ, Sarker A, Ram K et al (2023) Microplastic toxicity in aquatic organisms and aquatic ecosystems: a review. Water Air Soil Pollut 234:52. https://doi.org/10.1007/s11270-023-06062-9

Reichert J, Schellenberg J, Schubert P, Wilke T (2018) Responses of reef building corals to microplastic exposure. Environ Pollut 237:955–960. https://doi.org/10.1016/j.envpol.2017.11.006

Reichert J, Arnold AL, Hoogenboom MO et al (2019) Impacts of microplastics on growth and health of hermatypic corals are species-specific. Environ Pollut 254:113074. https://doi.org/10.1016/j.envpol.2019.113074

Reisser J, Shaw J, Wilcox C et al (2013) Marine plastic pollution in waters around Australia: characteristics, concentrations, and pathways. PLoS One 8:e80466. https://doi.org/10.1371/journal.pone.0080466

Rocha RJM, Rodrigues ACM, Campos D et al (2020) Do microplastics affect the zoanthid Zoanthus sociatus? Sci Total Environ 713:136659. https://doi.org/10.1016/j.scitotenv.2020.136659

Rotjan RD, Sharp KH, Gauthier AE et al (2019) Patterns, dynamics and consequences of microplastic ingestion by the temperate coral, *Astrangia poculata*. Proc R Soc B 286:20190726. https://doi.org/10.1098/rspb.2019.0726

Sabdono A, Ayuningrum D, Sabdaningsih A (2022) First evidence of microplastics presence in Coralsof Jepara coastal waters, Java Sea: a ComparisonAmong habitats receiving different Degreesof sedimentations. Pol J Environ Stud 31:825–832. https://doi.org/10.15244/pjoes/139376

Saliu F, Montano S, Garavaglia MG et al (2018) Microplastic and charred microplastic in the Faafu Atoll, Maldives. Mar Pollut Bull 136:464–471. https://doi.org/10.1016/j.marpolbul.2018.09.023

Saliu F, Montano S, Leoni B et al (2019) Microplastics as a threat to coral reef environments: detection of phthalate esters in neuston and scleractinian corals from the Faafu Atoll, Maldives. Mar Pollut Bull 142:234–241. https://doi.org/10.1016/j.marpolbul.2019.03.043

Saliu F, Veronelli M, Raguso C et al (2021) The release process of microfibers: from surgical face masks into the marine environment. Environ Adv 4:100042. https://doi.org/10.1016/j.envadv.2021.100042

Saliu F, Biale G, Raguso C et al (2022) Detection of plastic particles in marine sponges by a combined infrared micro-spectroscopy and pyrolysis-gas chromatography-mass spectrometry approach. Sci Total Environ 819:152965. https://doi.org/10.1016/j.scitotenv.2022.152965

Samanta P, Dey S, Kundu D et al (2022) An insight on sampling, identification, quantification and characteristics of microplastics in solid wastes. Trends Environ Anal Chem 36:e00181. https://doi.org/10.1016/j.teac.2022.e00181

Sandre F, Dromard CR, Le Menach K et al (2019) Detection of adsorbed Chlordecone on microplastics in marine sediments in Guadeloupe: a preliminary study. GCR 30:GCFI8–GCFI14. https://doi.org/10.18785/gcr.3001.14

Santana MFM, Dawson AL, Motti CA et al (2021) Ingestion and depuration of microplastics by a Planktivorous coral reef fish, Pomacentrus amboinensis. Front Environ Sci 9:641135. https://doi.org/10.3389/fenvs.2021.641135

Strafella P, Fabi G, Despalatovic M et al (2019) Assessment of seabed litter in the northern and Central Adriatic Sea (Mediterranean) over six years. Mar Pollut Bull 141:24–35. https://doi.org/10.1016/j.marpolbul.2018.12.054

Su Y, Zhang K, Zhou Z et al (2020) Microplastic exposure represses the growth of endosymbiotic dinoflagellate Cladocopium goreaui in culture through affecting its apoptosis and metabolism. Chemosphere 244:125485. https://doi.org/10.1016/j.chemosphere.2019.125485

Syakti AD, Jaya JV, Rahman A et al (2019) Bleaching and necrosis of staghorn coral (Acropora formosa) in laboratory assays: immediate impact of LDPE microplastics. Chemosphere 228:528–535. https://doi.org/10.1016/j.chemosphere.2019.04.156

Tan F, Yang H, Xu X et al (2020) Microplastic pollution around remote uninhabited coral reefs of Nansha Islands, South China Sea. Sci Total Environ 725:138383. https://doi.org/10.1016/j.scitotenv.2020.138383

Tang J, Ni X, Zhou Z et al (2018) Acute microplastic exposure raises stress response and suppresses detoxification and immune capacities in the scleractinian coral Pocillopora damicornis. Environ Pollut 243:66–74. https://doi.org/10.1016/j.envpol.2018.08.045

Tang J, Wu Z, Wan L et al (2021) Differential enrichment and physiological impacts of ingested microplastics in scleractinian corals in situ. J Hazard Mater 404:124205. https://doi.org/10.1016/j.jhazmat.2020.124205

Torres HR, Reynolds CJ, Lewis A et al (2019) Examining youth perceptions and social contexts of litter to improve marine debris environmental education. Environ Educ Res 25:1400–1415. https://doi.org/10.1080/13504622.2019.1633274

Utami DA, Reuning L, Konechnaya O, Schwarzbauer J (2021) Microplastics as a sedimentary component in reef systems: a case study from the Java Sea. Sedimentology 68:2270–2292. https://doi.org/10.1111/sed.12879

Vidyasakar A, Neelavannan K, Krishnakumar S et al (2018) Macrodebris and microplastic distribution in the beaches of Rameswaram Coral Island, Gulf of Mannar, Southeast coast of India: a first report. Mar Pollut Bull 137:610–616. https://doi.org/10.1016/j.marpolbul.2018.11.007

Wang F, Wong CS, Chen D et al (2018) Interaction of toxic chemicals with microplastics: a critical review. Water Res 139:208–219. https://doi.org/10.1016/j.watres.2018.04.003

Wang T, Zou X, Li B et al (2019) Preliminary study of the source apportionment and diversity of microplastics: taking floating microplastics in the South China Sea as an example. Environ Pollut 245:965–974. https://doi.org/10.1016/j.envpol.2018.10.110

Wilson SK, Graham NAJ, Pratchett MS et al (2006) Multiple disturbances and the global degradation of coral reefs: are reef fishes at risk or resilient? Glob Chang Biol 12:2220–2234. https://doi.org/10.1111/j.1365-2486.2006.01252.x

Yang Y, Liu W, Zhang Z et al (2020) Microplastics provide new microbial niches in aquatic environments. Appl Microbiol Biotechnol 104:6501–6511. https://doi.org/10.1007/s00253-020-10704-x

Zhang L, Zhang S, Wang Y et al (2019) The spatial distribution of microplastic in the sands of a coral reef Island in the South China Sea: comparisons of the fringing reef and atoll. Sci Total Environ 688:780–786. https://doi.org/10.1016/j.scitotenv.2019.06.178

Zhou Z, Wan L, Cai W et al (2022) Species-specific microplastic enrichment characteristics of scleractinian corals from reef environment: insights from an in-situ study at the Xisha Islands. Sci Total Environ 815:152845. https://doi.org/10.1016/j.scitotenv.2021.152845

Micro- and Nanoplastics—Impact on Marine Vertebrates and Invertebrates

14

Tandel Bhavika, Tandel Mayur, Dave Tushar, Tandel Vivek, Tandel Mitul, Tandel Nehal, Tandel Ritika, Timbadia Shriyas, Likhar Varsha, and Patel Bhoomi

Abstract

The effect of Micro-nano plastics (MNPs) is receiving attention in the marine environment because of their potential impact on marine organisms and humans. Future environmental hazards from these (MPs and NPs) pollutants cannot be neglected. The majority of the plastic trash discovered in the marine environment comes from land-based sources, and once plastic entered into the ocean, it can break down into smaller fragments and consumed by marine vertebrates and invertebrates directly or indirectly. Since these plastics have been discovered in microalgae, invertebrates like zooplankton and some benthonic creatures, vertebrates including fish and shellfish, seabirds, and amphibians, brine shrimp larvae and bivalves etc., assessing their toxicological effects on soil/marine sediments and aquatic life is a growing concern. Some marine vertebrates are impacted by MNPs, including birds, turtles (nutrient depletion) and whales (starvation). Small bits of plastic debris can be ingested by aquatic organisms, accumulating as they travel up the food chain and finally reaching humans. Marine vertebrates and invertebrates are important part of aquatic food chain and important as food, medicine, tourism for humans. Therefore, the chapter delineates impact of micro-nano plastics in vertebrates and invertebrates, which is being a major concern today.

T. Bhavika (✉) · D. Tushar · T. Nehal · T. Ritika
Department of Fisheries Resource Management, College of Fisheries Science, Kamdhenu University, Veraval, India
e-mail: tushar.dave@kamdhenuuni.edu.in

T. Mayur
Department of Fishery Engineering and Technology, College of Fisheries Manglore, KVAFSU University, Bidar, India

T. Vivek
Department of Aquatic Environment Management, College of Fisheries Science, Kamdhenu University, Veraval, India

T. Mitul
Department of Ocean Engineering and Marine Science, Florida Institute of Technology, Melbourne, FL, USA

T. Shriyas
Department o Biological Science, California State University, Fullerton, CA, USA

L. Varsha
Department of Fish Processing Technology, College of Fisheries Science, Kamdhenu University, Veraval, India

P. Bhoomi
Masters of Business Administration (Global), De Montfort University, Leicester, UK

S. Goswami et al. (eds.), *Micro-Nano Plastics Exposure, Environmental Degradation and Public Health Crisis*, Sustainable Development Goals Series, https://doi.org/10.1007/978-981-96-5047-7_14

Keywords

Microplastics · Nanoplastics · Vertebrates · Invertebrates · Food chain · Impact · Toxicity

14.1 Introduction

The most structurally complex organisms are marine vertebrates (MarineBio 2021a). Fish and other living resources from aquatic ecosystems have economic, social, and cultural significance as food and recreation sources (Cojocaru et al. 2022). Diverse organisms that can be found from the sea surface to its bottom and into the substrate are known as marine invertebrates. Invertebrates that are found in the sea comprise a wide variety of species, including starfish, crabs, sea cucumbers, worms, sea urchins, bluebottles, corals, and nudibranchs (DCCEEW 2023). Marine worms, clams, octopuses, crabs, and other invertebrates are essential to the health of the ocean ecosystems. Fish, marine life, and humans all rely on a wide variety of species as food. Others, including oysters and corals, provide vital habitat to marine life (NOAA 2023).

Plastic debris pollution is a persistent global environmental concern, with global plastic production reaching as high as 400.3 million tonnes in 2022 (Plastics Europe 2023). Plastic pollution is worsening globally due to rising plastic consumption, inadequate garbage management and recycling practices, and lack of environmental awareness (Rajpal et al. 2024). Microplastics significantly affect the marine ecosystem when they derived in the marine environment from marine and terrestrial-based sources (Choudhury et al. 2022; Arat 2024). Plastics are divided into three types based on their size: macroplastics (>25 mm), microplastics (<5 mm), and nanoplastics (<1 μm or 1000 nm) (Allen et al. 2022).

The toxicity of nanoplastics to marine creatures is influenced by pollutants, the availability of food, species, and stage of development and varies on particle size, concentration, and exposure time on marine creatures (Kogel et al. 2020). Nanoplastic toxicity to marine biota is related to decreased growth rate, energy and movement, stress, inflammation, and deformities (Kogel et al. 2020). These NPs' chance to pass through biological membranes increases with decreasing particle size, making marine species more vulnerable to their harmful effects (Goncalves and Bebianno 2021). The impacts of MNPs on marine biota are currently poorly understood. This chapter aims to compile all data to date on the toxicity of MNPs on marine biota.

14.2 Types of Marine Vertebrates and Invertebrates

14.2.1 Types of Marine Vertebrates

Among the most structurally complex creatures are marine vertebrates, which are categorized under the Kingdom Animalia, Phylum Chordata, and Subphylum Vertebrata (MarineBio 2021a). Owing to their roles in food, educational, and recreational systems, they are diverse and charismatic, capturing and inspiring societal interests. This is largely due to their visibility (Sydeman et al. 2015) which makes them ideal for studying the effects of micronanoplastics on marine ecosystems.

Vertebrata has four major marine superclasses and classes (MarineBio 2021a), which is described in Fig. 14.1. that shows the classification of marine vertebrates. It is divided into four superclasses: Agnatha, which includes jawless fishes like lampreys and hagfishes, and superclass Gnathostomata; Pisces, which includes bony fishes like ray-finned fishes, rays, chimaeras, and lungfish; and Tetrapoda, which includes sea birds, mammals like whales, and reptiles like sea turtles. Each class is illustrated with representative images of the species.

14.2.2 Types of Marine Invertebrates

Invertebrate animals are those without backbones. Invertebrates, which make up more than 98% of all organisms on Earth, depend on mechanisms for support other than a backbone, like

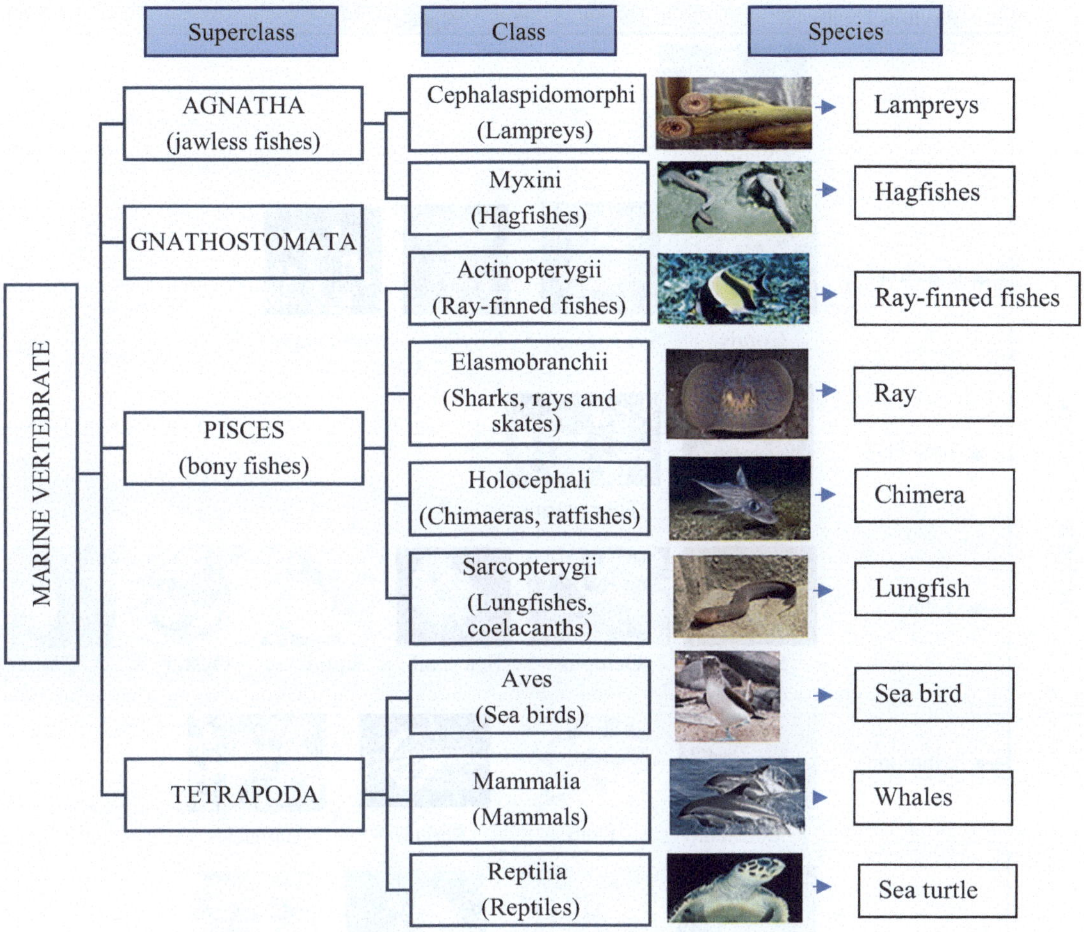

Fig. 14.1 Classification of marine vertebrates

hydrostatic pressure, exoskeletons, shells, and, occasionally, glass spicules (MarineBio 2021b).

Some invertebrate phyla have only one species, whereas others, such as Arthropoda, have over a million species and account for more than 83% of all described animal species. Sponge, cnidarian, worm, phosphorate, mollusc, arthropod, echinoderm, and hemichordate are the most prevalent marine invertebrates (MarineBio 2021b). Figure 14.2 shows the classification of marine invertebrates. It includes images and names of species under different phylum: Porifera (Sponges), Cnidaria (Corals, Sea anemones, Jellyfish, Hydrozoans), Annelida (Bristle worms, Tube worms), Mollusca (Sea Slug, Octopus, Sea snail, Squid, Oyster, Clam), Arthropoda (Lobsters, Crab, Shrimp, Barnacles), and Echinoderms (Sea urchins, Sea cucumber, Sealily, Starfish).

14.3 Sources of Micronanoplastics in Marine Vertebrates and Invertebrates

Microplastics and nanoplastics are mainly introduced into aquatic ecosystems by human activities (tourist, aquaculture, and home and industrial wastewaters) and natural erosion (Guo et al. 2020; Birch et al. 2020). Eight rivers in Asia (Amur, Ganges, Hai He, Indus, Mekong, Pearl, Yangtze, and Yellow) and two rivers in Africa (Nile and Niger) are the sources of 90% of the plastic waste that ends up in the ocean annually

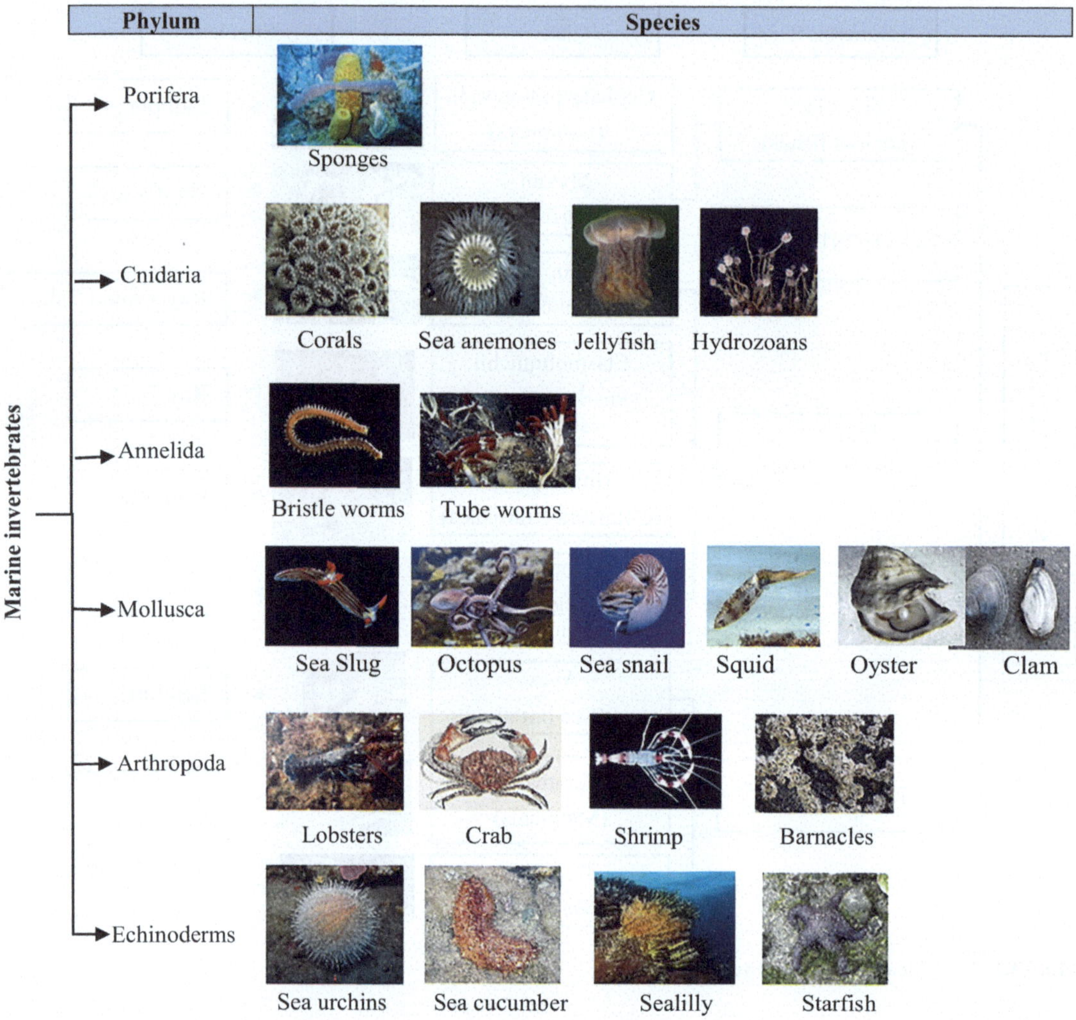

Fig. 14.2 Classification of marine invertebrates (MarineBio 2021b)

(Mariano et al. 2021). MNPs, which are present in most commonly used items (primary) or can emerge from the fragmentation of more considerable plastic trash (secondary), are emerging as significant concerns (Mariano et al. 2021; Kokalj et al. 2021). Microplastics flow significantly through filtration systems, while macroplastics are retained within oxidation ponds or wastewater treatment plants (Magalhaes et al. 2020).

Plastics and wastewater are frequently entered into water bodies (Wu et al. 2021). These polymers may remain in water bodies or disintegrate into smaller parts due to biological processes, temperature changes, physical stress, UV radiation, and wave impacts (Sarma et al. 2022). Every year, between 4.8 and 12.7 million tonnes of plastic wastes mostly from land-based sources are dumped into the ocean worldwide (Mofijur et al. 2021). Primary MPs are produced in enormous amounts in a variety of commercial and household uses, including exfoliating face scrubs, toothpaste, detergents, personal care products, abrasive cleaning agents, plastic powder for moulding, and synthetic clothing made of nylon and polyester (Wang et al. 2020; Birch et al. 2020). Paints, adhesives, electronics, and other substances have been accountable for NP release (Kihara et al. 2021). Just 15–30% of all plastic emitted into the environment is attributed to primary MNPs (Mariano et al. 2021), whereas sec-

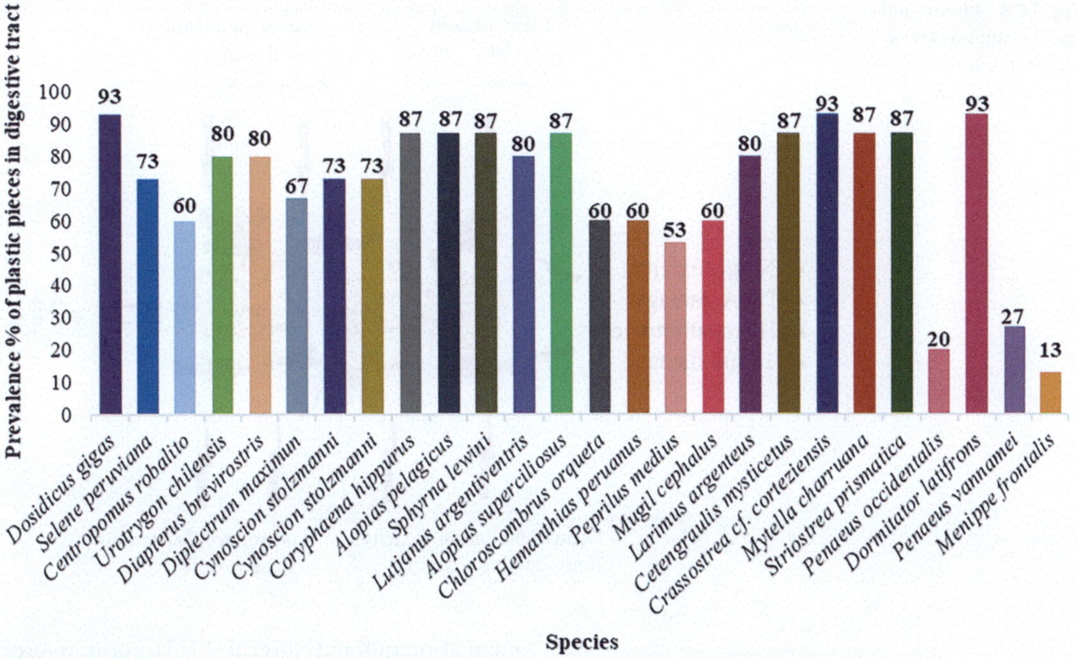

Fig. 14.3 Prevalence of microplastic particles in the digestive tract of marine species (Caceres-Farias et al. 2023)

ondary MP/NPs, which make up 70–80% of all plastic unrestricted into the environment, originate from the fragmentation of macroplastics through tensile pressures (Kiran et al. 2022).

14.4 Prevalence of Micronanoplastics in Marine Vertebrates and Invertebrates

MNPs can be consumed by aquatic biota indirectly (through trophic transfer) or directly from the water or sediment (Vethaak and Martínez-Gómez 2020). Most aquatic ecosystems include micro- and nanoplastics (MNPs), and their rising prevalence endangers aquatic organisms (Benson et al. 2022). Plastic waste interacts with a variety of marine organisms at all trophic levels as soon as it reaches the marine environment (Wang et al. 2020).

Figure 14.3 summarizes the prevalent status of MNPs in marine vertebrates and invertebrates. The highest percentage of microplastic particles are found in the digestive tracts of carnivorous (79%), planktivorous (74%) and detritivores (38%). Bivalve molluscs comprise a large majority of the food chain at 93%, resulting in accumulation processes at higher trophic levels (Caceres-Farias et al. 2023). Earlier research has revealed that 97% of the soft bodies of numerous mussel species contain microplastic contamination (Wakkaf et al. 2020). Molluscs such as the Pacific oyster *Crassostrea gigas* and the palmate oyster *Saccostrea palmula* were abundant in microplastic contamination, with 83% and 47%, respectively (Jones et al. 2021).

Figure 14.3 shows the prevalence percentage of microplastic pieces in the digestive tracts of various marine species. The highest prevalence is observed in *Dosidicus gigas, Pterois volitans*, and *Penaeus occidentalis*, each with 93%. Other species like *Selene peruviana* and *Lutjanus argentventr*is show lower percentages, while *Menippe frontalis* has the lowest at 13%. The chart highlights the widespread presence of microplastics across different species.

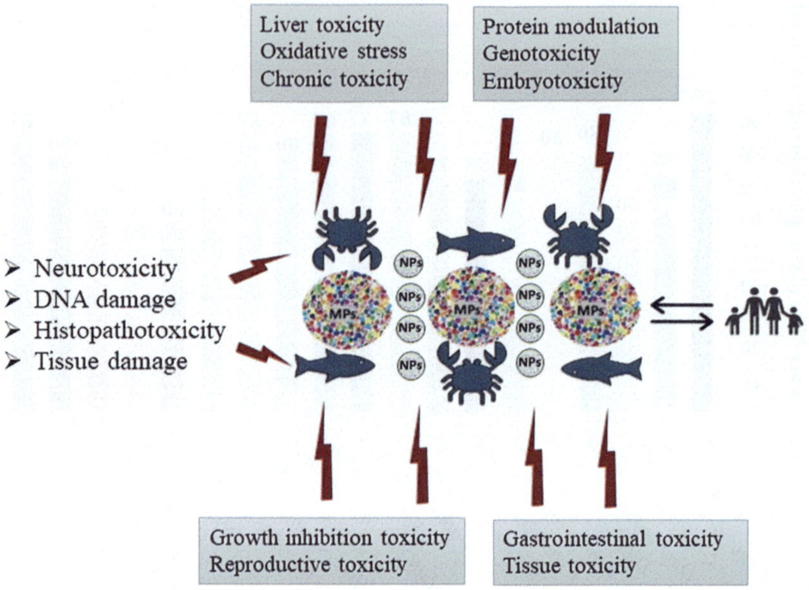

Fig. 14.4 Micro- and nanoplastic impact on marine organisms (Benson et al. 2022)

14.5 Impact on Marine Vertebrates and Invertebrates

NPs pose a greater risk to aquatic/marine creatures (Fig. 14.4) than MPs due to their smaller size, increased mobility, and more significant surface area (Wang et al. 2021). Figure 14.4 illustrates the impact of micro- and nanoplastics (MNPs) on marine organisms. Central image shows MNPs affecting aquatic organisms. Key impacts include neurotoxicity, DNA damage, histopathotoxicity, and tissue damage. Additional effects such as liver toxicity, oxidative stress, chronic toxicity, protein modulation, genotoxicity, embryotoxicity, growth inhibition, reproductive toxicity, gastrointestinal toxicity, and tissue toxicity are suggested potential impacts on humans. Because of their small size, benthos and zooplankton are more prone to ingest MPs and NPs, which might cause an accumulation of them in animals at higher trophic levels, including fish (Goncalves and Bebianno 2021). Micro- and nanoplastics are not just eliminated from the body but enter the bodies of animals through a variety of channels and then accumulate in the tissues. The build-up of micro- and nanoplastics in living tissue cells initially results in histopathological alterations (Niu et al. 2023). Furthermore, a significant element determining MNP toxicity and bioaccumulation is particle size. Due to their smaller size, higher surface-to-volume ratio, possible penetration, and higher bioavailability, NPs are often reported to be more hazardous than MPs (Gu et al. 2023). Li et al. (2020) discovered that benthic clams (*Corbicula fluminea*) were more severely affected by the gut mucosal barrier due to nanoplastic-induced genotoxicity than microplastics.

Goncalves and Bebianno (2021) have reported that marine taxa are adversely affected by MNP consumption. These effects include reduced feeding behaviour, impaired growth, inhibited growth, nutrient uptake, impaired immune response, oxidative stress, cellular toxicity, pseudo-satiety sensation, decreased fertility, malnutrition, and ultimately death. Subjected to 0.5 and 15 μm PS-NPs (190 μg/L), *Sebastes schlegelii* samples displayed reduced protein and fat contents, decreased swimming speed, increased oxygen consumption, and increased ammonia excretion (Jiang et al. 2023). Even after ingesting more than 90% of the microalgae containing polystyrene nanoparticles (PS-NPs), the brine shrimp (*Artemia franciscana*) showed no noticeable effects (Sendra et al. 2020). During laboratory testing and field surveys, Mohsen et al. (2022) discovered MPs in

Table 14.1 Effects of MNPs on marine vertebrates and invertebrates

Marine organisms	Species	Type	Size	Concen-tration	Effects	References
Fishes	*Sparus aurata*	PMMA	45 nm	0–10 μg.mL-1	Increase in plasma cholesterol and triglyceraldehyde	Brandts et al. (2021)
	Sparus aurata	PMMA	45 nm	0.001–10 mg.L-1	Transcriptional level of genes and antioxidant response inhibited; increase the anti-inflammatory response	Balasch et al. (2021)
	Larimichthys crocea	PS	100 nm	5.5 × 10–12 -5.50 × 10–7 mg.L-1	Oxidative stress increases, survival rate decreases	Gu et al. (2020)
Annelids	*Hediste diversicolor*	PS	100 nm	0.005–50 mg.L-1	Inhibit neurotransmission behaviour	Silva et al. (2020)
Echinoderms	*Paracentrotus lividus*	PS-NH2 and PS-COOH	60 nm	3.85 mg.L-1	Decreases cell viability, reduction of growth and development of its larva	Murano et al. (2021)
Molluscs	*Mytilus galloprovincialis*	PS-NPs	50 nm	1.5–150 mg.L-1	Rise in general stress and a weakening of neurological and immunological function	Capolupo et al. (2021)
	Mytilus galloprovincialis	PS-NH2	50 nm	2.5–10 mg.L-1	Embryonic development was restricted	Das (2023)

PMMA polymethylmethacrylate, *PS* polystyrene

the edible sections of *A. japonicus* (sea cucumber), including fibres, bits, films, and beads. Two polymer materials were isolated from the stomach of a sea cucumber (*Stichopus horrens*) in Pulau Pangkor, Malaysia. The majority of the MNPs diameters were found to be between 0.51 and 2 μm (Gu et al. 2023).

It has been observed that MNPs accumulate in the reproductive organs (Liang et al. 2021). Pacific oyster and Daphnia parents and offspring cannot reproduce when directly exposed to naturally aged MNPs (Liu et al. 2022). Furthermore, MNPs can alter the process of mitosis in sea urchins, leading to a significant rise in the number of deformities and, eventually, a significant decrease in fertilization rates (Trifuoggi et al. 2019). Much research has been done on the effects of MNPs on marine vertebrates and invertebrates, which is described in Table 14.1.

14.6 Mechanism of Toxicity

MNPs in aquatic environments have the ability to directly impact organisms through absorption, which can have toxicological consequences. Indirect effects may result from their ability to transfer pollutants and pathogens (Liu et al. 2023). By means of pinocytosis and phagocytosis, marine organisms absorb MP/NPs (Allen et al. 2020). According to Allen et al. (2020), they go by a variety of physiological routes, mostly to the stomach, gut, gills, and digestive tract, but also occasionally to the liver. Consuming MNPs has a variety of biological effects on an organism, including disturbances in oxidative balance and energy metabolism, as well as disturbances in brain function, immune response, and tissue integrity (Liu et al. 2023).

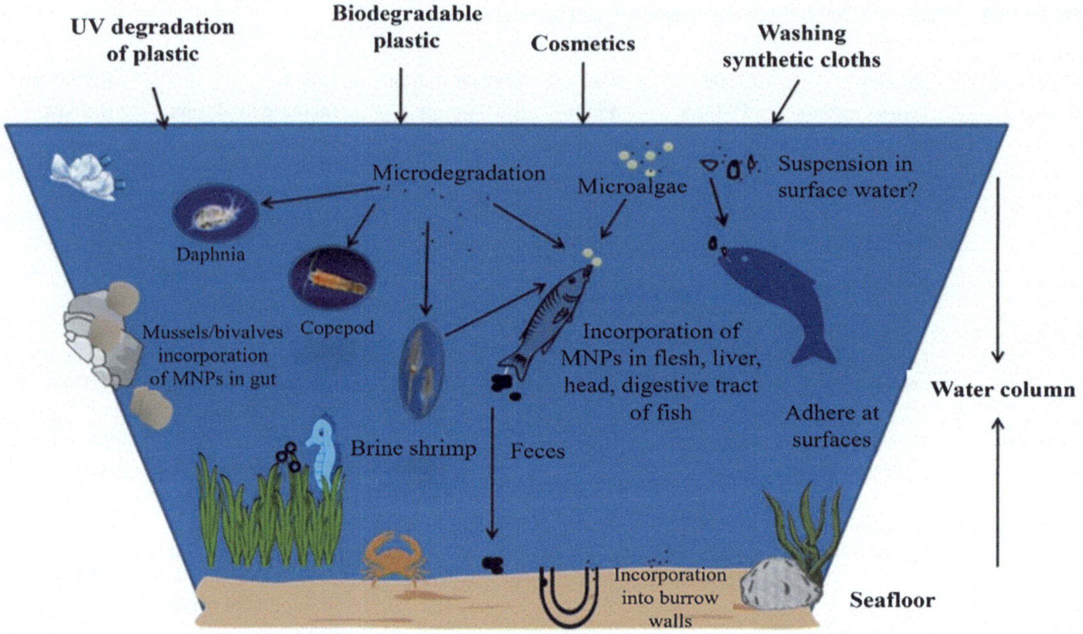

Fig. 14.5 Toxicity of MNPs in aquatic environment (Al-Thawadi 2020)

14.6.1 Developmental Toxicity and Digestive Tract Injury

MNPs accumulate in the digestive system of aquatic animals after ingestion, leading to gastrointestinal diseases such as intestinal blockage and injury, false satiety, starvation, and, ultimately, death (Guo et al. 2020). This phenomenon significantly impacts how aquatic species feed and maintain their energy balance, which hinders their ability to grow, develop, and maintain general health (Guo et al. 2020). It has been shown that MNP intake alters the shape and function of the fish larvae's small intestine, changing the glandular cells, epithelial cell structure, and villi height width (Mbugani et al. 2022).

14.6.2 Reproductive Toxicity and Endocrine Disorders

Since the harmful effects of aggregated MNPs in the gonads can be passed on to future generations, their presence can substantially affect the physiology and development of aquatic populations (Cormier et al. 2022). For example, MNPs in the paternal gonads, for instance, can be transferred to the offspring embryos, reducing body length and hatching rates in later generations, as research on teleost fish has shown (Cormier et al. 2022).

Endocrine disruption caused by MNPs is mainly observed as changes in levels of sex hormones, including testosterone, oestradiol, luteinizing hormone (LH), and lutein-stimulating hormone (LSH) (Liu et al. 2023). Furthermore, the disruption of the hypothalamic-pituitary-gonadal axis (HPG-axis) by MPs leads to altered gene expression, the creation of empty follicles, a rise in the prevalence of follicular atresia, and an increase in the toxicity of reproduction in marine medaka (Li et al. 2022).

Figure 14.5 illustrates the toxicity of micro- and nanoplastics (MNPs) in aquatic environments and shows various sources of MNPs, including UV degradation of plastic, biodegradable plastic, cosmetics, and washing synthetic cloths. MNPs undergo microdegradation and are consumed by aquatic organisms such as daphnia, copepods, and brine shrimp. The diagram highlights the incorporation of MNPs in the gut of mussels and bivalves, and in the flesh, liver, head, and digestive tract of fish. MNPs are shown in suspension in surface water, adhering to surfaces, and being incorporated into burrow walls on the seafloor.

14.7 Factors Influencing Marine Vertebrates and Invertebrates Toxicity

The following factors have been identified as the primary drivers of microplastic ingestion or assimilation by marine organisms and their associated toxicity: size, density, abundance, and colour. All of these factors increase the bioavailability of microplastics in organisms compared to other anthropogenic waste (Ugwu et al. 2021).

14.7.1 Size

MNP consumption in aquatic ecosystems reduces with increasing size. Organisms often mistake MNPs for natural prey or passively swallow them during normal eating. In fish, particles larger than 150 μm pass through the circulatory system due to their sorption threshold. All ages of Pacific oyster larvae consumed 1.84–7.3 μm polystyrene (PS) beads, while larger larvae consumed 20.3 μm beads (Suman et al. 2021).

14.7.2 Density

Particles with low densities naturally prefer to float near the water's surface (because they have a density lower than or comparable to that of water) and can spread quickly and widely horizontally (Peng et al. 2020). For example, foamed polystyrene particles with a 1.05 g/cm^3 density could travel across the Baltic Sea, around 250 km across, in a single day, even with considerable windage. On the other hand, heavier MNPs particles typically transfer almost vertically. For example, in the central Gotland basin, 5 mm polyoxymethylene particles with a 1.6 g/cm^3 density could sink through the water column of about 250 m in b18 h (Peng et al. 2020).

14.7.3 Abundance

The increased abundance of plastic particles in marine environments increases the likelihood of organisms encountering and ingesting them (Benson et al. 2022). Aquatic fishes living and feeding in shallow coastal waters are more likely to ingest microplastic particles than bottom-dwelling species directly or accidentally through their feeding behaviours. This is most likely due to the increased introduction of macro-, meso-, and microplastics into aquatic coastlines from land-based anthropogenic activities (Benson et al. 2022).

14.7.4 Colour

Colour increases MNP bioavailability, particularly for wild animals, due to the reflection of prey particles. The colours of observed MNPs differed between studies: black, clear, red or white, brown, blue, green, and yellow were the most commonly found colour particles (Suman et al. 2021). Several zooplankton species quickly consumed the pale-coloured MPs, while copepod and euphausiid species consumed the more colourful MNPs, primarily black, red, and blue. The digestive tracts of *Seriolella violacea* larvae contained 66% of the blue and black MPs (Suman et al. 2021).

14.8 Future Concerns and Recommendations

Due to the increasing demand for plastics and their introduction into the environment, the issue of micro- and nanoplastic plastics impacting marine vertebrates and invertebrates in marine ecosystems will never be reduced. Micro- to nanoplastics are more difficult to detect than large plastics, and characterizing their ecotoxicological effects is significantly more difficult. NPs have a greater ability to enter animal bodies and move between organs than MPs do. After being ingested by marine creatures, the fate and effects of the MNPs are still up for debate. In addition, there is a shortage of long-term data regarding the fate of ingested nanoparticles in aquatic animals, and other important questions need to be explored in further studies. It is unclear how organic compounds and metals interact mechanistically and how harmful they are to marine life. Consequently, we advise doing the following:

1. Investigating the animals at higher trophic levels, like marine mammals.
2. More investigation is necessary to understand the vertical distribution of MNPs in the water column.
3. Determining how long and how much microplastics accumulate in the digestive tract depends on our understanding of how they migrate between trophic levels.

14.9 Conclusion

Because of the increasing usage of plastics and their release into the environment, the presence of micro- and nanoplastics (MNPs) in marine ecosystems is a persistent problem. Compared to big plastics, MNPs are harder to detect and harder to characterize in terms of their ecotoxicological consequences, which can vary depending on the species, type, size, and dose. The marine ecosystem has seen an increase in the production of micro- and nanoplastics as a result of processes like hydrolysis, photodegradation, thermo-oxidative degradation, biodegradation, and heat. In marine ecosystems, primary MNPs constitute a substantial source of plastic pollution that impacts both vertebrates and invertebrates and also affects the trophic food chain in the ecosystem. To provides a more realistic picture of the destruction that MNPs cause to marine fauna, it is necessary to look into the ecotoxicological effects of these plastics at ambient concentrations as well as their interactions with other environmental contaminants. MNPs negatively impact marine biota, bacterial communities, top consumers, primary producers, and primary/secondary consumers.

References

Allen S, Allen D, Moss K et al (2020) Examination of the ocean as a source for atmospheric microplastics. PLoS One 15(5):e0232746

Allen S, Allen D, Karbalaei S et al (2022) Micro (nano) plastics sources, fate, and effects: what we know after ten years of research. J Hazard Mater Adv 6:100057

Al-Thawadi S (2020) Microplastics and nanoplastics in aquatic environments: challenges and threats to aquatic organisms. Arab J Sci Eng 45(6):4419–4440

Arat SA (2024) An overview of microplastic in marine waters: sources, abundance, characteristics and negative effects on various marine organisms. Desalin Water Treat 317:100138

Balasch J, Brandts I, Barría C et al (2021) Short-term exposure to polymethylmethacrylate nanoplastics alters muscle antioxidant response, development and growth in *Sparus aurata*. Mar Poll Bull 172:112918

Benson NU, Agboola OD, Fred-Ahmadu OH et al (2022) Micro (nano) plastics prevalence, food web interactions and toxicity assessment in aquatic organisms: a review. Front Mar Sci 291

Birch QT, Potter PM, Pinto PX et al (2020) Sources, transport, measurement and impact of nano and microplastics in urban watersheds. Rev Environ Sci Biotechnol 19:275–336

Brandts I, Barría C, Martins M et al (2021) Waterborne exposure of gilthead seabream (*Sparus aurata*) to polymethylmethacrylate nanoplastics causes effects at cellular and molecular levels. J Hazard Mater 403:123590

Caceres-Farias L, Espinoza-Vera M, Oros J et al (2023) Macro and microplastic intake in seafood variates by the marine organism's feeding behaviour: is it a concern to human health? Heliyon 9(5):e16452

Capolupo M, Valbonesi P, Fabbri E (2021) A comparative assessment of the chronic effects of micro- and nanoplastics on the physiology of the Mediterranean mussel *Mytilus galloprovincialis*. Nano 11(3):649

Choudhury M, Sharma A, Pervez A, Upadhyay P, Dutta J (2022) Growing menace of microplastics in and around the coastal ecosystem. In: Coastal ecosystems: environmental importance, current challenges and conservation measures, pp 117–137. https://doi.org/10.1007/978-3-030-84255-0_6

Cojocaru AL, Liu Y, Smith MD, Akpalu W, Chávez C et al (2022) The "seafood" system: aquatic foods, food security, and the global south. Rev Environ Econ Policy 16(2):306–326

Cormier B, Cachot J, Blanc M et al (2022) Environmental microplastics disrupt swimming activity in acute exposure in Danio rerio larvae and reduce growth and reproduction success in chronic exposure in D. Rerio and Oryzias melastigma. Environ Pollut 308:119721

Das S (2023) Impact of Nanoplastics on marine life: a review. Nat Environ Poll Tech 22(4):1983

DCCEEW (2023) Marine invertebrates. Department of Climate Change, Energy, the Environment and Water https://www.dcceew.gov.au/environment/marine/marine-species/marine-invertebrates. Accessed 11 Jan 2023

Goncalves JM, Bebianno MJ (2021) Nanoplastics impact on marine biota: a review. Environ Poll 273:116426

Gu H, Wang S, Wang X et al (2020) Nanoplastics impair the intestinal health of the juvenile large yellow croaker Larimichthys crocea. J Hazard Mater 397:122773

Gu Y, Xu D, Liu J et al (2023) Bioaccumulation of functionalized polystyrene nanoplastics in sea cucumber *Apostichopus japonicus* (Selenka, 1867) and their toxic effects on oxidative stress, energy metabolism and mitochondrial pathway. Environ Pollut 319:121015

Guo JJ, Huang XP, Xiang L et al (2020) Source, migration and toxicology of microplastics in soil. Environ Int 137:105263

Jiang N, Luo C, Shao M, Zheng Z et al (2023) Uncovering the role of ribosomal protein L8 in milk fat synthesis mechanisms in yak mammary epithelial cells. Pak J Zool:1–16

Jones JS, Porter A, Muñoz-Pérez JP et al (2021) Plastic contamination of a Galapagos Island (Ecuador) and the relative risks to native marine species. Sci Total Environ 789:147704

Kihara S, Köper I, Mata JP, McGillivray DJ (2021) Reviewing nanoplastic toxicology: it's an interface problem. Adv Colloid Interf Sci 288:102337

Kiran BR, Kopperi H, Venkata MS (2022) Micro/nanoplastics occurrence, identification, risk analysis and mitigation: challenges and perspectives. Rev Environ Sci Biotechnol 21(1):169–203

Kogel T, Bjorøy Ø, Toto B et al (2020) Micro- and nanoplastic toxicity on aquatic life: determining factors. Sci Total Environ 709:136050

Kokalj AJ, Hartmann NB, Drobne D et al (2021) Quality of nanoplastics and microplastics ecotoxicity studies: refining quality criteria for nanomaterial studies. J Hazard Mater 415:25751

Li Z, Feng C, Wu Y et al (2020) Impacts of nanoplastics on bivalve: fluorescence tracing of organ accumulation, oxidative stress and damage. J Hazard Mater 392:122418

Li Y, Yang G, Wang J, Lu L et al (2022) Microplastics increase the accumulation of phenanthrene in the ovaries of marine medaka (*Oryzias melastigma*) and its transgenerational toxicity. J Hazard Mater 424:127754

Liang B, Zhong Y, Huang Y et al (2021) Underestimated health risks: polystyrene micro-and nanoplastics jointly induce intestinal barrier dysfunction by ROS-mediated epithelial cell apoptosis. Part Fibre Toxicol 18(1):20

Liu Y, Zhang J, Zhao H et al (2022) Effects of polyvinyl chloride microplastics on reproduction, oxidative stress and reproduction and detoxification-related genes in Daphnia magna. Comp Biochem Physiol C Toxicol Pharmacol 254:109269

Liu W, Liao H, Wei M et al (2023) Biological uptake, distribution and toxicity of micro (nano) plastics in the aquatic biota: a special emphasis on size-dependent impacts. TrAC Trends Anal Chem 170:117477

Magalhaes S, Alves L, Medronho B et al (2020) Microplastics in ecosystems: from current trends to bio-based removal strategies. Molecules 25(17):3954

Mariano S, Tacconi S, Fidaleo M et al (2021) Micro and nanoplastics identification: classic methods and innovative detection techniques. Front Toxicol 3:636640

MarineBio (2021a, June 17) Marine Vertebrates. MarineBio Conservation Society. https://www.marinebio.org/creatures/marine-vertebrates/. Accessed 7 Jan 2023

MarineBio (2021b, June 12) Marine Invertebrates. MarineBio Conservation Society. https://www.marinebio.org/creatures/marine-invertebrates/. Accessed 8 Jan 2023

Mbugani JJ, Machiwa JF, Shilla DA et al (2022) Histomorphological damage in the small intestine of wami tilapia (*Oreochromis urolepis*)(Norman, 1922) exposed to microplastics remain long after depuration. Microplastics 1(2):240–253

Mofijur M, Ahmed SF, Rahman SA et al (2021) Source, distribution and emerging threat of micro-and nanoplastics to marine organism and human health: socio-economic impact and management strategies. Environ Res 195:110857

Mohsen M, Lin C, Liu S et al (2022) Existence of microplastics in the edible part of the sea cucumber *Apostichopus japonicus*. Chemosphere 287:132062

Murano C, Bergami E, Liberatori G et al (2021) Interplay between nanoplastics and the immune system of the mediterranean sea urchin *Paracentrotus lividus*. Front Mar Sci 8:647394

Niu H, Liu S, Jiang Y et al (2023) Are microplastics toxic? A review from eco-toxicity to effects on the gut microbiota. Meta 13(6):739

NOAA (2023) Invertebrates. National Oceanic and Atmospheric Administration. https://www.fisheries.noaa.gov/invertebrates. Accessed 9 Jan 2023

Peng L, Fu D, Qi H et al (2020) Micro-and nano-plastics in marine environment: source, distribution and threats—a review. Sci Total Environ 698:134254

Plastics Europe (2023) Plastics – the Fast Facts 2023: an Analysis of European Plastics Production, Demand and Waste Data. https://www.plasticseurope.org/en/resources/market-data

Rajpal A, Choudhury M, Goswami S, Chakravorty A, Raghavan V (eds) (2024) Waste management and treatment: advances and innovations, 1st edn. CRC Press. https://doi.org/10.1201/9781003258377

Sarma H, Hazarika RP, Kumar V et al (2022) Microplastics in marine and aquatic habitats: sources, impact, and sustainable remediation approaches. Environ Sustain 5(1):39–49

Sendra M, Sparaventi E, Blasco J et al (2020) Ingestion and bioaccumulation of polystyrene nanoplastics and their effects on the microalgal feeding of *Artemia franciscana*. Ecotoxicol Environ Saf 188:109853

Silva M, Oliveira M, Valente P et al (2020) Behavior and biochemical responses of the polychaeta Hediste diversicolor to polystyrene nanoplastics. Sci Total Environ 707:134434

Suman KH, Haque N, Uddin MJ et al (2021) Toxicity and biomarkers of micro-plastic in aquatic environment: a review. Biomarkers 26(1):13–25

Sydeman WJ, Poloczanska E, Reed TE et al (2015) Climate change and marine vertebrates. Science 350(6262):772–777

Trifuoggi M, Pagano G, Oral R et al (2019) Microplastic-induced damage in early embryonal development of sea urchin Sphaerechinus granularis. Environ Res 179:108815

Ugwu K, Herrera A, Gómez M (2021) Microplastics in marine biota: a review. Mar Pollut Bull 169:112540

Vethaak AD, Martínez-Gómez C (2020) Micro and nanoplastics in the aquatic environment with special reference to synthetic fibers. Synthetic nano-and microfibers 111

Wakkaf T, El Zrelli R, Kedzierski M et al (2020) Microplastics in edible mussels from a southern Mediterranean lagoon: preliminary results on seawater-mussel transfer and implications for environmental protection and seafood safety. Mar Pollut Bull 158:111355

Wang W, Ge J, Yu X (2020) Bioavailability and toxicity of microplastics to fish species: a review. Ecotoxicol Environ Saf 189:109913

Wang J, Zhao X, Wu F (2021) Characterization, occurrence, environmental behaviors, and risks of nanoplastics in the aquatic environment: current status and future perspectives. Fundam Res 1(3):317–328

Wu M, Tang W, Wu S et al (2021) Fate and effects of microplastics in wastewater treatment processes. Sci Total Environ 757:143902

Micro- and Nanoplastics: Impact on Crop Plants and Agricultural Practices

15

Mohd. Muzamil, Rizwan Ul Zama Banday, Sehreen Rasool, Danish Gul, Sana Rasool, Muzamil Hamid Wani, Suzhat Abbas, Masrat Mohiuddin, Kezia Rajan, Amir Ali, Muneeb Manzoor, Maliqa Majid, Usman Khan, and Boluram Gurjar

Abstract

Micro and nanoplastic debris in soil, aquatic, terrestrial and groundwater resources have jeopardized the food production system and human health and increased the vulnerability of the environmental ecosystem. The recent COVID pandemic has accelerated the pace of plastic debris infestation. The penetration of plastic debris through plastic mulch, fertilizer, sewage sludge, atmospheric deposition and illegal dumping ultimately culminates in the soil. The soil is the main reservoir due to its linkage with agricultural activities, microbial growth, food chain and human health. These particles interfere with the plant growth metabolism and development of beneficial soil microbes, significantly affecting their performance and susceptibility. Thus, it is imperative to understand the dynamics associated with the migration of micro and nanoplastics in soil, the environment, and the food chain. Most of the literature is directed to short-term lab-based studies, and the long-term field evaluation of plastic is essential to assess its impact on soil, plant metabolism, and human health. The remedial measures in the form of technological interventions or finding a biodegradable alternative to plastic can help to tackle the global problem of plastic pollution in general and micro/nano plastic pollution in particular. Accordingly, this chapter will address the impact of micro and nano plastics on soil, crops, the environment

M. Muzamil · R. U. Z. Banday · D. Gul · M. H. Wani
S. Abbas · K. Rajan · A. Ali · M. Manzoor
Division of Farm Machinery and Power Engineering, College of Agricultural Engineering and Technology, SKUAST-K, Shalimar, J&K, India

S. Rasool (✉)
Sericulture Production Department, Mirgund, J&K, India

S. Rasool
Department of Electrical Engineering, National Institute of Technology, Srinagar, J&K, India

M. Mohiuddin
Division of Agricultural Engineering, SKUAST-J, Chatha, Jammu, J&K, India

M. Majid
Project Scientist, CSIR, IIIM, Jammu, India

U. Khan
Bioproducts Science and Engineering Laboratory, Washington State University Tricities, Richland, WA, USA
e-mail: usman.khan@uaf.edu.pk

B. Gurjar
Department of Biological and Agricultural Engineering, Texas, A&M University, College Station, TX, USA

S. Goswami et al. (eds.), *Micro-Nano Plastics Exposure, Environmental Degradation and Public Health Crisis*, Sustainable Development Goals Series, https://doi.org/10.1007/978-981-96-5047-7_15

and associated agricultural practices, along with appropriate remedial measures in an articulate manner.

Keywords

Microplastic · Nanoplastic · Growth · Soil · Remediation · Plastic pollution

15.1 Introduction

The agricultural sector is witnessing tremendous application of plastics to improve productivity and reduce losses. Plastic is preferred as a low-cost, durable and flexible product for packaging food products, industrial manufacturing, pharmaceutical ingredients and the fashion industry (Jacques and Prosser 2021). The surge in the application has resulted in the production of 450 million metric tonnes (MMT), with the assertion that it will multiply by two to three times in the next four decades (OECD 2022). At a global level, about 12.5 MMT of plastics are used, with 80% of the total being attributed to agricultural crop and livestock production (FAO 2021). China accounts for 31% of the global production with 3, 10,000 tonnes devoted to crop production system (Plastics Europe 2020). However, the emergence of plastic is perceived as a global threat to the sustainability of the agricultural system and the viability of the environmental ecosystem (Krause et al. 2021). The agricultural and horticultural system is estimated to release 5400–23,500 tonnes of microplastics into the environmental ecosystem (FAO 2021; UNEP 2021). The recent COVID-19 pandemic boosted the prospectus of the application of plastics in protecting human health (Choudhury et al. 2022). Patricio-Silva et al. (2021) concluded that using PPE kits, facemasks, and gloves in a short duration jumped to 3.5% of global plastic waste production.

Dealing with plastic as small particles and migrating it to the environment pose serious challenges. It destroys soil structure and induces hindrances in the metabolic activities of plants,

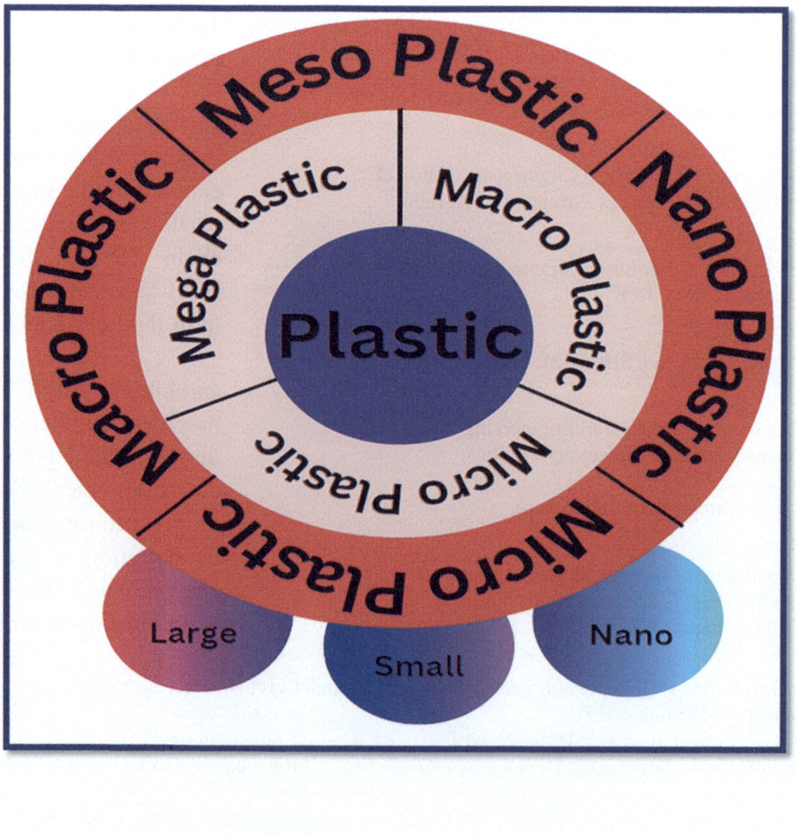

Fig. 15.1 Size based classification of plastic waste

thereby significantly influencing the growth and yield potential. The migration to the food chain directly or indirectly influences human health. Therefore, it is essential to understand the dynamics of plastic pollution to find a sustainable solution to this global problem. Moreover, the impact of micro- and nanoplastic on beneficial soil microbes and food security needs to be examined to take preventive measures (Krause et al. 2021). Dedicated efforts and energy are required to be invested in understanding the origin of the problem and the necessity to find a sustainable solution in order to protect soil, crops and the environment. This chapter will focus on those aspects to understand the impact of micro- and nanoplastic in changing soil attributes and metabolic processes in plants and the human body. It will also analyse the remedial measures to cure the system from the anomalies of plastic pollution and protect the environment from its harmful effects.

15.2 Classification of Plastic

Polyethylene, polypropylene and polyvinyl chloride are derived from fossil fuels and possess multiple applications. Polyethylene is the most used polymer, and its shape may vary as granular particles, spheres, fibres and films. The plastic waste is segregated based on particle size (Alimi et al. 2018) as macro plastics (>25 mm), mesoplastics (5–25 mm), microplastics (1–5 mm), and nanoplastics (< 1 μm) (Fig. 15.1). The microplastic and nanoplastic are known as primary and secondary particles and are emerging as significant threat to the environmental ecosystem (Mariano et al. 2021). These particles are used in cosmetics, food, industrial processing, and clothes. Lebreton et al. 2017 categorized plastic waste into metaplastic (>50 cm), macroplastic (5–50 cm) and microplastic (<0.5–5 cm). Ng et al. (2018) classified microplastic into large (1–5 mm), small (0.3–1 mm)

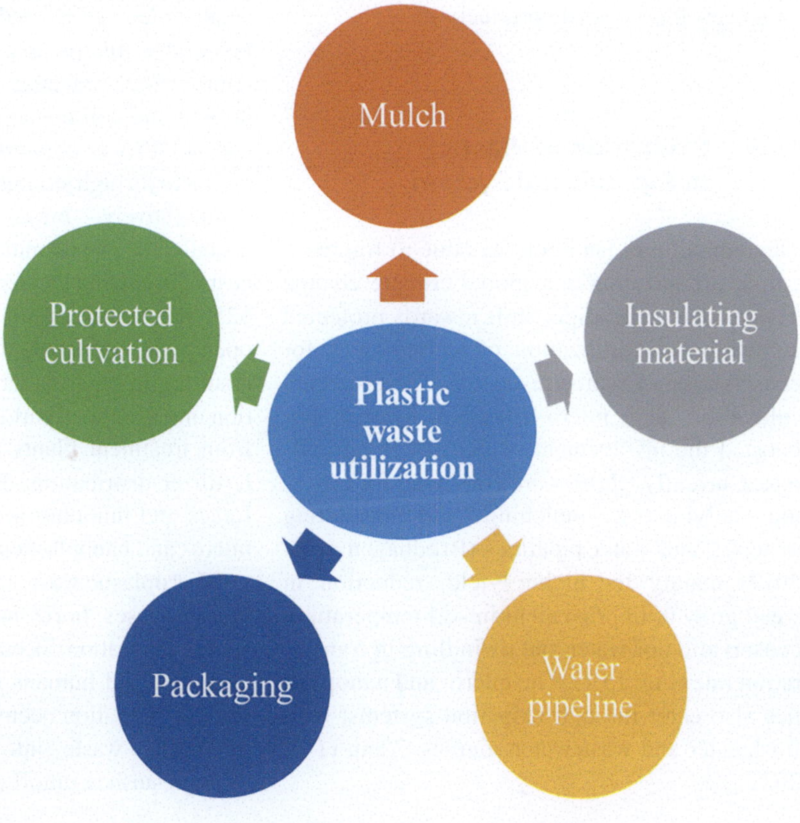

Fig. 15.2 Contribution of plastic waste to agricultural system

Fig. 15.3 Percentage distribution of plastic waste from different industries (Horton et al. 2017)

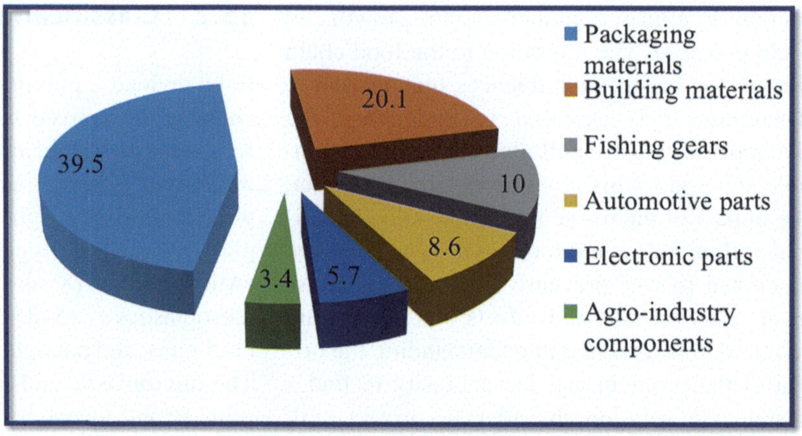

and nanoplastic (< 100 nm). The primary plastic waste undergoes physical, chemical and biological transformation into microplastic and nanoplastic due to exposure to solar ultraviolet radiations, mechanical abrasion, weathering and repeated utilization (Jacques and Prosser 2021; Meng et al. 2023). The breakdown of plastic mulch film in agricultural fields and farm machinery's rubber tyres often culminates in secondary plastic waste production.

15.3 Utilization of Plastic in Agricultural System

The reduction in land holdings due to fragmentation, urbanization and global climate change has ensured a paradigm shift towards protected cultivation and utilization of technologies for conservation of scarce resources. The protected cultivation and micro-irrigation system has boosted the involvement of plastics in agriculture. Currently, plastics are employed for covering polyhouses, mulching and insulating material, and water pipelines (Brodhagen et al. 2017) mainly for higher yield, reduction in weed growth, improvement in soil temperature, conservation of water and dwindling in transpiration rate (Fig. 15.2). The micro- and nanoplastics also enter through irrigation systems, both freshwater and wastewater sources (Yang et al. 2021).

15.4 Prevalence of MPs in Agricultural System

The agricultural industry accounts for 3.4%, with the highest proportion for plastic mulching, growing at a 5.6% compound annual growth rate (Hurley and Nizzetto 2018) (Fig. 15.3). The plastic mulch is laid on a well-prepared seedbed, and small holes are created to place the seeds. The favourable microclimatic structure regulates soil temperature, reducing evapotranspiration, weed growth and soil erosion. Low-density polyethylene (LDPE) is commonly used in plasticulture owing to its high strength, elongation abilities and comparatively lower gaseous permeability. Usually, the plastic mulch is left in the field as the cost of removing the degraded mulch film is about $250 per hectare. In some cases, LDPE is burnt in open fields or stockpiled in agricultural fields, resulting in environmental pollution and deposition into the soil. Apart from plastic mulch, sludge from treatment plants dealing with wastewater, fertilizer distribution, deposition by atmospheric forces and dumping in open areas also generate micro- and nanoplastics (Chai et al. 2020).

Microplastic waste migration generally occurs in two phases: horizontal and vertical. The horizontal migration occurs through soil, plants, microbes and humans (Yadav et al. 2022). The vertical migration occurs through the percolation of plastic waste into groundwater resources through surface runoff (Fig. 15.4).

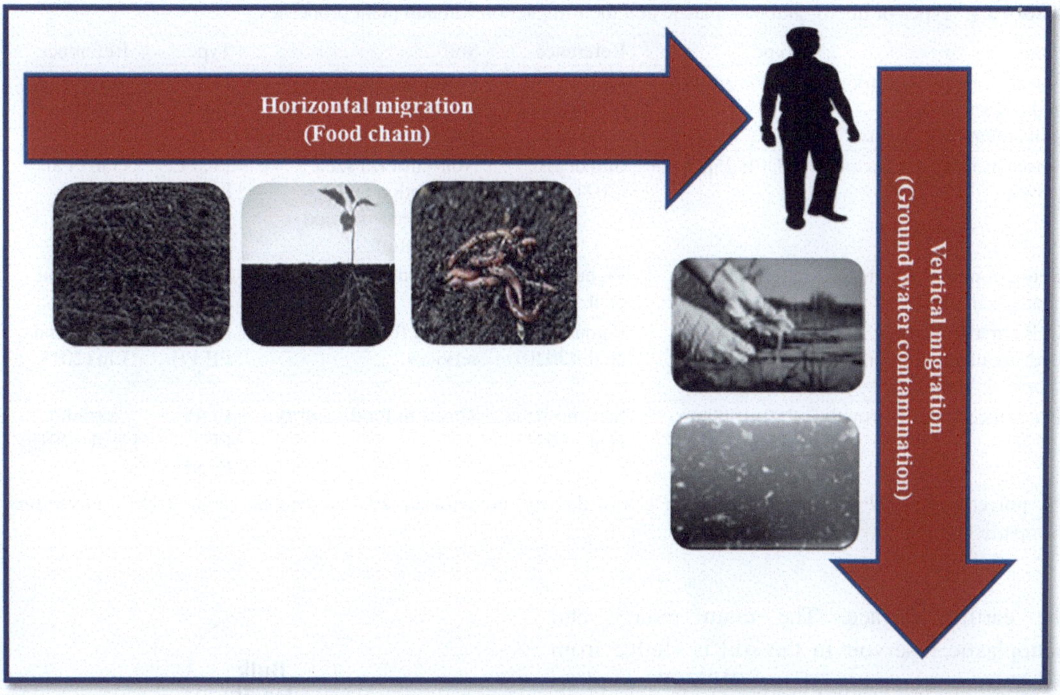

Fig. 15.4 Horizontal and vertical migration of microplastics

15.5 Consequences of MPs in Agricultural System

15.5.1 Impact of MPs on Plants

The plants' roots, tissues, leaves and edible parts absorb the micro and nano plastic monomers. The microparticles also enter the plant system through water and nutrient currents under the influence of transpiration, Table 15.1. Plastic transfer in the plant metabolism impacts growth performance and biochemical parameters and severely influences the growth of crops (Serrano-Ruiz et al. 2021). Many studies have confirmed the deformity in the growth process of corn, lettuce, onion, wheat and beans (Fan et al. 2022). In hydroponically grown plants, the damage to roots and cell bursting has also been recorded and documented (Pignattelli et al. 2020). The severity has alarmed the scientific community as the minute micro- and nanoplastic particles can enter the food chain from soil to microorganisms to humans (Beriot et al. 2021). Exposure to plastic-associated toxins can have a catastrophic impact on plants, including changes in growth and development phases, physiological disruptions and biochemical interruptions. The presence of microplastic and nanoplastic molecules can restrict chlorophyll synthesis in shoots and leaves, thereby inducing restrictions in plant photosynthesis. Therefore, the variation in the photosynthesis pigments can be used as an indicator to assess the presence of microplastic and nanoplastics in plants. The identification and monitoring of plastic waste are an intricate phenomenon. Therefore, the complex procedure of Fourier transformation, infra-red, and Raman spectroscopy can be the medium to identify and detect minute particles of plastic waste.

15.5.2 Impact of MPs on Soil

The distribution of the micro- and nanoplastic is widespread, encompassing the equator, poles, sea, mountains and deep underground layers of

Table 15.1 Types of micro- and nanoplastic and their impact on soil and plant properties

Plant	Type	Reference	Soil	Type	Reference
Absorption by plant roots and reduction in movement of nutrients	LDPE 250μm	Li et al. (2020)	Distortion of soil structure and soil-water dynamics	PE, PP and PS	Wang et al. (2022)
Absorption by tissues and leaves	LDPE 25μm	Sun et al. (2021)	Non-uniform seed distribution and interference in seed germination	PP, PE, LDPE	Guo et al. (2020)
Reduction in biochemical changes in plants	HDPE, PLA (100–154μm)	Pignattelli et al. (2020)	Destruction of soil ecology	LDPE, PE	Chen et al. (2020)
Cell bursting in hydroponically grown crops	100 mg/L	Pignattelli et al. (2020)	Loss of forest ecosystem services	LDPE, PP, PE	Zhang and Liu (2018)
Hindrance in plant growth	250–1000μm	Serrano-Ruiz et al. (2021)	Threat to food security	LDPE, PP, PSNP	Corradini et al. (2019)

PE polyethylene, *PP* polypropylene, *LDPE* low-density polyethene, *PLA* polylactic acid, *PSNP* polystyrene nanoparticles

the earth's surface. The main micro- and nanoplastic reservoir in the soil is sludge from wastewater treatment plants (WWTP) used as fertilizer, which results in 95% retention (O'Connor et al. 2019). Soil has low light and oxygen reserves, thereby providing a fertile ground for the micro- and nanoplastic particles to survive for generations. Over the years, agricultural soil is perceived as the sink of plastic waste, receiving 14% of the quantity discharged into the environment (Alimi et al. 2018; Horton et al. 2017). The contamination of soil ecosystems is 4–23 times more severe than aquatic ecosystems (Sridharan et al. 2021). The concentration of micro- and nanoplastic in agricultural soil (4400 pieces per kg) is three times higher than in average soil (1400 pieces per kg) (Tunali et al. 2023). Moreover, the waste management processes of landfilling, littering, atmospheric deposition and garden organic waste also result in the amassing of plastic waste in the agricultural soils (Rajpal et al. 2024; Mondal et al. 2023; Guo et al. 2020). The microplastic and nanoplastic distorts the soil's structure, influencing the metabolic activities of microorganisms and soil-water dynamics (Table 15.1).

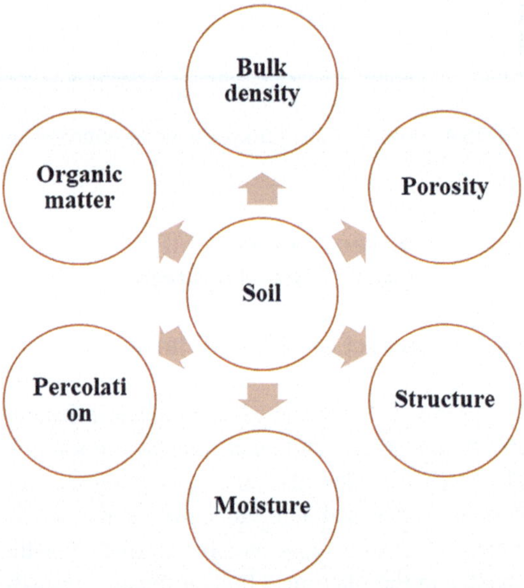

Fig. 15.5 Impact of MPs on soil properties (Wang et al. 2021)

The micro- and nanoplastic particles form aggregates as they possess lower density than soil minerals. This helps them alter the soil's physicochemical, biological and engineering properties. The properties of the soil in terms of bulk density, porosity, moisture content, structure, percolation and organic matter change in the presence of microplastics (Fig. 15.5). The presence of foreign particles in the range of 20–150 nm influences soil productivity and induces obstruction in the metabolic activities of microorganisms and soil fauna (Wahl et al. 2021). The presence of plastic leads to the non-uniform distribution of seeds and influences seed germination. Over the years, micro- and nanoplastic infestation has been detected in myriad situations: farmlands for crop production (Yu et al. 2021), Pastures for animal grazing (Corradini et al. 2021), forest soil for ecosystem services (Zhang and Liu 2018), Industrial area for commercial activities (Fuller and Gautam 2016) and floodplains for restoration of ecosystem (Scheurer and Bigalke 2018).

The contamination of vegetable farms in the sun-urban farming vicinity of China with microplastics was observed by Chen et al. (2020). Due to their long span and degradation rate, microplastics and nanoplastics are significant soil ecology and ecosystem hazards. Corradini et al. (2019) observed that the plastic fragments are retained in the agricultural fields for 15 years. The long-term impact of microplastic and nanoplastic waste can adversely affect food security. The accumulation of plastic mulch debris at the rate of 50–260 kg ha-1in the top 0–20 cm surface led to a reduction in the growth process and induced impediments in plant development in China (Liu et al. 2022). However, the research needs to be more balanced, with 47.02% focused on water contamination and only 7.01% directed to assess the influence of plastic on soil (Yang et al. 2021).

15.5.3 Impact of MPs on Soil Microbes

The soil microbes play an essential role in maintaining the equilibrium. The working of these microbes can be treated as an indicator of soil health. Microplastic and nanoplastic change the soil ecosystem and alter the habitat of microbes, including earthworms. These microbes usually ingest the micro- and nanoplastic particles, influencing their growth, mortality rate and ability to improve soil fertility status (Table 15.2). Moreover, nutrient recycling and oxygen flow inside the soil are also hampered. Huerta-Lwanga et al. (2017) detected the first evidence of microplastics' existence in the terrestrial food chain half a decade ago. Huerta-Lwanga et al. (2016) studied the impact of 150 μm ingested microplastic on the growth and development of *Lumbricus terrestris*. They concluded that the growth was severely affected and mortality amplified in different microplastic concentrations.

Table 15.2 Impact of MPs on soil microbes

Soil microbes	MPs Type	Reference
Alternation of habitat of beneficial soil microorganisms	PS (1–100μm)	Chen et al. (2022)
Microplastics damage immune system	PE (250–1000μm)	Feng et al. (2022)
Toxicity to soil microorganisms	LDPE (<150μm)	Aralappanavar et al. (2024)
Change in microbial properties towards nutrient recycling and gaseous emissions	80–250μm	Aralappanavar et al. (2024)

15.5.4 Soil Contamination Linked Transmission

Soil is essential for soil invertebrates and poultry to sustain their metabolism. The ingestion of the micro- and nanoplastic provides an easy pathway to reach the food chain and top echelon, that is, humans (Cox et al. 2019). The pathway of microplastics and their accumulation into agricultural soils, food systems, water, and air can seriously impact human health. Humans consume microplastics through skin contact, ingestion and inhalation (Prata et al. 2020). Moreover, plastic has the capacity to jeopardize the ecosystem by releasing toxic endocrine substances in the form of bisphenol A. Long-term exposure can result in the change in chromosomes, obesity, sterility and unwanted growth of cells (cancer) (Table 15.3).

Table 15.3 Presence of plastic in different organs of human body

Location	Size	Reference
Gonads, germ cells	50 nm	Ragusa et al. (2021); Ali et al. (2024)
Human faeces	50 nm	Zhang et al. (2021)
Cardiovascular system	40 nm and 200 nm	Ali et al. (2024)
Lungs	25 nm and 70 nm	Leslie and Depledge (2020)
Placenta	Up to 240 nm	Ragusa et al. (2021); Ali et al. (2024)
Blood (5–8μm)	≥700 nm	Leslie and Depledge (2020); Leslie et al. (2022)

15.5.5 Transmission of MPs in Aquatic System

The plastic waste migrates from soil to aquatic systems through point and diffuse sources. The point source of wastewater treatment plants and diffuse source of agricultural runoff or urban flooding has resulted in the entry of plastics into the food chain (Krause et al. 2021). The erosion of soil by wind energy and surface runoff transports the microplastic and nanoplastic waste to the water bodies, polluting the marine ecosystem (Alfaro-Nunez et al. 2021), groundwater reservoirs (Goeppert and Goldscheider 2021), drinking water and agricultural soils (Tian et al. 2022). Plastic waste penetrates deep into aquatic bodies with low temperatures and oxygen. This reduces the degradation rate and results in the accumulation of plastic waste (Garua and Sharma 2021). Lu et al. (2019) exposed a zebrafish to the micro- and nanoplastic media and observed the penetration into the gills, liver and gut within seven days of exposure. The plastic can accumulate in the water bodies and sediments to solidify its presence for hundreds of years.

15.6 Remedial Measures

The need for appropriate technology and high labour costs to tackle plastic waste have rendered the system unsustainable. The conventional remedial processes of landfills, incineration, and recycling (Fig. 15.6) still need to achieve the desired objectives. The landfills are least prioritized as it takes millions of years to degrade the plastic waste. Also, biodegradable plastics undergo degradation faster than conventional plastics, as carbon chain bonds are weak and friable. The interaction of the plastic results in the breakdown of the plastic into smaller particles, and more surface area prompts more microbial action. However, natural environment degradation is slow (Qin et al. 2021). The limitations associated with the conventional methods have prompted turning 76% of the plastic debris into waste (Ali et al. 2021), of which 79% is discarded into landfills, 12% incinerated and 9% recycled. Several processes have been tested, and it has been concluded that industrial composting is the best remedy for handling biodegradable plastic waste.

Moreover, the degradation rate is higher in composting, followed by the process inside the soil, freshwater sources, seawater and landfills. However, it has been observed that fertilizers (organic) and compost serve as pathways for micro- and nanoplastic particles. The density separation method to isolate the micro- and nanoplastic particles from waste water treatment plants (WWNP) can help to reduce the infestation (Schwinghammer et al. 2020). The primary and secondary treatments can partially remove the micro- and nanoplastic particles suspended in wastewater. Hale et al. (2020) observed that mechanical treatment could effectively reduce infestation by 35–59%, primary treatment by 23–53%, and pretreatment large-size microplastics. A procedure was developed to disintegrate polyethylene and polypropylene by hydrogen peroxide (H_2O_2) treatment. Fuller and Gautam (2016) used pressurized fluid extraction technology to extract plastic from soil and municipal waste and degrade polyethylene and polyvinyl chloride from soil samples.

15.6.1 Bioremediation

It involves microorganisms (Fan et al. 2022), earthworms (Prendergast-Miller et al. 2019) and fungal consortium to convert the biodegradable

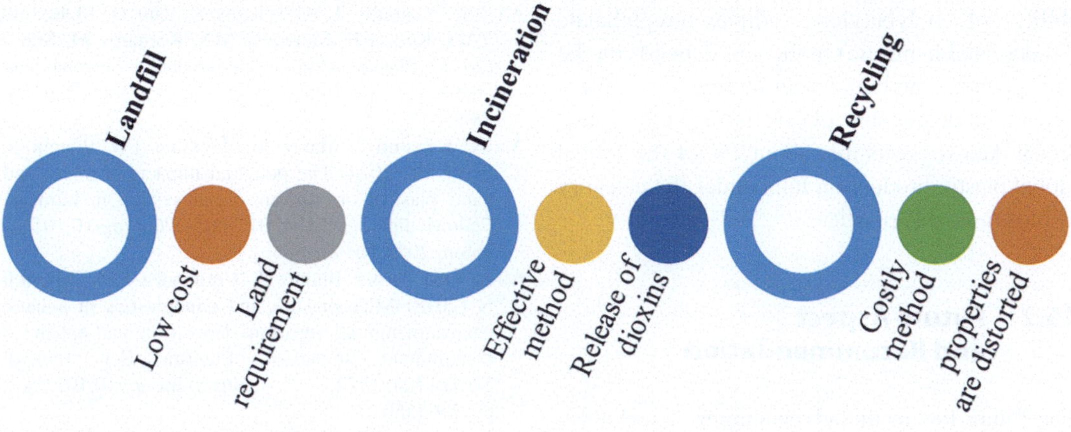

Fig. 15.6 Conventional methods of plastic disposal mechanism

plastics into environmentally benign products. The involvement of microbes is considered an environment-friendly and economical method for the remediation of environmental contaminants in the presence of biological and environmental factors (Paul et al. 2019). Initially, biodegradable plastic is subjected to molecular breakdown with the help of ultraviolet radiation, oxidation, reduction, hydrolysis, and degradation (Chowdhury et al. 2024). The products are decomposed and mineralized in the presence of microorganisms with the help of lipase, cutinase, carboxylesterases and laccases. The whole process is carried out in three stages: deterioration, fragmentation, and assimilation (Fig. 15.7). The process culminates in the production of carbon dioxide (CO_2), water (H_2O) and biomass.

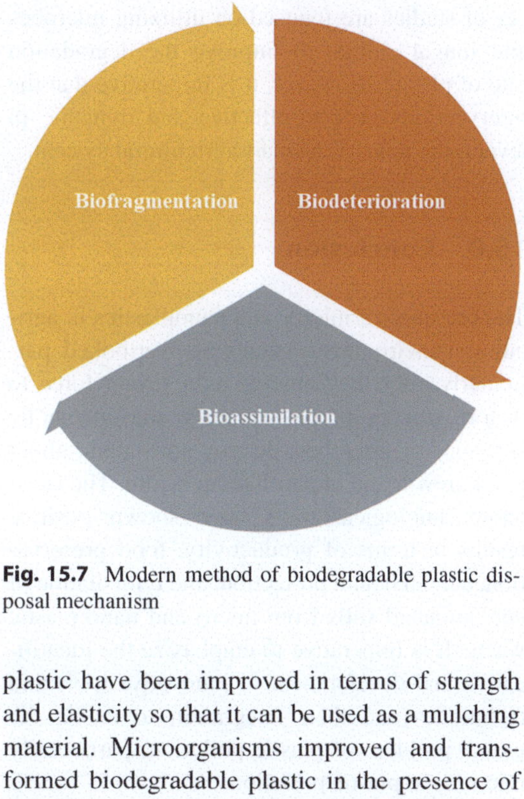

Fig. 15.7 Modern method of biodegradable plastic disposal mechanism

15.6.2 Alternative to Plastic Waste

Owing to the harmful effects of plastic on the food production system and human health, it is widely believed that biodegradable plastic can serve as an alternative (Roy and Choudhury 2022). The assertion can be ascertained that biodegradable plastic can be recycled and transformed into products without influencing the environment or human health. This is probably the reason that the biodegradable plastic market is exploding rapidly and expected to reach 2.9 million tonnes by 2025 (Meng et al. 2023). Moreover, over the years, the properties of biodegradable plastic have been improved in terms of strength and elasticity so that it can be used as a mulching material. Microorganisms improved and transformed biodegradable plastic in the presence of sufficient oxygen, appropriate temperature and moisture (Qin et al. 2021). The degradation of biodegradable plastic results in the generation of gaseous [carbon dioxide (CO_2, methane, CH_4), solid (biomass) or liquid (water, H_2O) products. Liu et al. (2022) degraded polybutylene adipate-terephthalate (PBAT) mulch after the harvesting operation and ploughed it with soil within a two-year time period. They concluded that the degrad-ability of polybutylene adipate-terephthalate (biodegradable plastic) in the soil depends on the soil type and microbial community. The utilization of biodegradable plastic is expanding, and recent data suggests that about 1% of the annual global plastic production falls under the category of biodegradable plastics.

15.7 Future Aspect and Recommendation

The future lies in the advancement of technologies to cure the problems of conventional plastic and the involvement of biodegradable plastic in agricultural production systems. In fact, a number of studies are focused on utilizing microbes and fungal culture to improve the degradation rate of plastic. However, it is imperative that the interventions be cost-effective and dynamic in leveraging towards a smart agricultural system.

15.8 Conclusion

The presence of micro- and nanoplastics in agricultural environments cannot be overlooked, particularly after they emerge as a strong force to distort soil properties, modify the metabolic activities of microbes, destroy soil fauna, affect plant growth and impair human health. The latest nanotechnological tools have shown positive results in terms of productivity, food preservation, animal health protection, and remediation of contaminated soils from micro and nano plastic waste. It is imperative to emphasize the identification and development of innovative tools and techniques and their integration to tackle the global problem of plastic pollution, particularly micro/nano plastic pollution.

References

Alfaro-Nunez A, Astorga D, Cáceres-Farías L, Bastidas L, Soto Villegas C, Macay KC, Christensen JH (2021) Microplastic pollution in seawater and marine organisms across the Tropical Eastern Pacific and Galápagos. Sci Rep 11:6424

Ali SS, Elsamahy T, Al-Tohamy R, Zhu D, Mahmoud YAG, Koutra E, Metwally MA, Kornaros M, Sun J (2021) Plastic wastes biodegradation: mechanism, challenges and future prospects. Sci Total Environ 780:1–18

Ali N, Katsouli J, Marczylo EL, Gant TW, Wright S, Serna JB (2024) The potential impacts of micro and nano plastics on various organ system in humans. EBioMedicine 99:104901. https://doi.org/10.1016/j.ebiom.2023.104901

Alimi OS, Farner Budarz J, Hernandez LM, Tufenkji N (2018) Microplastics and nanoplastics in aquatic environments: aggregation, deposition, and enhanced contaminant transport. Environ Sci Technol 52(4):1704–1724. https://doi.org/10.1021/acs.est.7b05559

Aralappanavar VK, Mukhopadhyaya R, Yu Y, Liu J, Bhatnagar A, Praveena SM, Li Y, Paller M, Adyel TM, Rinklebe J, Bolan NS, Sarkar B (2024) Effects of microplastics on soil microorganisms and microbial functions in nutrients and carbon cycling—a review. Sci Total Environ 924:171435. https://doi.org/10.1016/j.scitotenv.2024.171435

Beriot N, Peek J, Zornoza R, Geissen V, Huerta Lwanga E (2021) Low density microplastics detected in sheep faeces and soil: a case study from the intensive vegetable farming in Southeast Spain. Sci Total Environ 755:142653

Brodhagen M, Goldberger JR, Hayes D, Inglis DA, Marsh TL, Miles C (2017) Policy considerations for limiting unintended residual plastic in agricultural soils. Environ Sci Policy 69:81–84

Chai B, Wei Q, She Y, Lu G, Dang Z, Yin H (2020) Soil microplastic pollution in an e-waste dismantling zone of Chine. Waste Manag 118:291–301

Chen Y, Leng Y, Liu X, Wang J (2020) Microplastic pollution in vegetable farmlands of suburb Wuhan, Central China. Environ Pollut 257:113449

Chen M, Nan J, Xu Y, Yao J, Wang H, Zu X (2022) Efect of microplastics on the physical structure of cake layer for pre-coagulated gravity-driven membrane filtration. Sep Purif Technol 288:120632

Choudhury M, Sahoo S, Samanta P, Tiwari A, Tiwari A, Chadha U, Chakravorty A (2022) COVID-19: an accelerator for global plastic consumption and its implications. J Environ Public Health 1:1066350. https://doi.org/10.1155/2022/1066350

Chowdhury R, Barman S, Choudhury M, Kim KH, Bhattacharya SS (2024) Earthworm modifies microbial community and functional genes for lignocellulosic waste valorization: isolating plant-growth-promoting bacteria via next generation sequencing. Int Biodeterior Biodegradation 193:105854. https://doi.org/10.1016/j.ibiod.2024.105854

Corradini F, Meza P, Eguiluz R, Casado F, Huerta-Lwanga E, Geissen V (2019) Evidence of microplastic accumulation in agricultural soils from sewage sludge disposal. Sci Total Environ 671:411–420

Corradini F, Casado F, Leiva V, Huerta-Lwanga E, Geissen V (2021) Microplastics occurrence and frequency in

soils under different land uses on a regional scale. Sci Total Environ 752:141917
Cox KD, Covernton GA, Davies HL, Dower JF, Juanes F, Dudas SE (2019) Human consumption of microplastics. Environ Sci Technol 53(12):7068–7074
Fan P, Yu H, Xi B, Tan W (2022) A review on the occurrence and influence of biodegradable microplastics in soil ecosystems: are biodegradable plastics substitute or threat? Environ Int 163:1–12
FAO (2021) Assessment of agricultural plastics and their sustainability: a call for action. https://www.fao.org/3/cb7856en/cb7856en.pdf
Feng X, Wang Q, Sun Y, Zhang S, Wang F (2022) Microplastics change soil properties, heavy metal availability and bacterial community in a Pb-Zn-contaminated soil. J Hazard Mater. 424(Part A): 127364. https://doi.org/10.1016/j.jhazmat.2021.127364
Fuller S, Gautam A (2016) A procedure for measuring microplastics using pressurized fluid extraction. Environ Sci Technol 50:5774–5780
Garua B, Sharma JG (2021) Accumulation of plastics in terrestrial crop plants and its impact on the plant growth. J Appl Biol Biotechnol 9(6):25–33
Goeppert N, Goldscheider N (2021) Experimental field evidence for transport of microplastic tracers over large distances in an alluvial aquifer. J Hazard Mater 408:124844
Guo JJ, Huang XP, Xiang L, Wang YZ, Li YW, Li H, Cai QY, Mo CH, Wong MH (2020) Source, migration and toxicology of microplastics in soil. Environ Int 137:10526
Hale RC, Seeley ME, Guardia MJ, Mai L, Zeng EY (2020) A global perspective on microplastics. J Geophys Res Oceans 125(1):e2018JC014719
Horton AA, Walton A, Spurgeon DJ, Lahive E, Svendsen C (2017) Microplastics in freshwater and terrestrial environments: evaluating the current understanding to identify the knowledge gaps and future research priorities. Sci Total Environ 586:127–141
Huerta-Lwanga E, Gertsen H, Gooren H, Peters P, Salánki T, van der Ploeg M, Besseling E, Koelmans AA, Geissen V (2016) Microplastics in the terrestrial ecosystem: implications for Lumbricus terrestris (oligochaeta, lumbricidae). Environ Sci Technol 50(5):2685–2691. https://doi.org/10.1021/acs.est.5b05478
Huerta-Lwanga E, Mendoza Vega J, Ku Quej V, de Los Angeles Chi J, Sanchez Del Cid L, Chi C, Escalona Segura G, Gertsen H, Salánki T, van der Ploeg M, Koelmans AA, Geissen V (2017) Field evidence for transfer of plastic debris along a terrestrial food chain. Sci Rep 7:14071
Hurley RR, Nizzetto L (2018) Fate and occurrence of micro (nano) plastics in soils: knowledge gaps and possible risks. Curr Opin Environ Sci Health 1:6–11
Jacques O, Prosser RS (2021) A probabilistic risk assessment of microplastics in soil ecosystems. Sci Total Environ 757:143987
Krause S, Baranov V, Nel HA, Drummond JD, Kukkola A, Hoellein T, Smith GHS, Lewandowski J, Bonet B, Packman AI (2021) Gathering at the top? Environmental controls of microplastic uptake and biomagnification in freshwater food webs. Environ Pollut 268:115750
Lebreton LCM, Zwet JVD, Damsteeg J, Slat B, Andrady A, Reisser J (2017) River plastic emissions to the world's oceans. Nat Commun 8:15611
Leslie HA, Depledge MH (2020) Where is the evidence that human exposure to microplastics is safe? Environ Int 142:105807. https://doi.org/10.1016/j.envint.2020.105807
Leslie HA, Van-Velzen MJM, Brandsma SH, Vethaak AD, Garcia-Vallejo JJ, Lamoree MH (2022) Discovery and quantification of plastic particle pollution in human blood. Environ Int 163:107199. https://doi.org/10.1016/j.envint.2022.107199
Li L, Luo Y, Li R, Zhou Q, Peijnenburg WJGM, Yin N, Yang J, Tu C, Zhang Y (2020) Effective uptake of submicrometre plastics by crop plants via a crack-entry mode. Nat Sustain 3(11):929–937
Liu B, Guan T, Wu G, Fu Y, Weng Y (2022) Biodegradation behaviour of degradable mulch with poly (Butylene Adipate-co-Terephthalate) (PBAT) and poly (Butylene Succinate (PBS) in simulation marine environment). Polymers 14(8):1515
Lu L, Luo T, Zhao Y, Cai C, Fu Z, Jin Y (2019) Interaction between microplastics and microorganism as well as gut microbiota: a consideration on environmental animal and human health. Sci Total Environ 667:94–100
Mariano S, Tacconi S, Fidaleo M, Rossi M, Dini L (2021) Micro and nanoplastics identification: classic methods and innovative detection techniques. Front Toxicol 3:2
Meng K, Lwanga EH, Zee MVD, Munhoz DR, Geissen V (2023) Fragmentation and deploymerization of microplastics in the earthworm gut: a potential for microplastic bioremediation. J Hazard Mater 447:1–11
Mondal T, Choudhury M, Kundu D, Dutta D, Samanta P (2023) Landfill: an eclectic review on structure, reactions and remediation approach. Waste Manag 164:127–142. https://doi.org/10.1016/j.wasman.2023.03.034
Ng E-L, Lwanga EH, Eldridge SM, Johnston P, Hu H-W, Geissen V, Chen D (2018) An overview of microplastic and nanoplastic pollution in agroecosystems. Sci Total Environ 627:1377–1388
O'Connor D, Pan S, Shen Z, Song Y, Jin Y, Wu WM, Hou D (2019) Microplastics undergo accelerated vertical migration in sand soil due to small size and wet-dry cycles. Environ Pollut 249:527–534
OECD (2022) Global plastics outlook: economic drivers, environmental impacts and policy options. OECD Publishing, Paris
Patricio-Silva AL, Prata JC, Duarte AC, Barcelo D, RochaSantos T (2021) An urgent call to think globally and act locally on landfill disposable plastics under and after covid19 pandemic: pollution prevention and technological (bio) remediation solutions. Chem Eng 426:131201

Paul S, Choudhury M, Deb U, Pegu R, Das S, Bhattacharya SS (2019) Assessing the ecological impacts of ageing on hazard potential of solid waste landfills: a green approach through vermitechnology. J Clean Prod 236:117643. https://doi.org/10.1016/j.jclepro.2019.117643

Pignattelli S, Broccoli A, ve Renzi, M. (2020) Physiological responses of garden cress (L. sativum) to different types of microplastics. Sci Total Environ 727:138609

PlasticsEurope (2020) Plastics-the facts: an analysis of European plastics production, demand and waste data

Prata JC, da Costa JP, Lopes I, Duarte AC, Rocha-Santos T (2020) Environmental exposure to microplastics: an overview on possible human health effects. Sci Total Environ 702:134455

Prendergast-Miller MT, Katsiamides A, Abbass M, Sturzenbaum SR, Thorpe KL, Hodson ME (2019) Polyester-derived microfibre impacts on the soil dwelling earthworm Lumbricus terrestris. Environ Pollut 251:453–459. https://doi.org/10.1016/j.envpol.2019.05.037

Qin M, Chen CY, Song B, Shen MC, Cao WC, Yang HL, Zeng GM, Gong JL (2021) A review of biodegradable plastics to biodegradable microplastics: another ecological threat to soil environments? J Clean Product 312:127816

Ragusa A, Svelato A, Santacroce C, Catalano P, Notarstefano V, Carnevali O, Papa F, Rongioletti MCA, Baiocco F, Draghi S, D'Amore E, Rinaldo D, Matta M, Giorgini E (2021) Plasticenta: first evidence of microplastics in human placenta. Environ Int 146:10627

Rajpal A, Choudhury M, Goswami S, Chakravorty A, Raghavan V (eds) (2024) Waste management and treatment: advances and innovations, 1st edn. CRC Press. https://doi.org/10.1201/9781003258377

Roy P, Choudhury M (2022) Bioeconomy and green plastic production. In: Baskar C, Ramakrishna S, Daniela La Rosa A (eds) Encyclopedia of green materials. Springer, Singapore. https://doi.org/10.1007/978-981-16-4921-9_189-1

Scheurer M, Bigalke M (2018) Microplastics in Swiss floodplain soils. Environ Sci Technol 52:3591–3598

Schwinghammer L, Krause S, Schaum C (2020) Determination of large microplastics: wet-sieving of dewatered digested sludge, co-substrates, and compost. Water Sci Technol 84(2):384–392. https://doi.org/10.2166/wst.2020.582

Serrano-Ruiz H, Martin-Closas L, Pelacho AM (2021) Biodegradable plastic mulches: impact on the agricultural biotic environment. Sci Total Environ 750:141228

Sridharan S, Kumar M, Bolan NS, Singh L, Kumar S, Kumar R, You S (2021) Are microplastics destabilizing the global network of terrestrial and aquatic ecosystem services? Environ Res 198:111243

Sun H, Lei C, Xu J, Li R (2021) Foliar uptake and leaf-to-root translocation of nanoplastics with different coating charge in maize plants. J Hazard Mater 416:125854

Tian L, Jinjin C, Ji R, Ma Y, Yu X (2022) Microplastics in agricultural soils: sources, effects, and their fate. Curr Opin Environ Sci Health 25:100311

Tunali M, Adam V, Nowack B (2023) Probabilistic environmental risk assessment of microplastics in soils. Geoderma 430:116315. https://doi.org/10.1016/j.geoderma.2022.116315

UNEP (2021) Plastics in agriculture: sources and impacts. United Nations Environment Programme https://drive.google.com/file/d/1U0v-md500GvoCAss_YF3jLtdk3akEBai/view

Wahl A, Le-Juge C, Davranche M, El-Hadri H, Grassl B, Reynaud S, Gigault J (2021) Nanoplastic occurrence in a soil amended with plastic debris. Chemosphere 262:127784

Wang FY, Wang QL, Adams CA, Sun YH, Zhang SY (2021) Effects of microplastics on soil properties: current knowledge and future perspectives. J Hazard Mater 424:127531

Wang W, Yuan W, Xu EG, Li L, Zhang H, Yang Y (2022) Uptake, translocation, and biological impacts of micro (nano) plastics in terrestrial plants: progress and prospects. Environ Res 203:111867

Yadav V, Dhanger S, Sharma J (2022) Microplastics accumulation in agricultural soil: evidence for the presence, potential effects, extraction and current bioremediation approaches. J Appl Biol Biotechnol 10(2):38–47. https://doi.org/10.7324/JABB.2022.10s204

Yang L, Zhang YL, Kang SC, Wang ZQ, Wu C (2021) Microplastics in soil: a review on methods, occurrence, sources, and potential risk. Sci Total Environ 780:146546

Yu L, Zhang J, Liu Y, Chen L, Tao S, Liu W (2021) Distribution characteristics of microplastics in agricultural soils from the largest vegetable production base in China. Sci Total Environ 756:143860

Zhang G, Liu Y (2018) The distribution of microplastics in soil aggregate fractions in southwestern China. Sci Total Environ 642:12–20

Zhang N, Li YB, He HR, Zhang JF, Ma GS (2021) You are what you eat: microplastics in the faeces of young men living in Beijing. Sci Total Environ 767:144345

16 Micro- and Nanoplastics Impact on Food Sources of Marine Origin

Neha Kumari, Poonam Maurya, Neelesh Kumar Maurya, Latika Yadav, and Sanjay Kumar

16.1 Introduction

Worldwide waste plastics are becoming a large problem. Much talk goes on about pathways of this kind wasted and its great environmental impact. During the period of 1950–2015, over 8.3 billion metric tons were produced, and nearly four-fifths of those plastics ended up in landfills. Over 12,000,000,000 metric tons of plastic trash would most certainly be buried in the earth below us by 2050 (Geyer et al. 2017). Plastic debris accounts for 61–87% of plastic floating litter in our oceans, with around 8.75 million metric tons estimated to be entering the ocean per annum from land-based sources (Jambeck et al. 2015; Barboza et al. 2019; Tekman et al. 2019). On the other hand, when it breaks down into even smaller fragments, such as microplastics (MPs) and sub-micron nanoplastics (NPs) (<1 mm), the problem is exacerbated (Wright et al. 2020). Today it is estimated that there are over 5.25 trillion plastic particles floating on the world's oceans (Eriksen et al. 2014). More scientific research conducted by Uddin et al. (2020) revealed that deeper insights into the contribution of MPs to the aquatic world range around 1.471015 in terms of annual deposits of processed effluent from waste-water treatment plants, and MPs have an approximate value of 3.851016 annually due to untreated disposals (Uddin et al. 2020). Microplastics are found in a range of environmental components, including marine (rivers, estuaries, and inshore and pelagic waters) and terrestrial phases (Bao et al. 2023). Aerosols are now contributing a significant part of the MPs' total human exposure since they have recently gained a certain level of confidence among the general population (Dewika et al. 2023). A series of studies have pointed to MPs being transported up to 95 kilometers on a single journey. MPs are formed in the air through numerous processes and sources, including washing and tire abrasion, furniture

N. Kumari
Centre of Advanced Study in Botany, Institute of Science, Banaras Hindu University, Varanasi, India

P. Maurya
Shambhunath Institute of Pharmacy, Prayagraj, UP, India

N. K. Maurya (✉)
Department of Nutrition and Dietetics, School of Allied Health Science, Sharda University, Greater Noida, India

L. Yadav
Department of Home Science Faculty of Arts Vijay Singh Pathik Government Post Graduate Degree College Kairana, Shamli, Uttar Pradesh, India

S. Kumar
Division of Hematology, Mayo Clinic, Rochester, MN, USA
e-mail: sanjay.kumar7@sharda.ac.in

S. Goswami et al. (eds.), *Micro-Nano Plastics Exposure, Environmental Degradation and Public Health Crisis*, Sustainable Development Goals Series, https://doi.org/10.1007/978-981-96-5047-7_16

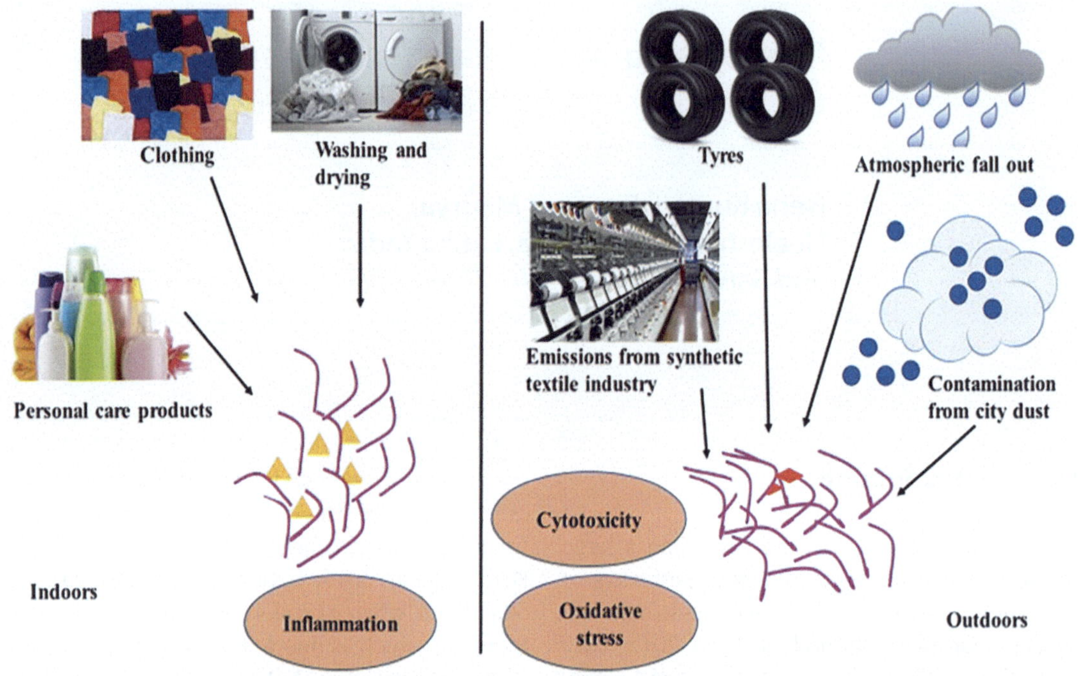

Fig. 16.1 Microplastics sources in the maritime environment and its consequences on human wellbeing

degradation, vinyl chloride and PVC production (Habibi et al. 2022) and city dust contamination (Fig. 16.1).

The most common type of marine plastic pollution is microplastics, which consist of tiny, invisible particles of less than 5 mm in size. The more popular polymer types are low-density polyethene (LDPE), high-density polyethene (HDPE), polypropylene (PP), polystyrene (PS), polyethene terephthalate (PET) and polyvinyl chloride. Types of other polymers present are nylon, acrylic, polyurethanes, polylactic acid (PLA—a biodegradable plastic), or polycarbonate, etc. (Bratovcic 2019). These materials made of plastic turn into artificial plastic scraps and finish as waste in the aquatic ecosystems' interior (Turra et al. 2014). Microplastics pose an unadorned threat to marine organisms, which, when consumed, lead to reduced growth rates, decreased reproductive ability, ulcers, abrasions, and oxidative stress (Fossi et al. 2016; Sutton et al. 2016; Choudhury et al. 2022), injuries due to abrasion, asphyxiation, and restricted movements (Alimba and Faggio 2019). Numerous examples have demonstrated that many of the studied aquatic fauna ingest microplastics (Carrasco et al. 2019; Hara et al. 2020; Naidoo et al. 2020). There exists an unknown risk of exposure, accumulation, and magnification of MPs or sorbed additives among food chains in freshwaters (Fig. 16.1) and the toxicities that MPs may carry out on higher-trophic level biota are not well studied. Several studies have investigated the negative effects of MPs on organisms from different groups; nevertheless, a single paper was missing that synthesizes all these impacts. Thus, in this chapter effect of microplastic on aquatic organisms and the environmental fate of microplastics is studied.

16.2 Sources of Micro- and Nanoplastics in Marine Environment

The sources of MPs in maritime environment are discussed under the following headings:

16.2.1 Land-Sourced and Ocean-Sourced MPs

Most MPs and NPs in the ocean and seas come from the land. Ineffective waste management, unlawful dumping, and inevitable discharges from construction, industrial, agriculture, home use, and recreation generate this problem (Kershaw et al. 2011). Plastic wastes can enter the water directly during fishing, boating, and aquaculture activities. They are harder to control than land-based discharges since they go directly into the ocean. For example, only 10% of plastic MPs come from marine sources, of which almost 80% of input comes from land-based ones (Sambandam et al. 2022).

16.2.2 Land-Based Plastic Waste Entering the Ocean

Three routes are involved in carrying land-based plastic waste into oceans–the wind, soil, and water pathways. A waterway most commonly accesses it. Sewer overflows, soil runoffs, air (dust) precipitation, and tourist trash are just a few of the origins of the plastic trash that eventually makes its way into rivers and oceans. The wind might blow plastic trash from the coast to distant locations where it could be deposited. Rain runoff and tidal washing can sweep soil-born plastic litter, including agricultural films and pesticide and fertilizer packaging, into the ocean.

16.3 Interactions of Microplastics with Food Web

At various trophic levels, MPs can move up and down the food chain from the bottom up. This gives them unique qualities that make them far easier to incorporate into the food chain. Because of some distinct characteristics shown in Fig. 16.2, much greater admission to the food chain is much increased due to the following reasons:

Size: When the size of MPs is the same as food criteria, they are preyed by low-trophic level organisms, and after ingestion, transfer them to higher-trophic levels by consumption (Wright et al. 2013; Anik et al. 2021).

Low weight: Low-density, lightweight plastics could be ingested by planktivorous filters and suspension feeders in saline before being diluted. Marine life, meanwhile, can gobble up the dense plastic particles as they fall down the water column to any depth.

More colorful: Plastic particles are often multi-colored, leading certain marine predators to mistake them for their natural prey (Peng et al. 2020).

Abundance: Enrichment of plastic particles in seawaters would double the likelihood of marine organisms consuming MPs (Wright et al. 2013).

16.3.1 MPs Translocation to Digestive Organs

MPs can enter animals by a variety of routes, including ingestion, but they can also stick to the skin, foot, mantle, and other non-gastrointestinal organs. Most MPs that are swallowed stay in the digestive tract. Microvilli activity or ciliate movement can maintain these particles in numerous species' gills, stomachs, intestines, and digestive tubules, including the pelagic fish *Platycephalus indicus* and the mussel *Mytilus edulis* (Khoshmanesh et al. 2023). After recovering the acid-digested tissue (individual) of Arenicola marina that had been exposed to the beach sediment, it was discovered that the MPs concentration approximately ranged from 1.2 to 2.8 p/g of tissue. Additionally, the initial abundance of PS microspheres per gram of beach sand was 110 pieces. After being subjected to PET fiber MPs ranging in length from 62–1400 m over 48 hours, the fibers were seen to have emerged from within the stomach of the *D. magna* (Kwak et al. 2022). Fibers that are ingested normally have a diameter of around 300 micrometers, but some can be over 1400 micrometers long and are severely twisted.

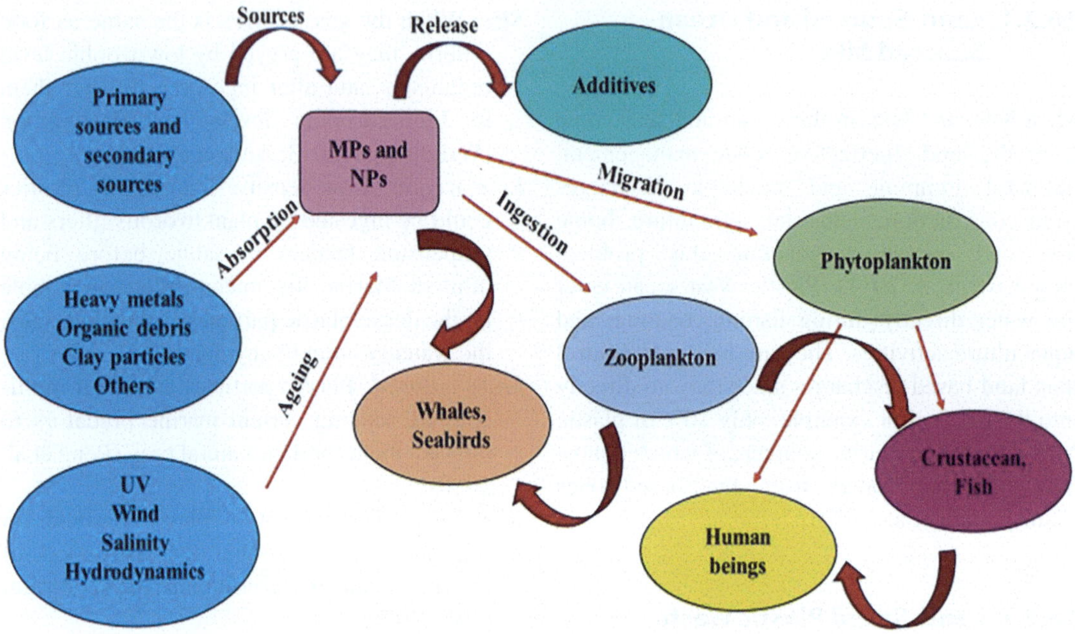

Fig. 16.2 MPs and NPs interactions with other environmental contaminants or organisms

16.3.2 MPs Translocation to Non-digestive Organs

MPs can damage tissues by translocating to and concentrating in them (Abbasi et al. 2018; Ribeiro et al. 2019). It was found that 3.0 μm and 9.6 μm MPs tracked in the hemolymph after consumption of the blue mussel M. edulis. They can travel from the digestive tract to the bloodstream within as little as three days, and researchers have found that 3.0 μm MPs are far more common in the bloodstream than 9.6 μm MPs (Browne et al. 2008).

16.4 Factors Influencing Food Web Interactions

The food web is a more complex representation of how various animals in an ecosystem obtain their nourishment. Energy from sunlight is captured by producers (e.g., plants and algae) and transferred to herbivores, which are then consumed by carnivores and other predators, forming a food chain. In this ecosystem, decomposers, producers, and primary, secondary, and tertiary consumers all play essential roles. The food web should remain balanced, with the ratio of biomass and energy gradually decreasing as it moves up the food chain. However, microplastics and persistent organic pollutants disrupt this balance as they accumulate at higher-trophic levels.

While some microplastics may be excreted by organisms, many find their way into vital organs such as the stomach, gills, liver, and even the brain. In mammals, organic pollutants tend to accumulate in fatty tissues. Both microplastics and toxic substances bioaccumulate within the food chain. Microplastics attached to phytoplankton and zooplankton are ingested by primary consumers like small fish, which are then consumed by larger predators. This process spreads microplastic contamination throughout the entire aquatic food web, affecting life from the bottom to the top of the chain. As microplastics cannot degrade, they pose long-term toxic risks, with chemical contamination potentially causing harm over time. In larger fish, the accumulation of microplastics may even lead to death.

The main sources of microplastics entering freshwater systems, such as rivers, lakes, and streams, include air deposition, wastewater treatment facilities, sewage and industrial effluents, urban runoff, and improper waste disposal (Rajpal et al. 2024). Historically, rivers were the primary conduit for plastic waste into marine ecosystems. The occurrence and dispersal of microplastic particles in freshwater environments are influenced by various factors, including climate and hydrodynamic characteristics like water current, waves, wind, and river discharge rates, along with geographical features.

Microplastics tend to concentrate in the water column, making them bioavailable to a variety of organisms, including planktonic, nektonic, and bottom-dwelling species. Many different types of organisms, including diatoms, planktonic crustaceans, fish, mussels, and zooplankton, are able to consume microplastics. Abiotic factors, such as temperature, biotic factors, such as food availability, diet composition, and interactions between predators and prey, as well as the kind of microplastic, feeding behavior of the species, and exposure route, all have a role in influencing the likelihood of ingestion.

16.4.1 Light

The abiotic degeneration of plastic litter in the sea is mainly hindered by oxygen and light restriction. The main degradation processes are photodegradation, thermo-oxidative degradation and hydrolysis (Gewert et al. 2015). Plastics are photo-oxidized when they are exposed to ultra violet (UV) light, and this changes the surface of the polymer at a chemical and physical level. This mechanism results in the production of oxidation products such as carbonyl groups at the surface, increases hydrophilicity and makes plastic biodegradable (Hakkarainen and Albertsson 2004). Plastic particles developed microcracks merely under extended UV-exposure and become brittle hastening physical breakdown (Andrady 2011). Nevertheless, in deep marine waters, where light is minimal but pressure and cold temperatures are maximized the controls on microbial/plastic aggregates are expected to deviate considerably from those in shallow or freshwater environments (Wagner and Lambert 2018).

16.4.2 Temperature

The degradation of plastic during its transportation processes accelerates as ambient temperatures rise. On coastal areas, the thermo-oxidative breakdown of plastic waste is driven by increasing temperatures. Prolonged exposure to sunlight causes plastics to degrade thermally and undergo faster photodegradation (Torikai et al. 1986; Kamweru et al. 2011). It has been estimated that ice melting in the Arctic Ocean could release up to one trillion microplastic particles into the water within the next decade (Obbard et al. 2014). A study on glacier melt-induced microplastic pollution in a remote mountainous lake basin on the Tibetan Plateau revealed that most of the discharged microplastics were small and lightweight (Dong et al. 2021). Additionally, due to physical abrasion and light exposure, the freeze-thaw cycles occurring annually would further break down microplastics into nano plastics (Horton and Dixon 2018).

16.4.3 Rain

According to Xia et al. In the case of Goel et al., the microplastic concentration in surface water rose from 7.4 microplastic per liter to 30 microplastic per liter during a heavy rainfall, the value increased fourfold. Rainfall would trap atmospheric microplastics causing them to be transported into aquatic habitats (Abbasi 2021). Furthermore, ARA has the potential of causing the acidification-induced coral reef bleaching and enhancing release of sequestered microplastics as redistribution of MPP relationships exist (O'Connor et al. 2019). It can also lead to the degradation of plastics and alter the pH in aquatic environments (Hocker et al. 2014; Krupa and Manning 1988).

16.4.4 Wind

The aquatic ecosystem can also serve as a sink for microplastics, depending on the nature of the plastic particles and their environment (i.e., in water column, sediments or bedrock). Influenced by those external factors, global warming would stimulate winds and rainfall, and the water circulation will be promoted, leading to transferring of microplastics transported all the way to the Arctic Sea. Glaciers in areas with higher wind speeds receive more microplastics via the atmosphere. Rapid melting of ice caps releases the stored microplastic into their natural ranges after unprecedented global warming (Haque and Fan 2023).

16.5 Distribution of MPs on Marine Organisms

Table 16.1 Summary of the types and size ranges of microplastics (MPs) found in various organisms collected at different locations based on a literature review. A number of marine species have been sampled and identified as already carrying MPs such as mussels, eastern oysters, shrimp and the fresh water flea, *Daphnia magna*. Turtles were found to have ingested soft floating MPs, which were probably ingested mechanically during feeding or as a result of entanglement with their food items (such as macroalgae or seagrass branches). Filter feeders and seabirds can accumulate MPs and macroplastics, which are then transferred to the next higher-trophic level. Marine animals—including whales can consume MPs directly from the water. MPs are ubiquitous in the marine environment, and MPs including polypropylene (PP), polyethylene terephthalate (PET), polyethylene (PE), nylon, and mesoplastics are present in large numbers in the marine environment. A range of microplastics, varying in size from 5 mm to 1200 mm, have been reported to be found in the intestine of humpback whales. A recent study conducted in Southeastern Australia revealed that 9 out of 30 seabirds tested positive for MPs in their stom-

Table 16.1 Micro/nanoplastics in organism samples

Site	Organisms	Type of MPs	Size of MPs (mm)	References
China	Dophin	Fibers, polyester, polyethylene acrylic acid	1–5	Zhu et al. (2019)
Arctic and sub-Arctic regions	Benthos	Microfibers, PE, PET, PA	0.01–5	Naji et al. (2018)
	Mollusca	Northern part of the Persian Gulf	–	
Sanggou Bay, China	Oysters	Fiber	0.05–5	Wang et al. (2019)
Great Barrier Reef	Sea turtles	Acrylic paint, polyvinyl, copolymer	0.45–2.95	Agm et al. (2018)
Southern Labrador Sea (Canada)	Seabird	Pooled and fragment	˂42.4	Avery-Gomm et al. (2018)
East Greenland Current	Seabird	Filaments	–	Avery-Gomm et al. (2018)
–	Mussel	–	30 nm	Wegner et al. (2012)
–	Mussel	–	50 nm	Canesi et al. (2015)
–	Mussel	–	51 mm	Pitt et al. (2018)
–	Zebrafish	PS-NPs (fluorescently labeled)	1000 nm	Catarino et al. (2019)
–	Zooplankton, fish	PS-NPs (fluorescently labeled)	26–330 nm	Mattsson et al. (2017)
Australian urban wetlands	Fish	Polyester, fibers	–	Su et al. (2019)
Yellow Sea and Bohai Sea of China	Finless porpoises	PP, fibers	0–1200	Xiong et al. (2018)
French, Belgian and Dutch North Sea coast	Lugworm	LDPE, HDPE, fiber, PS	˂1	Van Cauwenberghe et al. (2015)

achs. Additionally, *Fulmarus glacialis* (a northern fulmar) from the southern Labrador Sea in Canada was reported to have an average of 11.6 particles per individual (0.151 grams per bird) of MPs (Peng et al. 2020).

16.6 Impact on Marine Food Sources and Associated Health Risk

16.6.1 Inhibit Growth and Development

Polystyrene microplastics suppress the growth of Arenicola marina in sediment, with the level of suppression increasing with microplastic concentration. On the contrary, polyethylene microplastics can affect on the growth and feeding behavior of *Tripneustes gratilla* (Li et al. 2021). Meanwhile, polystyrene microplastics greatly reduce energy reserves of *Sebastes schlegelii* (Liu Qiang et al. 2017), so they consume less food which directly results in lower feeding rates or active metabolic suppression; hence ingesting through marine life make this organism less nutritious. In bivalve molluscs which were exposed to larger plastic particles, no differences in lipid and protein content were found. On the other hand, their total energy reserves improved when being exposed to higher concentrations of microplastics. For instance, the exposure of *Mytilus galloprovincialis* to polyethylene microplastics have resulted a disturbed energy balance and hence, an increased energy demand with delayed growth.

16.6.2 Cause Toxicological Effects

Microplastics can affect the standards of reproductive health among marine life. For example, polystyrene microplastics can lead to a decrease in ovulated egg cells and lowered sperm motility level in *Crassostrea gigas*, where microplastics would interfere with the reproductive ability of *C. gigas* on significant extent (Li et al. 2021). Microplastics are then absorbed into the biological tissues and organs of marine fishery where they generate immune responses. For instance, polyvinyl chloride (PVC) and polyethene (PE) microplastics with a particle size of 40–150 μm can induce oxidative stress in the systemic erythrocytes of *Sparids verata* and sea bass *Laotarox*, ultimately leading to calcium toxicity depending on immunosuppression (Liu Qiang et al. 2017).

16.6.3 Cause Genetic Damage

It is known that microplastics affect the genetic integrity of marine organisms and cause damage to those who eat them. Moreover, *Magallana galloprovincialis* can be exposed with several microplastics that absorb polycyclic aromatic hydrocarbons and induce immunotoxicity, neurotoxicity and genotoxicity, whereas based on the few data available indicate also possible genetic damage in mussels. Nevertheless, few studies have evaluated the genotoxicity of microplastics in marine biota (Du Jing et al. 2018).

16.6.4 Digestive Disorders

MPs and microplastics have been found in the stomachs of dead or stranded seabirds such as Sooty Shearwaters (*Ardenna grisea*) and Northern Fulmars (*F. glacialis*) (Robuck et al. 2022). These plastics produced ventricular overloading and clogged the bird's digestive system, eventually killing it. Furthermore, it has been revealed that plastic garbage harms or obstructs fish and seabirds' digestive tracts. When fish larvae were exposed to fluorescent PE-MPs with diameters ranging from 1 to 5 m, their growth was halted, and early juvenile fish (*P. microps*) were more severely harmed than more significant adults (Yuan et al. 2022). *Mytilus*, a type of oyster, showed signs of increased lysozyme activity and cytotoxicity and decreased phagocytic and feeding activity. Lugworms exposed to conventional (PE and PVC) and biodegradable (PLA) MPs in beach sands ate less, moved less, and kept food in their stomachs longer (Green 2016).

16.6.5 Inflammation

The persistent MPs can injure the organs, tissues, cells, molecules of the digestive system, and other organs if they travel there. These injuries may cause tissue damage and inflammation. Plastic chemical leaching, such as phthalates, bisphenol A, and brominated flame retardants, could be to blame for the inflammation in M. edulis in response to MPs (Green 2016).

16.6.6 Decreased Reproduction

MPs would cause a long-distance migrant seabird to delay breeding. *Calanus helgolandicus* copepods experienced a considerable decline in fecundity (reproductive production) after exposure to 20 m polystyrene MPs (Bhuyan et al. 2021). Plasticizers may have contributed to the endocrine disturbance that prevented aquatic animals and non-mammalian species from reproducing normally.

16.6.7 Effects on Metabolisms

Fish have been reported to inhibit the enzymatic activity of acetylcholinesterase (AChE), e.g., Japanese medaka (*Pomatoschistus microps*), *Oryzias latipess*. The primary cholinesterase, AChE, breaks down the neurotransmitter's acetylcholine and numerous other choline esters. When fin whales were subjected to MP pollution, substantial lipid peroxidation (LPO) and oxidative stress were seen in the animals. Bisphenol A caused energy disturbances in rainbow trout Oncorhynchus mykiss oocytes, such as changing how energy was allocated from somatic growth to vitellogenesis (BPA) (Minaz et al. 2022). In addition to this, a few genotoxic outcomes were reported.

16.6.8 Other Detrimental Effects

The copepod *C. helgolandicus* was negatively impacted by MPs of size 7.3 μm (Piliyan et al. 2023). Plastic extracts such as phthalates, brominated flame retardants, phthalates and bisphenol A are just a few examples of those that have been demonstrated to leach into food sources and raise the risk of toxicity. Adding MPs to beach sands influenced the nitrogen cycle, primary production, and the metabolism of the lugworm *A. marina.*

16.7 Future Concerns and Recommendations

The release of MPs, NPs, and other secondary pollutants into the ocean is associated with human activity. While most originate on land, it travels to the sea by wind, rain, and rivers. MPs have been found in water, soil, and animal samples from oceans and places worldwide, whereas NPs have only been found in animals due to their subtlety. MPs and NPs can infiltrate the marine ecosystem, food chain, and animal bodies due to their small size. Consequently, they pose a danger to the maritime environment.

In contrast to macro plastics, MPs and NPs are small enough to pass through the tissues of numerous organs, potentially causing damage to those organs through physical, chemical, and biological toxicity. Plastics are pretty stable, meaning they could accumulate in animals and cause long-term harm. No in-vivo research has yet been done on the impact of MPs and NPs on human health. However, they are known to be ingested, and in vitro studies have shown that they harm human health. It is crucial to keep an eye on the fate of microplastics in different habitats as the effects of climate change on the ecosystem grow. Knowing the mechanisms controlling the dispersion of microplastics in the aquatic environment is essential since the aquatic habitat serves as the primary sink for these tiny plastic particles. Since glacier snow has been recognized as another ultimate sink for microplastics, glacier melting and the release of microplastics are still unavoidable due to rising global temperatures. As a result, it is essential to keep an eye on the redistribution of microplastic caused by ice caps melting. Under various climate change scenarios, investigate the possible consequences of microplastics on ecosystem functioning and services, including implications on nutrient cycling, carbon sequestration, and overall ecosystem resilience. Develop frameworks for predictive modeling that consider both direct and indirect effects when evaluating the combined effects of climate change and microplastic pollution on specific ecosystems and species at risk. Investigate innovative technologies for detecting, monitoring, and removing microplastics in various environmental compartments, considering the influence of climate change on the effectiveness of these methods.

In contrast to MPs, nanoscales (NPs) have a far easier time penetrating animal cells and moving from one organ to another. Not only are they more efficient pollution adsorbents and transporters, but they are also known to be harmful to human and animal health. However, most studies have ignored them because of their evasiveness, the limitations of available sampling and analysis methods, and the need for standard units for necessary quantities like particle abundance.

16.8 Conclusion

In nature, the marine environment has been seriously polluted by huge amounts of microplastics (MPs) and nano plastics (NPs), which come from the arbitrary use and mishandling of plastic that has caused ecological damage on an upscale with larger marine organisms affected. They halt the effects of particles on sea inhabitants and could increase morbidity and mortality rates for a vital actor in the food chain. It is highly probable that marine organisms will ingest microplastics, and its biological consequence of this seems to be significant on the marine food chain. To examine for how long these particles are present in the gut of marine organisms, and to what degree they accumulate, it is crucial to investigate microplastic movement between trophic levels. Try to collect more quantitative data with better resolution on how common micro- and nanoplastics are in the air, how frequent they go in and come out of organisms, and how often freshwater and marine species take up hydrophobic pollutants. Such fish would include but not be entirely limited to a suite of demersal and pelagic species at intermediate and higher-trophic levels. Besides, there is the demand for studying metabolic pathways Modulation readily available in marine species after ingest MNPs taking into consideration the MNPs concentration exposure time and physical and chemical properties of particles.

References

Abbasi S (2021) Microplastics washout from the atmosphere during a monsoon rain event. J Hazard Mater Adv 4:100035

Abbasi S, Soltani N, Keshavarzi B, Moore F, Turner A, Hassanaghaei M (2018) Microplastics in different tissues of fish and prawn from the Musa Estuary, Persian Gulf. Chemosphere 205:80–87

Agm C, Thomas CR, Kle B, Motti CA, Arieil E, Brodie JE (2018) Ingestion of microplastic debris by green sea turtles (*Chelonia mydas*) in the Great Barrier Reef: validation of a sequential extraction protocol. Mar Pollut Bull 127:743–751

Alimba CG, Faggio C (2019) Microplastics in the marine environment: current trends in environmental pollution and mechanisms of toxicological profile. Environ Toxicoll Pharmacol 6689(18):30571–30574

Andrady AL (2011) Microplastics in the marine environment. Mar Pollut Bull 62:1596–1605

Anik AH, Hossain S, Alam M, Sultan MB, Hasnine MT, Rahman MM (2021) Microplastics pollution: a comprehensive review on the sources, fates, effects, and potential remediation. Environ Nanotechnol Monitor Manag 16:100530

Avery-Gomm S, Provencher JF, Liboiron M, Poon FE, Smith PA (2018) Plastic pollution in the Labrador Sea: an assessment using the seabird northern fulmar *Fulmarus glacialis* as a biological monitoring species. Mar Pollut Bull 127:817–822

Bao M, Xiang X, Huang J, Kong L, Wu J, Cheng S (2023) Microplastics in the atmosphere and water bodies of coastal agglomerations: a mini-review. Int J Environ Res Public Health 20(3):2466

Barboza LGA, Cózar A, Gimenez BCG et al (2019) Macroplastics pollution in the marine environment. World seas: an environmental evaluation. Academic Press, London, pp 305–328

Bhuyan MS, Nabi MR, Alam MW, Islam MN, Cáceres-Farías L, Bat L, Musthafa MS, Chung SY, Senapathi V, Alfaro-Núñez A (2021) Impacts of microplastics on marine organisms and in human health

Bratovcic A (2019) Degradation of micro- and nanoplastics by photo catalytic methods. J Nanosci Nanotechnol Appl 3:206

Browne MA, Awantha D, Galloway TS, Lowe DM, Thompson RC (2008) Ingestedmi croscopic plastic

translocates to the circulatory system of the mussel, *Mytilus edulis* (L). Environ Sci Technol 42:5026–5031
Canesi L, Ciacci C, Bergami E, Monopoli MP, Dawson KA, Papa S, Canonico B, Corsi I (2015) Evidence for immunomodulation and apoptotic processes induced by cationic polystyrene nanoparticles in the hemocytes of the marine bivalve *Mytilus*. Mar Environ Res 111:34–40
Carrasco A, Pulgar J, Quintanilla-Ahumada D, Perez-Venegas D, Quijon PA, Duarte C (2019) The influence of microplastics pollution on the feeding behavior of a prominent sandy beach amphipod, Orchestoidea tuberculata (Nicolet, 1849). Mar Pollut Bull 145:23–27
Catarino AI, Frutos A, Henry TB (2019) Use of fluorescent-labelled nanoplastics (NPs) to demonstrate NP absorption is inconclusive without adequate controls. Sci Total Environ 670:915–920
Choudhury M, Sharma A, Pervez A, Upadhyay P, Dutta J (2022) Growing menace of microplastics in and around the coastal ecosystem. In: Coastal ecosystems: environmental importance, current challenges and conservation measures, pp 117–137. https://doi.org/10.1007/978-3-030-84255-0_6
Dewika M, Markandan K, Irfan NA, Abdah MA, Ruwaida JN, Sara YY, Khalid M (2023) Review of microplastics in the indoor environment: distribution, human exposure and potential health impacts. Chemosphere 4:138270
Dong H, Wang L, Wang X, Xu L, Chen M, Gong P, Wang C (2021) Microplastics in a remote Lake Basin of the Tibetan plateau: impacts of atmospheric transport and glacial melting. Environ Sci Technol 55:12951–12960
Du Jing Y, Mingxi SG et al (2018) Research advances on monitoring and toxicology of microplastic pollution in marine as indicated by bivalves [J]. Chinese J Ecol 37(7):2205–2212
Eriksen M, Lebreton LC, Carson HS et al (2014) Plastic pollution in the world's oceans: more than 5 trillion plastic pieces weighing over 250,000 tons afloat at sea. PLoS One 9:e111913
Fossi MC, Marsili L, Baini M, Giannetti M, Coppola D, Guerranti C, Caliani I, Minutoli R, Lauriano G, Finoia MG, Rubegni F, Panigada S, Berub M, Urban Ramírez J, Panti C (2016) Fin whales and microplastics: the Mediterranean Sea and the sea of Cortez scenarios. Environ Pollut 209:68e78
Gewert B, Plassmann MM, Macleod M (2015) Path ways for degradation of plastic polymers floating in the marine environment. Environ Sci Process Impacts 17:1513–1521
Geyer R, Jambeck JR, Law KL (2017) Production, use, and fate of all plastics ever made. Sci Adv 3:e1700782
Green DS (2016) Effects of microplastics on European flat oysters, *Ostrea edulis* and their associated benthic communities. Environ Pollut 216:95–103
Habibi N, Uddin S, Fowler SW, Behbehani M (2022) Microplastics in the atmosphere: a review. J Environ Expos Assess 1:6
Hakkarainen M, Albertsson A-C (2004) Environmental degradation of polyethylene. Adv Polym Sci 169:177–199
Haque F, Fan C (2023) Fate of microplastics under the influence of climate change. Iscience 26:107649
Hara J, Frias J, Nash R (2020) Quantification of microplastic ingestion by the decapod crustacean Nephrops norvegicus from Irish waters. Mar Pollut Bull 152:110905
Hocker S, Rhudy AK, Ginsburg G, Kranbuehl DE (2014) Polyamidehydrolysis accelerated by small weak organic acids. Polymer 55:5057–5064
Horton AA, Dixon SJ (2018) Microplastics: an introduction to environmental transport processes. WIREs Water 5:e1268
Jambeck JR, Geyer R, Wilcox C et al (2015) Marine pollution. Plastic waste inputs from land into the ocean. Science 347:768–771
Kamweru PK, Ndiritu FG, Kinyanjui TK, Muthui ZW, Ngumbu RG, Odhiambo PM (2011) Study of temperature and UV wavelength range effects on degradation of photo-irradiated polyethylene films using DMA. J Macromol Sci, Part B 50(7):1338–1349
Kershaw P, Saido K, Sangjin L (2011) Plastic debris in the ocean. I Gland, Switzerland. IUCN, pp 1–45. https://doi.org/10.2305/IUCN.CH.2014.03.en
Khoshmanesh M, Sanati AM, Ramavandi B (2023) Co-occurrence of microplastics and organic/inorganic contaminants in organisms living in aquatic ecosystems: a review. Mar Pollut Bull 187:114563
Krupa SV, Manning WJ (1988) Atmospheric ozone: formation and effects on vegetation. Environ Pollut 50:101–137
Kwak JI, Liu H, Wang D, Lee YH, Lee JS, An YJ (2022) Critical review of environmental impacts of microfibers in different environmental matrices. Comp Biochem Physiol C Toxicol Pharmacol 251:109196
Li Y et al (2021) Research on the influence of microplastics on marine life IOP Conf. Ser.: earth environmental science 631:012006
Liu Qiang X, Xudan HW, Xiaoqun X, Shou L, Jiangning Z (2017) Research advances on the ecological effects of microplastic pollution in the marine environment [J]. Acta Ecol Sin 37(22):7397–7409
Mattsson K, Johnson EV, Malmendal A, Linse S, Hansson LA, Cedervall T (2017) Brain damage and behavioural disorders in fish induced by plastic nanoparticles delivered through the food chain. Sci Rep 7(1):11452
Minaz M, Er A, Ak K, Nane İD, İpek ZZ, Kurtoğlu İZ, Kayış Ş (2022) Short-term exposure to bisphenol A (BPA) as a plastic precursor: hematological and behavioral effects on *Oncorhynchus mykiss* and *Vimba*. Water Air Soil Pollut 233(4):122
Naidoo T, Sershen, Thompson RC, Rajkaran A (2020) Quantification and characterization of microplastics ingested by selected juvenile fish species associated with mangroves in KwaZulu-Natal, South Africa. Environ Pollut 257:113635

Naji A, Nuri M, Vethaak AD (2018) Microplastics contamination in molluscs from the northern part of the Persian Gulf. Environ Pollut 235:113–120

O'Connor D, Pan S, Shen Z, Song Y, Jin Y, Wu WM, Hou D (2019) Microplastics undergo accelerated vertical migration in sand soil due to small size and wet-dry cycles. Environ Pollut 249:527–534

Obbard RW, Sadri S, Wong YQ, Khitun AA, Baker I, Thompson RC (2014) Global warming releases microplastic legacy frozen in Arctic Sea ice. Earth's Future 2:315–320

Peng L, Fu D, Qi H, Lan CQ, Yu H, Ge C (2020) Micro- and nano-plastics in marine environment: source, distribution and threats-a review. Sci Total Environ 698:134254

Piliyan R, Meril D, Karuppaiyaa ND, Selvaraju A, Ayyasamy G, Gunaseelan RJ, Mathiazhagan S, Manickam M, John R, Pachiyappan P (2023) Dietary effects of microplastics on the physiological and biochemical profiles of keystone secondary producers *Oithona dissimilis* (Lindberg, 1941)

Pitt JA, Kozal JS, Jayasundara N, Massarsky A, Trevisan R, Geitner N, Wiesner M, Levin ED, Di Giulio RT (2018) Uptake, tissue distribution, and toxicity of polystyrene nanoparticles in developing zebrafish (*Danio rerio*). Aquat Toxicol 194:185–194

Rajpal A, Choudhury M, Goswami S, Chakravorty A, Raghavan V (eds) (2024) Waste management and treatment: advances and innovations, 1st edn. CRC Press. https://doi.org/10.1201/9781003258377

Ribeiro FG, O'Brien J, Galloway TS, Thomas KV (2019) Accumulation and fate of nano and micro-plastics and associated contaminants in organisms. Trends Anal Chem 111:139–147

Robuck AR, Hudak CA, Agvent L, Emery G, Ryan PG, Perold V, Powers KD, Pedersen J, Thompson MA, Suca JJ, Moore MJ (2022) Birds of a feather eat plastic together: high levels of plastic ingestion in Great Shearwater adults and juveniles across their annual migratory cycle. Front Marine Sci 8:1754

Sambandam M, Dhineka K, Sivadas SK, Kaviarasan T, Begum M, Hoehn D, Sivyer D, Mishra P, Murthy MR (2022) Occurrence, characterization, and source delineation of microplastics in the coastal waters and shelf sediments of the central east coast of India, Bay of Bengal. Chemosphere 303:135135

Su L, Nan B, Hassell KL, Craig NJ, Pettigrove V (2019) Microplastics biomonitoring in Australian urban wetlands using a common noxious fish (*Gambusia holbrooki*). Chemosphere 228:65–74

Sutton R, Mason SA, Stanek SK, Willis-Norton E, Wren IF, Box C (2016) Microplastic contamination in the San Francisco Bay, California, USA. Mar Pollut Bull 109:230e235–230e235

Tekman MB, Gutow L, Macario A, Haas A, Walter A, Bergmann M (2019) Alfred-wegener-institut helmholtz-zentrum für Polar- und meeresforschung. Available from: https://litterbase.awi.de/litter_detail. Last accessed on 20 Jan 2022

Torikai A, Takeuchi A, Nagaya S, Fueki K (1986) Photodegradation of polyethylene: effect of crosslinking on the oxygenated products and mechanical properties. Polym Photochem 7:199–211

Turra A, Manzano AB, Dias RJS, Mahiques MM, Barbosa L, Balthazar Silva D, Moreira FT (2014) Three-dimensional distribution of plastic pellets in sandy beaches: shifting paradigms. Sci Rep 4:4435

Uddin S, Fowler SW, Behbehani M (2020) An assessment of microplastic inputs into the aquatic environment from wastewater streams. Mar Pollut Bull 160:111538

Van Cauwenberghe L, Claessens M, Vandegehuchte MB, Janssen CR (2015) Microplastics are taken up by mussels (*Mytilus edulis*) and lugworms (*Arenicola marina*) living in natural habitats. Environ Pollut 199:10–17

Wagner M, Lambert S (eds) (2018) Freshwater micro plastics: emerging environmental contaminants? SpringerOpen

Wang J, Lu L, Wang M, Jiang T, Liu X, Ru S (2019) Typhoons increase the abundance of microplastics in the marine environment and cultured organisms: a case study in Sanggou Bay,China. Sci Total Environ 667:1–8

Wegner A, Besseling E, Foekema EM, Kamermans P, Koelmans AA (2012) Effects of nanopolystyrene on the feeding behavior of the blue mussel (*Mytilus edulis* L.). Environ Toxicol Chem 31:2490–2497

Wright SL, Thompson RC, Galloway TS (2013) The physical impacts of microplastics on marine organisms: a review. Environ Pollut 178:483–492

Wright SL, Ulke J, Font A, Chan KLA, Kelly FJ (2020) Atmospheric microplastic deposition in an urban environment and an evaluation of transport. Environ Int 136:105411

Xiong X, Chen X, Zhang K, Mei Z, Hao Y, Zheng J, Wu C, Wang K, Ruan Y, Lam PK, Wang D (2018) Microplastics in the intestinal tracts of East Asian finless porpoises (*Neophocaena asiaeorientalis sunameri*) from Yellow Sea and Bohai Sea of China. Mar Pollut Bull 136:55–60

Yuan Z, Nag R, Cummins E (2022) Human health concerns regarding microplastics in the aquatic environment-from marine to food systems. Sci Total Environ 7:153730

Zhu J, Yu X, Zhang Q, Li Y, Tan S, Li D, Yang Z, Wang J (2019) Cetaceans and microplastics: first report of microplastic ingestion by a coastal delphinid, *Sousa chinensis*. Sci Total Environ 659:649–654

Nano and Microplastic—Impact on Human Life

17

Jige Sandipan B. Sandip,
Mohamed Abdel-Raheem,
and Subhangi H. Fulamarkar

17.1 Introduction

The word plastic comes from the Greek word 'plastikos', which means to be shaped or molded. The first man-made plastic was created by Alexander Parkes in 1862; its material is known as Parkesine. It was organic and derived from the cellulose on the heating and, after being molded, retained its shape when cooled. Plastic is made from petroleum, and this petroleum contributes to deforestation, so the use of more plastic cuts many trees for oil drilling and refining facilities. The plastic has link with cancer, reproductive problems, and other health problems for humans and animals. The biological, chemical, and environmental elements will break down plastics into more amount of microplastic. Microplastic refers to plastic fragments and particles that have a diameter of less than 5 mm. The plastic particle with a diameter of less than 1 μm is known as nanoplastics. In the twenty-first century, plastic pollution is one of the most serious environmental problems; microplastics are ubiquitous in the global environment. The microplastics has become a ubiquitous pollutant in our environment and contaminates water, soil, and air (Adhikari 2018). Humans ingest the micro plastic through consumption of fish, food, water, and inhalation. In the atmosphere the evaporation of ponds and used equipment and tools degraded, the honeybees also collect microplastic from the air and interact with plants, soils, and water (Chauhan 2012). In humans, through food nano- and microplastic enters into body, causing internal damage such as gastrointestinal blockage, reduced feeding, inflammatory response. According to the WWF report on normal water and food consumption, it estimates that every person consumes about 5 grams of plastic each week. The nanoplastic fragments are invisible to the human eye; these filaments, granules, and flakes are also called secondary nanoplastic. All forms of nanoplastics are formed by the decomposition of larger plastic pieces and packing (Rajak 2019). Microplastics cause physical harm to the human body such as gastrointestinal tract damage, intestinal blockages, and damage to

J. S. B. Sandip (✉)
Sant Ramdas College Ghansawangi Dist, Jalna, Maharashtra, India

M. Abdel-Raheem
Agricultural and Biological Research Institute, National Research Centre, Cairo, Egypt

S. H. Fulamarkar
Johnson Ave, San Jose, CA, USA

S. Goswami et al. (eds.), *Micro-Nano Plastics Exposure, Environmental Degradation and Public Health Crisis*, Sustainable Development Goals Series, https://doi.org/10.1007/978-981-96-5047-7_17

organs like liver, kidney, lungs, and different tissues. The chemicals present in microplastics affect on reproductive and developmental issues. It also increases the risk of cancer, immune system suppression, cardiovascular disease, neurodegenerative diseases and neurological damage (La Nasa 2022). Nevertheless, extremely few studies have been done on the prevalence and number of microplastics in different organs/cells of the human body. Accordingly, this chapter will address the following issues:

- To study the prevalence of nano and microplastic in the human body
- To understand potential risks associated with exposure to nano and microplastic particles
- To develop strategies to minimize exposure and mitigate nano and microplastic effects on human health
- To aware new generation about effect of nano and microplastic on living organisms

17.2 Prevalence of Microplastics in Human

The microplastic damages the human cell, decreases human reproductive health, and disrupts the endocrine system. It also acts as a vessel for harmful substances to enter the body as they absorb chemicals linked to cancer and a weakened immune system. Microplastics are found in the human placenta, colon, lung, liver, sputum, feces, blood, and breast milk. In the lung, cotton, rayon, and polyethylene microplastics are observed. The tumor tissues show 38 particles and normal tissues 27 particles per 50 samples (Chen et al. 2021).

In the sputum, the HNO, NaOH Fourier transform-infrared spectroscopy detection method detects the polyurethane, polyvinyl chloride, and polyethylene types of microplastics 18.75 to 91.75 particles per 10 ml samples (Hung et al. 2022). In human blood, the proteinase K pyrolysis-gas chromatography-mass spectrometry detection method detects the polyethylene terephthalate and polyethylene 1.6 μg/ml. The human liver, kidney, and spleen showed 11, 3, and 3 microplastic particles, respectively, of polyethylene terephthalate, polyvinyl chloride, and polystyrene types of microplastics by KOH Raman digestion method, which is equivalent to 1.4 particles per gram (John Toedt 2005). In the human placenta, the 10% KOH Raman digestion method detects polyethylene, polystyrene, film fiber, and fragments types of microplastics (Rajak 2019). The human breast milk shows 38% polyethylene, 21% polyvinyl chloride, and 17% polypropylene by the digestion method KOH Raman and fragments of 26 microplastic particles per 34 samples (Chen et al. 2021). The human urine contains polyethylene, polyvinyl chloride, polypropylene, polyvinyl alcohol of microplastic by KOH Raman digestion method, and fragments of microplastics of 4–15 particles per sample (Adhikari 2018).

17.3 Impact of Microplastics on Human Life

The nanoplastic in the human body crosses the cell membrane and enters the bloodstream, and it also interacts with biomolecules such as protein and DNA. It also damages cells and disrupts cellular function and the immune system (Rajak 2019). This plastic has a significant threat to human health due to its unique properties and ability to interact with biological system of body. The microplastics found in blood vessels and increase the risk of heart attack and stroke. It also presents in higher concentration in people with inflammatory bowel disease. It also causes internal blockages and tissue damage; it also interferes with the reproductive system of human and marine animals. The inhaling microplastic leads to respiratory issues such as asthma and chronic obstructive pulmonary diseases. It also triggers autoimmune response which leads to diseases like rheumatoid arthritis and lumps (Chen et al. 2021).

The brain infiltration by MPs has caused a decrease in glial fibrillary acidic protein GFAP in support of cell process in brain. The decrease in GFAP is related to early stages of neurodegenerative diseases like depression and Alzheimer's disease. The GFAP signaling is altered by the microplastics (Kids and Andrus 2020). The blood of human makes up to 6–7% of the body weight of human; it transports oxygen and nutrients for different tissues and organs. The microplastics eliminated by renal filtration or binary excretion or it deposited in liver and spleen via filtered capillaries and sinusoids. Vein tissue was also contaminated by microplastics. The microplastics detected in human stool, lung, heart, liver, and placenta also affect circulatory, respiration, nervous, and reproductive system of human (Bergmann et al. 2022).

In the respiration process, microplastics enter in upper sputum, nasal cavity and lower alveoli, lung tissue of the tract detected in the human. It raises concerns about their health effect on the respiration system of human. The direct link between microplastics and human respiratory disease does not occur, but current research shows microplastics alter endogenous surfactants of lung. The lung cell impairs and increases its susceptibility to lung disorders like asthma, pulmonary fibrosis, and pulmonary frosted glass nodules, so microplastics can cause lung diseases. The circulatory system supplies nutrients and oxygen to different tissues in the human body and removes waste products. The microplastics identified in human thrombus raise concerns about potential impact on the human circulatory system (Jacopo La Nasa 2022). The human reproductive recently paid attention to the potential impact of microplastics in human placenta.

The wokers of coal miners, oil and gas fields extrct fossil carbon feedstock for the plastic production, at time of work they suffermortality due to traumatic injury. The coal worker has also been infected by lung cancers, silicosis, pneumoconiosis, and cardiovascular diseases. The plastic producer worker shows the risk of brain cancer, breast cancer, leukemia, lymphoma, hepatic angiosarcoma, neurotoxin injury, and mesothelioma with decreased fertility (Bingham 2019). The workers of plastic textile production are at increased risk of bladder cancer, intestinal cancer, lung cancer, and mesothelioma. Plastic recycling workers face increased risk of cardiovascular diseases, neuropathy, lung cancer, and toxic metal poisoning (Karapanagioti 2019). In the environment, the microplastic also acts as a carrier of heavy metals, and it is absorbed on the surface by the electrostatic interaction and complication. It releases magnesium, zinc, arsenic, chromium, copper, nickel, and lead in gastrointestinal solution. The inhalation process also absorbs plastic particles, and it damages the lung depending upon the sizes of the particles potentially embedded into the lung and stay on the alveolar surface and other parts of the body (Aves et al. 2022).

17.4 Mechanism of Microplastic Toxicity in Human Life

A set of adverse outcome pathways (AOPs), which outline the biological mechanisms that result in negative health effects upon exposition, can be utilized to comprehend the damaging effects of microplastics on humans. These processes aid in clarifying how microplastics trigger cellular and molecular reactions that lead to toxicity by virtue of their chemical and physical attributes. Key AOPs linked to cytotoxicity caused by microplastics are summarized below in sequential steps:

Uptake and Bioavailability: The human system can absorb microplastic particles through the mouth, nose, or skin. Small enough microplastics, especially nanoplastics, can penetrate biological barriers and enter the bloodstream through injured skin, the gastrointestinal epithelium, or the lungs' alveolar membranes (Vethaak and Legler 2021).

Disruption of Membranes and Cellular Uptake: Microplastics are capable of being undertaken via endocytosis and pass-through cell membranes, especially those that have specific surface charges. By interacting with lipid bilayers, microplastics' hydrophobic and charged properties can either damage membranes or improve cellular absorption. Cellular internationalization might occur from interactions between positively charged substances and negatively charged cellular membranes (Wegner et al. 2012).

Oxidation and Inflammation: Reactive oxygen species (ROS) undergo formation in response to the accumulation of microplastics in cells, particularly in organs such as the liver or lungs. The harmful substances that contain MPs have the capability to produce an excessive amount of reactive oxygen species (ROS), overpowering the cell's antioxidant defenses (Aves et al. 2022).

Cellular constituents like proteins, lipids, and DNA may sustain serious harm as the outcome of this oxidative stress. Activation of the immune system: The body's defenses identify microplastic particles as foreign intruders. Immunological cells (such as dendritic cells or macrophages) launch a defense mechanism when they come into contact with microplastics. The immune system might phagocytes bigger particles, but it might additionally overpower or miss smaller fragments (Karapanagioti 2019).

Release of harmful substances: As soon as microplastics infiltrate the body, they unleash chemical additives or absorb environmental contaminants. Microplastics frequently absorb contaminants from the surroundings, such as permanent organic pollutants (POPs) and polycyclic aromatic hydrocarbons (PAHs), or include hazardous additions like heavy metals, phthalates, and bisphenol A (BPA). Cellular functions may be affected when these compounds are released into the bloodstream (Blom and Gabrielson 2020).

Genotoxicity and Damage to DNA: Direct or indirect DNA damage results from exposure to microplastics. Exposure to microplastics can produce reactive oxygen species, which can directly oxidative damage DNA. Furthermore, DNA might become mutated by the association of harmful compounds generated by microplastics (Rellan et al. 2023).

Endocrine Disturbances: The body absorbs endocrine-disrupting chemicals (EDCs) from microplastics. Substances that imitate or disrupt the body's hormonal processes include BPA and phthalates, which are frequently employed as plasticizers. Regular biological signaling might be altered by their ability to attach to hormone-binding sites (Aves et al. 2022).

Respiratory and Gastrointestinal Toxicology: Microplastics build up in the breathing passages or digestive system. Microplastics that are breathed can settle in the lung tissue, while those that are consumed can build up in the stomach. Normal functioning in this machinery might be interfered with by the tangible presence of these foreign particles and the leaching of harmful chemicals (Ye et al. 2022).

Neurotoxicity: The blood-brain barrier (BBB) is crossed by microplastics, especially nanoplastics. Damage to neurons may result from nanoplastics that are small enough to penetrate the blood-brain barrier and infiltrate the central nervous system (CNS) (Nollet and Siddhiqi 2020).

17.5 Factors Affecting the Microplastic Associated Human Toxicity

Numerous factors affect how these fragments become acquainted with human tissue and boost a possible health hazard, which in turn affects the toxicity of microplastics connected with humans. A list of significant variables is provided below, along with citations to studies in this new area.

Size One important factor influencing the toxicity of microplastics is their size. The rain, liver, kidneys, and other organs can be reached by tiny particles such as nanoplastics (less than 1 μm), which can pass through biological obstacles like cell membranes and infiltrate the bloodstream. Although they are less likely to go through these obstacles, larger microplastics can build up in the lungs and digestive system and may result in tissue destruction and inflammatory responses (Blom and Gabrielson 2020).

Shape The toxic nature of microplastics is influenced by various forms, including fibers, pieces, and spheres. Because they could stick to the respiratory system and result in persistent inflammation and respiratory disorders, fibrous microplastics are especially dangerous. On the other hand, pieces with irregular shapes have a higher potential to mechanically harm structures than smoother fragments, which raises the possibility of cellular disruptions (Blom and Gabrielson 2020).

MP's Chemical Makeup One important aspect influencing cytotoxicity is the chemical composition of microplastics, particularly the kind of polymer and their inclusion of additives. The risk for hazardous chemical releases varies across common plastics, such as polyethylene (PE), polypropylene (PP), and polystyrene (PS). Upon reaching the human system, additives like phthalates, bisphenol A (BPA), and heavy metals can leak out and cause reproductive issues, endocrine disruption, and possibly even cancer.

Microplastics' hydrodynamic properties and their electrical charge can affect how they connect to biological tissues. The possibility of cellular uptake is increased by the strongly charged particles' propensity to attach to negatively charged membranes on cells. This might result in increased toxicity, particularly in cells in the immune system, which may react by causing inflammation or cell mortality (Hung et al. 2022).

Duration and Route of Exposure The main ways individuals become accustomed to microplastics are by eating or drinking them, breathing in airborne fragments, and coming into touch with them through cosmetics. Exposure route affects the degree of poisoning, with inhaling causing respiratory disorders and ingesting causing gastrointestinal irritation. The buildup of microplastics in tissues over time makes chronic exposure, even at low levels, especially alarming (Vethaak and Legler 2021).

One important consideration is bioavailability, or the extent to which microplastics and the compounds they contain are metabolized by the tissues of humans. Because they are more accessible than bigger particles, nanoplastics have a higher propensity to penetrate biological membranes, gather in internal organs, and increase their systemic damage (Kids and Andrus 2020).

Presence of Contaminants The contaminants in the environment are spread by microplastics. They absorb harmful elements from the environment, such as heavy metals, persistent organic pollutants, and polycyclic aromatic hydrocarbons (PAHs). These substances have the ability to become absorbed and infiltrate the body after ingestion, which can lead to cellular damage, immunological disorder, and the development of oxidative stress. Combining exposure to microplastics with other environmental pollutants may have cumulative or synergistic consequences, increasing the dangers to one's general health (Rellan et al. 2023). Microplastics, for example, may increase the uptake of co-exposed toxicants, including organic contaminants or heavy metals, hence escalating inflammation-related and oxidative stress reactions (Bergmann et al. 2022).

Biological Attributes People with already occurring medical disorders and kids are among the vulnerable groups most at risk from microplastic-associated cytotoxicity. Since their immunity and organs are still growing, children may be more vulnerable to prolonged exposure. Furthermore, the consequences of microplastic buildup might be more severe for those who suffer from gastrointestinal or respiratory illnesses (Bingham 2019).

Environmental Attributes Environmental triggers that might boost the harmful effect of microplastics include weathering and disintegration. They become more responsive and have a broader surface area as they decompose into smaller fragments, which enables them to absorb more environmental contaminants. Additionally, temperature and pH conditions affect how quickly toxic compounds discharge from microplastics (Chen et al. 2021).

Trophic Translocations Trophic transmission results from the building up of microplastics in marine species and other living things that humans eat. Contamination is increased throughout the food chain by this technique, especially in seafood. A greater level of microplastics and related contaminants in humans can be caused via accumulation in lower trophic-level species (Wani and Ariana 2019).

17.6 Future Aspects and Recommendation

Today every person has come in contact with plastic due to more use of plastice for daily needs like textiles, personal care products and medical items like solution containers, transfer tubing, and transfusion sets, devices, and drug packing. The personal products include hand cleaner, toothpaste, shaving foam, soap, sunscreen, bubble bath, facial scrub and shampoo. It is used by every person in home; these are microplastics spreading products. In the future, the use of these products will become harmful to human health. The microplastic and nanoplastics spreading articles and products use need to be banned for better human health (Crawford and Quinn 2016). The research on effect of microplastic and nanoplastics on human health is necessary to be carried out in every country. The government has imposed strict regulations for the use of plastic articles also, people's awareness about the plastic waste impact on the environment is necessary. The government also strictly follows law and regulation related to environment and plastic ban. For the control of plastic pollution, humans also use the four R, in which the first R means refuse plastic products, and the second R indicates reduce the use of plastic instead of using bioplastic. The third R indicates reuse it before getting discarded, and the fourth R indicates recycle the plastic and create other products for a better future for the planet (Chintan Environment Research and Action Group 2021).

17.7 Conclusion

In human life, plastic plays an important role, and it is used by every person due to its lightweight and ease of carrying. But plastic use increases its pollution because it is not degraded easily in the environment. It contains different types of chemicals that are deposited in different ecosystems, such as terrestrial and marine, and disturb living organisms. Today in the world, many tons of plastic products occur in landfills and oceans which produces pollution. The use of plastic decreases day by day, but its recycling rate is very small, so it occurs in every place of the world. The entry of plastic into animal and plant bodies affects human health, so today, the human body also shows the amount of content in blood and other parts of the human body. For a better future for the planet and living organisms, more research and control of plastic are necessary.

References

Adhikari (2018) Environment: pollution and protection. Narosa, Delhi

Aves AR, Revell LE, Gaw S, Ruffell H et al (2022) First evidence of micro plastics in Antarctic snow. Cryosphere J EGU 16:2127–2145

Bergmann M, Collard F, Fabres J, Gabrielsen GW et al (2022) Plastic pollution in the Arctic. Nat Rev Earth Environ J 3:323–337. https://doi.org/10.1038/s43017-022-00279-8

Bingham EC (2019) An investigation of the laws of plastic flow. Forgotten Books, London

Blom K, Gabrielson GW (2020) Plastic Sea: a bird's-eye view. Cornell Lab, New York

Chauhan AS (2012) Society and environment. Jain Brother's, New Delhi

Chen Q, Gao J, Yu H, Su H, Yang Y, Cap Y, Zhang Q, Ren Y et al (2021) An emerging role of micro plastics in the etiology of lung ground glass nodules. J Environ Sci 34:25

Chintan Environment Research & Action Group (ed) (2021) Plastic: a threat to mankind. National Book Trust, New Delhi

Crawford CB, Quinn B (2016) Micro-plastic pollutants. Elsevier Science, Wilmington

Hung H, Hung X, Bi R, Guo Q, Yu X, Zeng Q, Hung Z, Liu T et al (2022) Detection and analysis of micro plastics in human septum. J Environ Sci Technol 168:107459. https://doi.org/10.1016/j.envint.2022.107459

Karapanagioti H (2019) Micro-plastics in water and wastewater, Glendale Heights

Kids L, Andrus A (2020) The plastic problem. Lonely planet, Sydney

La Nasa J (2022) Micro-plastics degradation and characterization. MDPI, Basel

Nollet LML, Siddhiqi K (2020) Analysis of Nano plastics and micro-plastics in food. CRC Press

Rajak A (2019) Micro-plastic pollution. Notion Press, Chennai

Rellan AG, Ares DV, Lopez CV, Bugallo PM et al (2023) Sources sinks and transformation of plastics in our

oceans: review, management strategies and modeling. Sci Total Environ J 854:158745. https://doi.org/10.1016/j.scitotenv.2022.158745
Toedt J (2005) Chemical composition of everyday products. Greenwood, New York
Vethaak AD, Legler J (2021) Micro plastic and human health. Sci J 371(6530):672–674. https://doi.org/10.1126/science.abe5041
Wani K, Ariana L (eds) (2019) Environmental and human health impacts of plastic pollution. IGI Global, USA
Wegner A, Besseling E, Foekema EM, Kamermans P, Koelmans AA (2012) Effect of nano polystyrene on the feeding behavior of the blue mussel (Mytilus edulis L.). Environ Toxicity Chem J 123:18–25. (10.1016)
Ye Y, Yu K, Zhao Y (2022) The development and application of advanced analytical methods in micro plastics contamination detection: a critical review. Sci Total Environ J 818:151851. https://doi.org/10.1016/j.scitotenv.2021.151851

A Comprehensive Review on Micro and Nanoplastics-Mediated Cardiovascular Complications

18

Jashabir Chakraborty, Dibyendu Shil, Saumendu Deb Roy, Amit Samanta, Diptendu Bhowmik, Sharbani Roy, Manab Deb Adhikari, and Pankaj Kanti Jodder

18.1 Introduction

In the modernized world, society produces hundreds of millions of plastics, including their use in consumer commodities, packaging, and dressing. An estimate currently provided states that 364 million tons of plastics are generated yearly and that this may increase threefold by the year 2050. (Rubio et al. 2020). Plastics may also be categorized based on key characteristics relevant to their manufacturing process, product designing, and end-of-use, like thermoplastics and thermosets. Thermoplastics, basically, are melted by heat and solidified on cooling but thermosets got a chemical change with temperature and modified into a three-dimensional network that is impossible to be re-melted or reformed. Thermoplastics include polyethylene (PE) and polyamides, polypropylene (PP), polycarbonate (PC), polystyrene (PS), such as expanded polystyrene; polyethylene terephthalate (PET); or the case of polyvinyl-chloride (PVCs), among others. In the case of the thermoset polymer class, a choice for prototype has been polyurethane (PUR) as in the case of an injection molding application, along with silicones, a composite application for epoxy as for phenolic resins as also an acrylic example for completion (Da Costa et al. 2020).

After that, once these materials come into the environment, they are prone to degrade abiotically or biotically and end up in microplastics. Generally, primary and secondary are the two types of micro and nanoplastics, basically differentiated on the basis of their entry into the environment. One of them, the primary ones, enter the environment in final forms while the second kind get its forms by weathering and wearing away of plastics. (Boucher 2020). The generation of microplastics (MPs, <5 mm) and nanoplastics (NPs, <1000 nm) poses significant health risks. These pollutants can induce various toxicological effects in living creatures mostly including cyto-

J. Chakraborty (✉) · D. Bhowmik · S. Roy
Department of Pharmacology, Mata Gujri College of Pharmacy, Mata Gujri University, Kishanganj, Bihar, India

D. Shil · S. D. Roy
Department of Pharmacognosy, Mata Gujri College of Pharmacy, Mata Gujri University, Kishanganj, Bihar, India

A. Samanta
Department of Chemistry, Mata Gujri College of Pharmacy, Mata Gujri University, Kishanganj, Bihar, India

M. D. Adhikari
Department of Biotechnology, University of North Bengal, Siliguri, India

P. K. Jodder
University of North Carolina, Chapel Hill, NC, USA

S. Goswami et al. (eds.), *Micro-Nano Plastics Exposure, Environmental Degradation and Public Health Crisis*, Sustainable Development Goals Series, https://doi.org/10.1007/978-981-96-5047-7_18

toxicity, oxidative stress, enzyme disruptions, lysosomal damage, immune dysfunction, genetic alterations, and proteomic/transcriptomic changes. Cardiovascular effects are a primary focus of current research (Galloway et al. 2017; Liang et al. 2022; Matthews et al. 2021).

Cardiac disorders are basically caused by pathological changes in microvasculature in the myocardium and changes in the myocardium due to the effects of different pollutants. Plenty of predisposing factors like unhealthy diet, lifestyle, obesity, sedentary lifestyle, and air pollution have been associated with CVDs. Recent reported experimental data on micro- and nanoplastics (MNPs) depicted that high exposure to these types of pollutants can cause abnormalities in blood fluid dynamics, cardiac output, heart rate (HR), fibric changes in myocardial tissue, and cardiometabolic disorder. Research on Microplastics and Nanoplastics (MNPs) has demonstrated their capacity to traverse the subcutaneous barrier and enter the circulatory system. Subclinical cardiac toxicity associated with MNP exposure includes arrhythmias, cardiac malfunctioning, effusion into pericardium, fibromyocardial tissue formation, and other cardiac pathologies. As well, MNPs have been shown to induce toxicity at the microvascular level, manifesting as hemolysis, thrombosis, coagulopathy, and endothelial dysfunction. Still, there is a dearth of information available regarding MP-induced cardiovascular disorders and explaining its mode of action. Accordingly, in this chapter, a brief description of mechanisms for etiopathogenesis of these diseases induced by MPs is discussed in a comprehensive way. Additionally, the chapter examines the potential cardiac risks, including arrhythmia, atherosclerosis, myocardial infarction, and other CVDs, associated with MP exposure.

18.2 Molecular Basis of Cardiovascular Disease Etopathogenesis by Micro and Nanoplastics Materials

In recent human's studies, scientists have been able to make an estimate an intake of 80 g per day of micro and nanoplastics (MNPs) that accumulate via plants (fruits and vegetables) through uptake as food from polluted soil (Claudia Campanale et al. 2020). Micro and nanoplastic materials can cause severe damage to the cardiovascular system, which ends up in acute or chronic cardiovascular complications like impairments in cardiac functions like heart rate, dyslipidemia, myocardial fibrosis, pericardial edema, induced apoptosis and pyroptosis of the myocardial cell, induce hemolysis, thrombosis and vascular endothelial damage (Fig. 18.1).

18.2.1 Induction of Apoptosis

Polystyrene microplastics (PS-MPs) were observed to exert cardiovascular toxicity by initiating cardiac fibrosis. This was achieved by promoting apoptosis in myocardial cells. The probable mechanism involved oxidative stress, which resulted in increased collagen deposition within the heart. MPs induced an initial state of myocardial oxidative stress in rats that promotes cardiomyocyte apoptosis. Subsequently, cardiac fibrosis started, which was driven by the augmentation of the Wnt/β-catenin signaling cascade (Wei et al. 2021; Li et al. 2020a). PS-MPs were implicated in inflammatory processes and a specific type of cell death known as pyroptosis, both of which were mediated by oxidative stress. This oxidative stress, in turn, resulted in the development of myocardial fibrosis. The molecular mechanisms underlying these effects involved the production of two key upstream mediators. One of these was reactive oxygen species (ROS), generated by polymorphonuclear neutrophils, and the activation of NF-κB. PS-MPs also activated the NLRP3/Caspase-1/GSDMD pathway. These events led to a specific increase in malondialdehyde levels and a noticeable reduction in antioxidant enzymes, prominently identified superoxide dismutase, glutathione peroxidase, and catalase, primarily due to the generation of ROS (Wei et al. 2021). It is observed that oral administration of polystyrene microplastics (PS-MPs, 5 μm, 1–100 mg/L) to chickens for a span of 42 days resulted in noticeable histopathological changes in the heart-like disrupted myocardial fibers, infiltration of inflammatory cells, and damaged mito-

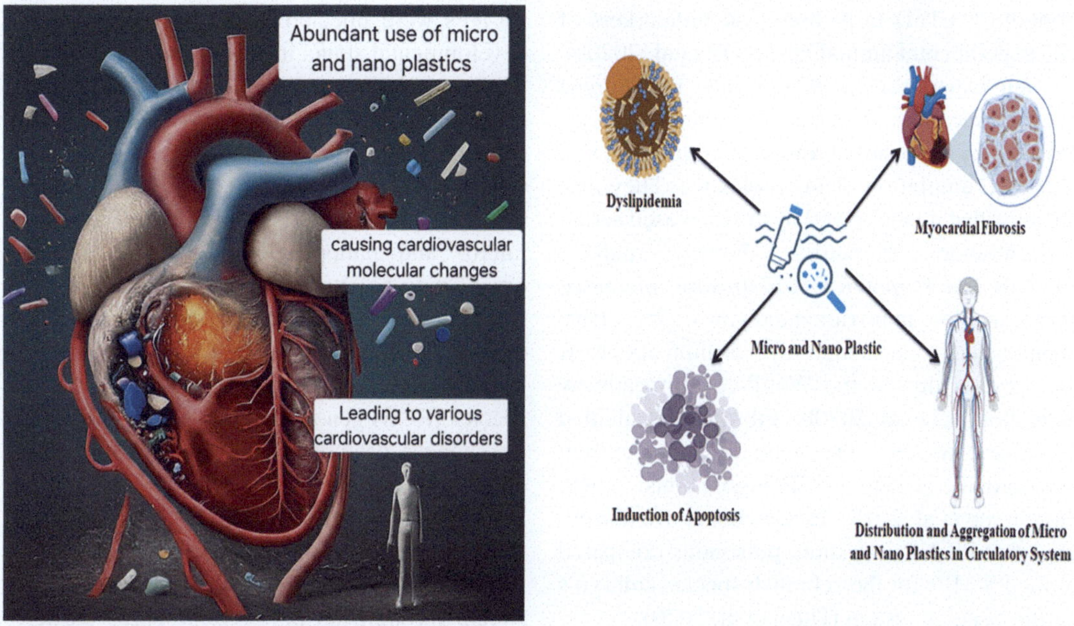

Fig. 18.1 Graphical illustration of the molecular basis of various cardiovascular disorders in micro and nanoplastics poisoning

chondria. Additionally, the researchers found that PS-MP exposure mediated oxidative stress by altering antioxidant enzyme levels and elevating reactive oxygen species (ROS) production. Thus, oxidative stress, in turn, activated the NF-κB-NLRP3-GSDMD cascade pathway, leading to pyroptotic cell death and worsening myocardial inflammation (Zhang et al. 2022). The Carp exposed to polystyrene nanoparticles (PS-NPs) of various sizes (50–400 nm) in a concentration of 1000 μg/L exhibited myocardial inflammation and cell death (apoptosis). Smaller PS-NPs were associated with increased cardiac damage and disrupted immune responses (Th1/Th2 imbalance) (Wu et al. 2022).

18.2.2 Changes in the Lipid Profile or Dyslipidemia

Dyslipidemia is a major cause of atherosclerosis, which ends in the development of major CVS and ischemic heart disease. Polystyrene microplastic (PS-MPs) dietary exposure in males causes dyslipidemia. Rats exposed to microplastics orally experienced a downfall in high-density lipoprotein cholesterol (HDL-C) and significant elevation in low-density lipoprotein cholesterol (LDL-C) (Nnoruka et al. 2022). The micro and nanoplastics (MNPs) were shown to induce lipid accumulation in human macrophages. This effect was associated with increased oxidative stress within mitochondria, leading to a loss of mitochondrial membrane potential. These cellular changes are indicative of foam cell transformation and progression into atherosclerotic plug formation (Florance et al. 2022).

18.2.3 Damage to the Myocardium Fibrosis

Consumption of Microplastics has been found to disrupt the structural and functional integrity of the heart muscles. The maximum damage was observed in the mitochondria of cardiomyocytes. Death of these cells was also observed. Elevated amount of Creatine kinase-MB and Cardiac troponin I (cTnI) in pathological biomarkers of this experimental animal model (Li et al. 2020b). In their experiment in Wistar rats, polystyrene microplastic polysaccharide was found to be cardiotoxic by myocardial vascular congestion due to the accumulation of microplastics. They also reported that it is followed by the increased serum concentration of cardiac damage markers CK-MB and Troponin I at high dose, increased MDA, and low antioxidant enzymes. These have been known to cause high myocardium apoptosis and fibrosis due to the Wnt/β-catenin pathway activation. (Li et al. 2020b). Fibroid and calcified adaptive changes cause hypoxia and recurrent myocardial ischemia. (Frangogiannis 2019; Hutcheson et al. 2015). Researchers have discovered that PS-NPs are more poisonous compared to the PS-MPs for the zebrafish species embryo's cardiovascular system (Duan et al. 2020).

18.3 Distribution In Vivo and Aggregation of Micro and Nanoplastics in the Circulatory System

Across various animal models, including mussels, zebrafish, and mice, ingested MNPs were observed to accumulate significantly in the circulatory system. In mice, this accumulation was particularly evident in specific tissues such as kidneys, blood vessels, heart, and other highly vascular parts of organs (Browne et al. 2008). It was found that polystyrene microplastics with diameters of 3.0 and 9.6 micrometers, respectively, traversed the gut and entered the circulation within a 3-day period. These microplastics persisted in the bloodstream for more than 48 days. (Browne et al. 2008). In a zebrafish study, researchers examined how the distribution of 700 nm PS-NPs within the body changes when injected into the yolk of embryos at various stages of their development (Veneman et al. 2017). PS-NPs injected into blastula-stage embryos were observed to concentrate around the developing heart and within the circulatory system 3 days post-injection. However, when PS-NPs were injected in the later phase of the developmental stage, its traces were not found in the ducts of Cuvier. Researchers found bioaccumulation and diverse allocations of MNPs within the circulation influenced by their particle size (Duan et al. 2020; Pitt et al. 2018a). Scientists have postulated that the wide biodistribution of micro- and nanoplastics (MNPs) to various organs may be attributed to the signaling induction of epithelial cell death (apoptosis) and the disruption of the intestinal barrier performance, both of which are triggered by reactive oxygen species (ROS) generated by MNPs. These cellular changes promote the absorption and subsequent systemic distribution of MNPs (Liang et al. 2021). In some studies, MNPs can traverse the protective layers of the intestines and lungs. Once they breach these barriers, they enter the bloodstream through the portal vein. These particles can then be carried throughout the body, eventually accumulating near the heart. Larger particles, however, face greater obstacles. Microplastics exceeding 0.5 millimeters in size are less likely to penetrate through the intestinal wall, and those bigger than 400 nanometers are unable to enter into the delicate lung tissue. (Lusher et al. 2020).

18.3.1 Cardiac Toxicity of MNPs and Prognosis of Cardiac Abnormalities

18.3.1.1 Abnormality in Heart Rate (HRs)

Irregular heart rhythms, including excessively fast (tachycardia) or slow (bradycardia) heart rates, can signal underlying heart problems and related disorders (Reil and Böhm 2007; Tadic et al. 2018). Pitt et al. (2018a, b) in their experiment found that the impact of MNP exposure on heart rate is complex and can vary. While some studies have linked MNP exposure to increased heart rate, others have reported a lowering effect on heart rate. In a specific study, zebrafish exposed to MNPs (34.5 ± 10.8 nm; 0.1–10 mg/L) displayed a dose-dependent reduction in heart rate, known as bradycardia. This effect was also observed in case of the offspring of zebrafish exposed to PS-NPs (30.67 ± 8.97 nm; 1 mg/g BW) during maternal and co-parental stages. (Pitt et al. 2018a, b).

Zebrafish embryos exposed to PS-NPs (100 nm, 400 items/mL) exhibited increased heart rate 24 h after fertilization (hpf) (Duan et al. 2020). Some studies specifically focused on fish to identify how different types of metal nanoparticle (MNP) exposure can alter heart development. In case of embryos collected from marine medaka exposed to polystyrene-coated MNPs (0.05–6.00 micrometers in size, at concentrations of 0.1–1 million particles per milliliter) experienced significant changes in heart rate. These changes were linked to altered expression of genes (BMP4, GATA4, and NKX2.5) known to play a key role in heart development (Chen et al. 2020). In an experiment, they have found that when isolated adult zebrafish hearts were exposed to PS-MPs (8 μm; 26 and 260 mg/L) for 2 h, a significant downfall in heart rate was observed. This effect was more pronounced at higher concentrations of PS-MPs, indicating a concentration-dependent response (Dimitriadi et al. 2021). A direct interaction between A and MNPs was identified by Geiser et al., suggesting a mechanism through which sarcomeres can influence HR (Geiser et al. 2005). Heart rate variability (HRV) was reported to be a significant inference of cardiovascular disease (CVD) prognosis, reflecting autonomic nervous dysfunction (Malik et al. 1996). Individuals with lower HRV are at increased risk of heart-related illnesses and death, highlighting its importance as a prognostic indicator for cardiovascular diseases. (Bigger Jr et al. 1992; Tsuji et al. 1996).

18.3.1.2 Impairment of Cardiac Function

The heart's primary function is to efficiently pump blood throughout the body. Several factors regulate this pumping action, including heart rate, which means the speed at which the heart contracts, and preload, which is the amount of blood filling the heart chambers before contraction. Afterload means the resistance that the heart must overcome to eject blood. Contractility means the forcefulness of the ventricular contraction (Muir and Hamlin 2020). Heart function is evaluated by heart rate and the strength of heart muscle contractions (Chico et al. 2008; Kitambi et al. 2012). When adult forms of zebrafish were isolated and given exposure to PS-MPs (8 μm; 26 and 260 mg/L) for a duration of 2 h, a significant frequency downfall found in atrial systole. However, ventricular contractility remained unaffected. These findings indicate that PS-MPs preferentially target pacemaker cells over cardiac myocytes in ventricle (Dimitriadi et al. 2021). In Young's study, by interacting with β-adrenergic receptors on ventricular cardiomyocytes, MNPs could disrupt the heart's electrical activity, increasing the risk of cardiac toxicity. (Yong et al. 2020). In vitro study on neonatal stage rat ventricular cardiomyocytes (NRVMs) were treated with amine- or carboxylate-PS-NPs (50 nm; 25 μg/mL) for a period of 60 min prominently revealed that the amine form of PS-NPs was more readily internalized. This differential uptake resulted in a more significant lowering effect in NRVM contractility. The early phase of contractile impairment was characterized by a sharp fall in intracellular Ca^{2+} amount and alterations in electrophysiological activity. Subsequently, in the late phase, mitochondrial membrane potential and cellular metabolism were significantly compromised, further exacerbating contractile dysfunction. Given the heightened sensitivity of developing hearts to MNP exposure, newborn offspring may be particularly vulnerable during the infant feeding process of infants (Roshanzadeh et al. 2021). A 24-h exposure of human-induced pluripotent stem cells (hiPSCs) to polyethylene nanoparticles (PE-NPs) of 40 nm and 200 nm diameter, at a concentration of 1 × 10^9 particles/mL, resulted in significant adverse effects on ventricular valve development (Bojic et al. 2020). Changes in cardiac output (CO) can affect cardiac function (Muir and Hamlin 2020). PE-NPs (191.10 ± 3.13 nm) at a concentration of 200 μg/mL significantly compromised the cardiovascular health of zebrafish at 96 hpf. This was manifested by a decrease in CO and fluid dynamics of blood flow, indicative of elevated cardiac afterload (Sun et al. 2021). NP-induced thrombus causes alteration in arterial blood pressure leading to an effect on myocardial contractility and CO levels (Diamond 2016). In several research on zebrafish, scientists observed significant pericardial edema after an aqueous exposure to PS-NPs (50 nm; 1 mg/L). Likewise, zebrafish embryos, when injected with PS-NPs (20 ± 3.2 nm; 0.0318 μg/animal), apart from HR increase and changes in flow rate of blood developed pericardia region edema (Bhagat et al. 2021).

18.3.1.3 Pericardial Edema

MNPs trigger pericardial edema, a frequent indicator of general cardiovascular harm in zebrafish research. As exposure to MNPs increases, so does toxicity, which is associated with the recruitment of neutrophils and subsequent acute inflammatory responses (Bhagat et al. 2021; Sun et al. 2021).

18.3.2 MNPs Toxicity on Microvasculature Sites

18.3.2.1 Hemolysis

Micro and nanoplastics (MNPs) can damage human red blood cells (RBCs) in laboratory conditions. When RBCs are exposed to polystyrene nanoparticles (PS-NPs) of various sizes (50–250 nm) and concentrations (50–500 μg/mL) in a solution devoid of protein, they break down. However, adding a small amount of albumin (0.01–0.5%) to the solution can protect the RBCs from this damage (Barshtein et al. 2011). Hwang et al. (2020). Research reported that high concentrations of irregular PE-MP particles (1–100 micrometers in size range 10–500 micrograms per milliliter in mass) with a rough surface produce hemolysis in the RBC of sheep (Choi et al. 2021). The hemolytic effects of MNPs are a complex interplay of various factors, likely to be particle size and structure, exposure amount, interactions with blood proteins, and surface modifications (Zhu et al. 2023).

18.3.2.2 Effect of Micro and Nanoplastics on Thrombosis

A few researchers have reported that when human venous blood was exposed to carboxylate-PLNPs (50 nm, 260–750 μg/mL), aggregation of platelets was observed. This effect was accompanied by an upregulation of CD62P and PAC-1, key markers of thrombocyte activation, in both cases of whole blood and isolated thrombocyte samples. Notably, unmodified and amine-PLNPs were involved in similar platelet activation. McGuinnes et al. (2011) hypothesized that the thrombocytes or platelet aggregation caused by amine-PS-NPs was pinpointly a result of membrane damage.

18.3.2.3 Micro and Nanoplastic Materials Affecting Blood Coagulation

MNPs can alter the delicate balance between blood clotting and anticoagulation processes. Amine-PS-NPs, sized 57 and 330 nanometers at concentrations of 0.06 to 0.5 milligrams per milliliter, bound to clotting factors VII and IX, hindering the extrinsic pathway and reducing thrombin formation. Conversely, carboxylate-PS-NPs, sized 24 and 220 nanometers at a concentration of 0.5 milligrams per milliliter, stimulated the intrinsic pathway, promoting coagulation. Sulfonated PS-NPs, sized 23 and 200 nanometers at a concentration of 0.5 milligrams per milliliter, did not impact either the intrinsic or extrinsic pathways due to their negative charge (Oslakovic et al. 2012).

18.3.2.4 Micro and Nanoplastic Materials Causing Vascular Endothelial Damage

In the zebrafish model, PE-NPs compromised the integrity of vascular endothelial cells, thereby establishing a hypoxic microenvironment (Duan et al. 2020). Circulating MNPs may trigger RBC and monocyte activation, leading to endothelial cell interactions and subsequent circulatory disorders, mostly atheroma formation (Barshtein et al. 2016). Treatment of human red blood cells with PS-NPs, sized between 50 and 250 nanometers at concentrations ranging from 0.05 to 0.5 milligrams per milliliter, promoted the clumping of erythrocytes and their attachment to endothelial cells (Barshtein et al. 2016). Tiny plastic particles, both micro and nano-sized, can directly interact with cells in our body, particularly immune cells and cells lining blood vessels. This interaction can trigger inflammation and contribute to the development of atherosclerosis, a condition that hardens arteries. When exposed to specific plastic particles (carboxylate-PS-MPs) in laboratory experiments, researchers observed an increase in certain molecules (VCAM-1 and ICAM-1) on the surface of blood vessel cells, which can further promote inflammation. Surprisingly, the release of proinflammatory mediators (IL-1β and TNF-α) in immune cells decreased under the same conditions (Pober and Sessa 2007).

In an experiment of in vivo studies in C57BL/6 N mice intravenously injected with 2.5 mg of 1 μm polystyrene microplastics (PS-MP) showed an increased expression of IL-1β and ICAM-1 in the cells of aortic tissue (Vlacil et al. 2021). When Zhao et al., in their experiment in a 12-week study, C57BL/6 mice were orally administered Polystyrene micro and nanoplastics (PS-MNPs) at doses of 0.5 and 5 μm, and 0.36 and 3.6 μg/animal, respectively. The researchers found no significant impact on aortic contractility or endothelial function. However, in a study with a minute sensitization of smooth muscle in the aorta by α-adrinomimetic agents and subsequent exposure to PS-MNPs causes perivascular adipose tissue (PVAT) dysfunctioning and inflammation. The potential mechanism behind these effects may involve some alteration in the gene expression, particularly within the *Wnt gene* superfamily, which has been very prominently implicated in cardiovascular disorders and atherosclerotic plug formation (Zhao et al. 2022). Scientists have suggested that PS-MNP intake may increase the risk of cardiovascular disorders by promoting a cardiometabolic disease phenotype (Ahmadieh et al. 2020). Autophagy is one mechanism contributing to MNP-induced endothelial dysfunction, though its role requires further investigation (Nussenzweig et al. 2015).

18.4 Combination Toxicity of Micro and Nanoplastics and Other Environmental Contaminants on Cardiovascular System

18.4.1 Synergistic Toxicity

In an experiment, zebrafish embryos were treated with 50 nm polystyrene nanoparticles (PS-NPs) at a concentration of 1 mg/L, either alone or in association with antifungal drug ketoconazole (KCZ) or fluconazole (FCZ) at 1 mg/L. All exposed groups exhibited decreased heart rate, while significant pericardial and yolk sac edema were observed compared to the control group. Additionally, the study revealed enhanced toxicity of KCZ and FCZ in the presence of PS-NPs, potentially due to increased reactive oxygen species (ROS), lipid peroxidation, and alteration in the expression of oxidative stress and apoptosis-related genes in co-exposed groups (Bhagat et al. 2021). In another observation by Lee et al., the presence of PS-NPs was found to significantly increase the toxicity of Au ions. This synergistic effect is believed to be driven by proinflammatory responses, which ultimately lead to damaging effects inside mitochondria (Lee et al. 2019).

18.4.2 Antagonistic Effects of Micro and Nanoplastics (MNPs)

MNPs can also mitigate CVS toxicity of different natural contaminants, like polycyclic aromatic hydrocarbons (PAHs) and ionic metals, by acting either antagonistically or protectively. Exposure of zebrafish larvae to Elizabeth River Sediment Extract (ERSE, 0.1–5%) containing PAHs (5.07–25.36 μg/L) causes deformed heart, impairments in cerebrovascular blood vessel formation. The cardiovascular toxic effect of PAHs was dramatically moderate in case of co-exposure PAHs with PS-NPs (44 nm; 0.1–10 μg/L) together. Its protective effect is attributed to the reduced bioavailability and bioaccumulation of the hydrocarbons in the presence of PS-NPs. Cationic interactions reduced PAH sorption to PS-NPs, increasing their lipophilicity and enhancing PAH retention and aggregation (Trevisan et al. 2019). In summary, Co-exposure to MNPs reduces the bioavailability and bioaccumulation of contaminants, leading to antagonistic effects.

18.4.3 Biphasic Effects Found in the Exposure of Micro and Nanoplastics

Biphasic effects mean that when two or more pollutants are mixed together, their combined effect can change over time. Sometimes, they work together to cause more harm (synergistic), and other times, they cancel each other out (antagonistic). This change can be influenced by how long the exposure lasts, the number of pollutants, and other factors. Zhang et al. (2020) have reported that embryonic zebrafish exhibited biphasic responses to MPs + $CdCl_2$ treatment (Zhang et al. 2020). When zebrafish embryos were exposed to $CdCl_2$ (0.01 mg/L), their heart rate (HR) decreased. However, when these embryos were simultaneously exposed to $CdCl_2$ and PS-MP, the effect on HR depended on the concentration of PS-MP. When Co-exposed with a lower amount of PS-MP (0.05 and 0.1 mg/L), it resulted in a hike of HR as compared to $CdCl_2$ exposure alone. Conversely, co-exposure to higher concentrations of PS-MP (1–10 mg/L) led to a downfall in HR (Zhang et al. 2020).

18.5 Conclusion

The current chapter explores the cardiovascular risks posed by microplastics and nanoplastics (MNPs). We have discussed how these materials can exacerbate or enhance the toxicity of other pollutants. The underlying pathogenesis of MNP-induced cardiovascular toxic effect includes oxidative stress, inflammatory response, apoptosis, pyroptosis, and direct cellular interactions. Despite growing concerns, our understanding of how MNPs affect the human cardiovascular system remains limited. The lack of direct human evidence and experimental data hinders our ability to fully assess the health risks posed by MNP exposure. Researchers must account for a multitude of factors, such as particle size, exposure amount or dose and duration; surface modifications and particle structural conformations; the presence of proteins, the life stage, genetic modifications, and mutagenic properties of the tested organism; and the potential for gene activation or silencing. This research could bridge the gap between scientific findings and environmental reality, improving our understanding of human cardiovascular disorders. However, the potential health risks of MNPs, especially cardiovascular toxicity, from various exposure routes (airborne, waterborne, soilborne, and direct contact) should be thoroughly assessed. Future research efforts should focus on identifying biochemical indicators of MNP-mediated cardiovascular damage and cellulaotoxicity, with the goal of developing strategies to mitigate cardiovascular risk and reduce related mortality rates.

References

Ahmadieh S, Kim HW, Weintraub NL (2020) Potential role of perivascular adipose tissue in modulating atherosclerosis. Clin Sci. 134(1):3–13

Barshtein G, Arbell D, Yedgar S (2011) Hemolytic effect of polymeric nanoparticles: role of albumin. IEEE Trans Nanobiosci 10(4):259–261

Barshtein G, Livshits L, Shvartsman LD, Shlomai NO, Yedgar S, Arbell D (2016) Polystyrene nanoparticles activate erythrocyte aggregation and adhesion to endothelial cells. Cell Biochem Biophys 74:19–27

Bhagat J, Zang L, Nakayama H, Nishimura N, Shimada Y (2021) Effects of nanoplastic on toxicity of azole fungicides (ketoconazole and fluconazole) in zebrafish embryos. Sci Total Environ 800:149463

Bigger JT Jr, Fleiss JL, Steinman RC, Rolnitzky LM, Kleiger RE, Rottman JN (1992) Frequency domain measures of heart period variability and mortality after myocardial infarction. Circulation 85(1):164–171

Bojic S, Falco MM, Stojkovic P, Ljujic B, Gazdic Jankovic M, Armstrong L, Markovic N, Dopazo J, Lako M, Bauer R, Stojkovic M (2020) Platform to study intracellular polystyrene nanoplastic pollution and clinical outcomes. Stem Cells 38(10):1321–1325

Boucher J (2020) Primary microplastics in the oceans: a global evaluation of sources. IUCN

Browne MA, Dissanayake A, Galloway TS, Lowe DM, Thompson RC (2008) Ingested microscopic plastic translocates to the circulatory system of the mussel, Mytilus edulis (L.). Environ Sci Technol 42(13):5026–5031

Campanale C, Massarelli C, Savino I, Locaputo V, Uricchio VF (2020) A detailed review study on potential effects of microplastics and additives of concern on human health. Int J Environ Res Public Health 17(4):1212

Chen JC, Chen MY, Fang C, Zheng RH, Jiang YL, Zhang YS, Wang KJ, Bailey C, Segner H, Bo J (2020) Microplastics negatively impact embryogenesis and modulate the immune response of the marine medaka Oryzias melastigma. Mar Pollut Bull 158:111349

Chico TJ, Ingham PW, Crossman DC (2008) Modeling cardiovascular disease in the zebrafish. Trends Cardiovas Med. 18(4):150–155

Choi D, Hwang J, Bang J, Han S, Kim T, Oh Y, Hwang Y, Choi J, Hong J (2021) In vitro toxicity from a physical perspective of polyethylene microplastics based on statistical curvature change analysis. Sci Total Environ 752:142242

Da Costa JP, Rocha-Santos T, Duarte AC (2020) The environmental impacts of plastics and micro-plastics use, waste and pollution: EU and national measures

Diamond SL (2016) Systems analysis of thrombus formation. Circ Res 118(9):1348–1362

Dimitriadi A, Papaefthimiou C, Genizegkini E, Sampsonidis I, Kalogiannis S, Feidantsis K, Bobori DC, Kastrinaki G, Koumoundouros G, Lambropoulou DA, Kyzas GZ (2021) Adverse effects polystyrene microplastics exert on zebrafish heart–molecular to individual level. J Hazard Mater 416:125969

Duan Z, Duan X, Zhao S, Wang X, Wang J, Liu Y, Peng Y, Gong Z, Wang L (2020) Barrier function of zebrafish embryonic chorions against microplastics and nanoplastics and its impact on embryo development. J Hazard Mater 395:122621

Florance I, Chandrasekaran N, Gopinath PM, Mukherjee A (2022) Exposure to polystyrene nanoplastics impairs lipid metabolism in human and murine macrophages in vitro. Ecotoxicol Environ Saf 238:113612

Frangogiannis NG (2019) Cardiac fibrosis: cell biological mechanisms, molecular pathways and therapeutic opportunities. Mol Asp Med 65:70–99

Galloway TS, Cole M, Lewis C (2017) Interactions of microplastic debris throughout the marine ecosystem. Nat Ecol Evol 1(5):0116

Geiser M, Rothen-Rutishauser B, Kapp N, Schürch S, Kreyling W, Schulz H, Semmler M, Hof VI, Heyder J, Gehr P (2005) Ultrafine particles cross cellular membranes by nonphagocytic mechanisms in lungs and in cultured cells. Environ Health Perspect 113(11):1555–1560

Hutcheson JD, Goettsch C, Rogers MA, Aikawa E (2015) Revisiting cardiovascular calcification: a multifaceted disease requiring a multidisciplinary approach. In: Seminars in cell & developmental biology, vol 46. Academic, pp 68–77

Hwang J, Choi D, Han S, Jung SY, Choi J, Hong J (2020) Potential toxicity of polystyrene microplastic particles. Sci Rep 10(1):7391

Kitambi SS, Nilsson ES, Sekyrova P, Ibarra C, Tekeoh GN, Andäng M, Ernfors P, Uhlén P (2012) Small molecule screening platform for assessment of cardiovascular toxicity on adult zebrafish heart. BMC physiology Dec 12:1–7

Lee WS, Cho HJ, Kim E, Huh YH, Kim HJ, Kim B, Kang T, Lee JS, Jeong J (2019) Bioaccumulation of polystyrene nanoplastics and their effect on the toxicity of Au ions in zebrafish embryos. Nanoscale 11(7):3173–3185

Li B, Ding Y, Cheng X, Sheng D, Xu Z, Rong Q, Wu Y, Zhao H, Ji X, Zhang Y (2020a) Polyethylene microplastics affect the distribution of gut microbiota and inflammation development in mice. Chemosphere 244:125492

Li Z, Zhu S, Liu Q, Wei J, Jin Y, Wang X, Zhang L (2020b) Polystyrene microplastics cause cardiac fibrosis by activating Wnt/β-catenin signaling pathway and promoting cardiomyocyte apoptosis in rats. Environ Pollut 265:115025

Liang B, Zhong Y, Huang Y, Lin X, Liu J, Lin L, Hu M, Jiang J, Dai M, Wang B, Zhang B (2021) Underestimated health risks: polystyrene micro-and nanoplastics jointly induce intestinal barrier dysfunction by ROS-mediated epithelial cell apoptosis. Particle Fibre Toxicol 18(1):20

Liang B, Huang Y, Zhong Y, Li Z, Ye R, Wang B, Zhang B, Meng H, Lin X, Du J, Hu M (2022) Brain single-nucleus transcriptomics highlights that polystyrene nanoplastics potentially induce Parkinson's disease-like neurodegeneration by causing energy metabolism disorders in mice. J Hazard Mater 430:128459

Lusher AL, Welden NA, Sobral P, Cole M (2020) Sampling, isolating and identifying microplastics ingested by fish and invertebrates. In: Analysis of nanoplastics and microplastics in food. CRC Press, pp 119–148

Malik M, Bigger JT, Camm AJ, Kleiger RE, Malliani A, Moss AJ, Schwartz PJ (1996) Heart rate variability: standards of measurement, physiological interpretation, and clinical use. Eur Heart J 17(3):354–381

Matthews S, Mai L, Jeong CB, Lee JS, Zeng EY, Xu EG (2021) Key mechanisms of micro-and nanoplastic (MNP) toxicity across taxonomic groups. Comp Biochem Physiol C: Toxicol Pharmacol 247:109056

McGuinnes C, Duffin R, Brown SL, Mills N, Megson IL, MacNee W, Johnston S, Lu SL, Tran L, Li R, Wang X (2011) Surface derivatization state of polystyrene latex nanoparticles determines both their potency and their mechanism of causing human platelet aggregation in vitro. Toxicol Sci 119(2):359–368

Muir WW, Hamlin RL (2020) Myocardial contractility: historical and contemporary considerations. Front Physiol 11:222

Nnoruka UC, Okonkwo CJ, Ilechukwu I, Okonkwo CJ, Belonwu DC (2022) Impact of polystyrene microplastic exposure on lipid profile and oxidative stress status of male and female Wistar rats. Environ Anal Health Toxicol 37(3):e2022024-0

Nussenzweig SC, Verma S, Finkel T (2015) The role of autophagy in vascular biology. Circ Res 116(3):480–488

Oslakovic C, Cedervall T, Linse S, Dahlbäck B (2012) Polystyrene nanoparticles affecting blood coagulation. Nanomedicine 6:981–986

Pitt JA, Kozal JS, Jayasundara N, Massarsky A, Trevisan R, Geitner N, Wiesner M, Levin ED, Di Giulio RT (2018a) Uptake, tissue distribution, and toxicity of polystyrene nanoparticles in developing zebrafish (Danio rerio). Aquat Toxicol 194:185–194

Pitt JA, Trevisan R, Massarsky A, Kozal JS, Levin ED, Di Giulio RT (2018b) Maternal transfer of nanoplastics to offspring in zebrafish (Danio rerio): a case study with nanopolystyrene. Sci Total Environ 643:324–334

Pober JS, Sessa WC (2007) Evolving functions of endothelial cells in inflammation. Nature Rev Immunol 7(10):803–815

Reil JC, Böhm M (2007) The role of heart rate in the development of cardiovascular disease. Clin Res Cardiol 96(9):585–592

Roshanzadeh A, Oyunbaatar NE, Ganjbakhsh SE, Park S, Kim DS, Kanade PP, Lee S, Lee DW, Kim ES (2021) Exposure to nanoplastics impairs collective contractility of neonatal cardiomyocytes under electrical synchronization. Biomaterials 278:121175

Rubio L, Marcos R, Hernández A (2020) Potential adverse health effects of ingested micro-and nanoplastics on humans. Lessons learned from in vivo and in vitro mammalian models. J Toxic Environ Health B 23(2):51–68

Sun M, Ding R, Ma Y, Sun Q, Ren X, Sun Z, Duan J (2021) Cardiovascular toxicity assessment of polyethylene nanoplastics on developing zebrafish embryos. Chemosphere 282:131124

Tadic M, Cuspidi C, Grassi G (2018) Heart rate as a predictor of cardiovascular risk. Eur J Clin Invest 48(3):e12892

Trevisan R, Voy C, Chen S, Di Giulio RT (2019) Nanoplastics decrease the toxicity of a complex PAH mixture but impair mitochondrial energy production in developing zebrafish. Environ Sci Technol 53(14):8405–8415

Tsuji H, Larson MG, Venditti FJ, Manders ES, Evans JC, Feldman CL, Levy D (1996) Impact of reduced heart rate variability on risk for cardiac events: the Framingham Heart Study. Circulation 94(11):2850–2855

Veneman WJ, Spaink HP, Brun NR, Bosker T, Vijver MG (2017) Pathway analysis of systemic transcriptome responses to injected polystyrene particles in zebrafish larvae. Aquat Toxicol 190:112–120

Vlacil AK, Bänfer S, Jacob R, Trippel N, Kuzu I, Schieffer B, Grote K (2021) Polystyrene microplastic particles induce endothelial activation. PLoS One 16(11):e0260181

Wei J, Wang X, Liu Q, Zhou N, Zhu S, Li Z, Li X, Yao J, Zhang L (2021) The impact of polystyrene microplastics on cardiomyocytes pyroptosis through NLRP3/Caspase-1 signaling pathway and oxidative stress in Wistar rats. Environ Toxicol 36(5):935–944

Wu H, Guo J, Yao Y, Xu S (2022) Polystyrene nanoplastics induced cardiomyocyte apoptosis and myocardial inflammation in carp by promoting ROS production. Fish Shellfish Immunol 125:1–8

Yong CQ, Valiyaveettil S, Tang BL (2020) Toxicity of microplastics and nanoplastics in mammalian systems. Int J Environ Res Public Health 17(05):1509

Zhang R, Wang M, Chen X, Yang C, Wu L (2020) Combined toxicity of microplastics and cadmium on the zebrafish embryos (Danio rerio). Sci Total Environ 743:140638

Zhang Y, Yin K, Wang D, Wang Y, Lu H, Zhao H, Xing M (2022) Polystyrene microplastics-induced cardiotoxicity in chickens via the ROS-driven NF-κB-NLRP3-GSDMD and AMPK-PGC-1α axes. Sci Total Environ 840:156727

Zhao J, Gomes D, Jin L, Mathis SP, Li X, Rouchka EC, Bodduluri H, Conklin DJ, O'Toole TE (2022) Polystyrene bead ingestion promotes adiposity and cardiometabolic disease in mice. Ecotoxicol Environ Saf 232:113239

Zhu X, Wang C, Duan X, Liang B, Xu EG, Huang Z (2023) Micro-and nanoplastics: a new cardiovascular risk factor? Environ Int 171:107662

Impact of Micro-nanoplastics on Gut and Immune Function

19

Chanchal Singh, Raghubir Singh,
Apoorva Shekhar, and Indranil Das

19.1 Introduction

The digestive systems of aquatic and terrestrial invertebrates and vertebrates host different categories of microorganisms known as the gut microbiota or microbiome. This includes bacteria, fungi, protozoa, and viruses. The term "microbiota" refers to the community of these microorganisms, while "microbiome" represents the collection of their genomes in a given environment. These microorganisms engage in a symbiotic relationship with their host, playing critical roles in digestion and immune function. Moreover, the gut microbiota helps preserve the integrity of the intestinal lining, metabolizes dietary components and xenobiotics, influences behavior via the gut-brain axis, and regulates immune responses.

C. Singh (✉) · A. Shekhar
Department of Veterinary Physiology and Biochemistry, COVS, Guru Angad Dev Veterinary and Animal Sciences University, Ludhiana, Punjab, India

R. Singh
Department of Veterinary Public Health and Epidemiology, College of Veterinary Sciences and Animal Husbandry (Central Agricultural University-Imphal), Jalukie, Nagaland, India

I. Das
MPH, University of Eastern Finland & Team member eith HMS Host, Joensuu, Finland

When micro and nanoplastics (MNPs) are ingested, most are excreted through the digestive system, but some can remain in the intestines for several days (Dawson et al. 2018). MNPs are absorbed by intestinal cells through endocytosis (Fig. 19.1), facilitating their entry into the bloodstream (Ma et al. 2021b). Once in the bloodstream, MNPs may spread to other tissues, where they can persist and potentially induce long-term damage and inflammation (Volkheimer 1975).

Exposure to MNPs prompts significant transcriptional changes in immune cells, altering enzyme levels and cytokine production. For instance, Lehner et al. (2020) used an in vitro 3D intestinal model composed of human intestinal epithelial cells (Caco-2 and HT29-MTX-E12), dendritic cells, and macrophages derived from blood monocytes. This study examined the effects of MPs, specifically 50–500 μm polymers from tyre wear and polyolefins, at concentrations between 823.5 and 1380.0 μg/cm^2. Although changes in inflammatory cytokines such as IL-8, TNFα, and IL-1β and alterations in intestinal barrier integrity were observed, these changes were not statistically significant. Han et al. investigated the immune response to polyvinyl chloride (PVC) and acrylonitrile butadiene styrene (ABS), finding that both materials suppressed histamine release while increasing IL-6 and TNF-α levels. However, this study focused on short-term expo-

S. Goswami et al. (eds.), *Micro-Nano Plastics Exposure, Environmental Degradation and Public Health Crisis*, Sustainable Development Goals Series, https://doi.org/10.1007/978-981-96-5047-7_19

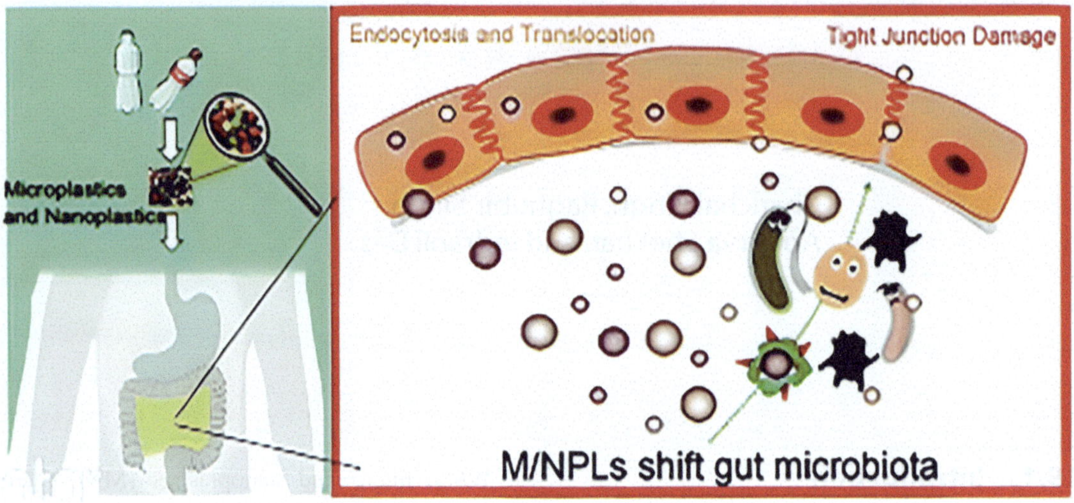

Fig. 19.1 MPs and gut microbiota alteration. (*Courtesy*: Jiyan Oiao et al.)

sure, leaving the long-term effects of MNP exposure on the immune system unresolved.

19.2 MPs and Microbiota: Dysbiosis

The complex community of gut microbiota is often regarded as an "organ" due to its essential role in health and its direct exposure to environmental changes. As the body's first line of defense, the microbiota plays a crucial role in protecting the host from external threats (Tremaroli and Bäckhed 2012). The microbiota composition varies across individuals, families, and species—particularly between carnivores, omnivores, and herbivores. Microbiota protects against pathogen colonization through direct interactions or by stimulating the immune system and strengthening the mucosal barrier (Eberl 2010).

Micro and nanoplastics (MNPs) can penetrate the gut directly, accumulate in the intestines, and disrupt the gut microbiota, leading to health concerns (Cheng et al. 2020). Numerous studies have explored the impact of microplastic exposure on gut microbiota across various species, consistently showing shifts in microbial diversity and composition (Lu et al. 2018). While specific changes vary by species, the common result is dysbiosis. This imbalance can impair immune functionality, weakening defenses against infections, increasing inflammation, and raising the risk of developing chronic diseases (Lehner et al. 2020).

The dysbiosis caused by MNP exposure is often the result of mechanical damage to the gut environment, ingestion of pathogens, or exposure to harmful chemicals that adhere to microplastics (Wright and Kelly 2017). Dysbiosis typically manifests as a decline in microbial diversity and a reduction in beneficial bacterial species that produce short-chain fatty acids, which are vital for protecting the mucosal lining and stimulating the immune system (Gao et al. 2021). This imbalance in the gut microbiota can have significant consequences for the immune system, including contributing to chronic diseases, increasing vulnerability to infections, and altering gene expression within the gut microbiome (Khan et al. 2020). In this context, MNPs represent a serious threat, with the potential to undermine gut health and, by extension, immune defenses.

19.2.1 Impact of Dysbiosis on the Immune System

- Altering cytokine production: The microbiome is key in regulating cytokine production, signaling molecules critical for immune responses. Dysbiosis can alter cytokine production, enhancing or suppressing immune functions (Zmora et al. 2018).

- Modifying T-cell responses: The microbiome influences the development and function of T-cells, a type of white blood cell crucial for immune defense. Dysbiosis can impair T-cell responses, weakening the immune system's ability to fight pathogens (Belkaid and Harrison 2017).
- Changing gut permeability: A healthy microbiome maintains the integrity of the gut lining, which serves as a barrier against harmful substances. Dysbiosis can increase gut permeability, allowing greater exposure to pathogens and toxins (Al Nabhani and Eberl 2020).
- Affecting the gut-liver axis: The gut microbiome plays a role in the gut-liver axis, a communication pathway between the gut and liver. Dysbiosis can disrupt this axis, negatively affecting liver function and overall health (Tripathi et al. 2018).
- Reducing antimicrobial peptide production: The microbiome regulates the production of antimicrobial peptides, which help fight infections. Dysbiosis can reduce the production of these peptides, increasing the host's vulnerability to pathogens (Riley et al. 2018).
- The adaptive immune system, which has long-term memory, allows the host to respond more effectively to pathogens after initial exposure. It comprises humoral and cell-mediated responses, primarily driven by lymphocytes (B- and T-cells). B-cells secrete antibodies (immunoglobulins), while T-cells manage cell-mediated immunity (Murphy and Weaver 2016).

19.2.2 In Vivo Studies of MNP Impact on Immunity

In vivo studies have demonstrated that exposure to micro and nanoplastics (MNPs) can lead to immunotoxic effects in both invertebrates and vertebrates. These effects include a wide range of immune disruptions, such as altered immune cell function, changes in gene expression, and disturbances in various immune parameters. Such disruptions can impair an organism's ability to respond effectively to pathogens, heightening vulnerability to infections and contributing to immune-related diseases (Rist et al. 2021). MNP exposure has been shown to significantly impact the health of various species, including humans.

Research investigating the effects of MNP exposure on the gut epithelium in organisms such as nematodes, zooplankton, mussels, fish, and mice has revealed several harmful outcomes. These include intestinal oxidative damage, cellular deformations, inflammation, altered enzyme activity, and changes in gene expression, all pointing to potential damage to the gut lining (Choi et al. 2021; Xia et al. 2020).

19.3 Microbiota and Dysbiosis in Aquatic Animals

Microplastic (MP) and nanoplastic (NP) exposure has been linked to significant microbial disruptions in aquatic animals. For example, zebrafish exposed to polystyrene (PS) microplastics exhibited a reduction in bacterial diversity, with an increase in the Firmicutes/Bacteroidetes ratio—a change often associated with disease development (Wan et al. 2019). In the same species, nanoplastics (NPs) caused even more severe dysbiosis and inflammation than MPs. Similarly, large croakers and medaka showed an increase in Firmicutes and Bacteroidetes levels, while Proteobacteria were reduced. These microbial shifts correlated with increased juvenile mortality and gut inflammation (Yang et al. 2022). In carp, PS MPs were found to promote the growth of pathogenic bacteria such as Shewanella, contributing to intestinal barrier dysfunction (Ouyang et al. 2021).

Further research revealed that grass carp and largemouth bass (*Micropterus salmoides*) experienced alterations in microbiota composition and function after exposure to polystyrene NPs and environmental pollutants (Chen et al. 2022b; Liao et al. 2022). Marine medaka exposed to PS NPs and sulfamethazine displayed moderate microbial shifts, while juvenile guppies exhibited an increase in *Proteobacteria* and a decrease in *Actinobacteria*, further highlighting the potential harm of MP exposure on aquatic microbiomes (Zhang et al. 2021; Huang et al. 2020a, b). Additionally, blue mussels exposed to MPs showed an increase in human pathogens, indicating the broad and harmful impacts of plastic pollution on marine microbiomes (Li et al. 2020c; Maszczyk et al. 2022).

MP exposure has also been linked to oxidative stress and inflammation in aquatic animals. Oxidative stress occurs when the levels of reactive oxygen species (ROS) exceed the body's antioxidant defenses, leading to cellular and tissue damage. Inflammation, which is the immune system's response to injury or infection, often involves the recruitment of immune cells to the affected area. In aquatic animals, exposure to MPs can disrupt gut microbiota, leading to dysbiosis, reduced bacterial diversity, and shifts in beneficial gut bacteria. Since the gut microbiome plays a critical role in modulating immune responses and preventing the overgrowth of harmful bacteria, dysbiosis caused by MPs can heighten the risk of oxidative stress and inflammation (Huang et al. 2020a, b). Moreover, MP exposure can facilitate the transfer of toxic substances, pollutants, and pathogens from the environment to aquatic organisms, amplifying health risks (Wan et al. 2019). These findings highlight the complex and detrimental effects of plastic pollution on aquatic ecosystems, particularly through its disruption of microbiota and immune system responses.

19.4 Dysbiosis in Terrestrial Animals

In terrestrial animals, exposure to microplastics (MPs) and nanoplastics (NPs) has been shown to disrupt gut microbial communities, leading to dysbiosis. For example, in *Folsomia candida* (springtail), exposure to polyvinyl chloride (PVC) microspheres (80–250 μm, 1 μg/kg dry soil, 56 days) resulted in altered microbial diversity, characterized by an increase in Firmicutes and a decrease in Bacteroidetes (Ju et al. 2019). Similarly, polystyrene (PS) NPs exposure in earthworms caused an increase in Proteobacteria and Bacteroidetes, further demonstrating the effects of plastics on gut microbial balance (Yang et al. 2022).

In *Caenorhabditis elegans*, MPs disrupt the digestive tract, with the toxicity exacerbated by chemical additives. However, fecal microbiota transplantation helped mitigate these adverse effects, suggesting potential therapeutic interventions (Jewett et al. 2022; Chu et al. 2021a, b).

In mammalian models, such as mice, pre-exposure to NPs (18 and 180 μg/kg/day for 90 days) worsened the toxicity of cyclophosphamide (CTX) by disturbing gut microbial abundance (Wen et al. 2022). Studies also showed that polyethylene (PE) MPs increased the abundance of bacteria such as *Actinobacteria, Lactobacillus, Adlercreutzia*, and *Parabacteroides*, indicating possible immune disruptions due to microbial changes (Deng et al. 2020a, b). Additionally, different PS MP sizes and charges resulted in significant gut dysbiosis (Qiao et al. 2021). PE MPs also led to a rise in Staphylococcus, associated with gut inflammation, and a decline in Parabacteroides (Li et al. 2020a, b, c).

In studies involving human gut microbiota cultures, PE MPs enhanced the growth of Clostridium, Bacteroides, and Escherichia, disrupting metabolic pathways and promoting the proliferation of potentially pathogenic bacteria (Huang et al. 2021; Fackelmann and Sommer 2019). In macrophages, MPs induced metabolic changes, including a reduction in α-Proteobacteria, Actinobacteria, and Firmicutes (Merkley et al. 2021). In C57BL/6 mice, exposure to micro and nanoplastics (MNPs) led to similar dysbiosis although some studies reported an increase in Proteobacteria (Zhu et al. 2022; Qiao et al. 2021).

MP exposure has also been linked to alterations in bacterial metabolism, including increased amino acid metabolism (Qiao et al. 2021). In honey bees, PS MPs decreased microbial diversity, reduced Lactobacillus and Bifidobacterium populations, activated immune inhibitory genes, and heightened susceptibility to Hafnia alvei, a pathogen, which ultimately increased mortality rates (Wang et al. 2022).

19.5 Effect of MNPs on Immune Dysfunction of Vertebrates and Invertebrates

In humans, exposure to microplastics and nanoplastics (MNPs) occurs through ingestion, inhalation, and skin contact, which can trigger immune responses and inflammation, and carry harmful chemicals. Studies indicate that MNPs increase IL-6 and IL-8 production in macrophages as well as IL-1-mediated inflammation in PE-exposed cells (Lehner et al. 2020). MNPs may cause oxidative stress and hemolysis, potentially leading to autoimmune diseases (Zarus et al. 2020). Disruptions in microbiota composition can alter immune responses (Santos et al. 2022), with irregular NPs inducing stronger effects than spherical ones (Weber et al. 2022). S-NPs lead to lipid droplet accumulation in macrophages, causing oxidative stress and impaired lysosomal clearance (Florance et al. 2021). Despite these findings, the full impact of MNPs on human health remains unclear, necessitating further research.

In mice, exposure to polyethylene (PE) microplastics reduced white blood cells and lymphocytes while increasing neutrophils. These particles accumulate in the stomach, affecting mitochondria and spleen organelles and decreasing dendritic cells. PE exposure (10–150 μm) altered cytokine levels, increasing IL-1α, G-CSF, and Th17 cells while reducing regulatory T-cells in the spleen (Li et al. 2020a, b, c). Prolonged exposure (7 μm, 90 days) elevated neutrophils and IgA levels, altered spleen lymphocytes, and increased intestinal inflammation, enhancing pro-inflammatory genes such as TLR4 and IRF5 (Park et al. 2020). Polystyrene nanoplastics (10 μm, 5 μm, and 80 nm) caused hematopoietic injury, reducing colony-forming capacity and differentiation, likely disrupting gut microbiota-cytokine pathways (Jing et al. 2022).

In small-spotted catsharks (*Scyliorhinus canicula*), gut accumulation of macroplastics increased T-cell receptor (TCR β and δ) and immunoglobulin M (IgM) levels in the spleen (Mancia et al. 2020). Neutrophil death ranged from 50 to 97% with 7.3 μm microplastics (MPs), while 49–880 nm particles caused 49–53% death; 880 nm MPs did not affect neutrophil function. Zebrafish larvae showed liver inflammation, marked by increased neutrophil aggregation and macrophage apoptosis, with liver-type inflammatory binding protein (fabp10a) expression rising in a dose- and size-dependent manner. Nanoplastics (NPs) also stimulated amino acid and purine catabolism and steroid synthesis linked to immune disorders (Cheng et al. 2022).

In carp, 30-day exposure to polyethylene (PE) microparticles (250–500 μg/L) disrupted complement systems and plasma immunity-related enzymes (Banaee et al. 2019). Zebrafish exposed to high-density PE and polystyrene (PS) particles (100–1000 μg/L for 20 days) showed decreased leukotriene B4 receptor (ltb4r) and interferon-induced transmembrane protein (ifitm1) expression (Limonta et al. 2019). In marine medaka (*Oryzias melastigma*) larvae, exposure to relevant PS NP concentrations (102, 104, and 106 particles/L for 14 days) upregulated mucin genes (muc7-like and muc13-like) but downregulated others (muc2 and muc5AC-like). Genes involved in metabolism, immunity, and inflammation (jak, stat-3, IL-6, IL-1β, TNF-α, ccl-11, nf-κb, and sod) were significantly induced at 104 and 106 particles/L, indicating that PS NP-induced inflammation may impair immune function (Chen et al. 2022a, b).

In bivalve molluscs, PS exposure decreased hemocyte counts and phagocytic activity, modulating inflammation, apoptosis, and oxidative stress (Shi et al. 2020; Tang et al. 2020). In *Mytilus galloprovincialis,* amino-modified PS MPs caused hemocyte degranulation, lysosome acidification, and mitochondrial disruption, while PS exposure in Mediterranean Sea urchins increased coelomocyte counts and ROS/RNS levels (Auguste et al. 2020; Murano et al. 2020). In sea cucumber (*Apostichopus japonicus*), exposure to PS NPs (100 nm and 20 μm at 100 mg kg − 1 diet for 60 days) indicated size-dependent alterations in immune response and ammonia stress tolerance, with NPs proving more toxic than MPs (Liu et al. 2022). In juvenile *Macrobrachium nipponense* (shrimp), NPs showed an inverse relationship with antioxidant enzymes, impacting antioxidant-related gene expression. High NP concentrations (10, 20, and 40 mg/L) had inhibitory and toxic effects on antioxidant and immune enzyme activities (Li et al. 2020a, b, c).

Polystyrene (PS) nanoparticles (NPs) reduced phagocytosis and lysosomal membrane stability (LMS) in hemocytes, with smaller particles (50 nm) causing more damage than larger ones (100 nm). In *Mytilus hemocytes*, PS NPs (5 μg/mL) led to cellular damage and altered morphology (Auguste et al. 2021; Canesi et al. 2015). In European sea bass (*Dicentrarchus labrax*), PS NP exposure increased viral replication and reduced antiviral response, potentially affecting disease resistance (González-Fernández and Cuesta 2022). In Chinese mitten crab (*Eriocheir sinensis*), MPs altered immune responses by changing hemocyanin (Hc) and lysozyme (LSZ) gene expression, disrupting calcium signaling and apoptosis pathways (NF-κB), affecting T cell differentiation, and increasing infection risk (Liu et al. 2019; Tang et al. 2020).

Zebrafish larvae showed liver inflammation, marked by increased neutrophil aggregation and macrophage apoptosis, with liver-type inflammatory binding protein (fabp10a) expression rising in a dose- and size-dependent manner.

19.6 Future Aspects and Recommendations

Since the 1950s, the rapid increase in plastic production and use across all industries and inadequate recycling efforts has led to significant environmental contamination. Animal studies have shown potential harm from MNPs to various aspects of health, including gut microbiota, immune response, and gut epithelium. These findings underscore the necessity for comprehensive strategies that minimize future plastic waste and address existing environmental contamination to mitigate ongoing exposure to MNPs and their associated health risks.

Micro and nanoplastics (MNPs) are pervasive in the environment, leading to frequent human exposure from various sources. Studies indicate that MNP exposure can result in oxidative stress, inflammation, impaired immune function, alterations in cellular and energy metabolism, inhibition of cell proliferation, tissue degeneration, abnormal organ development, biochemical changes, and even genotoxicity and carcinogenicity. While numerous animal and cell culture studies have demonstrated these potential biological effects, the exact mechanisms by which MNPs affect human health remain unclear. Moreover, whether long-term exposure to MNPs increases susceptibility to human diseases is still unknown. Future research should focus on quantifying the health impact of MNPs on humans and elucidating the underlying mechanisms of their pathogenesis. This understanding is critical for summarizing current knowledge and identifying research gaps.

An emerging area of investigation is the potential for probiotics to mitigate the harmful effects of MNPs on gut health. Modulating the gut microbiota with probiotics may offer protective benefits against MNP-induced damage through several mechanisms, including:

- Reinforcing the intestinal barrier: Probiotics can enhance the integrity of the gut lining by promoting the production of tight junction proteins, which help seal gaps between intestinal cells. This prevents harmful substances, including MNPs, from crossing into the bloodstream and causing systemic damage.
- Promoting the production of cytokines, chemokines, and IgA-secreting cells: Probiotics stimulate the immune system to produce cytokines and chemokines, which are signaling molecules that regulate immune responses. They also boost the production of IgA-secreting cells, which release immunoglobulin A (IgA), a key antibody that neutralizes

pathogens and toxins in the gut, offering protection against MNP-induced damage.
- Normalizing microbial communities: Probiotics help restore balance to the gut microbiome by promoting the growth of beneficial bacteria and inhibiting harmful bacteria that may proliferate due to MNP exposure. This balance is crucial for maintaining a healthy gut ecosystem.
- Producing beneficial fatty acids: Probiotics can enhance the production of short-chain fatty acids (SCFAs) such as butyrate, acetate, and propionate. These SCFAs nourish gut cells, strengthen the intestinal barrier, and have anti-inflammatory properties, which may counteract the effects of MNP-induced inflammation.
- Modulating anti-inflammatory responses: Probiotics can reduce inflammation by inhibiting the production of pro-inflammatory cytokines and promoting anti-inflammatory cytokines. This modulation of the immune response may help protect the gut from the inflammatory damage caused by MNPs.
- Potentially influencing the brain-gut axis: The gut microbiota interacts with the brain through the brain-gut axis. By maintaining a healthy microbiome, probiotics could positively affect mood, behavior, and overall brain health, potentially counteracting the neurotoxic effects that may arise from MNP exposure.

19.7 Conclusion

In conclusion, addressing the challenges posed by micro and nanoplastics (MNPs) requires a multifaceted strategy. This approach should include reducing plastic production, improving waste management to limit environmental contamination, and exploring therapeutic interventions such as probiotics. Probiotics offer a promising avenue to mitigate the harmful effects of MNPs on gut health by reinforcing the intestinal barrier, modulating immune responses, and restoring microbial balance. Together, these efforts are essential to safeguarding both environmental and human health from the pervasive threat of plastic pollution.

References

Al Nabhani Z, Eberl G (2020) Imprinting of the immune system by the microbiota early in life. Mucosal Immunol 13(2):183–189

Auguste M, Lasa A, Balbi T, Pallavicini A, Vezzulli L, Canesi L (2020) Impact of nanoplastics on hemolymph immune parameters and microbiota composition in *Mytilus galloprovincialis*. Mar Environ Res 159:105017

Auguste M, Mayall C, Barbero F, Hočevar M, Alberti S, Grassi G et al (2021) Functional and morphological changes induced in Mytilus hemocytes by selected nanoparticles. Nano 11(2):470

Banaee M, Soltanian S, Sureda A, Gholamhosseini A, Haghi BN, Akhlaghi M et al (2019) Evaluation of single and combined effects of cadmium and microplastic particles on biochemical and immunological parameters of common carp *Cyprinus carpio*. Chemosphere 236:124335

Belkaid Y, Harrison OJ (2017) Homeostatic immunity and the microbiota. Immunity 46(4):562–576

Canesi L, Ciacci C, Bergami E, Monopoli MP, Dawson KA, Papa S, Canonico B, Corsi I (2015) Evidence for immunomodulation and apoptotic processes induced by cationic polystyrene nanoparticles in the hemocytes of the marine bivalve *Mytilus*. Mar Environ Res 111:34–40

Chen JC, Fang C, Zheng RH, Chen ML, Kim DH, Lee YH et al (2022a) Environmentally relevant concentrations of microplastics modulated the immune response and swimming activity and impaired the development of marine medaka *Oryzias melastigma* larvae. Ecotoxicol Environ Saf 241:113843

Chen L, Liu Z, Wang L, Peng X, Zhang Y (2022b) Combined effects of microplastic and nanoplastic exposure on aquatic animals: potential risks to the immune system. Environ Sci Pollut Res 29:48385–48395

Cheng Y, Wang Y, Wu J (2020) Effects of microplastics and nanoplastics on aquatic ecosystems and organism health. Environ Sci Technol 54(19):11454–11462

Cheng H, Duan Z, Wu Y, Wang Y, Zhang H, Shi Y et al (2022) Immunotoxicity responses to polystyrene nanoplastics and their related mechanisms in the liver of zebrafish *Danio rerio* larvae. Environ Int 161:107128

Choi D, Hwang J, Bang J, Han S, Kim T, Oh Y, Hwang Y, Choi J, Hong J (2020) In vitro toxicity from a physical perspective of polyethylene microplastics based on statistical curvature change analysis. Sci Total Environ 752:142242

Choi Y, Lee S, Park J (2021) Effects of nanoplastics on aquatic environments and organisms: current understanding and future perspectives. Environ Pollut 270:116086

Chu X, Peng J, Yin J, Chen W, Xu H, Zhang L, Ju L (2021a) Toxicity and stress response of microplastics and nanoplastics on *Caenorhabditis elegans*. Ecotoxicol Environ Saf 207:111340

Chu Q, Zhang S, Yu X, Wang Y, Zhang M, Zheng X (2021b) Fecal microbiota transplantation attenuates nano-plastics induced toxicity in *Caenorhabditis elegans*. Sci Total Environ 779:146454

Dawson AL, Kawaguchi S, King CK et al (2018) Turning microplastics into nanoplastics through digestive fragmentation by Antarctic krill. Nat Commun 9:1001

Deng Y, Yan Z, Shen R, Wang M, Huang Y, Ren H et al (2020a) Microplastics release phthalate esters and cause aggravated adverse effects in the mouse gut. Environ Int 143:105916

Deng Y, Zhang Y, Lemos B, Ren H (2020b) Tissue accumulation of microplastics in mice and biomarker responses suggest widespread health risks of exposure. Sci Rep 10:21322

Eberl G (2010) A new vision of immunity: homeostasis of the superorganism. Mucosal Immunol 3(5):450–460

Fackelmann G, Sommer S (2019) Microplastics and the gut microbiome: how chronically exposed species may suffer from gut dysbiosis. Mar Pollut Bull 143:193–203

Florance I, Ramasubbu S, Mukherjee A, Chandrasekaran N (2021) Polystyrene nanoplastics dysregulate lipid metabolism in murine macrophages. In Vitro Toxicol 458:152850

Gao F, Peng Y, Yang C (2021) Impact of microplastics on gut microbiota. Microorganisms 9(3):543

Gigault J, Ter Halle A, Baudrimont M, Pascal PY, Gauffre F, Phi TL, Reynaud S (2018) Current opinion: what is a nanoplastic? Environ Pollut 235:1030–1034

González-Fernández C, Cuesta A (2022) Nanoplastics increase fish susceptibility to nodavirus infection and reduce antiviral immune responses. Int J Mol Sci 23(3):1483

Huang H, Ren Z, Qiao Z, Wang X, Liu Y, Feng J (2020a) Gut microbiota dysbiosis and alterations of metabolic profiles in juvenile guppies exposed to microplastics. Environ Sci Technol 54(18):11220–11231

Huang JN, Wen B, Zhu JG, Zhang YS, Gao JZ, Chen ZZ (2020b) Exposure to microplastics impairs digestive performance stimulates immune response and induces microbiota dysbiosis in the gut of juvenile guppy *Poecilia reticulate*. Sci Total Environ 733:138929

Huang Y, Xia X, Cao M, Zheng M, Luo C, Wang T (2021) Microplastic exposure alters human gut microbiota. Environ Sci Technol 55(13):8639–8648

Jewett LE, Xu X, Denison MS, Cheng L, Ho J (2022) Microplastic impacts on gut microbiota and health: lessons from *Caenorhabditis elegans*. Sci Total Environ 813:151901

Jin Y, Lu L, Tu W, Luo T, Fu Z (2019) Impacts of polystyrene microplastic on the gut barrier microbiota and metabolism of mice. Sci Total Environ 649:308–317

Jing J, Zhang L, Han L, Wang J, Zhang W, Liu Z, Gao A (2022) Polystyrene micro−/nanoplastics induced hematopoietic damages via the crosstalk of gut microbiota metabolites and cytokines. Environ Int 161:107131

Ju L, Zhang H, Huang X, Fan L, Sun J (2019) Exposure of polyvinyl chloride microplastics influences the gut microflora in terrestrial springtail *Folsomia candida*. Sci Total Environ 667:942–948

Khan FR, Shashoua Y, Crawford A, Drury A, Sheppard K, Stewart K, Thompson RC (2020) Impacts of plastic pollution on freshwater ecosystems: a review of the known unknowns. Sci Total Environ 730:138884

Lehner R, Wohlleben W, Septiadi D, Landsiedel R, Petri-Fink A, Rothen-Rutishauser B (2020) A novel 3D intestine barrier model to study the immune response upon exposure to microplastics. Arch Toxicol 94:2463–2479

Li B, Ding Y, Cheng X, Sheng D, Xu Z, Rong Q, Wu Y (2020a) Polyethylene microplastics affect the distribution of gut microbiota and inflammation development in mice. Chemosphere 244:125492

Li Y, Liu Z, Li M, Jiang Q, Wu D, Huang Y, Zhao Y (2020b) Effects of nanoplastics on antioxidant and immune enzyme activities and related gene expression in juvenile *Macrobrachium nipponense*. J Hazard Mater 398:122990

Li J, Zhang H, Zhang S (2020c) Microplastics in mussels along the coastal waters of China. Environ Pollut 257:113569

Li W, Tian Y, Chen L, Zhao Y (2023a) Effects of microplastic exposure on fish gut microbiota and immune responses. Mar Pollut Bull 189:114820

Li Z, Yan L, Junaid M, Chen X, Liao H, Gao D et al (2023b) Impacts of polystyrene nanoplastics at the environmentally relevant and sub-lethal concentrations on the oxidative stress immune responses and gut microbiota to grass carp *Ctenopharyngodon idella*. J Hazard Mater 441:129995

Liao H, Liu S, Junaid M, Gao D, Ai W, Chen G, Wang J (2022) Di-(2-ethylhexyl) phthalate exacerbated the toxicity of polystyrene nanoplastics through histological damage and intestinal microbiota dysbiosis in freshwater Micropterus salmoides. Water Res 219:118608

Limonta G, Mancia A, Benkhalqui A, Bertolucci C, Abelli L, Fossi MC, Panti C (2019) Microplastics induce transcriptional changes immune response and behavioral alterations in adult zebrafish. Sci Rep 9(1):1–11

Liu Z, Yu P, Cai M, Wu D, Zhang M, Chen M, Zhao Y (2019) Effects of microplastics on the innate immunity and intestinal microflora of juvenile *Eriocheir sinensis*. Sci Total Environ 685:836–846

Liu J, Xu D, Chen Y, Zhao C, Liu L, Gu Y et al (2022) Adverse effects of dietary virgin (nano) microplastics on growth performance immune response and resistance to ammonia stress and pathogen challenge in juvenile sea cucumber *Apostichopus japonicus* (Selenka). J Hazard Mater 423:127038

Lu L, Wan Z, Luo T, Fu Z, Jin Y (2018) Polystyrene microplastics induce gut microbiota dysbiosis and hepatic lipid metabolism disorder in mice. Sci Total Environ 631:449–458

Ma C, Li L, Chen Q, Lee JS, Gong J, Shi H (2021a) Application of internal persistent fluorescent fibers in tracking microplastics in vivo processes in aquatic organisms. J Hazard Mater 401:123336

Ma C, Chen Q, Li J et al (2021b) Distribution and translocation of micro- and nanoplastics in fish. Crit Rev Toxicol 51:740–753

Mancia A, Chenet T, Bono G, Geraci ML, Vaccaro C, Munari C et al (2020) Adverse effects of plastic ingestion on the Mediterranean small-spotted catshark *Scyliorhinus canicula*. Mar Environ Res 155:104876

Maszczyk P, Gosciniak G, Dymek J (2022) Microplastic pollution in aquatic ecosystems: toxicity interactions with gut microbiota and human health. Mar Pollut Bull 174:113235

Merkley SD, Bradley S, Lamendella R (2021) Impact of polyethylene microplastics on gut bacterial communities in zebrafish and their health. Ecotoxicology 30:1035–1047

Merkley SD, Moss HC, Goodfellow SM, Ling CL, Meyer-Hagen JL, Weaver J et al (2022) Polystyrene microplastics induce an immunometabolic active state in macrophages. Cell Biol Toxicol 38(1):31–41

Murano C, Agnisola C, Caramiello D, Castellano I, Casotti R, Corsi I, Palumbo A (2020) How sea urchins face microplastics: uptake tissue distribution and immune system response. Environ Pollut 264:114685

Murphy K, Weaver C (2016) Janeway's Immunobiology, 9th edn. Garland Science

Ouyang X, Ma Z, Wang Z, Zhuang Y (2021) PS microplastics induce intestinal barrier dysfunction in common carp *Cyprinus carpio* via bacterial dysbiosis and inflammatory responses. J Hazard Mater 414:125535

Park EJ, Han JS, Park EJ, Seong E, Lee GH, Kim DW et al (2020) Repeated-oral dose toxicity of polyethylene microplastics and the possible implications on reproduction and development of the next generation. Toxicol Lett 324:75–85

Qiao R, Deng Y, Zhang S, Wolosker MB, Zhu Q, Ren H, Zhang Y (2021) Accumulation of different shapes of microplastics initiates intestinal injury and gut microbiota dysbiosis in the marine organism *Girella laevifrons*. Mar Pollut Bull 173:113076

Rahman A, Sarkar A, Yadav OP, Achari G, Slobodnik J (2021) Potential human health risks due to environmental exposure to nano-and microplastics and knowledge gaps: a scoping review. Sci Total Environ 757:143872

Riley K, Eberl G, Oh J (2018) Gut microbiota regulates intestinal immune system. Immunity 48(6):964–976

Rist S, Baun A, Hartmann NB (2021) Ingestion of micro- and nanoplastics by aquatic organisms. Environ Pollut 275:116659

Santos AL, Rodrigues CC, Oliveira M, Rocha TL (2022) Microbiome: a forgotten target of environmental micro (nano) plastics? Sci Total Environ 822:153628

Sendra M, Saco A, Yeste MP, Romero A, Novoa B, Figueras A (2020) Nanoplastics: from tissue accumulation to cell translocation into Mytilus galloprovincialis hemocytes resilience of immune cells exposed to nanoplastics and nanoplastics plus *Vibrio splendidus* combination. J Hazard Mater 388:121788

Shi W, Han Y, Sun S, Tang Y, Zhou W, Du X, Liu G (2020) Immunotoxicities of microplastics and sertraline alone and in combination to a bivalve species: size-dependent interaction and potential toxication mechanism. J Hazard Mater 396:122603

Tang Y, Rong J, Guan X, Zha S, Shi W, Han Y et al (2020) Immunotoxicity of microplastics and two persistent organic pollutants alone or in combination to a bivalve species. Environ Pollut 258:113845

Tremaroli V, Bäckhed F (2012) Functional interactions between the gut microbiota and host metabolism. Nature 489(7415):242–249

Tripathi A, Debelius JW, Lozupone CA (2018) The gut-liver axis and the role of the gut microbiome in liver disease. Nat Rev Gastroenterol Hepatol 15(9):524–537

Volkheimer G (1975) Hematogenous dissemination of ingested polyvinyl chloride particles. Ann N Y Acad Sci 246:164–171

Wan Z, Wang C, Zhou J, Shen M, Wang X, Fu Z, Jin Y (2019) Effects of polystyrene microplastics on the composition of the microbiome and metabolism in larval zebrafish. Chemosphere 217:646–658

Wang K, Zhu L, Rao L, Zhao L, Wang Y, Wu X et al (2022) Nano-and micro-polystyrene plastics disturb gut microbiota and intestinal immune system in honeybee. Sci Total Environ 842:156819

Weber A, Schwiebs A, Solhaug H, Stenvik J, Nilsen AM, Wagner M et al (2022) Nanoplastics affect the inflammatory cytokine release by primary human monocytes and dendritic cells. Environ Int 163:107173

Wen S, Zhao Y, Liu S, Chen Y, Yuan H, Xu H (2022) Polystyrene microplastics exacerbated liver injury from cyclophosphamide in mice: insight into gut microbiota. Sci Total Environ 840:156668

Wright SL, Kelly FJ (2017) Plastic and human health: a micro issue? Environ Sci Technol 51(12):6634–6647

Xia Y, Wang J, Ren L (2020) Toxicity and uptake of nanoplastics: a review of the impact on human health. Nanotoxicology 14(9):1231–1247

Xu M, Halimu G, Zhang Q, Song Y, Fu X, Li Y et al (2019) Internalization and toxicity: a preliminary study of effects of nanoplastic particles on human lung epithelial cell. Sci Total Environ 694:133794

Yan W, Hamid N, Deng S, Jia PP, Pei DS (2020) Individual and combined toxicogenetic effects of microplastics and heavy metals (Cd Pb and Zn) perturb gut microbiota homeostasis and gonadal development in marine medaka *Oryzias melastigma*. J Hazard Mater 397:122795

Yang S, Cheng Y, Chen Z, Liu T, Yin L, Pu Y, Liang G (2021) In vitro evaluation of nanoplastics using human lung epithelial cells microarray analysis and co-culture model. Ecotoxicol Environ Saf 226:112837

Yang S, Wei X, Cheng Y, Li Q (2022) The interaction between microplastics and gut microbiota in aquatic organisms: a key role in mediating immune responses and gut health. Front Microbiol 13:810943

Yong CQY, Valiyaveettil S, Tang BL (2020) Toxicity of microplastics and nanoplastics in mammalian system. Int J Environ Res Public Health 17(5):1509

Zarus GM, Muianga C, Hunter C, Pappas RS (2020) A review of data for quantifying human exposures to micro and nanoplastics and potential health risks. Sci Total Environ 756:144010

Zhang YT, Chen H, He S, Wang F, Liu Y, Chen M et al (2021) Subchronic toxicity of dietary sulfamethazine and nanoplastics in marine medaka *Oryzias melastigma:* insights from the gut microbiota and intestinal oxidative status. Ecotoxicol Environ Saf 226:112820

Zheng S, Wang WX (2023) Disturbing ion regulation and excretion in medaka *Oryzias melastigma* gills by microplastics: insights from the gut-gill axis. Sci Total Environ 857:159353

Zhu Z, Yang Y, Zhang Y, Shen C, Yang Y (2022) Nanoplastics cause gut microbiota dysbiosis affecting metabolic and immune systems in zebrafish. Environ Pollut 310:119810

Zmora N, Levy M, Elinav E (2018) Microbiome and cytokines: a dialogue between the gut microbiota and innate immunity. Cytokine 113:1–8

Micro and Nanoplastics: Impact on Cellular Health and Neuro-Function

20

Chanchal Singh, Raghubir Singh, Apoorva Shekhar, and Indranil Das

20.1 Introduction

The micro-nanoplastics (MNPs) are minute plastic particles ubiquitously distributed in the environment, affecting air, water, soil, and living organisms. They result from the degradation of larger plastics which break down into smaller particles due to physical, chemical, and biological processes such as sunlight exposure, weathering, and microbial action (Andrady 2011). Nanoplastic particles are released during the production or application of plastic materials, leading to their dissemination into air, soil, and water (Geyer et al. 2017). The effects of MNPs have raised significant concerns due to their prevalence and ability to infiltrate and accumulate within living organisms (Cole et al. 2011).

C. Singh (✉) · A. Shekhar
Department of Veterinary Physiology and Biochemistry, COVS, Guru Angad Dev Veterinary and Animal Sciences University, Ludhiana, Punjab, India

R. Singh
Department of Veterinary Public Health and Epidemiology, College of Veterinary Sciences and Animal Husbandry (Central Agricultural University-Imphal), Jalukie, Nagaland, India

I. Das
MPH, University of Eastern Finland & Team member eith HMS Host, Joensuu, Finland

The plastic-habituated humans are exposing themselves to MNPs through inhalation, ingestion, and dermal contact. Once inside the body, these particles infatuate to host cells; however, the interaction depends on several factors, viz., size of particles, surface properties, and molecules they encounter, such as carbohydrates, proteins, and phospholipids (Fadhl et al. 2021). Following exposure, MNPs can interact with tissues in ways that may enhance their toxicity potential. Studies indicate that micro-nanoplastics have the potential to damage organs and tissues causing oxidative stress and the development of inflammation. MNPs exposure has been linked to DNA damage and apoptosis in human beings (Li et al. 2020). However, much investigation has been undertaken regarding the harmful impacts of microplastics, and a large portion has been conducted utilizing experimental strategies such as animal studies and cell cultures. These studies have clarified some potentially toxic pathways, including oxidative stress, DNA damage, and inflammatory responses. Still, there continues to be a big knowledge vacuum on the underlying molecular pathways. Accordingly, in the present chapter, a detailed discussion regarding MPs' exposure to humans and related health risks has been carried out.

S. Goswami et al. (eds.), *Micro-Nano Plastics Exposure, Environmental Degradation and Public Health Crisis*, Sustainable Development Goals Series, https://doi.org/10.1007/978-981-96-5047-7_20

20.2 Micro and Nanoplastics' Impact on Cellular Health and Neuro-Function

Micro and nanoplastics (MNPs) have been shown to cause a range of negative impacts in laboratory and field studies, including oxidative stress, cellular toxicity, and developmental and reproductive toxicity in various organisms. Numerous studies have investigated the effects of MNPs, particularly focusing on oxidative stress, cytotoxicity, and neurotoxicity.

20.2.1 Oxidative Stress

Oxidative stress occurs when there is an imbalance between the production of reactive oxygen species (ROS) and the body's ability to detoxify these reactive intermediates or repair the damage they cause. ROS, which includes free radicals such as superoxide (O_2^-) and hydroxyl radicals (OH^-) as well as non-radical species such as hydrogen peroxide (H_2O_2), are byproducts of normal cellular metabolism, especially in the mitochondria.

MNPs can generate reactive oxygen species (ROS), causing oxidative damage to cells, tissues, and organs. This may induce inflammation and activate apoptosis pathways, resulting in cell death (Pérez-Guevara et al. 2020). Furthermore, MNPs adsorb environmental pollutants, such as heavy metals and organic compounds, which contribute to oxidative stress by generating ROS that damage cellular components such as proteins, lipids, and DNA (Hwang et al. 2022; Tiwari et al. 2021). Additionally, the physical characteristics of MNPs enable them to act as pro-oxidants, transferring electrons and generating free radicals (Liu et al. 2022). In vitro studies on MNPs also authenticate that these particles induce oxidative stress leading to cellular damage and chronic diseases by a mechanism to generate ROS either directly or indirectly and carrying toxic contaminants and therefore contributing to cellular damage, inflammation, and potential long-term health effects, including cancer, neurodegenerative diseases, and metabolic disorders. The endogenous antioxidant potential of cells is regulated by enzymatic pathways involved in free radical scavenging, encompassing enzymes such as superoxide dismutase (SOD), catalase (Cat), glutathione reductase (GR), and glutathione peroxidase (GPx) (Halliwell and Gutteridge 2015). However, excessive free radical production can lead to an oxidant-antioxidant imbalance, resulting in oxidative stress and neuronal damage, thus leading to neurotoxicity and neurodegeneration. Studies have also shown that MNPs are able to inhibit the activity of acetyl-cholinesterase (AChE), an enzyme crucial for neurotransmitter regulation, impairing neuronal network function (Feng et al. 2020).

In aquatic invertebrates, MNPs exposure has been shown to produce ROS and impair antioxidant defenses resulting in oxidative stress and DNA damage. This oxidative stress negatively impacts the growth, development, and reproduction of species such as snails, mussels, and crustaceans, potentially disrupting aquatic ecosystems (Browne et al. 2011; Rist and Sanz-Lázaro 2019). The ingestion of MNPs by filter-feeding organisms, such as mussels, can lead to plastic accumulation in tissues, affecting physiology and ecosystem interactions (Prata et al. 2019). Aquatic vertebrates, including fish, also suffer from MNP-induced oxidative stress, which impairs growth, development, and immune function. Studies on zebrafish and medaka have demonstrated that MNP exposure increases ROS production, weakens antioxidant defenses, and raises disease susceptibility (Liu et al. 2021; Rist et al. 2017). These effects can lead to broader ecological consequences, such as altered nutrient cycling and disrupted food webs in aquatic environments (Zhang et al. 2021).

20.2.2 Cytotoxicity

Cytotoxicity is defined as the ability of substances to harm or kill cells. MNPs exhibit cytotoxic properties which have been demonstrated in in vitro studies, thus resulting in cellular damage and apoptosis (programmed cell death) in a dose-dependent manner. Furthermore, studies on aquatic animals, fishes, and invertebrates have shown that MNPs exposure can lead to cell death, alteration of gene expression, stress, inflammation, and reduced antioxidant enzyme activity leading to oxidative stress (Rist and Sanz-Lázaro 2019; Browne et al. 2011). Research on mammals, such as rats and mice, and human cell cultures, has demonstrated that MNPs can cause cytotoxicity through physical damage, toxic chemical release, and particle accumulation. In liver, neuronal, and immune cells, MNPs lead to increased cell death and changes in gene expression linked to inflammation and oxidative stress (Zhang et al. 2021; Tiwari et al. 2021). The size, shape, and surface chemistry of MNPs influence their cytotoxic effects. The mechanisms of MNP-induced cytotoxicity are still in the infancy stage but are believed to involve several factors:

- Oxidative Stress: MNPs can induce oxidative stress by generating ROS and depleting antioxidant defenses, leading to cellular damage and death (Hwang et al. 2022; Tiwari et al. 2021).
- Inflammation: MNPs can trigger low-level inflammation, further contributing to cell damage (Liu et al. 2022).
- Physical Damage: MNPs can physically disrupt cellular membranes, decreasing cell viability and increasing apoptosis (Zhang et al. 2021).
- Release of Toxic Substances: MNPs may release harmful substances, such as heavy metals and persistent organic pollutants, into cells, exacerbating cytotoxic effects (Prata et al. 2019).

20.2.3 Neurotoxicity

Neurotoxicity is the ability of a substance to cause damage to the nervous system. MNPs have the ability to cross the blood-brain barrier and accumulate in the brain, thus leading to neurotoxic effects. The extent and mechanisms of MNP-induced neurotoxicity, however, are not well understood and are the subject of ongoing research.

Few studies on mice and rats suggest that exposure to MNPs may lead to neurotoxicity in these animals, with potential implications for human health as well (Hwang et al. 2022). MNPs enter the bodies of mice and rats through ingestion, inhalation, and dermal absorption, leading to significant increases in ROS production and decreased activity of antioxidant enzymes. This exposure is associated with neuro-inflammation, a condition linked to various neurodegenerative diseases (Liu et al. 2022). In aquatic animals, similar studies suggest MNPs exposure may result in neurotoxicity and impact the health of aquatic ecosystems. Micro and nanoplastics enter the bodies of aquatic animals through ingestion, absorption through the gills and skin, and direct injection by predators that consume contaminated prey. Furthermore, research has shown that exposure to polystyrene microplastics caused behavioral changes in zebrafish, including decreased locomotor activity and increased anxiety-like behavior. These studies also indicated that microplastic exposure leads to changes in the expression of genes related to the nervous system and neurotransmitter signaling (Zhang et al. 2023; Tiwari et al. 2021). Exposure to MNPs can affect the development and morphology of the nervous system in aquatic animals, causing neuro-inflammation and oxidative stress (Hwang et al. 2022; Liu et al. 2021).

20.3 Empirical Evidence of MNPs Toxic Effects

Zebrafish Studies Zebrafish exposure to polystyrene nanoplastics (PS-NPs) and nonylphenol (4-NP) indicates neurotoxicity linked to oxidative stress and is evidenced by a decrease in enzymes catalase and glutathione reductase activities. PS-NPs & 4-NP together as combined exposure has intensified these effects, while individual exposures by zebrafishes has significantly inhibited acetylcholinesterase (AChE) activity, suggesting antagonistic interactions. Additionally, increased glutamine synthetase activity and disruption of energy metabolism indicate the reduced α-ketoglutarate dehydrogenase (α-KGDH) activity and loss of neural cells (Aliakbarzadeh et al. 2023). In another study, Zebrafish exposure to PS micro and nanoparticles (MNPs) show size-dependent cellular distribution, with 50 nm PS NPs being distributed evenly in the cytoplasm and 1 μm PS NPs forming aggregates. The smaller particles were internalized via dynamin-dependent endocytosis, while the larger ones were phagocytized. These particles were degraded inside lysosomes, leading to alkalinization and altered expression of cathepsin genes (Sendra et al. 2021).

Caenorhabditis elegans In *C. elegans*, exposure to polystyrene nano and microplastics (100, 500 nm, 1.0, 2.0, and 5.0 μm) affects the survival rates, lifespan, motor behavior, and neuron function and 1.0 μm particles having the most pronounced effects. Downregulation of unc-17 and unc-47 indicated damage to cholinergic and GABAergic neurons, while increased gst-4 expression signals oxidative stress. Antioxidants such as oligomeric proanthocyanidins and curcumin mitigate these effects (Lei et al. 2018).

Clams Studies In *Mactraveneriformis* clams, exposure to phenanthrene (Phe) and polystyrene (PS) induced oxidative stress, and their combination significantly elevates oxidative stress markers (SOD and GST) and inhibits AChE activity in both gills and digestive glands, indicating potential neurotoxicity (Zhang et al. 2021). In *Donaxtrunculus* clams, exposure to a microplastic mixture (0.06 g/Kg of sand; 1:1 ratio of polyethene (PE) to polypropylene (PP)) significantly induces oxidative stress in all organs and inhibits AChE activity in both gills and the digestive gland, suggesting potential neurotoxicity (Tlili et al. 2020).

Mouse Cell Studies Studies involving mouse cell lines, such as embryonic fibroblasts and mixed neuronal cells, have revealed that exposure to PS NPs (100 nm) significantly reduces neuronal cell viability, as indicated by increased cleaved caspase-3, a marker for apoptosis. PS NPs also induced astrocytosis, as shown by elevated lipocalin-2 and pro-inflammatory cytokine levels, potentially leading to apoptosis (Jung et al. 2020).

Human Neuroblastoma Cells Exposure of cultured human neuroblastoma cells (SH-SY5Y) to PS NPs resulted in cytotoxic effects, promoting differentiation into a neuronal phenotype similar to the effects of the neurotoxin acrylamide. This exposure caused neurite shrinkage, morphological alterations, nuclear swelling, and leakage of intracellular components (Ban et al. 2021). In humans, Li et al. (2020) report that exposure to microplastics results in oxidative stress, DNA damage, and apoptosis in human liver cells. The evidence underscores that exposure to MNPs can harm various human organ systems.

Bioaccumulation Studies The bioaccumulation of five types of nanoplastics—polyvinyl chloride (PVC), polystyrene (PS), polylactic acid (PLA), polypropylene (PP), and polyethene terephthalate (PET)—on dipalmitoyl phosphatidylcholine (DPPC) bilayers showed a reduction in lipid bilayer thickness (Yuan et al. 2023).

Recent Studies Rats treated with PS NPs (1, 3, 6, and 10 mg/kg-day) for 5 weeks have exhibited increased ROS production and altered antioxidant enzyme levels, suggesting neurobehavioral and energy metabolism dysfunctions (Babaei et al. 2022). In freshwater prawns (*Macrobrachiumnipponense*), MNPs exposure significantly has reduced moulting rates and induced oxidative stress (Fan et al. 2022). Additionally, adult *Procambarusclarkii* shows altered gene expression relating to oxidative stress and lipid metabolism due to dietary exposure to 100 nm carboxylated polystyrene nanoparticles (Capanni et al. 2021). Rochman et al. (2015) found that ingesting microplastics by fish resulted in liver and gill tissue damage leading to reduced growth and survival. Van der Meeren et al. (2015) have demonstrated that zooplankton ingesting microplastics experienced oxidative stress, reduced survival, and behavioral alterations. Setälä et al. (2014) have reported that earthworms exposed to microplastics in soil exhibited reduced growth, reproduction, and behavioral changes. Similarly, Chen et al. (2015) observed oxidative stress and decreased survival in ants exposed to microplastics. Li et al. (2020) have found that microplastics induced neurotoxicity in zebrafish by altering the expression of genes associated with nervous system development and function.

20.4 Mechanisms of MNPs Toxicity

Protein Corona Formation When MNPs come into contact with biological fluids, they are rapidly coated by proteins, forming a "protein corona." This layer alters the particles' surface properties, influencing their interactions with cells and tissues (Nel et al. 2009). The protein corona composition affects the behavior of MNPs in the body, including their distribution, cellular uptake, and toxicity. For instance:

- Translocation Rates: Depending on the type and quantity of proteins involved, a protein corona can facilitate the translocation of nanoparticles across cellular barriers.
- Altered Toxicity: The protein corona can modify nanoparticles' physical and chemical properties, potentially increasing their interaction with cells and leading to heightened toxicity (Zhang et al. 2019).

Cellular Uptake Mechanisms MNPs can be internalized by cells through various pathways, with endocytosis being the primary route. Endocytosis involves engulfing particles by the cell membrane, leading to their internalization. There are several types of endocytosis:

- Phagocytosis: Primarily employed by immune cells to engulf large particles or pathogens.
- Micropinocytosis: Involves the uptake of extracellular fluid and its contents, including small particles.
- Clathrin-mediated endocytosis: A well-studied pathway where particles are internalized through clathrin-coated pits on the cell membrane.
- Caveolae-Mediated Endocytosis: Involves the formation of small, flask-shaped invaginations in the cell membrane known as caveolae.

After exposure, bioavailable MNPs that enter the circulatory system can translocate to secondary organs, potentially accumulating to levels that result in cellular damage. The xenobiotic effects of MNPs may disrupt biological membranes through lipid peroxidation and induce cellular damage by generating free radicals (Kumar et al. 2021). Figure 20.1 demonstrates the MNP toxicity mechanism.

20.5 Factors Affecting MNPs Toxicity

The toxic effects of MNPs on living organisms have been extensively studied, revealing various mechanisms through which they interact with biological systems. These interactions are influenced by multiple factors, including size, shape, surface charge, weathering/ageing processes, and the capacity to adsorb other pollutants. Below is an overview of these factors and their implications for toxicity:

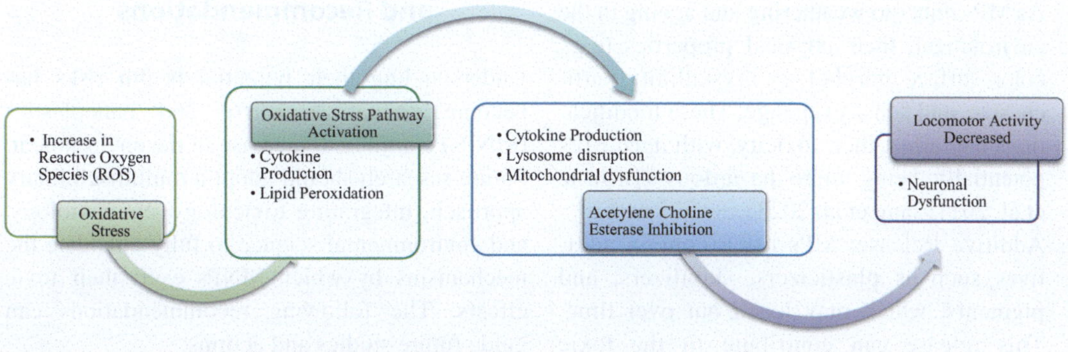

Fig. 20.1 Oxidative stress, cytotoxicity, and neurotoxicity mechanism

Size

- Cellular Uptake: Smaller particles, particularly nanoplastics (NPs), are more likely to penetrate cells and tissues than larger microplastics (MPs). Larger MPs are generally excluded from cellular entry, leading to lower oxidative stress and potentially reduced toxicity (Khan et al. 2020; Rist et al. 2017).
- Oxidative Stress: The plastic particles' size influences the extent of oxidative stress, with smaller particles inducing greater stress due to their higher surface area-to-volume ratio. This increased oxidative stress can lead to cellular damage and inflammation (Tiwari et al. 2021; Hwang et al. 2022).

Shape

- Physical Effects: The shape of MPs significantly affects their interactions with cells and tissues. Spherical MPs tend to cause less physical damage compared to irregularly shaped fragments, which can result in more severe mechanical irritation and tissue damage (Liu et al. 2021; Prata et al. 2019).

Surface Charge

- Cellular Uptake Efficiency: The surface charge, often expressed as zeta potential, plays a crucial role in determining the internalization efficiency of MPs by cells. A higher positive surface charge is typically associated with increased uptake, which can lead to greater toxicity (Chen et al. 2022; Aruoja et al. 2021).

Weathering/Aging Process

- Changes in Physical and Chemical Properties: As MPs undergo weathering and ageing in the environment, their physical properties (e.g., color, surface morphology, crystallinity, particle size, and density) change. These modifications can affect their toxicity, with aged MPs potentially being more hazardous (Browne et al. 2011; Jang et al. 2023).
- Additive Release: MPs often contain additives such as plasticizers, stabilizers, and pigments, which may leach out over time. This release can contribute to the toxic effects of MPs as they degrade (Gonçalves et al. 2021; Zhang et al. 2021).

Adsorption of Other Pollutants

- Pollutant Vector: MPs can adsorb various environmental pollutants, including heavy metals and hydrophobic organic chemicals (HOCs), due to being small in size but possessing a larger surface area. This capability significantly enhances their toxicity by acting as carriers for these harmful substances (Huang et al. 2020; Rist and Sanz-Lázaro 2019).
- Heavy Metals: MPs in marine and coastal environments have the ability to adsorb heavy metals at 10^{-1}–10^{4} μg/g concentrations range during various existing environmental conditions there (Wang et al. 2023).
- Hydrophobic Organic Chemicals (HOCs): MPs' ability to adsorb HOCs further increases their toxic potential, particularly in aquatic ecosystems (Liu et al. 2022; Lee et al. 2022).

Nanoplastics and Biomolecule Interaction

- Protein Corona Formation: As MPs degrade into NPs, they can interact with biological molecules such as proteins, forming a protein corona. This corona can alter the particles' persistence, bioavailability, and ecotoxicity. The composition of the protein corona may change due to aging processes, which can further modify the particles' interactions with biological systems and their toxic effects (Zhang et al. 2023; Accinelli et al. 2021).

20.6 Future Aspects and Recommendations

Understanding their potential health risks has become crucial as micro and nanoplastics (MNPs) continue to increase in the environment. Future research should adopt a multidisciplinary approach, integrating toxicology, epidemiology, and environmental science to fully elucidate the mechanisms by which MNPs exert their toxic effects. The following recommendations can guide future studies and actions:

- Longitudinal Human Studies: There is an urgent need for long-term epidemiological studies that track the health outcomes of populations exposed to MNPs. These studies should focus on various exposure routes—ingestion, inhalation, and dermal contact—and assess their correlations with chronic diseases, including respiratory, cardiovascular, and metabolic disorders.
- Mechanistic Studies: Research should prioritize uncovering the molecular pathways involved in MNP toxicity. Understanding how MNPs interact with cellular structures, induce oxidative stress, and contribute to inflammation will provide insights into their health impacts. In vitro studies using advanced models that mimic human tissues can be instrumental in this endeavor.
- Standardized Assessment Methods: Developing standardized protocols for measuring MNP exposure and assessing their effects will enhance the comparability of studies. This includes establishing methods for quantifying MNPs in biological samples and environmental matrices.
- Risk Assessment Frameworks: Regulatory bodies should develop frameworks for assessing the risks associated with MNP exposure. This includes setting safety thresholds and guidelines for acceptable levels of MNPs in food, water, and air, particularly in immune-compromised populations such as children and pregnant women.
- Public Awareness and Education: It is vital to increase public awareness about MNPs and their potential health impacts. Educational campaigns can inform communities about minimizing exposure, promoting practices such as reducing single-use plastics and participating in local clean-up efforts.
- Interdisciplinary Collaboration: Collaboration is essential for developing comprehensive strategies to address MNP pollution mainly among scientists, policymakers, industry stakeholders, and public health officials. Such partnerships can foster innovative solutions to reduce MNP production and enhance waste management practices.

20.7 Conclusion

Micro and nanoplastics are emerging as significant environmental pollutants with a range of toxic effects on living organisms, including potential adverse impacts on human health. There is an urgent need for comprehensive research to fully understand the extent of MNP toxicity, the underlying biological mechanisms, and the long-term health implications of chronic exposure. By focusing on these areas, the scientific community can better inform public health strategies and regulatory policies to mitigate the risks associated with MNPs. Addressing these challenges requires a concerted effort from researchers, policymakers, and society to ensure a healthier future in an increasingly plastic-polluted world.

References

Accinelli C, Sacca ML et al (2021) Interactions of microplastics and Nano plastics with biological systems: an overview. Environ Pollut. https://doi.org/10.1016/2020.115642

Aliakbarzadeh F, Rafiee M, Khodagholi F, Khorramizadeh MR et al (2023) Adverse effects of polystyrene Nano plastic and its binary mixtures with nonylphenol on zebrafish nervous system: from oxidative stress to impaired neurotransmitter system. Environ Pollut 317:120587. https://doi.org/10.1016/2023.120587

Andrady AL (2011) Microplastics in the marine environment. Mar Pollut Bull 62(8):1596–1605

Aruoja V et al (2021) Impact of microplastics on aquatic organisms: ecotoxicological effects. Mar Pollut Bull. https://doi.org/10.1016/2021.112134

Babaei AA, Rafiee M, Khodagholi F, Ahmadpour E, Amereh F (2022) Nanoplastics-induced oxidative stress, antioxidant defense, and physiological response in exposed Wistar albino rats. Environ Sci Pollut Res 29:82331–82342

Ban M, Shimoda R, Chen J (2021) Investigation of Nano plastic cytotoxicity using SH-SY5Y human neuroblastoma cells and polystyrene nanoparticles. Tox In Vitro 76:105225. https://doi.org/10.1016/2021.105225

Browne MA, Crump P, Niven S et al (2011) Accumulation of microplastic on shorelines worldwide: sources and sinks. Environ Sci Technol 45(21):9175–9179

Capanni F, Greco S, Tomasi N, Giulianini PG, Manfrin C (2021) Orally administered nano-polystyrene caused vitellogenin alteration and oxidative stress in the red swamp crayfish (Procambarus clarkii). Sci Total Environ 791:147984

Chen Y, Li Y, Li S (2015) Toxicity of microplastics to four species of terrestrial invertebrates. Environ Pollut 196:22–28

Chen Q, Wang X, Li Y, Zhang T (2022) Influence of surface charge on microplastic uptake in marine organisms. Environ Sci Technol 56(12):8552–8560

Cole M, Lindeque P, Fileman E, Halsband C, Goodhead R, Moger J, Galloway TS (2011) Microplastics as contaminants in the marine environment a review. Environ Sci Technol 45(22):9746–9755

Fadhl BM, Li J, Lin L, Zhao X (2021) Mechanisms of microplastics toxicity in aquatic organisms a review. Environ Toxicol Pharmacol 85:103640

Fan W, Yang P, Qiao Y, Su M, Zhang G (2022) Polystyrene nanoplastics decrease molting and induce oxidative stress in adult *Macrobrachium nipponense*. Fish Shellfish Immunol 122:419–425

Feng X, Li D, Ma Y, Gao X, Zhao L, Wei Q (2020) Inhibition of acetylcholinesterase activity in the nervous system of goldfish (*Carassius auratus*) by microplastics. Aquat Toxicol 228:105634

Geyer R, Jambeck JR, Law KL (2017) Production use and fate of all plastics ever made. Sci Adv 3(7):e1700782

Gonçalves C, Lima A, Ribeiro F, Da Costa JP (2021) Microplastics and their additives toxicological effects in aquatic organisms. Environ Res 197:111147

Halliwell B, Gutteridge JM (2015) Free radicals in biology and medicine. Oxford University Press, London

Huang Y, Li J, Zhang H, Zhan Y, Li L (2020) Microplastics as vectors for environmental pollutants a review. Environ Sci Technol 54(1):98–110

Hwang HM, Lee S, Choi J, Kim K (2022) Nanoplastics induce oxidative stress and altered gene expression in aquatic organisms. Environ Pollut 295:118653

Jang M, Lee S, Yang Y, Kim K (2023) Environmental aging of microplastics implications for ecotoxicology. Sci Total Environ 871:162169

Jung BK, Han SW, Park SH, Bae JS, Choi J, Ryu KY (2020) Neurotoxic potential of polystyrene nanoplastics in primary cells originating from mouse brain. Neurotoxicology 81:189–196

Khan FR, Chubarenko I, Bagaev A (2020) Toxicity of microplastics and their effects on aquatic life. Mar Pollut Bull 160:111528

Kumar R, Yadav K, Singh R, Singh S (2021) Microplastics and human health a comprehensive review. J Hazard Mater 404:124079

Lee Y, Kim MS, Park JJC, Lee YH, Lee JS (2022) Oxidative stress–mediated synergistic deleterious effects of nano-and microplastics in the hypoxia-conditioned marine rotifer *Brachio nusplicatilis*. Mar Pollut Bull 181:113933

Lei L, Liu M, Song Y, Lu S, Hu J, Cao C, He D (2018) Polystyrene (nano) microplastics cause size-dependent neurotoxicity, oxidative damage and other adverse effects in *Caenorhabditis elegans*. Environ Sci Nano 5(8):2009–2020

Li Y, Zhao H, Xu X, Li Y, Liu Y (2020) Microplastics induce neurotoxicity in zebrafish through altering the expression of genes related to nervous system development. Aquat Toxicol. https://doi.org/10.1016/2019.105453

Liu P, Huang M, Zhang J et al (2021) Shape-dependent toxicity of microplastics to aquatic organisms. Environ Sci Technol 55(20):13785–13794

Liu X, Chen L, Wang Y et al (2022) The role of microplastics in environmental pollution: a comprehensive review. Water Res. https://doi.org/10.1016/2022.118505

Nel AE, Madler L, Velegol D (2009) Understanding biophysics of the protein corona on nanoparticles: a comprehensive overview. J Phys Chem B113(12):3884–3894

Pérez-Guevara F, González-Fernández C, Muniain-Mujika I (2020) The impact of microplastics on aquatic systems a review. Aquat Toxicol 224:105496

Prata JC, da Costa JP, Lopes I (2019) Microplastics an overview of their sources effects and solutions. Environ Sci Pollut Res 26(19):18954–18965

Rist S, Sanz-Lázaro C (2019) Microplastics and their pollutants: the role of environmental factors. Environ Pollut 252:338–346

Rist S, Jansen M, Pahl S (2017) Microplastics in the marine environment an overview of the effects on marine organisms. Environ Pollut 230:29–36

Rochman CM, Tahir A, Williams SL, Baxa DV, Lam R, Miller JT, Teh SJ (2015) Anthropogenic debris in seafood: plastic debris and fibers from textiles in fish and bivalves sold for human consumption. Sci Rep 5(1):1–10

Sendra M, Pereiro P, Yeste MP, Mercado L, Figueras A, Novoa B (2021) Size matters: zebrafish (*Daniorerio*) as a model to study toxicity of nanoplastics from cells to the whole organism. Environ Pollut 268:115769

Setälä O, Reinikainen M, Kukkonen JV, Tiainen J, Paavola R (2014) Microplastics in soil environments a review of their occurrence, effects and fate. Soil Biol Biochem 75:180–187

Tiwari P, Zubair M, Bhatia M (2021) Nanoplastics induced toxicity in aquatic organisms mechanisms and mitigation strategies. J Hazard Mater 415:125817

Tlili S, Jemai D, Brinis S, Regaya I (2020) Microplastics mixture exposure at environmentally relevant conditions induce oxidative stress and neurotoxicity in the wedge clam *Donax trunculus*. Chemosphere 258:127344

Van der Meeren T, Amundsen PA, Zettler EL (2015) Microplastics in zooplankton and filter-feeding fish from the pelagic ecosystem of the Norwegian Sea. Environ Sci Technol 49(18):10913–10920

Wang X, Ren XM, He H, Li F, Liu K, Zhao F, Pan X (2023) Cytotoxicity and pro-inflammatory effect of polystyrene nano-plastic and micro-plastic on RAW264.7 cells. Toxicology 484:153391

Yuan S, Zhang H, Yuan S (2023) Understanding the transformations of nanoplastic onto phospholipid bilayers mechanism microscopic interaction and cytotoxicity assessment. Sci Total Environ 859:160388

Zhang Y, Zhang X, Wang L, Guo L (2019) The influence of protein corona on the biological effects of nanoparticles. J Nanobiotechnol 17(1):1–15

Zhang C, Wu S, Song X, Wang Z (2021) The release of additives from microplastics environmental implications. Environ Sci Technol 55(18): 12320–12330

Zhang X, Wu J, Wang H, Chen Y, Liu J, Zhang L (2023) Biological interactions of nanoplastics understanding the protein corona. Nat Commun 14:2218

Micro-nano Plastics: Impact on Gastrointestinal System

21

Shareen Fatima Rizvi, Syed Khalida Izhar, Uzma Afaq, Mohammed Kuddus, Danish Iqbal, and Roohi

21.1 Introduction

Aside from the numerous technological advantages of living in the "plastic age," the sheer abundance of plastic products, their unsustainable use and disposal, and their great durability in the environment all contribute to pollution dangers, raising significant environmental and public health concerns (Fackelmann and Sommer 2019). Growing concerns have been raised that waste management needs to be equipped to deal with the vast quantities of plastics being produced and disposed of via the several available channels (Rajpal et al. 2024). Microplastics (MPs) are tiny fragments of plastic released into the environment when consumer or industrial plastic goods are discarded or degraded. Microplastics are categorized based on their origin into two distinct groups. The first ones are primary microplastics. These are manufactured in shapes such as pellets, nurdles, and microfibers for cosmetics, toothpaste, pharmaceutical drugs, and textiles. Secondary microplastics are the second category of microplastics. These originate when larger plastic materials are fragmented into tinier pieces in nature (Weber et al. 2022).

Microplastic pollution is a widespread problem ranging from soil to aquatic environment (Choudhury et al. 2022). Microplastic contamination has been documented in agricultural, industrial, floodplain, and coastal soils (Kumar et al. 2020; Zhou et al. 2018; Wang et al. 2021; Nematollahi et al. 2022; Weber and Opp 2020 and Weber et al. 2022). Microplastic pollution sourced from shipping, fishing, oil, and gas extraction has a major impact on aquatic ecosystems, whether they are freshwater or marine (Wang et al. 2021; Sharma et al. 2023). Figure 21.1 shows the breakdown (fragmentation) and the variables that lead to the degradation of plastic trash into smaller pieces of varying sizes.

While once believed to be harmless and inactive substances, the adverse biological effects of microplastic and nanoplastic pollution have lately been proven in several studies (Jiang et al. 2020; Yee et al. 2021; Hong et al. 2023). From a

S. F. Rizvi · S. K. Izhar · U. Afaq
Department of Biosciences, Integral University, Lucknow, Uttar Pradesh, India
e-mail: uzmaafaq@iul.ac.in

M. Kuddus
College of Medicine University of Hail, Hail, Saudi Arabia

D. Iqbal
Department of Health Information Management, College of Applied Medical Sciences, Buraydah Private Colleges, Buraydah, Saudi Arabia

Roohi (✉)
Department of Bioengineering, Integral University, Lucknow, Uttar Pradesh, India

S. Goswami et al. (eds.), *Micro-Nano Plastics Exposure, Environmental Degradation and Public Health Crisis*, Sustainable Development Goals Series, https://doi.org/10.1007/978-981-96-5047-7_21

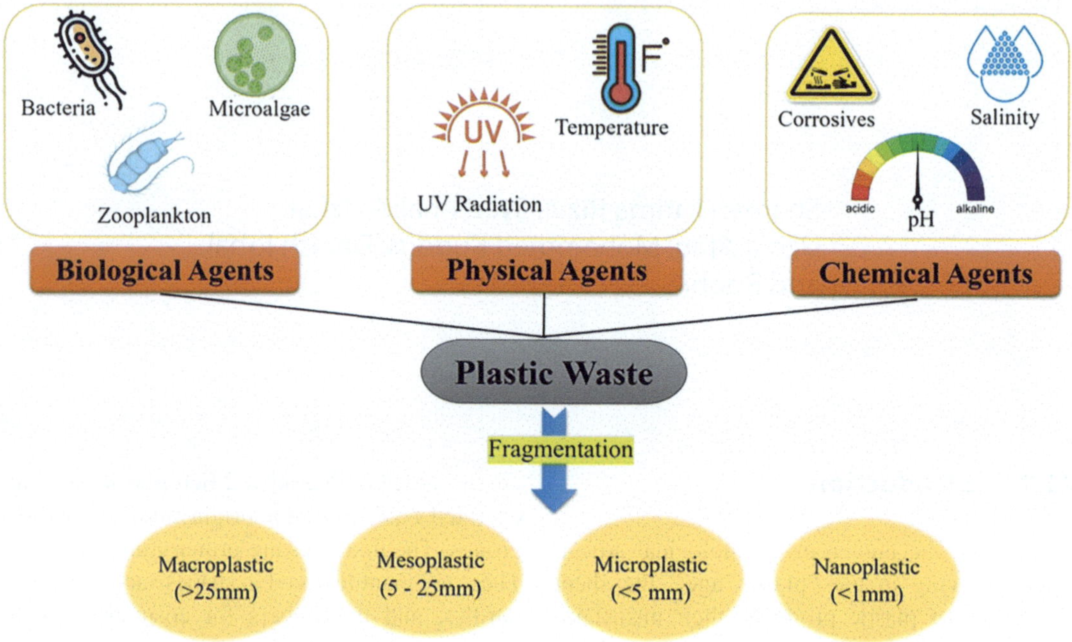

Fig. 21.1 Size-wise characterization of plastics

study by Schwabl et al. (2019), it is well-established that the gut is a primary location for exposure to microplastics (MPs) in the human body; where Schwabl et al. (2019) analyzed samples from eight healthy volunteers from various locations worldwide to determine the presence of MPs. All the subjects' stool samples included microplastics (MPs), ranging from 50 mm (the smallest detectable size) to 500 mm. On average, the samples had 20 micrograms of MPs per 100 grams of stool (Schwabl et al. 2019). Clinical trials evaluating the risks of MNPs to human health are currently not feasible due to ethical concerns. Therefore, it is still not known how MNPs affect human health. We have little insight into the full capacity of humans to absorb and accumulate MNPs.

Additionally, this includes the precise mechanisms governing their movement inside the human body and the biological response of the human body to these substances. Accordingly, this chapter aims to examine the potential impacts of microplastics on the gastrointestinal system, drawing from pertinent studies recorded in the literature. The research findings present today lack sufficient clinical data, but they provide a good overview of the problem and can thus serve as a springboard for future research.

21.2 MP/NPs Exposure by Ingestion

Tainted food or water consumption is the key way of exposure to microplastic and nanoplastic particles (Lehner et al. 2019). There is undeniable evidence of a rising prevalence of microplastics in edibles and drinkables. The prevalent usage of plastic and the tenacity of these particles have increased the likelihood that humans will come into contact with microplastics. In recent investigations, plastic particles have been found in several everyday foods (Galafassi et al. 2019). Disturbingly, a solitary plastic tea bag soaked in a drink could expose individuals to billions of microplastics (MPs) and nanoplastics (NPs) that are released into the beverage (Rubio et al. 2019). Nanoparticles (NPs) have been identified in table salts (Zhang et al. 2020) and seafood (Garrido Gamarro et al. 2020), hence elevating the potential for oral exposure. Human exposure by ingestion is thus highly probable, requiring research that examines the production and consumption phases of the food supply chain. The consumption of microplastics constitutes a significant pathway for human exposure to these pollutants. You can find polymers such as polystyrene (PS) (Ahmad and Bajahlan 2007), polyvinyl chloride (PVC), polyethylene (PE), and polyethylene terephthalate (PET) (Pivokonsky et al. 2018) in a lot of edible and drinkable substances. Although they are identified less commonly, polymethyl methacrylate (PMMA) and polyamide (PA) (Jung et al. 2022) are also detected. In regards to the consumption of MNPLs through alcohol, as stated by the WHO, in 2016, the global per capita alcohol consumption for those aged 15 and up was 6.4 liters. There is a dearth of research on MNPLs in alcoholic drinks, with the few that do exist primarily addressing beer (Li et al. 2022). It is unclear how micro and nanoplastics are distributed upon ingestion. Multiple micro and nanoplastics pathways can be taken in depending on their size. There are several types of uptakes that cells can undergo, such as microfold cell transcytosis, enterocyte endocytosis, villus crevice persorption, and paracellular uptake (Covello et al. 2024).

21.3 Translocation of NMPS in Gastrointestinal Tract

Nanoplastics and microplastics are transported and absorbed into cells through active or passive mechanisms. Passive transport is a concentration-dependent process that occurs only when the size of MNPs aligns with the size of surface pores on the cell membrane. Conditions (such as cancer) that enlarge the cell membrane surface pores enhance the penetration and translocation of MNPs into the cell. The active mode of translocation is an ATP-dependent process and occurs against the concentration gradient of MNPs inside and outside the cell. The type of corona compounds attached to the MNPs is the major factor in active translocation (Yee et al. 2021; Hong et al. 2023). Other factors, such as the type of cell in which MNPs are entering and the size and shape of the MNPs, are directly influenced by the corona compound attached to the MNPs. Figure 21.2 shows the infiltration of microplastics and nanoplastics in biological systems by studying how polystyrene nanoparticles' protein corona reacts to in vitro gastrointestinal digestion and how they traverse an in vitro intestinal barrier. Walczak et al. (2015) determined that particle size and surface chemistry affected translocation. Both undamaged and broken down 50 nm polystyrene nanoparticles with varying electrical charges (positive and negative) were added to a co-culture of intestinal Caco-2 and HT29-MTX cells utilizing an artificial gastrointestinal digestion model in a laboratory setting. The digestion time for "neutral," positively charged and negatively charged PS-NPs was 4.8, 3.5, and 1.8 times, respectively, of protein in the corona. Hence, it is well-established that NMPs are not inert and capable of interacting well with the gastrointestinal tract on the cellular level (Covello et al. 2024).

21.4 GI Injuries Caused by MP/NPs Exposures

Given that most individuals acquire microplastics (MPs) and nanoplastics (NPs) through water and food consumption, the gastrointestinal tracts could be particularly vulnerable to their harmful effects. There is limited knowledge concerning the effects of MPs/NPs on the stomach even though a recent study found them in many human organs. It still needs to be completely known how MNPs affect the gut microbiota and the digestive system in humans and other mammals. In a comparative systematic research investigation, Park et al. (2023) examined the impact of microplastics (microscopic particles, 50–100 nm in size) using organoids derived from the human colon, human colonocytes, and a murine model of colonic inflammation. At 5 mg/mL doses, the scientists found colon organoids lost more than 20% of their viability. Inflammation, apoptosis, and immunity-related genes were also upregulated after this exposure. In addition, evidence col-

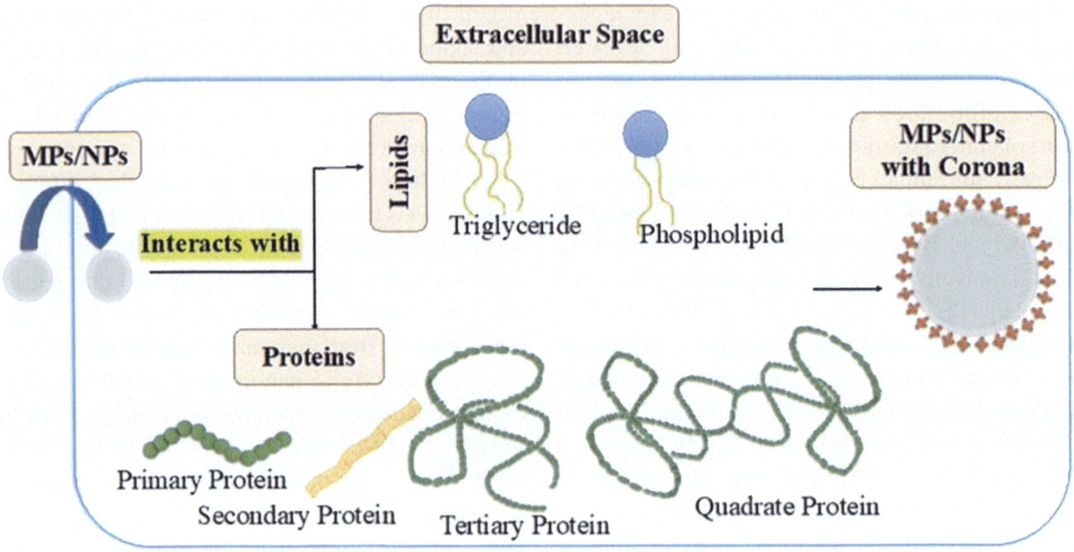

Fig. 21.2 Conceptual model of the possible entry of MPs/NPs into the extracellular space

lected from a murine model showed that after seven (7) days of treatment, 50 nm MPs accumulated in many organs of the mice, including the colon. Adsorption by polyvinyl chloride microplastics (PVC-MPs) altered the protein structure and functionality of digestive enzymes (α-amylase, pepsin, and trypsin) (Liu et al. 2024).

Additionally, they looked into how PVC-MPs undergo biotransformation in the intestines. Due to increased hydrophilicity upon digestion, PVC-MPs' mobility within organisms will be enhanced. The cytotoxicity of PVC-MPs—the evaluated parameters encompassed cellular uptake, viability, membrane integrity, ROS, and mitochondrial membrane potential—is influenced by the altered physicochemical qualities that occur after digestion, according to cell tests. Analysis of transcriptomes provided additional evidence that digesting treatment amplified the biotoxic effects of PVC-MPs. Disruption of intestinal barriers, cytotoxicity, dysbiosis of gut microbiota, immunotoxicity, and oxidative reactive oxygen species (ROS) mediated apoptosis are the major potential toxicological effects of MP/NPs on the gastrointestinal system (Lu et al. 2018). Table 21.1 summarizes the overall research on MPs' impact on gastrointestinal injuries.

21.4.1 MP/NPs and Oxidative Stress and Cytotoxicity

Oxidative stress can arise from an overabundance of antioxidant responses. Microplastics may have stress due to their large surface area, allowing them to adsorb oxidizing chemicals such as metals or generate reactive oxygen species (Table 21.1) during the inflammatory response (Marciano and Vajro 2017). For instance, researchers have shown oxidative stress in zebrafish (*Danio rerio*) and mice following exposure to microplastics (Lu et al. 2018). This triggers a positive free radical production feedback loop and exposes potential plastic removal methods (Bhuyan 2022). In addition, MPs with smaller dimensions might be more readily absorbed by cells and had enhanced physicochemical characteristics, likely attributed to greater surface area. Ding et al. (2022) examined the effects of pH, PS-MP particle sizes, initial nonylphenol (NP) concentration, and metal ion content on batch adsorption. The study found that PS-MP particle size (0.1, 1, 10, 50, and 100 μm) significantly affected their NP adsorption ability. Both PS-MPs and NP considerably impacted cell proliferation, as seen by decreased cell viability, hindered progression through the G1 phase of the cell cycle,

Table 21.1 Overview of research on the impact of microplastics (MPs) on gastrointestinal injuries

Organism/cell line	MP type/size/ concentration	Exposure duration	Observed effects	Reference
Zebrafish, mice	Microplastics (not specified)	Acute exposure (short-term)	Oxidative stress due to ROS generation	Lu et al. (2018)
Zebrafish, mice	Microplastics (not specified)	Acute exposure (short-term)	Oxidative stress due to ROS generation	Ding et al. (2024)
Human cells (unspecified)	PS-MP (0.1–100 μm), nonylphenol	Short-term (unspecified)	Decreased cell viability, G1 phase hindrance, apoptosis	
HT29 human intestinal cell line	PS-MP (3 μm and 10 μm)	24 h	Cytotoxicity: 3 μm (6.1%–29.6% mortality), 10 μm (6.7%–21.6% mortality); ROS production	Jin et al. (2019)
Mice	5 μm fluorescent polystyrene microplastics	6 weeks	Decrease in mucus secretion, intestinal barrier impairment	
Coastal and Highland human populations	Microplastics (not specified)	Not specified	Altered gut microbial community composition, association with specific bacterial taxa	Nugrahapraja et al. (2022)
Female mice	PET microplastics (2–631 μm) at 500 mg/kg	28 days	Decreased Bacteroidetes, increased Firmicutes and lactobacillus	Liu et al. (2024)
Caco-2 and HT29-MTX-E12 cell co-culture	Carboxy-modified PS-MP (500 nm and 50 nm)	Short-term (unspecified)	Minor cytotoxicity, except at high concentrations, suggesting mucus layer protection	Hesler et al. (2019)
Mice	PS-MP (500 mg/L)	Short-term (unspecified)	Imbalance in intestinal immunity, increased TNF-α, IL-1β, IFN-γ	Liu et al. (2022)
Mice	PE microbeads (36 and 116 μm)	6 weeks	Intestinal disruptions, gene expression changes, immune cell recruitment impairment	Djouina et al. (2022)
Murine immune cells (PBMCs, raw 264.7, HMC-1)	PP microplastics (20–200 μm)	Short-term (unspecified)	Cytokine release, increased proinflammatory responses at high concentrations and smaller particle size	Hwang et al. (2019)
Mice	PE microplastics (10–150 μm), doses of 2, 20, 200 μg/g	5 weeks	Colonic inflammation, elevated AP-1, IRF5, worsened histological damage in the colonic mucosa	Li et al. (2020)

and increased apoptosis by disrupting fundamental cellular processes. In this study, Visalli et al. (2021) investigate the effects of polystyrene microplastics (PS-MPs) with sizes of 3 and 10 μm on the HT29 human intestinal cell line et al. (2021) investigate the effects of polystyrene microplastics (PS-MPs) with sizes of 3 and 10 μm on the HT29 human intestinal cell line. According to their findings, both PS suspensions showed substantial cytotoxicity after 24 h; the mortality rates for the 10 μm PS range from 6.7 to 21.6%, while the mortality rates for the 3 μm PS range from 6.1 to 29.6%. The particle concentration in the suspensions ranged from 100 to 1600 particles per mL. HT29 cells treated to PS 3 μm exhibited more significant lysosomal membrane permeabilization, as observed under a microscope. Cells exposed to PS particles measuring 10 μm showed increased reactive oxygen species production levels.

21.4.2 MP/NPs and Intestinal Barrier Disruption and Changes in Gut Microbiota

The human gastrointestinal tract harbors many microbial communities, commonly known as the gut microbiota. Research on mammals has revealed that acute and chronic exposure to MNPs can alter microbial community composition, highlighting that the gut microbiota is an integral part of many physiological and pathological processes in human life. This could cause systemic and localized outbreaks of metabolic disorders and infectious diseases (Table 21.1). According to variations observed in research findings regarding the impact of MP/NPs on intestinal barriers, it can be concluded that the effect of MP on gut cells is heavily influenced by the features of the MP (such as size and polymer type), the concentration used, and the period of exposure. Furthermore, it is crucial to consider the specific cell line utilized and the experimental methodology when examining the uptake of MP and its effects on the gut. A 2019 study by Jin et al. demonstrated that mice fed a diet containing 5 μm fluorescent polystyrene microplastics for 6 weeks exhibited accumulation of these microplastics in their intestines, leading to a reduction in intestinal mucus secretion and subsequently compromising the efficacy of the intestinal barrier. Using next-generation sequencing to study the gut microbial makeup in stool samples from coastal and highland populations in Indonesia, Nugrahapraja et al. (2022) compare samples from populations with and without microplastic contamination. The study revealed that microplastic pollution does not impact the general assortment of gut microbes and the variety of genes associated with gut health.

They identified both negative and positive correlations between certain categories of microplastics and specific bacterial taxa, notably from the genera Roseburia, Clostridium, and Prevotella. In their investigation, Liu et al. (2022), the researchers subjected female mice to a concentration of 500 mg/kg of polyethylene terephthalate (PET) microplastics, which ranged in size from 2 μm to 631 μm, for 28 days. The study documented a decline in the prevalence of Bacteroidetes and Parabacteroides, with a rise in the population of Firmicutes and Lactobacillus. In their work, Hesler et al. (2019) examined the effects of carboxy-modified polystyrene microparticles, specifically those with a size of 500 nm and smaller particles measuring 50 nm, using sophisticated in vitro co-culture models of Caco-2 cells and mucus-secreting HT29-MTX-E12 cells to study the human intestinal and placental barriers. Cells exposed to PS particles measuring 10 μm showed increased reactive oxygen species production levels. It was shown that these particles did not inflict any substantial damage to the cells, except when used at extremely high concentrations. This suggests that the mucus layer may have a protective effect. No significant difference in absorption was detected in a separate investigation involving polystyrene microplastics (2 μm) on Caco-2 and Caco-2/Raji-B models. Overall, all the studies investigating the intestinal microbiota in animals exposed to microplastics have consistently shown dysbiosis. Although the specific characteristics of this dysbiosis may differ depending on the situation, the observed changes in the variety and composition of microflora are likely to result in functional deficiencies in the immune system.

21.4.3 MP/NPs and Immunology of Gut

Intestinal microbes and food-borne antigens are constant threats to the intestinal immune system and tolerance to these threats must be developed. The gut immune system, however, must also be ready to deal with harmful bacteria and other foreign substances (Table 21.1). Intestinal B- and T-lymphocytes, myeloid cells, and innate lymphocytes all play a role in the delicate balance between pro- and anti-inflammatory stimuli maintained by the mesenteric lymph nodes. In response to MNP exposure, immune cells undergo a substantial transcriptional modification that impacts enzyme levels and cytokine release. Nano and microplastics have been found to have an immuno-toxic effect on the immunity of the digestive system, according to various research conducted on invertebrates and vertebrates. Liu et al. (2022) exposed mice to 500 mg/L of PS-MPs. The analysis found that this exposure led to an imbalance in intestinal immunity, which was related to increased inflammatory damage characterized by elevated TNF-α, IL-1 β, and IFN-γ levels. Djouina et al. (2022) conducted an experiment in which mice were given PE microbeads of 2 diameters, namely, 36 and 116 μm orally, for 6 weeks. Both sizes were detected in mouse stools. Intestinal disruptions were noted following separate exposure to each size of polyethylene microbead, with the most significant harmful effects observed in mice that had been exposed to multiple sizes simultaneously. There were quantifiable changes in gene expression associated with epithelial permeability and inflammatory indicators; there was also a noticeable impairment in the recruitment of some immune cells throughout the intestinal tract, specifically from the upper part of the small intestine to the colon. In their study, Hwang et al. (2019) examined the cellular reactions of secondary polypropylene microplastics (PP particles) measuring around 20 μm and 25–200 μm under various circumstances and sizes in relation to normal cells, analysis of immune cells, blood cells, and murine immune cells using cytokine assessment, reactive oxygen species test, polarization assay, and proliferation assay. Their research demonstrated that PP particles displayed a moderate cytotoxic effect contingent upon size and concentration. Nevertheless, when exposed to small PP particles at high concentrations, the immune system was stimulated, leading to a rise in cytokine levels and the possibility of hypersensitivity to these particles. In a study by Li et al. (2020), mice were subjected to varying quantities of polyethylene microplastics (ranging from 10 to 150 μm) at doses of 2, 20, and 200 μg/g of chow over 5 weeks. The colon and duodenum of the infected mice displayed obvious histological signs of inflammation, and they showed increased levels of several factors, including interferon regulatory factor 5 (IRF5) and the proinflammatory transcription factor activator protein 1 AP-1 (c-Jun). In their 2022 study, Lui et al. compared the effects of 500 μg/L polystyrene microplastics (PSMPs) on mice with normal immune systems versus animals with intestinal immunological imbalance. The effects were evaluated using a range of techniques, including colon images and histological research, measurement of inflammatory cytokines, analysis of PSMPs distribution, examination of the microbial community, and metabolomics analysis; the findings indicated that there was an aggravation of the colonic response to PSMPs due to an imbalance in the intestinal immune system. The exposure to PSMPs greatly enhanced the expression of inflammatory mediators (TNF-α, IL-1β, and IFN-γ) in mice exhibiting intestinal immunological dysregulation. Furthermore, PSMPs worsened the histological damage to the colonic mucosa in mice with an imbalance in their intestinal immune system. It also significantly disrupted the composition and functioning of the microbial community in the colon. Although research has shown that MNPs can affect the immune system, their influence on the adaptive immune response still needs to be discovered because most studies have only examined the innate immune response.

21.5 Future Prospective and Recommendation

Every year, the world produces millions of tonnes of plastic waste, which breaks down into tiny particles and spreads over the land, causing pollution. Consumption is a common method by which animals and people are exposed to MP/NPs, which can build up in the digestive system to an extent and in a way that is not fully understood. The intestinal lining is exposed to a wide variety of plastic materials. Given the intricate nature of the situation, our current knowledge regarding the quantities and characteristics of consumed plastics still needs to be improved. Identifying particles measuring a few micrometers is subject to analytical restrictions, meaning the environmental exposure data only pertains to micro-sized plastics.

Moreover, interstudy comparisons must be revised due to the absence of standardized technical data collection and processing protocols. The current data at our disposal provide us with limited insights into the presence of plastic contaminants in drinking water and various food items. More comprehensive investigations on the presence of plastics in the total diets of the general population still need to be conducted, which would provide a more accurate estimate of the extent of oral contamination by plastics. Research should primarily concentrate on the presence of microplastics in human feces. This method is especially valuable for identifying the polymers that cause harm by directly contacting the gut mucosa. Ensuring the standardization of NP characterization is essential to ensure that toxicological evaluations are reproduced and compatible. Since animal models may not perfectly reflect human traits and reactions to NPs and microplastics, human organoids could provide a more accurate toxicity evaluation tool. Identifying possible exposure pathways and developing evaluation procedures might benefit from a deeper comprehension of NPs' behavior. Nanoparticles (NPs) can interact with soil, water, and biota differently depending on their size, surface properties, and aggregation state. The microbiome is essential to human health and the proper functioning of ecosystems; thus, it is important to evaluate the possible toxicity of NPs in this community. A thorough risk management plan should be developed to lessen the dangers of NPs. This plan should include cutting down on emissions of nanoplastics, creating substitute materials, and enhancing waste management techniques.

21.6 Conclusion

In conclusion, there have been inconsistent results from in vitro studies performed on human cell lines for investigating the toxicity of MNPs in the digestive system. One possible explanation for the differences is that the particle doses utilized in the various studies were varied. Another possible explanation for the contradictory results is the wide range of treatment times and particle concentrations used. Additionally, an investigation has been conducted into the potential long-term consequences to bees of MNPs; all of the referenced research solely evaluates their short-term impacts on various endpoints. Results from mammalian research point to the potential toxicity, immunotoxicity, and dysbiosis of MNPs on intestinal cells. Observational and biomarker-based studies will provide valuable insights into the actual detrimental impacts of these particles on human gut health. Therefore, additional research and analytical approaches are necessary to fully understand the actual toxicological impacts of MPNs on the human intestinal system and the specific involvement of the gut microbiota in this regard.

References

Ahmad M, Bajahlan AS (2007) Leaching of styrene and other aromatic compounds in drinking water from PS bottles. J Environ Sci 19(4):421–426

Bhuyan M (2022) Effects of microplastics on fish and in human health. Front Environ Sci 10:827289

Choudhury M, Sharma A, Pervez A, Upadhyay P, Dutta J (2022) Growing menace of microplastics in and around the coastal ecosystem. In: Coastal ecosystems: environmental importance, current challenges and conservation measures, pp 117–137. https://doi.org/10.1007/978-3-030-84255-0_6

Covello C, Di Vincenzo F, Cammarota G, Pizzoferrato M (2024) Micro (nano) plastics and their potential impact on human gut health: a narrative review. CIMB 46(3):2658–2677

Ding F, Zhao Q, Wang L, Ma J, Song L, Huang D (2022) Adsorption Behavior of Nonylphenol on Polystyrene Microplastics and Their Cytotoxicity in Human Caco-2 Cells. Water 14(20):3288. https://doi.org/10.3390/w14203288

Ding R, Chen Y, Shi X, Li Y, Yu Y, Sun Z, Duan J (2024) Size-dependent toxicity of polystyrene microplastics on the gastrointestinal tract: oxidative stress related-DNA damage and potential carcinogenicity. Sci Total Environ 912:169514

Djouina M, Vignal C, Dehaut A, Caboche S, Hirt N, Waxin C, Body-Malapel M (2022) Oral exposure to polyethylene microplastics alters gut morphology, immune response, and microbiota composition in mice. Environ Res 212:113230

Fackelmann G, Sommer S (2019) Microplastics and the gut microbiome: how chronically exposed species may suffer from gut dysbiosis. Mar Pollut Bull 143:193–203

Galafassi S, Nizzetto L, Volta P (2019) Plastic sources: a survey across scientific and grey literature for their inventory and relative contribution to microplastics pollution in natural environments, with an emphasis on surface water. Sci Total Environ 693:133499

Garrido Gamarro E, Ryder J, Elvevoll EO, Olsen RL (2020) Microplastics in fish and shellfish–a threat to seafood safety? J Aquat Food Prod Technol 29(4):417–425

Hesler M, Aengenheister L, Ellinger B, Drexel R, Straskraba S, Jost C, Kohl Y (2019) Multi-endpoint toxicological assessment of polystyrene nano-and microparticles in different biological models in vitro. In Vitro Toxicol 61:104610

Hong Y, Wu S, Wei G (2023) Adverse effects of microplastics and nanoplastics on the reproductive system: a comprehensive review of fertility and potential harmful interactions. Sci Total Environ 903:166258

Hwang J, Choi D, Han S, Choi J, Hong J (2019) An assessment of the toxicity of polypropylene microplastics in human derived cells. Sci Total Environ 684:657–669

Jiang B, Kauffman AE, Li L, McFee W, Cai B, Weinstein J, Xiao S (2020) Health impacts of environmental contamination of micro-and nanoplastics: a review. Environ Health Prev Med 25:1–15

Jin Y, Lu L, Tu W, Luo T, Fu Z (2019) Impacts of polystyrene microplastic on the gut barrier, microbiota and metabolism of mice. Sci Total Environ 649:308–317

Jung JW, Kim S, Kim YS, Jeong S, Lee J (2022) Tracing microplastics from raw water to drinking water treatment plants in Busan, South Korea. Sci Total Environ 825:154015

Kumar M, Xiong X, He M, Tsang DC, Gupta J, Khan E, Bolan NS (2020) Microplastics as pollutants in agricultural soils. Environ Pollut 265:114980

Lehner R, Weder C, Petri-Fink A, Rothen-Rutishauser B (2019) Emergence of nanoplastic in the environment and possible impact on human health. Environ Sci Technol 53(4):1748–1765

Li B, Ding Y, Cheng X, Sheng D, Xu Z, Rong Q, Zhang Y (2020) Polyethylene microplastics affect the distribution of gut microbiota and inflammation development in mice. Chemosphere 244:125492

Li Y, Peng L, Fu J, Dai X, Wang G (2022) A microscopic survey on microplastics in beverages: the case of beer, mineral water and tea. Analyst 147(6):1099–1105

Liu S, Li H, Wang J, Wu B, Guo X (2022) Polystyrene microplastics aggravate inflammatory damage in mice with intestinal immune imbalance. Sci Total Environ 833:155198

Liu G, Jiang Q, Qin L, Zeng Z, Zhang P, Feng B, Qing T (2024) The influence of digestive tract protein on cytotoxicity of polyvinyl chloride microplastics. Sci Total Environ 945:174023

Lu L, Wan Z, Luo T, Fu Z, Jin Y (2018) Polystyrene microplastics induce gut microbiota dysbiosis and hepatic lipid metabolism disorder in mice. Sci Total Environ 631:449–458

Marciano F, Vajro P (2017) Oxidative stress and gut microbiota. In: Gastrointestinal tissue. Academic, pp 113–123

Nematollahi MJ, Keshavarzi B, Mohit F, Moore F, Busquets R (2022) Microplastic occurrence in urban and industrial soils of Ahvaz metropolis: a city with a sustained record of air pollution. Sci Total Environ 819:152051

Nugrahapraja H, Sugiyo PWW, Putri BQ, Huang L, Hafza N, Götz F, Luqman A (2022) Effects of microplastic on human gut microbiome: detection of plastic-degrading genes in human gut exposed to microplastics preliminary study. Environment 9(11):140

Park SB, Jung WH, Choi KJ, Koh B, Kim KY (2023) A comparative systematic analysis of the influence of microplastics on colon cells, mouse and colon organoids. Tissue Eng Regen Med 20(1):49–58

Pivokonsky M, Cermakova L, Novotna K, Peer P, Cajthaml T, Janda V (2018) Occurrence of microplastics in raw and treated drinking water. Sci Total Environ 643:1644–1651. https://doi.org/10.1016/j.scitotenv.2018.08.102

Rajpal A, Choudhury M, Goswami S, Chakravort A, Raghavan V (eds) (2024) Waste management and treatment: advances and innovations, 1st edn. CRC Press. https://doi.org/10.1201/9781003258377

Rubio L, Marcos R, Hernández A (2019) Potential adverse health effects of ingested micro- and nanoplastics on humans. Lessons learned from in vivo and in vitro mammalian models. J Toxicol Environ Health B 23(2):51–68. https://doi.org/10.1080/10937404.2019.1700598

Schwabl P, Koppel S, Konigshofer P, Bucsics T, Trauner M, Reiberger T, Liebmann B (2019) Detection of various microplastics in human stool: a prospective case series. Ann Intern Med 171(7):453–457

Sharma S, Bhardwaj A, Thakur M, Saini A (2023) Understanding microplastic pollution of marine ecosystem: a review. Environ Sci Pollut Res 31:41402–41445

Visalli G, Facciolà A, Pruiti Ciarello M, De Marco G, Maisano M, Di Pietro A (2021) Acute and sub-chronic effects of microplastics (3 and 10 μm) on the human intestinal cells HT-29. Int J Environ Res Public Health 18(11):5833. https://doi.org/10.3390/ijerph18115833

Walczak AP, Kramer E, Hendriksen PJ, Helsdingen R, Van Der Zande M, Rietjens IM, Bouwmeester H (2015) In vitro gastrointestinal digestion increases the translocation of polystyrene nanoparticles in an in vitro intestinal co-culture model. Nanotoxicology 9(7):886–894

Wang Z, Zhang Y, Kang S, Yang L, Shi H, Tripathee L, Gao T (2021) Research progresses of microplastic pollution in freshwater systems. Sci Total Environ 795:148888

Weber CJ, Opp C (2020) Spatial patterns of mesoplastics and coarse microplastics in floodplain soils as resulting from land use and fluvial processes. Environ Pollut 267:115390

Weber CJ, Hahn J, Opp C (2022) Spatial connections between microplastics and heavy metal pollution within floodplain soils. Appl Sci 12(2):595

Yee MSL, Hii LW, Looi CK, Lim WM, Wong SF, Kok YY, Leong CO (2021) Impact of microplastics and nanoplastics on human health. Nano 11(2):496

Zhang Q, Xu EG, Li J, Chen Q, Ma L, Zeng EY, Shi H (2020) A review of microplastics in table salt, drinking water, and air: direct human exposure. Environ Sci Technol 54(7):3740–3751

Zhou Q, Zhang H, Fu C, Zhou Y, Dai Z, Li Y, Luo Y (2018) The distribution and morphology of microplastics in coastal soils adjacent to the Bohai Sea and the Yellow Sea. Geoderma 322:201–208

Micro-nano Plastics: Development of Carcinogenicity

22

Ayan Chatterjee, Rajdip Goswami, Parna Chatterjee, Ashi Ramavat, and Jewelson Santos

22.1 Introduction

Plastics are integral to modern life, from packaging and household items to medical devices and automotive components. Their versatility, durability, and cost-effectiveness have made them indispensable (Llorca et al. 2018). However, widespread use and environmental disasters have resulted from the incorrect disposal of plastics, particularly through the proliferation of microplastics and nanoplastics—tiny particles pervasive in ecosystems and even in human bodies (Rajpal et al. 2024). Plastics do not biodegrade traditionally; microplastics, particles smaller than 5 mm, and nanoplastics, or particles smaller than 100 nm, are the last products of their breakdown (Yee et al. 2021). These particles are concerning due to their ability to persist in the environment, build up inside the food chain, and interact with biological systems in ways larger plastic debris cannot. Microplastics originate from diverse sources, including the breakdown of larger plastic debris, synthetic fibres from clothing, and personal care items that incorporate tiny, spherical particles known as microbeads [Bouwmeester et al. 2015]. They are now ubiquitous in oceans, rivers, soils, and even the air we breathe (Choudhury et al. 2022). Nanoplastics, though less studied, pose an even greater threat due to their smaller size and higher surface area-to-volume ratio, making them more reactive and capable of interacting with biological tissues at the cellular and molecular levels. The ability of nanoplastics to cross biological barriers, such as the blood-brain barrier and cellular membranes, raises significant concerns about their potential to cause harm in ways that are not yet fully understood. One of the most pressing concerns associated with microplastics and nanoplastics is their potential role in carcinogenicity (Cox et al. 2019). The complex transformation of healthy cells into malignant ones, termed carcinogenesis, encompasses a multifaceted process involving genetic mutations, persis-

A. Chatterjee (✉)
School of Health Science and Technology, Medhavi Skills University, Sikkim, India

R. Goswami
Department of Pharmaceutical Technology, Adamas University, Kolkata, West Bengal, India

P. Chatterjee
School of Healthcare, Rishihood University, Sonipat, Haryana, India

A. Ramavat
Department of Biotechnology, SRM University, Sonipat, Haryana, India

J. Santos
Faculty Member, University of Batangas, Batangas City, Philippines
e-mail: jewelson.santos@deped.gov.ph

S. Goswami et al. (eds.), *Micro-Nano Plastics Exposure, Environmental Degradation and Public Health Crisis*, Sustainable Development Goals Series, https://doi.org/10.1007/978-981-96-5047-7_22

tent inflammation, oxidative stress, disruptions to cellular communication networks, and other contributing mechanisms. Emerging evidence suggests that exposure to microplastics and nanoplastics could contribute to these processes, potentially increasing the risk of cancer. Microplastics and nanoplastics might contribute to carcinogenesis through several pathways (Cox et al. 2019). Their physical properties, such as size, shape, and surface chemistry, enable them to uniquely interact with cells and tissues. For example, nanoplastics can penetrate deep into tissues and cells, potentially inducing mechanical stress, disrupting cellular structures, and interfering with normal cellular functions. This interaction could damage DNA, oxidative stress, and other cellular changes that increase cancer risk (Sendra et al. 2019). Moreover, microplastics and nanoplastics can act as vectors for other harmful substances. Plastics are known to adsorb and eliminate heavy metals, other hazardous substances, and persistent organic pollutants (POPs) from the environment. When these contaminated particles are ingested or inhaled, they can introduce hazardous substances into the body, where they may contribute to carcinogenesis. This is particularly concerning in aquatic environments, where marine organisms readily consume these particles, causing dangerous compounds to bioaccumulate and biomagnify in the food chain (Llorca et al. 2018). The fact that oxidative stress and chronic inflammation—both of which are known to be important factors in the development of cancer—can be brought on by microplastics and nanoplastics raises additional concerns. Reactive oxygen species (ROS) generation and the body's capacity to eliminate these reactive intermediates or repair the harm they cause are out of balance, leading to oxidative stress. Chronic inflammation can lead to a sustained immune response that damages tissues and alters the cellular environment, creating conditions that favour cancer development. Research has demonstrated that exposure to microplastics and nanoplastics can cause cells to produce more inflammatory cytokines and ROS, suggesting that these particles may contribute to carcinogenesis through these pathways. Furthermore, the potential for microplastics and nanoplastics to disrupt endocrine function adds another concern. Many plastics contain additives, such as bisphenol A (BPA) and phthalates, which are recognized to be hormone disruptors (Bouwmeester et al. 2015). These substances can obstruct hormone signalling pathways. Leading to various health problems, including an elevated risk of cancers linked to hormones, including prostate and breast cancer. When microplastics and nanoplastics containing these additives enter the body, they may release these harmful chemicals, potentially contributing to carcinogenesis. The expanding corpus of studies on the effects of nanoplastics and microplastics on health is still in its infancy, and much remains to be understood about the specific mechanisms through which these particles might contribute to cancer development. However, the evidence that has emerged so far underscores the need for further investigation into this critical issue. Understanding the carcinogenic potential of microplastics and nanoplastics is essential for assessing the full scope of the risks posed by plastic pollution and for developing effective strategies to mitigate these risks. This chapter will explore the emerging evidence on the carcinogenic potential of microplastics and nanoplastics, highlighting key studies, mechanisms, and areas of ongoing research. It will also discuss the broader implications of these findings for public health, environmental policy, and future research directions. As the world continues to confront the challenges of plastic pollution, understanding the potential role of microplastics and nanoplastics in carcinogenesis is critical for ensuring a sustainable future (Bouwmeester et al. 2015).

22.2 Mechanisms of Carcinogenesis Due to Micro and Nanoplastics

Microplastics can be exposed to the human body in three ways: ingestion, contact with the skin, and inhalation. The elements of the human body, as well as carbohydrates, lipids, and proteins, combine with nanoplastics to generate coronated nanoplastic particles. These coronated nanoplastic particles are absorbable (Lai et al. 2022). Plastic exposure inside an individual's body may occur in three ways: food, ingestion, and absorption (Yee et al. 2021).

22.2.1 Gateway to Toxicity: Ingestion, Uptake, and Translocation of Micro and Nanoplastics

The unwitting consumption of micro and nanoplastics (MNPs) through contaminated food and water has become a pressing concern, with far-reaching consequences for human health and the environment. These microscopic pollutants, measuring mere micrometres in size, can effortlessly breach the gut-blood barrier, embarking on a perilous journey through the circulatory system, organs, and tissues. As MNPs accumulate in the liver, kidneys, and brain, they orchestrate a cascade of cellular damage, disrupting the intricate balance of physiological processes (Wang et al. 2022). The diminutive size of MNPs belies their colossal impact, as they insidiously infiltrate the very fabric of our tissues and cells, leaving an indelible mark on our well-being. The image below reveals the shocking pathway of MNPs into the human body (Fig. 22.1), a journey that can lead to a toxic takeover of our cells, tissues, and organs (Yee et al. 2021).

22.2.2 Cellular Chaos: Cell Uptake, Inflammation, and Oxidative Stress

The internalization of micro and nanoplastics (MNPS) by cells sets off a catastrophic chain reaction, culminating in inflammation and oxidative stress induction. The physical presence of these microscopic pollutants within cellular confines wreaks havoc on delicate cellular components, including DNA, proteins, and lipids (Del Rio et al. 2005). This destructive process triggers the production of reactive oxygen species (ROS), which in turn orchestrates a vicious cycle of oxidative damage, targeting DNA and other vital cellular constituents (Yee et al. 2021). As the cell struggles to maintain homeostasis, the ensuing inflammation and oxidative stress precipitate a cascade of changes in cellular behaviour, ultimately paving the way for cancer development (Panizzolo et al. 2023). The internalization of micro and nanoplastics (MNPS) by cells sets off a catastrophic chain reaction, culminating in inflammation and oxidative stress induction. The physical presence of these microscopic pollutants within cellular confines wreaks havoc on delicate cellular components, including DNA, proteins, and lipids (Del Rio et al. 2005). This destructive process triggers the production of reactive oxygen species (ROS), which in turn orchestrates a vicious cycle of oxidative damage, targeting DNA and other vital cellular constituents (Yee et al. 2021). As the cell struggles to maintain homeostasis, the ensuing inflammation and oxidative stress precipitate a cascade of changes in cellular behaviour, ultimately paving the way for the development of cancer (Panizzolo et al. 2023).

22.2.3 Genetic Mayhem: Genotoxicity, Epigenetic Changes, and Immune System Suppression

The insidious presence of MNPs within cells triggers a complex interplay of genetic and epigenetic alterations, ultimately culminating in the development of cancer and other diseases. MNPs can directly induce DNA damage and mutations, leading to genotoxicity and influencing gene expression through epigenetic modifications (Llorca et al. 2018). The immunosuppressive properties of MNPs further compromise the body's ability to mount an effective defence against cancer cells, creating a permissive environment for tumorigenesis. The genetic scars inflicted by MNPs can be transmitted to future generations, perpetuating a cycle of toxicity that poses a significant threat to human well-being (Mohapatra et al. 2021).

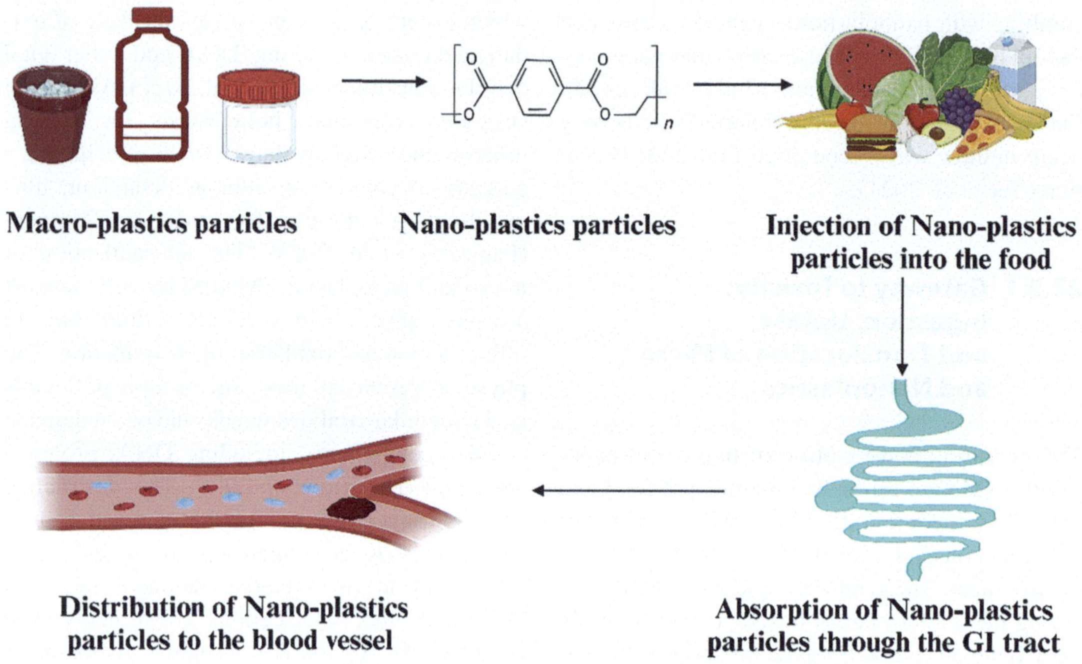

Fig. 22.1 Toxic invasion: the sinister journey of micro and nanoplastics into the human body

22.2.4 Metabolic Mayhem: Mitochondrial Dysfunction, Telomere Shortening, and Metabolic Disruption

MNPs can also directly induce mitochondrial dysfunction, leading to changes in cellular metabolism that increase the risk of cancer. Additionally, MNPs can accelerate telomere shortening, precipitating cellular ageing and further increasing the risk of cancer. The resulting metabolic changes can profoundly impact multiple organ systems, unleashing many health problems that can have devastating consequences (Shammas 2012).

22.3 Factors Influencing the Carcinogenicity of Micro and Nanoplastic Particles

The size and shape of micro and nanoplastic particles are crucial determinants of their carcinogenic potential. These characteristics significantly affect how these particles interact with biological systems, particularly their capacity to penetrate tissues, bypass immune defences, and cause cellular damage. Their small size and irregular shapes enable micro and nanoplastics to cross biological barriers that larger particles cannot, potentially leading to prolonged exposure and accumulation in sensitive tissues. These particles' minute dimensions and irregular morphology enable them to traverse biological barriers with greater ease, thereby amplifying their potential to cause adverse effects. In contrast, larger, more symmetrical particles are hindered in this regard. Below are some key points highlighting the toxicological implications of micro and nanoplastics (Bouwmeester et al. 2015). This accumulation may create conditions favourable for the initiation and progression of cancer. Below are key points that highlight the factors influencing the carcinogenic potential of micro and nanoplastics: the dimensions and morphology of micro and nanoplastic particles play a pivotal role in determining their toxicological impacts, with varying sizes and shapes contributing to distinct effects (Llorca et al. 2018).

22.3.1 Size-Dependent Toxicity

22.3.1.1 Increased Surface Area

Smaller particles, particularly nanoplastics, possess a greater ratio of surface area to volume. This increased surface area enhances their reactivity, allowing them to interact more readily with cells and tissues, potentially leading to toxic effects (Llorca et al. 2018).

22.3.1.2 Cellular Penetration

Nanoplastics, due to their minuscule size, can penetrate cellular membranes, enter the bloodstream, and even cross critical biological barriers such as the blood-brain barrier. This ability to infiltrate sensitive areas within the body raises concerns about their potential to disrupt cellular functions and cause damage at the molecular level (Bouwmeester et al. 2015).

22.3.2 Shape-Dependent Toxicity

22.3.2.1 Irregular Shapes

Micro and nanoplastics often possess irregular shapes with jagged edges, which can cause mechanical damage to cells and tissues upon contact. This physical interaction may trigger inflammatory responses or cellular stress, leading to adverse health effects (Llorca et al. 2018).

22.3.2.2 Fibre-Like Structures

Microplastics that resemble fibres or elongated shapes are particularly concerning as they can be inhaled and lodge in the respiratory system, potentially leading to respiratory issues or even systemic effects through inhalation exposure (Bouwmeester et al. 2015).

22.4 Microplastics and Carcinogenicity in Human

Microplastics (MPs) and nanoplastics (NPs) have emerged as a significant global concern, driven by their pervasive presence, profound ecological repercussions, and the stealthy yet ominous threats they pose to human health and the environment. The small-sized or minute plastic particles are regarded as environmental contaminants due to their intricate complexity, vast diversity, and remarkable persistence, ultimately contributing to the degradation of our ecosystem (Alimba et al. 2021). The United States took a pioneering step in 1987 by categorizing plastics as hazardous substances under the Resource Conservation and Recovery Act, thereby advocating for their controlled use in industrial settings to mitigate their environmental footprint (Bean 1987).

As per the hazard-ranking model of the United Nations' Globally Harmonized System of Classification and Labelling of Chemicals, a significant proportion of more than 50% of plastic monomers (such as polystyrene, polyurethane, and polycarbonate), additives (including PVC, PCBs, and PAH), and other chemical ingredients are categorized as hazardous and carcinogenic, falling under categories 1A and 1B (Lithner et al. 2011). The gastrointestinal tract (GTI) is subject to significant daily exposure to micro and nanoplastics (MNPs) due to the escalating global plastic pollution crisis. In addition to inhalation, the GIT has emerged as the primary portal of entry for MNPs into the body. A groundbreaking pilot study by Schwabl et al. in 2019 revealed the pervasive presence of diverse plastic particles in every human stool sample examined (Schwabl et al. 2019). Carcinogenesis can occur through two distinct pathways: genotoxic and non-genotoxic. The former involves direct DNA damage inflicted by mutagenic agents. Notably, microplastics can serve as vectors for environmentally acquired, surface-bound toxic chemicals, including persistent organic pollutants (POPs), polycyclic aromatic hydrocarbons (PAHs), and hydrophobic organic chemicals (HOCs). Regardless of their origin, microplastic absorption and bioaccumulation in the human body are increasingly implicated in disrupting the body's natural repair mechanisms. This can trigger a cascade of detrimental effects, including elevated apoptosis, necrosis, inflammation, and oxidative stress, and the consequences of microplastic exposure may be far more sinister than we currently understand. Subsequent studies have shed light on the extensive biodistribution of nanoplastics across multiple organs, including the lungs, heart, excretory organs, and even the brain (Yee et al. 2021).

Polyvinyl chloride (PVC), a widely used plastic, is synthesized from vinyl chloride monomer (VCM), a substance with well-documented carcinogenic properties. VCM exposure has been strongly linked to the development of hepatic angiosarcoma, a rare but aggressive form of liver cancer. While direct evidence linking PVC microplastics to cancer in humans is still emerging, the release of VCM and other toxic additives from degrading PVC microplastics presents a significant risk. These particles can act as carriers, delivering carcinogenic substances to human tissues (Fedeli et al. 2019). A significant correlation was found between rising vinyl chloride exposure and elevated liver cancer risk (across all types) among European workers with cumulative exposure levels below 1500 ppm-years (Ward et al. 2001).

22.4.1 Microplastics Exposure: A Novel Contributor to Skin Cancer Aggressiveness

An in vitro assessment of microplastic-induced effects on skin cancer progression by Wang et al. revealed that the microplastics were found to stimulate the proliferation of A431 skin cancer cells in a dose-dependent manner, as evidenced by MTT assay results. The MTT (3-(4, 5-dimethyl thiazolyl-2)-2, 5-diphenyltetrazolium bromide) assay is widely recognized as a sensitive and reliable indicator of cellular metabolic activity, offering a distinct advantage over alternative methods, such as ATP and 3H-thymidine incorporation assays, which involve the use of radioactivity. Furthermore, time-course experiments revealed that microplastics induced proliferative effects on SCL-1 skin cancer cells, with both dose- and time-dependent responses observed. These findings collectively suggest that microplastics exert a stimulatory effect on the growth of various skin cancer cell lines (Wang et al. 2023).

22.4.2 Microplastics: A Potential Trigger for Premature Colorectal Cancer Development

Globally, there is a rising trend of colorectal cancer cases among individuals under 50 years old, a phenomenon known as early-onset colorectal cancer (EOCRC). Notably, this increase is primarily driven by a surge in distal colonic and rectal cancers, which have biological similarities with late-onset colorectal cancer (LOCRC). However, EOCRC is characterized by a higher prevalence of mucinous or signet ring histology and poorly differentiated cancers. Various research studies suggest that the development of early-onset colorectal cancer (EOCRC) and sporadic colorectal cancer, in general, may be linked to the complex interplay between the gut microbiota and the gut mucosa (Purcell et al. 2017). Colonizing colonic bacteria in the gut mucosa is associated with an elevated risk of developing colorectal cancer (CRC). The ingestion of microplastics and its subsequent arrival in the colon may have far-reaching consequences, perturbing the intricate balance between the gut microbiome and the mucus layer. This disruption can expose colon cells to a toxic cocktail of harmful microbiome components, potentially fuelling the development of colorectal cancer. As we grapple with the complexities of this emerging health threat, one thing is clear: "The impact of microplastics on gut health demands urgent attention" (Wisniowska and Wlodarczyk-Makula 2018). The aggregations of microplastics near the inner mucus layer lining the colonic epithelium may be facilitated by forming a biofilm. In the proximity, microplastics laden with carcinogens may transfer these toxins to the colonocytes or undergo pinocytotic uptake, thereby exposing these cells to carcinogenic substances. This potential mechanism raises concerns about the role of microplastics in enhancing the risk of colorectal cancers. Microplastics may also serve as conduits for delivering carcinogenic bacterial toxins to the colonic epithelium. A prime example is the colonic carriage of *Escherichia coli*, which has been linked to an increased risk of colorectal cancer (CRC) through the expression of genotoxins (Nougayrede et al. 2006). While *E. coli* typically inhabits the colonic lumen, studies have demonstrated that these bacteria can bind to microplastics in aquaculture models, suggesting a similar phenomenon may occur in the colon. If so, microplastics with adherent pks + *E. coli* could potentially act as a vector for delivering these genotoxic bacteria to the surface of the colonic epithelium. However, this process may be contingent upon disrupting the intact inner mucus layer (Reuter et al. 2018). Non-genotoxic carcinogenesis can arise from errors during cell division, and microplastics may play a pivotal role in this process. By inducing inflammation and delivering estrogenic compounds, microplastics can interact with estrogen receptors on the surface of cells, leading to enhanced cell proliferation and signalling non-genotoxic carcinogenesis. This phenomenon adds significance because specific intestinal epithelial cells express estrogen receptors, and dietary estrogens and estrogen mimics have been implicated as promoters of colorectal cancer. Furthermore, some plasticizers, such as 4-nonyl phenol, have been identified as estrogen mimics, underscoring the potential for microplastics to contribute to non-genotoxic carcinogenesis (Graham and Shaw 2011). Table 22.1 summarizes the links between the widespread use of microplastics and the development of cancer in humans:

22.5 Future Prospect and Recommendation

The prospects of micro and nanoplastics (MNPs) about carcinogenicity highlight a critical area of ongoing research. As MNPs become increasingly prevalent in the environment, their potential to contribute to cancer development is gaining attention. Advancements in analytical techniques are expected to enhance our ability to detect and study these particles at the nanoscale, enabling a deeper understanding of their carcinogenic mechanisms. Future research will likely focus on the biochemical interactions of MNPs with cellular systems, are exploring how they may induce oxidative stress, chronic inflammation, and genetic damage in all cancer pathways. Epidemiological studies are needed to establish direct links between MNP exposure and cancer incidence, which could inform regulatory and policy measures to limit exposure. Public awareness, education, and innovations in safer, biodegradable plastics will be crucial in mitigating the risks associated with MNPs. Collaborative, interdisciplinary research efforts will play a key role in addressing the complex challenges of MNP-induced carcinogenicity, aiming to protect public health from the growing threat of plastic pollution.

22.6 Conclusion

The pervasive presence of micro and nanoplastics (MNPs) has emerged as a significant environmental concern, marking them as a newly recognized category of contaminants. These minute particles have infiltrated nearly every corner of our environment, exposing humans to their potentially harmful effects through various sources. MNPs are found in food, cosmetics, drinking water, and industrial products, leading to conscious and unconscious exposure. The complexity of the environmental systems—comprising terrestrial, aquatic, and atmospheric compartments—facilitates the transfer and spread of these particles across different ecosystems. This intricate web of environmental interactions is crucial in how MNPs move and disperse. They are not confined to a single medium but migrate between land, water, and air, making their presence ubiquitous and their impact far-reaching. The various pathways through which humans are exposed to MNPs differ in their effects on different body parts, but the full extent of these impacts remains poorly understood.

However, what is becoming increasingly clear is that these plastics can accumulate within the body and pose serious health risks. Research has shown that the presence of micro and nanoplastics in the human body can lead to detrimental effects on several critical components of the endocrine system of mammals, comprising the

Table 22.1 Summary of the links between the widespread use of microplastics and the development of cancer in humans

Aspect	Description	Cancer-related implications	MPs size	MPs types	References
Physical characteristics	Microplastics are small in size with irregular shapes, allowing them to penetrate biological barriers	Penetration into tissues can cause physical damage, cellular stress, and chronic inflammation, which are precursors to cancer	<5 mm	PE, PP, PVC, PS	Bouwmeester et al. (2015)
Chemical additives	Microplastics contain additives such as plasticizers and flame retardants that are not chemically bonded to the polymer matrix	These additives can leach out, many of which are carcinogenic or disrupt hormonal systems, increasing cancer risk	<5 mm	PVC, PS, PET	Llorca et al. (2018)
Environmental contaminants	Microplastics can adsorb environmental pollutants such as heavy metals and persistent organic pollutants (POPs)	These adsorbed contaminants can be introduced into the body, contributing to mutagenesis and cancer development	<5 mm	PE, PP, PVC, PS	Cox et al. (2019)
Oxidative stress	Interaction of microplastics with cells can lead to the generation of reactive oxygen species (ROS)	ROS cause oxidative stress and DNA damage, initiating cancerous transformations	<1 mm	PE, PP, PS	Yee et al. (2021)
Chronic inflammation	Accumulation of microplastics in tissues can lead to persistent inflammation	Chronic inflammation can lead to DNA damage and create a pro-tumorigenic environment	<1 mm	PE, PP, PVC, PS	Bouwmeester et al. (2015)
Immune system disruption	Microplastics may impair immune function	Disrupted immune surveillance allows potentially cancerous cells to evade detection and proliferate	<5 mm	PE, PP, PVC	Llorca et al. (2018)
Bioaccumulation	Microplastics can accumulate in the body over time, especially in the liver, kidneys, and lungs	Long-term accumulation increases exposure to toxic substances and the potential for carcinogenic effects	<1 mm	PE, PP, PVC, PS	Bouwmeester et al. (2015)
Inhalation of microplastics	Microplastics can be inhaled, particularly in urban environments	Inhaled particles can lodge in the lungs, causing inflammation and cellular damage that may lead to lung cancer	<5 mm	PE, PP, PVC, PS	Yee et al. (2021)
Consumption of contaminated food and water	Microplastics are found in drinking water and food sources such as seafood and produce	Ingested microplastics can release harmful chemicals into the body, raising the risk of gastrointestinal cancers	<5 mm	PE, PP, PVC, PS	Cox et al. (2019)
Synergistic toxicity	The combination of microplastics and the contaminants they carry may have amplified effects	Synergistic interactions between microplastics and absorbed toxins could heighten their carcinogenic potential	<5 mm	PE, PP, PVC, PS	Llorca et al. (2018)
Toxic chemical leaching	Chemicals such as phthalates and bisphenol A (BPA) leach from microplastics as they degrade	These chemicals are known carcinogens and increase cancer risk when they enter the body	<5 mm	PVC, PS, PET	Yee et al. (2021)

PE Polyethylene, *PP* Polypropylene, *PVC* Polyvinyl chloride, *PS* Polystyrene, *PET* Polyethylene terephthalate testes, ovaries, thyroid, pituitary, adrenal, hypothalamus, and thyroid gland. One of the primary concerns is the role of endocrine-disrupting chemicals (EDCs) associated with MNPs. These chemicals, either embedded in or carried by MNPs, have structural similarities to natural hormone receptors in the body. Due to these similarities, EDCs can interfere with the normal functioning of these receptors, leading to disruptions in hormonal activities and potentially causing long-term health effects. The unique physicochemical properties of MNPs make them particularly hazardous to the environment and living organisms. Their small size allows them to penetrate biological barriers and accumulate in various tissues, leading to different health issues. MNPs are now recognized as global stressors, given their widespread presence and the growing body of evidence linking them to adverse health outcomes. Studies conducted in vivo (within living organisms) and in vitro (in controlled laboratory environments) have demonstrated that microplastics can harm the digestive, respiratory, immune, and reproductive systems. In the digestive system, microplastics have been shown to cause inflammation, disrupt the gut microbiota, and impair nutrient absorption. Inhalation of these particles can lead to respiratory issues, including inflammation of lung tissues and potentially chronic respiratory conditions. The immune system is also not spared; MNPs can induce immune responses that may lead to chronic inflammation or autoimmunity.

Moreover, the reproductive system appears particularly vulnerable, with studies suggesting that exposure to microplastics can affect fertility and reproductive outcomes, potentially leading to developmental abnormalities in offspring. The interconnected nature of environmental compartments and the pervasive presence of MNPs underscore the urgency of addressing this emerging threat. As research continues to uncover the full extent of MNPs' effects on the environment and human health, it is becoming increasingly apparent that mitigation measures are required to lessen exposure and stop MNPs from spreading. This may include regulatory measures to limit the use of microplastics in consumer products, improved waste management practices to prevent their release into the environment, and ongoing public health initiatives to raise awareness about the risks associated with MNPs.

In conclusion, micro and nanoplastics represent a significant and growing environmental challenge with far-reaching implications for human health. These particles' ability to permeate various ecosystems and biological systems, coupled with their potential to disrupt endocrine functions and harm multiple organ systems, makes them a critical area of concern for researchers, policymakers, and the public. As we continue to understand and address this issue, we must prioritize strategies that minimize exposure and mitigate the harmful effects of these pervasive pollutants.

References

Alimba CG, Faggio C, Sivanesan S et al (2021) Micro(nano)-plastics in the environment and risk of carcinogenesis: insight into possible mechanism. J Hazard Mater 416:126143

Bean MJ (1987) Legal strategies for reducing persistent plastics in the marine environment. World Meritime University 18(6):357–360

Bouwmeester H, Hollman PCH, Peters RJB (2015) Potential health impact of environmentally released micro- and Nano plastics in the human food production chain: experiences from Nanotoxicology. Environ Sci Technol 49(15):8932–8947

Choudhury M, Sharma A, Pervez A, Upadhyay P, Dutta J (2022) Growing menace of microplastics in and around the coastal ecosystem. In: Coastal ecosystems: environmental importance, current challenges and conservation measures, pp 117–137. https://doi.org/10.1007/978-3-030-84255-0_6

Cox KD, Covernton GA, Davies HL et al (2019) Human consumption of microplastics. Environ Sci Technol 53:7068–7074

Del Rio D, Stewart AJ, Pellegrini N (2005) A review of recent studies on malondialdehyde as toxic molecule and biological marker of oxidative stress. Nutr Metab Cardiovasc Dis 15(4):316–328

Fedeli U, Girardi P, Mastrangelo G (2019) Occupational exposure to vinyl chloride and liver diseases. World J Gastroenterol 25(33):4885–48891

Graham L, Shaw I (2011) Does the oestrogen receptor encourage oestrogenicity in environmental pollutants? The case of 4-nonylphenol. SAR QSAR Environ Res 22(3):329–350

Lai H, Liu X, Qu M (2022) Nanoplastics and human health: Hazard identification and biointerface. Nano 12:1298

Lithner D, Larsson A, Dave G (2011) Environmental and health hazard ranking and assessment of plastic polymers based on chgemical composition. Sci Total Environ 409(18):3309–3324

Llorca M, Schirinzi G, Martínez M et al (2018) Adsorption of perfluoroalkyl substances on microplastics under environmental conditions. Environ Pollut 235(1):680–691

Mohapatra A, Sathiyamoorthy P, In-Kyu P (2021) Metallic nanoparticle-mediated immune cell regulation and advanced cancer immunotherapy. Pharmaceutics 13(11):1867

Nougayrede JP, Homburg S, Taieb F et al (2006) Escherichia coli induces DNA double-strand breaks in eukaryotic cells. Science 313(5788):848–851

Panizzolo M, Martins VH, Ghelli F et al (2023) Biomarkers of oxidative stress, inflammation, and genotoxicity to asses exposure to micro- and nanoplastics. A literature review. Ecotoxicol Environ Saf 267:115645

Purcell RV, Pearson J, Aitchison A et al (2017) Colonization with enterotoxigenic Bacteroides fragilis is associated with early-stage colorectal neoplasia. PLoS One 12(2):e0171602

Rajpal A, Choudhury M, Goswami S, Chakravorty A, Raghavan V (eds) (2024) Waste management and treatment: advances and innovations, 1st edn. CRC Press. https://doi.org/10.1201/9781003258377

Reuter C, Alzheimer M, Walles H et al (2018) An adherent mucus layer attenuates the genotoxic effect of colibactin. Cell Micobiol 20(2)

Schwabl P, Koppel S, Konigshofer P et al (2019) Detection of various microplastics in human stool: a prospective case series. Ann Intern Med 171(7):453–457

Sendra M, Stafferi E, Yeste MP et al (2019) Are the primary characteristics of polystyrene nano plastics responsible for toxicity and ad/absorption in the marine diatom Phaeodactylumtricornutum? Environ Pollut 249(1):610–619

Shammas MA (2012) Telomers, lifestyle, cancer, and aging. Curr Opin Clin Nutr Metab Care 14(1):28–34

Wang W, Yuan W, Xu EG et al (2022) Uptake, translocation, and biological impacts of micro(nano)plastics in terrestrial plants: Progress and prospects. Environ Res 203:111867

Wang Y, Xu X, Jiang G (2023) Microplastics exposure promotes the proliferation of skin cancer cells but inhibits the growth of normal skin cells by regulatig the inflammatory process. Ecotoxicol Environ Saf 267(15):115636

Ward E, Boffetta P, Anderson A et al (2001) Update of the follow-up of mortality and cancer incidence among European workers employed in the vinyl chloride industry. Epidemiology 12:710–718

Wisniowska E, Wlodarczyk-Makula M (2018) The effect of selected acidic or alkaline chemical agents amendment on leachability of selected heavy metals from sewage sludge. Sci Total Enviorn 633:463–469

Yee MSL, Hii LW, Looi CK et al (2021) Impact of microplastics and Nanoplastics on human health. Nano 11(2):496

From Plastics to Bioplastics: Technological Innovations and Nature based Solutions to achieve Global Sustainability

23

Rangabhatla Saishree, Suman Kumar Das, Nivedita Panda, B. Anjan Kumar Prusty, and Chandan Mahata

Abstract

Rising global plastic demand, dominated by synthetic plastic, has increased plastic waste pollution, negatively impacting human health and the environment. This chapter discusses bioplastics and nature-based solutions as promising alternatives for global sustainability. The chapter establishes a solid foundation through a literature review and successful case studies on biodegradable plastics. This chapter investigates the classification, characterization, and challenges linked to bioplastics in moving towards nature-based solutions. Ultimately, the aim is to increase awareness and understanding of bioplastics as a viable means for obtaining greater sustainability. Further, it provides an overview of a case study that holistically investigates the utilization of biodegradable plastics in different industries. This work emphasizes real-world applications, such as Novamont's Mater-Bi used in packaging and agriculture, NatureWorks' Ingeo in various sectors, and BASF's Ecoflex and TIPA's compostable packaging. These diverse applications highlight the improved characteristics and end-of-life possibilities of certain bioplastics, as the global market is expected to rise for PHA (poly-hydroxy alkanoates) and PLA (poly-lactic acid) based plastics. This chapter provides deeper insights into the importance of standards and certifications for ensuring bioplastic quality and effectiveness. To examine the significant impacts of plastics derived from biological mass to environmental stability, this chapter discusses the environmental impacts of bioplastics compared to traditional plastics. Furthermore, the findings discuss various solutions for producing parts and products using I4.0 technology, such as additive manufacturing for sustainable production. This chapter delivers valuable information for policymakers, researchers, and the public

R. Saishree · N. Panda · B. A. K. Prusty (✉)
Department of Environmental Science, Berhampur University, Bhanja Bihar, Berhampur, Odisha, India

S. K. Das
Department of Mechanical and Industrial Engineering, RISE– Research & Innovation for Smart Enterprises, University of Brescia, Brescia, Italy
e-mail: s.das001@unibs.it

C. Mahata
Department of Agricultural and Biological Engineering, University of Illinois Urbana-Champaign, Urbana, IL, USA
e-mail: cmahata2@illinois.edu

S. Goswami et al. (eds.), *Micro-Nano Plastics Exposure, Environmental Degradation and Public Health Crisis*, Sustainable Development Goals Series, https://doi.org/10.1007/978-981-96-5047-7_23

through critically analyzing current knowledge and assessing research gaps.

Keywords

Biodegradable plastics · Circular economy · Nature-based solutions · Plastic waste management

23.1 Introduction

Plastics' adaptability, longevity and affordability have increased their demand worldwide. Plastic has a broad range of applications, from packaging and manufacturing of domestic goods to automobile parts, which results in serious environmental consequences. Over one-third of the plastic is used in packaging (e.g. containers and polythene bags), and the rest is used in manufacturing products such as vinyl cladding and plastic pipes. Reportedly, the packaging sector consumes 42% of resin (Mutha et al. 2006). A significant volume of plastic is used by automobiles and manufacturing industries. Plastic is a high resource-efficient material because of its stiffness, toughness, electrical and thermal insulation, corrosion resistance, ductility, high strength-to-weight ratio, non-toxicity and amazing longevity at a comparatively low lifetime cost (Andrady and Neal 2009). Utilization of plastic throughout developing countries is accelerating due to reduced manufacturing costs, and the enhanced performance requirements persistently advocate for its replacements with materials such as paper, metals, wood and glass. For cloth production, around 40 million tons of plastics are transferred to textile fibres, including polyester, acrylics and nylon (Borrelle et al. 2020). According to the European Environmental Agency (EEA) 2020, per capita generation of textile waste was 16 kg, and 60–70% of this is plastics. Synthetic fibres such as polyester and nylon constitute 60% of clothing and 70% of house textiles. High-performance clothes are mainly made of plastic types, including ploy cotton clothing containing a great amount of Polyethylene Terephthalate (PET), while fleece garments are also made up of PET (Andrady and Neal 2009).

Components of footwear, e.g. footbeds and outsoles, are made up of polyurethane and composed of vinyl or synthetic polymers. In medical sectors, plastic is used to manufacture medical equipment such as surgical tools, aseptic packaging and medication packing (Czuba 2014). In food preservation, plastics help reduce wastage through innovations such as modified atmosphere packaging, extending perishable items' shelf life (Opara et al. 2020).

Plastic reduces transportation costs (Awoyera and Adesina 2020) and associated CO_2 emissions due to its lightweight nature which is due to low fuel consumption during transport (Costa et al. 2023). Modern vehicles' average weight contains up to 18% plastics in various components, while some aircraft, such as the Boeing Dreamliner, use up to 50% plastics. Plastics improve performance and lower the prices of elements such as insulation, window and door frames and lightweight fasteners. Additionally, energy is saved by plastics in many uses and enhances the standard of recreational activities by being used in sports equipment (Vieyra et al. 2022).

Despite their benefits, plastics pose substantial environmental challenges. Up to, 04% of gas and oil produced is utilized for plastic production. A considerable amount of energy is consumed in this process. Although plastic stores carbon in its structure, this becomes a challenge rather than a benefit, as it contributes to long-term environmental pollution and can release greenhouse gases (GHGs) upon degradation. Recycling plastics can significantly reduce energy consumption as compared to manufacturing new plastics from virgin fossil fuels (Kabeyi and Olanrewaju 2023), underscoring significant environmental challenges and sustainability issues associated with plastic production. The heavy dependence on fossil fuels in producing plastic and energy contributes to the depletion of resources and also leaves a huge carbon footprint. In 2022, fossil fuels released 36.6 billion tons of CO_2 globally, up by 0.8% from the previous year (Rana 2022). However, while driving home the point that plastics store carbon and that there is potential to retain some energy by recycling, it is ensured that recycling emerged as one of the most vital practices in curtailing those very environmental effects and thus a step forward towards sustainability. According to a new publication by the Organization for Economic Cooperation and Development (OECD), this is a fact that plastics from fossil fuel-based produce comprise 90% of global GHG emissions, amounting to 3.4% globally (OECD 2023). The extraction of raw materials and the manufacture and distribution of plastics are multiple phases through which the environment becomes contaminated. Each stage generates GHG emissions, energy use and soil, water and air pollution.

Ecosystems and wildlife are threatened by single-use plastics, significantly affecting marine habitats (Pinheiro et al. 2023). These plastics break down into microplastics and enter into food chain, and have been found in human beings, i.e. in kidneys, livers, placenta, etc. (de Oliveira et al. 2024). Furthermore, the International Union for Conservation of Nature (IUCN) confirms the possibility of leaching of carcinogenic chemicals from plastic items into water and resulting in various health impacts (illnesses related to the immunological system, nervous system, development and reproduction; IUCN 2024). Thus, they are known to play a significant role in aquatic ecosystem degradation and contamination (Marcharla et al. 2024). Microplastics are small plastic particles with size <05 mm in diameter (Nyadjro et al. 2023) that emerged as an environmental disaster as a breakdown product of plastics. These microplastics are pervasive throughout the environment, in oceans, rivers and soil, and to a great extent, they like to accumulate in trophic levels within the food chain. They have been found in human organs, including the kidney, liver and placenta, with grave health implications (de Oliveira et al. 2024). They may also have a bioaccumulation effect and be transferred along the food chain, making their impact worse on the environment (Adekanmbi et al. 2023).

Recycling plastic and plastic-associated products is promoted as a solution to mitigate its environmental impact. However, contamination, poor infrastructure and low recycling rates hinder its effectiveness. Recycling can also release emissions and contaminants, albeit less than primary production. Plastic waste's low degradation rate and higher environmental persistence lead to issues such as "White Pollution" (Hameed and Ahmed 2019), making the critical need for sustainable alternatives to traditional plastics quite evident. A potential remedy to the abovementioned problem is provided by bioplastics, derived from biologically renewable resources such as sugarcane, cellulose, corn and starch (Chen and Yan 2020). Production stages of bioplastics are given in Fig. 23.1. According to a report by Statista Research Department, 2024, the bioplastics production capacity that existed in 2022 was 2.2 million metric tons and can reach up to 7.4 million metric tons by 2028. The market value for bioplastic in 2023 was around 96.6 billion US dollars. Global market value is expected to touch 1353.3 billion US dollars. Bioplastics reduce the adverse impacts caused by conventional plastic and reliance on fossil fuels. Unfortunately, the manufacturing techniques limit bioplastic use because of high cost, scalability and material properties.

Recently, additive manufacturing (AM) or 3D printing offers a revolutionary plastic alternative to bioplastic. Complex geometries are difficult to achieve with traditional manufacturing and are possible with AM. Additionally, it minimizes material waste while rendering on-demand manufacture possible, which improves supply chains and lessens their adverse impacts on the environment. The performance and sustainability of bioplastics are enhanced by using the advantages of AM. Accordingly, the current chapter explores the strength of bioplastics as a sustainable substitute for plastics made from fossil fuels, highlighting the benefits of AM in overcoming the limitations of traditional bioplastic production techniques.

23.2 Classification of Plastics

On the basis of source, plastic is classified into fossil-based (derived from fossil fuel) and bioplastics (made from renewable biomass such as microorganisms and plants). The classification of

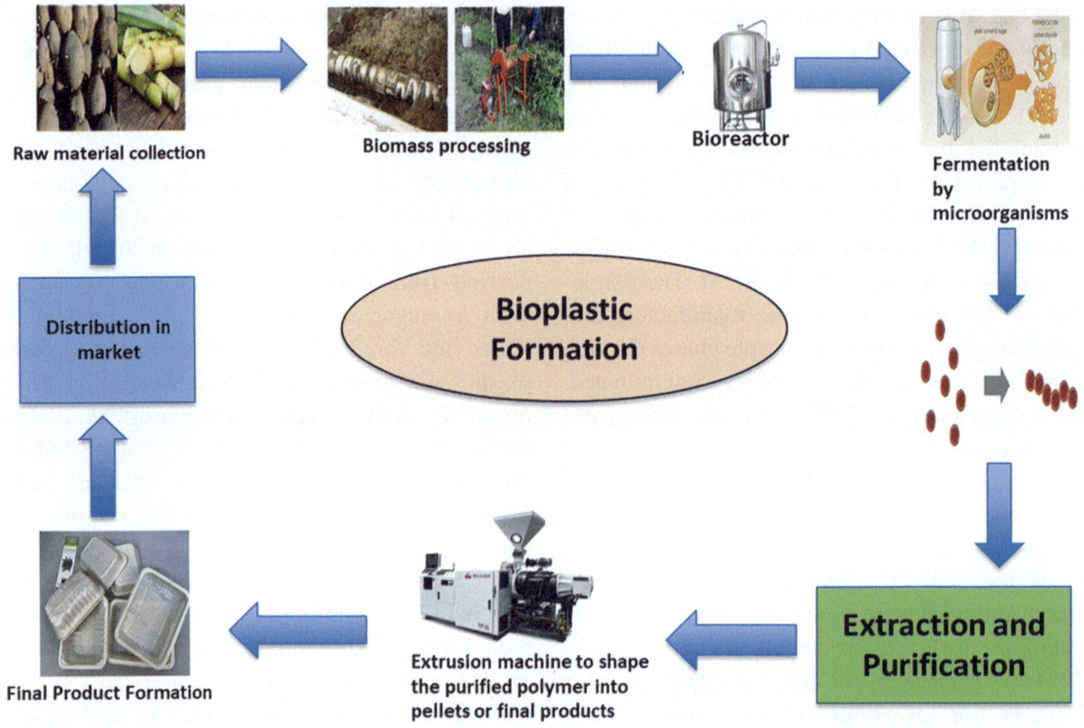

Fig. 23.1 Stages of bioplastic production. (*Source*: Ali et al. (2023a, b))

plastics with examples is given in Fig. 23.2. Bioplastics are the biologically unrelated products derived from renewable biomass such as microbes and agricultural products (Dilshad et al. 2021) and can be recycled by biological processes that help to minimize the usage of fossil fuels and protect the environment. Biopolyesters (biodegradable polymers derived from renewable sources like starch and microbial fermentation) and agropolymers are the two types of biodegradable plastics (Roy and Choudhury 2022). The characteristic differences between bio-based and fossil-based plastics are described in Table 23.1.

Agropolymers contain starch, chitin and protein, while biopolyesters encompass substances such as polyhydroxyalkanoates (PHA) and polylactic acid (PLA). Fruits and vegetables are packed using bioplastics with a short shelf life. Microorganisms' enzyme activities quicken the breakdown of bioplastic polymers and release CO_2, H_2O and other inorganic compounds. This decomposition underscores the environmentally friendly nature of bioplastics.

23.3 Plastic Production Status and its Environmental Consequences

The production of plastic in the past 10 years, has been much more lavish as compared to the last century. India produces 5.6 million tons of plastic each year (Laskar and Kumar 2019). The volume of the plastic output worldwide from 1950 to 2017 with a forecast figure of 2050 is provided in Fig. 23.3. Marine microplastic levels are predicted to be more than double by 2050 and projected a 50-fold increase by 2100, even if plastic debris inputs to marine ecosystems ceased today. Recently, it was estimated that approximately 11% of global plastic waste generated in 2016 (about 19–23 million metric tons) found its way into aquatic ecosystems. This is estimated to rise to 53 million metric tons annually by 2030 (Borrelle et al. 2020). According to Beaumont et al. (2019), plastic pollution causes loss in ecosystem services such as fisheries productivity, tourism, and water quality, ranges from US$3300

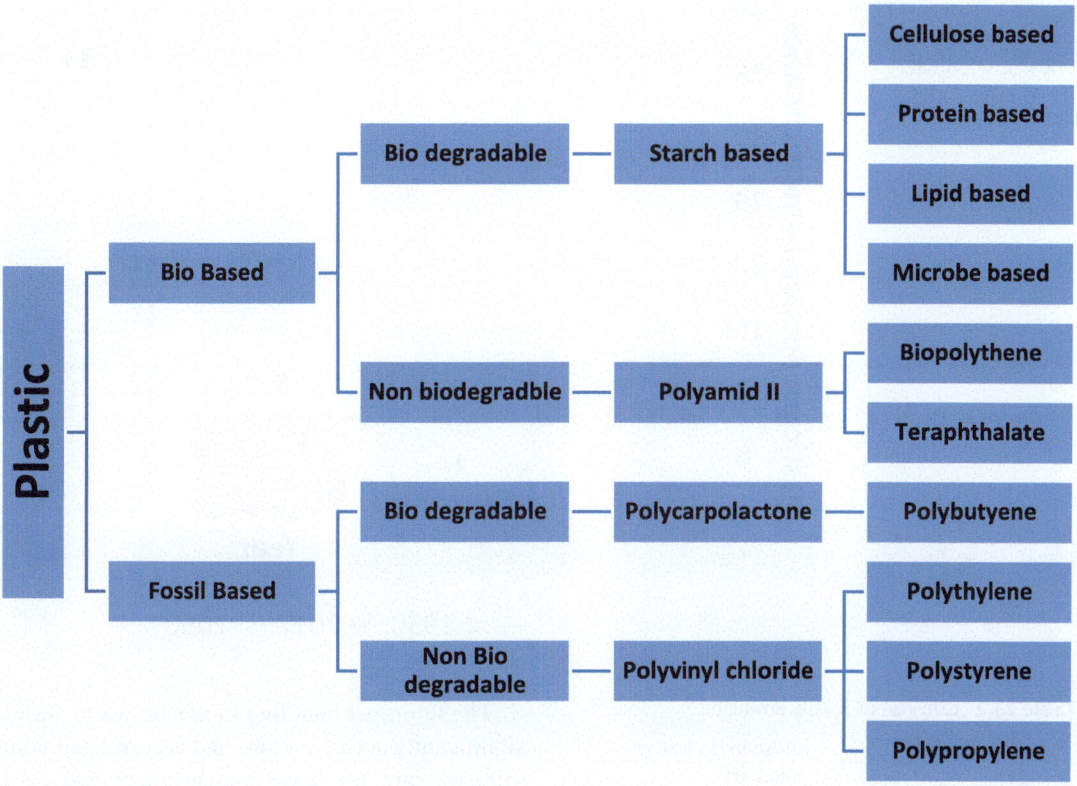

Fig. 23.2 Classification of plastics

Table 23.1 Conventional plastic versus bio-based plastic

Characteristics	Bioplastic	Conventional plastic
Source	Natural and renewable resources	Fossil fuels and petroleum-based products
Example	Polyhydroxyalkanoate, polylactic acid, polyhydroxy urethanes	Polyvinyl chloride, polystyrene, polyethylene terephthalate
Biodegradability	Degradable	Non-degradable
Environmental impact	Environment friendly	Leads to pollution
Applications	Food packaging, automotive, weather coating	Construction, bags, textile

to $33,000 for every ton of plastic that enters the marine ecosystem annually. Consequently, the total economic impact of marine plastic pollution is expected to cost several billion US dollars annually.

In the year 2022, global plastic production was 400.3 million metric tons (Statista Research 2024). Around 32% of global plastic materials are produced in China. The monthly plastic manufacturing in China is 06–12 million metric tons, the world's largest producer. The rest of Asian countries contribute 19% of plastic production. North America takes a 17% share of global production. According to the European Environmental Agency, the use of plastics increased drastically in China between 2010 and 2020. Worldwide plastic manufacturing expanded yearly between 2013 and 2019, generally exceeding population growth and plastic production increased by 2.5% in 2019, compared to 1.08%

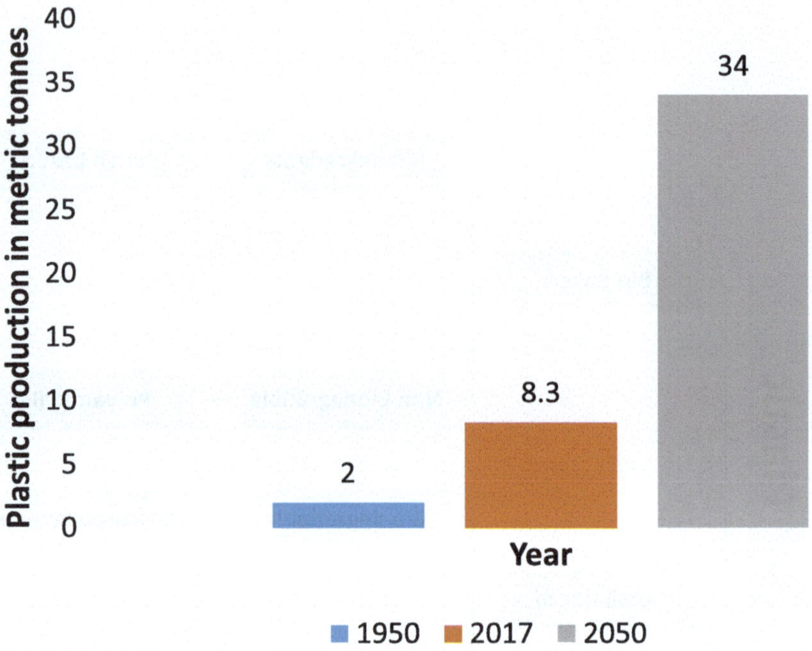

Fig. 23.3 Volume of plastic production globally (in billion metric tons). (*Source*: Geyer et al. (2017))

Table 23.2 Lifespan of plastic products

Plastic products	Lifespan (in years)
Packaging	Below 01
Consumer product	Less than 05
Textile	Less than 10
Electrical	Less than 10
Transportation	Greater than 10
Construction	20
Industrial machinery	Greater than 30

Source: OECD (2022)

of population growth in the same year. During COVID-19 challenges, plastic production increased by 0.9%. From the highest mountains to the deepest ocean trenches, plastic is omnipresent. It causes health risks to humans and harms ecosystems and wildlife (Choudhury et al. 2022). Since the dawn of the century, plastic production has doubled (400 million metric tons in 2021). Plastics take approximately 500 years to decompose since their components have ten (10) years of lifespan. The average lifespan of plastic products is given in Table 23.2. The global cleanup activities on International Coastal Cleanup Day in 2022 reveal plastic beverage bottles to be the second highest category of waste collected (after food wrappers).

The improper handling of plastic waste varies significantly across nations and is correlated with infrastructure for waste management and economic standing. As per the United Nations Development Programme, high levels of plastic pollution in natural ecosystems are caused by poor waste management in developing countries. Whereas developed countries export their plastic garbage to developing countries, they still need help with recycling. In 2016, the United States mismanaged 9.1% of its plastic garbage, while <1% was mismanaged in European nations such as Germany and the Netherlands (Law et al. 2020). Plastic is consumed either directly by ingestion or indirectly by entanglement of prey, trophic transfer and adherence of organisms to exterior surfaces of plastics (Eriksen et al. 2023). Research conducted by Alfred Wegener Institute Helmholtz Centre for Polar and Marine Research (AWI) records the most comprehensive account to date of the global impact of plastic pollution on marine ecosystems and its species, providing findings such as nearly every marine species group has encountered plastic pollution, with adverse effects observed in almost 90% of assessed species (Tekman et al. 2022). Large plastic debris causes death through entanglement, strangulation, choking and starvation, while micro and nanoplastics disrupt physiological processes such as feeding behaviour, reproductive outputs and growth (Tuuri and Leterme 2023). Plastic pollution has entered the marine food web, affecting critical ecosystems such as coral reefs and mangroves and leading to bioaccumulation and biomagnification. The Arctic Sea ice, East China, Mediterranean and Yellow seas have already surpassed ecological risk thresholds beyond which significant ecological risks may occur as a result of plastic pollution (Tekman et al. 2022).

23.3.1 Emergence of Bioplastics

Biodegradable plastics perform plastic functions without negative externalities, leading to a new and sustainable economy. Conventional plastic, derived from petrochemicals, persisting in the environment for 100 years, leads to pollution and harms living organisms. For example, plastic pollution affects marine organisms and even the food chain. In that respect, the demand for bioplastics has gone up, particularly in the substitution of single-use plastics. Bioplastics require less energy during production, create fewer waste materials and take a shorter time to break down. In contrast, bioplastics offer a promising solution as it degrades naturally and decomposes into natural resources. Bioplastic generation is based on renewable resources such as corn starch, sugarcane and cellulose. They are even more environmentally friendly compared to traditional plastics as they emit lesser amounts of GHG and fossil fuel energy. Bioplastics contain similar properties to traditional plastics, with additional benefits such as lower adverse environmental impacts in terms of reduced GHG emissions and appropriate waste management practices. This is an inexpensive, eco-friendly plastic material that helps cut pollution (Rajpal et al. 2024a).

It is an alternative for most food packaging applications, with a low carbon footprint and adverse environmental impacts (Zhao et al. 2023). As a part of the circular economy, biopolymers are derived from recycled or renewable materials, and the products are either recycled or reused after their useful lives. However, carbon neutrality is the goal of manufacturing these goods (Rosenboom et al. 2022). Harmful additives such as phthalates or bisphenol A are not present in bioplastics. They are versatile and suitable for the packaging of food and medical devices. Agropolymers contain starch, chitin and protein, while biopolyesters encompass substances such as Polyhydroxyalkanoates (PHA) and Polylactic Acid (PLA). Various types of bioplastics produced from different sources are given in Table 23.3. Foodstuffs such as vegetables and fruits are mainly packed with bioplastics with limited shelf life. Microorganisms cause bioplastic polymers to decay into CO_2, H_2O and other inorganic compounds through enzyme actions. This decomposition highlights the environmentally friendly nature of bioplastics. The global production capacity of bioplastics in 2022 by material is listed in Table 23.4.

Life cycle assessment (LCA) gives a detailed analysis of the impacts of bioplastics at each stage of their life cycle, from raw material extraction to disposal, so that their benefits can be appreciated fully (Barbhuiya and Das 2023). Bioplastics have lower environmental impacts and are sustainable alternatives to conventional plastics. The key stages of LCA are presented in Fig. 23.4.

Bioplastics also decrease the reliance on finite fossil resources but have a downside in that arable land and water are utilized, displacing food crops and sometimes biodiversity. Industrial-scale composting bioplastics minimizes landfills and returns valuable nutrients to the soil, dependent on appropriate management and facilities for environmental impacts to accrue (Chowdhury et al. 2024; Paul et al. 2019). Anaerobic digestion process bioplastics are into the production of biogas and digestate, which in turn can be used to generate renewable energy and applied as fertilizer. LCA studies demonstrate that there are substantial GHG emission reductions if bioplastics are managed by anaerobic digestion, especially when byproducts substitute conventional energy and fertilizers (Bishop et al. 2021).

Table 23.3 Various types of bioplastics

Type	Source	Category	Description	References
Starch-based plastics	Corn, potato	Biodegradable	Packaging, disposable cutlery, agricultural films	do Val Siqueira et al. (2021)
Polylactic acid	Plant starch	Biodegradable	Packaging, textiles	Ali et al. (2023a, b)
Polyhydroxy alkanoates	Bacterial fermentation of sugars, lipids	Biodegradable	Medical devices, packaging	Raza et al. (2018)
Polybutylene succinate	Synthesized from succinic acid and 1,4-butanediol	Biodegradable	Food packaging, disposable cutlery	Kato et al. (2023))
Cellulose-based plastics	Derived from cellulose of wood	Biodegradable	Films, coating, textile fibres	Sarkar et al. (2023)
Polyethylene	Derived from sugarcane and other biomass	Non-biodegradable	Bottles, bags, pipes and packaging	Burelo et al. (2023)
Polyethylene terephthalate	Produced from bio-based ethylene glycol and terephthalic acid	Non-biodegradable	Beverage bottles, food containers	Tiso et al. (2021)
Polytrimethylene terephthalate	Derived from bio-based 1,3-propanediol	Non-biodegradable	Textiles, carpet	Rahman and Bhoi (2021)

Table 23.4 Global production capacity of bioplastics in 2022

Category of bioplastic	Name	Global production capacity
Biodegradable	PBAT	4.5%
	PBS	0.9%
	PLA	20.7%
	PHA	3.9%
	Starch blends	17.9%
	Cellulose films	3.6%
Non-degradable	PE	14.8%
	PET	4.2%
	PA	11.1%
	PTT	13.3%
	PP	3.9%
	Others	1.1%

Source: https://www.statista.com/topics/8744/bioplastics-industry-worldwide/#topicOverview n.d. (viewed on 07.07.2024)

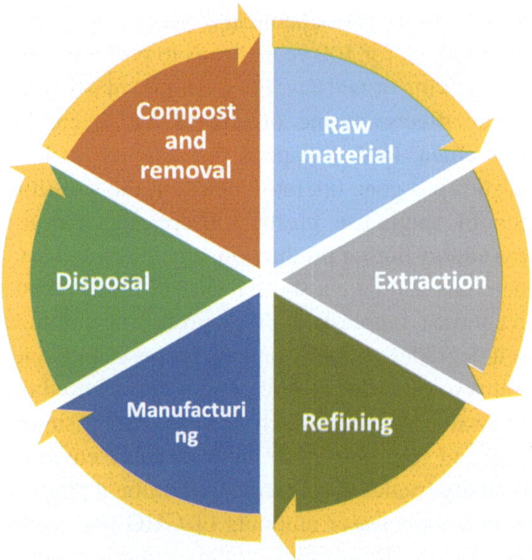

Fig. 23.4 Phases of life cycle assessment of bioplastic. (*Source*: Redrawn from Shah et al. (2021))

23.4 Case Studies: Real-World Applications of Bioplastics

A potential substitute for conventional petroleum plastics is bioplastic made from renewable resources, namely, vegetable oil, fats, corn starch, wood pulp, herbaceous fibre and microorganisms (Zhao et al. 2023). Their application and expansion across multiple industries demonstrate significant rises in reducing environmental impacts, enhancing waste management and improving sustainable practices (Rajpal et al. 2024b). Well-known bioplastics across the globe are described below. According to Clare Goldsberry's report in 2019, an Italian company called Novamont won an award for Innovation in Bioplastic in 2019. Mater-Bi is a family of bioplastics developed by the company Novamont that can be used for consumers, packaging and even agriculture. They are produced from sustainable resources such as plant-based feedstock. Mater-Bi products compostable are biodegradable. With Mater-Bi's implementation, the packaging industry has used more biodegradable materials, which has increased the bioplastics market in Europe (Mater-Bi – biodegradable and compostable bioplastics – Novamont). Coca-Cola, derived from 30% plant-based materials from sugarcane, introduced the PET bottle, which contains plant-based elements. This reduces the carbon impact of Coca-Cola's packaging. Plant-based bottles, used for beverage packing, help to popularize bioplastics in the mainstream market and encourage other companies to explore sustainable packaging options (Bottles Made From 100% Plant Plastic | The Coca-Cola Company). Durabio is a bioplastic developed by Mitsubishi Chemical made from plant-derived isosorbide. Because of its optical properties and impact resistance, it became renowned. Its applications include automotive interiors, mobile phone screens and durable, transparent materials. The success of Durabio highlights the versatility of bioplastics in high-performance applications and promotes in-depth research into it (New Bio-based Engineering Plastic "DURABIO" | Products | Mitsubishi Chemical Corporation (m-chemical.co.jp). Braskem, a Brazilian petrochemical company, manufactured green polyethylene, a type of polyethene based on ethanol derived from sugarcane. It has a significantly lower carbon footprint than conventional polyethene, as sugarcane absorbs CO_2 during its growth. The two global brands that have embraced green polythene to improve sustainability are Tetra Pak and Procter & Gambler. (News – Braskem developed world's first renewable-source polyethylene wax).

Compostable bioplastics, which include agricultural biofilms, compostable eggs and food packaging, are made from a combination of Ecoflex, a polymer derived from fossil fuels and PLA, as manufactured by German chemical company BASF. The positive environmental effects of utilizing Ecovio in agriculture induce reduced packaging waste and improved soil health through biodegradable mulch films. PTT Global Chemical, a Thai company, produces BioPBS, a biodegradable and compostable polymer from renewable resources such as sugarcane and cassava. Agricultural films, packaging and disposable goods are the uses of BioPBS. Its implementation in Thailand and other nations has reduced plastic waste and promoted sustainable farming methods (Kato et al. 2023). Carbios, a French company, developed an enzymatic recycling technology for PET plastics, including bioplastics. The process uses enzymes to disintegrate PET into its monomers, which can be reused to make new plastic. This technology resolves bioplastic end-of-life issues, offering a viable recycling option. Carbios' technology promotes a circular economy for plastics (DeFrancesco, 2020). PEF is a bioplastic developed by a Dutch company, Avantium, and is composed of plant-based sugars. PEF has the capability to replace PET because it has better barrier properties and a lower carbon footprint. PEF can be used in packaging, especially beverage packaging, as an alternative eco-friendly material to conventional plastics. Commercialization of PEF can change the entire landscape of the packaging industry into a more sustainable and high-performance material. Kingfa, a Chinese company, has been increasing production of PLA, a bioplastic produced from corn starch used for packaging, disposable tableware and medical devices. Due to large-scale production in China, the country has been fulfilling the global demand for biodegradable plastics, thereby reducing dependency on fossil-based plastics. Kingfa's efforts have helped the world supply PLA, making it widely used and reducing its impact on the environment. Studies created biodegradable plastics that were produced from corn starch as well as other renewable materials. These bioplastics are used in agriculture, packaging and disposable products. Products from Biotec reduce plastic pollution and support green agricultural practices in India. Degradable plastics have increased awareness of environmental sustainability and waste management technologies (Sharmila et al. 2021).

23.5 Challenges and Considerations

Bioplastics is ought to overcome various crucial challenges to achieve their full effectiveness in lowering global plastic pollution. Uncertainty regarding their long-term effectiveness is one of the primary reasons of concern. The degradation of bioplastics depends mainly on certain environmental factors, which are typically absent in regular disposal situations. The actual effectiveness of bioplastics in reducing plastic pollution comes into doubt due to its variability. Additionally, human behaviour significantly affects the acceptance and appropriate disposal of bioplastics, making public education and awareness campaigns important. Enhancing the feasibility and expandability of plastics also poses obstacles. The competitiveness of bioplastics in the market is lowered because their existing production methods are more expensive and inefficient than those for traditional plastics. Bioplastics are typically too large for the existing waste management infrastructures, necessitating significant investments in new facilities and technology for efficient recycling and composting. Furthermore, to effectively promote bioplastics' development and widespread use, assessing and reinforcing the current policy frameworks is necessary.

23.6 Conclusion and Future Directions for Bioplastics

From this chapter, a comprehensive understanding of the multifaceted role of plastics from packaging and automotive to medical and food preservation because of its remarkable adaptability, longevity and cost-effectiveness has been explored. However, the environmental role of plastics, characterized by their persistence, pollution and adverse effects on ecosystems and human health, underscores the urgent need for more sustainable practices. Increase in plastic production over recent decades are illustrating the increasing load on waste management systems and ecosystems. Solving these issues requires a multifaceted approach, integrating advanced waste management practices, improved recycling systems and innovative material alternatives. Bioplastics emerge as an alternative to fossil fuels and reduce the adverse effects of conventional plastics. In particular, Novamont's Mater-Bi, Coca-Cola's plant-based PET bottles and Carbios' enzymatic recycling technology stand as case studies discussed in this chapter that demonstrate various uses and success of bioplastics in the different industries. The examples discussed here go beyond replacing single-use plastics but rather lead to sustainable practices across the different sectors. Furthermore, AM or 3D printing increases its potential by enabling the production of complex geometries and reducing material waste. The combined use of AM and bioplastics holds promise for overcoming many limitations of traditional plastic production techniques, making bioplastics a more practical and effective solution. Bioplastics must overcome various crucial challenges to achieve their full effectiveness in lowering global plastic pollution. Uncertainty regarding their long-term effectiveness is one of the primary reasons for concern. The degradation of bioplastics depends mainly on certain environmental factors, which are typically absent in regular disposal situations. The actual effectiveness of bioplastics in reducing plastic pollution comes into doubt due to its variability. Enhancing the feasibility and expandability of plastics also poses obstacles. Bioplastics are now produced using more costly and insufficient processes than conventional plastics, so their competitiveness is reduced in the market. Since bioplastics are typically too large for current waste management systems, significant investments are needed in new facilities and technology for efficient recycling and composting.

Furthermore, to effectively promote bioplastics' development and widespread use, assessing and reinforcing the current policy frameworks is necessary. This chapter explores the benefits and challenges of bioplastics, emphasizing their role in addressing environmental problems caused by conventional plastics. Innovations in AM, improved recycling procedures, supportive legislation and manufacturing technology development will influence the future of bioplastics and encourage their wider use.

References

Adekanmbi T, Wang XX, Basheer S, Liu S, Yang A, Cheng H (2023) Climate change impacts on global potato yields: a review. Environ Res Clim 03(01):12001. https://doi.org/10.1088/2752-5295/ad0e13

Ali SS, Abdelkarim EA, Elsamahy T, Al-Tohamy R, Li F, Kornaros M, Zuorro A, Zhu D, Sun J (2023a) Bioplastic production in terms of life cycle assessment: a state-of-the-art review. Environ Sci Ecotechnol 15:100254

Ali W, Ali H, Gillani S, Zinck P, Souissi S (2023b) Polylactic acid synthesis, biodegradability, conversion to microplastics and toxicity: a review. Environ Chem Lett 21(3):1761–1786

Andrady AL, Neal MA (2009) Applications and societal benefits of plastics. Philos Trans R Soc Lond B Biol Sci 364(1526):1977–1984

Awoyera PO, Adesina A (2020) Plastic wastes to construction products: status, limitations and future perspective. Case Stud Construct Mater 12:e00330

Barbhuiya S, Das BB (2023) Life cycle assessment of construction materials: methodologies, applications and future directions for sustainable decision-making. Case Stud Construct Mater 19:e02326

Beaumont NJ, Aanesen M, Austen MC, Börger T, Clark JR, Cole M, Hooper T, Lindeque PK, Pascoe C, Wyles KJ (2019) Global ecological, social and economic impacts of marine plastic. Mar Pollut Bull 142:189–195

Bishop G, Styles D, Lens PN (2021) Environmental performance comparison of bioplastics and petrochemical plastics: a review of life cycle assessment (LCA) methodological decisions. Resour Conserv Recycl 168:105451

Borrelle SB, Ringma J, Law KL, Monnahan CC, Lebreton L, McGivern A, Murphy E, Jambeck J, Leonard GH, Hilleary MA, Eriksen M (2020) Predicted growth in plastic waste exceeds efforts to mitigate plastic pollution. Science 369(6510):1515–1518

Burelo M, Hernández-Varela JD, Medina DI, Treviño-Quintanilla CD (2023) Recent developments in bio-based polyethylene: degradation studies, waste management and recycling. Heliyon 9(11):e21374

Chen X, Yan N (2020) A brief overview of renewable plastics. Mater Today Sustainab 7:100031

Choudhury M, Sahoo S, Samanta P, Tiwari A, Tiwari A, Chadha U et al (2022) COVID-19: an accelerator for global plastic consumption and its implications. J Environ Public Health 2022(1):1066350. https://doi.org/10.1155/2022/1066350

Chowdhury R, Barman S, Choudhury M, Kim KH, Bhattacharya SS (2024) Earthworm modifies microbial community and functional genes for lignocellulosic waste valorization: isolating plant-growth-promoting bacteria via next generation sequencing. Int Biodeterior Biodegradation 193:105854. https://doi.org/10.1016/j.ibiod.2024.105854

Costa A, Encarnação T, Tavares R, Todo Bom T, Mateus A (2023) Bioplastics: innovation for green transition. Polymers 15(3):517

Czuba L (2014) Application of plastics in medical devices and equipment. In: Handbook of polymer applications in medicine and medical devices. William Andrew Publishing, pp 9–19

DeFrancesco L (2020) Closing the recycling circle. Nat Biotechnol 38(6):665–668. https://doi.org/10.1038/s41587-020-0541-0

de Oliveira RB, Pelepenko LE, Masaro DA, Lustosa GM, Cassani M, Roza NA, Marciano MA, dos Reis LM, Kamel S, Louvet L, Mazon T (2024) Effects of microplastics on the kidneys: a narrative review, vol 106. Kidney Int, p 400

Dilshad E, Waheed H, Ali U, Amin A, Ahmed I (2021) General structure and classification of bioplastics and biodegradable plastics. In: Bioplastics for sustainable development. Springer, pp 61–82

do Val Siqueira L, Arias CILF, Maniglia BC, Tadini CC (2021) Starch-based biodegradable plastics: methods of production, challenges and future perspectives. Curr Opin Food Sci 38:122–130

Eriksen M, Cowger W, Erdle LM, Coffin S, Villarrubia-Gómez P, Moore CJ, Carpenter EJ, Day RH, Thiel M, Wilcox C (2023) A growing plastic smog, now estimated to be over 170 trillion plastic particles afloat in the world's oceans—urgent solutions required. PLoS One 18(3):e0281596

Geyer R, Jambeck JR, Law KL (2017) Production, use, and fate of all plastics ever made. Sci Adv 3(7):e1700782

Hameed AM, Ahmed BAF (2019) Employment the plastic waste to produce the light weight concrete. Energy Procedia 157:30–38

https://www.braskem.com/news-detail/braskem-developed-worlds-first-renewable-source-polyethylene-wax

https://www.novamont.com/eng/mater-bi

https://www.statista.com/topics/8744/bioplastics-industry-worldwide/#topicOverview. Accessed on 07.07.2024, n.d.

IUCN (2024) Issues brief, plastic-pollution-issues-brief-may-2024-update.pdf (iucn.org)

Kabeyi MJB, Olanrewaju OA (2023) The levelized cost of energy and modifications for use in electricity generation planning. Energy Rep 9:495–534

Kato S, Ueda T, Aoshima T, Kosaka N, Nitta S (2023) BioPBS™(polybutylene succinate). In: Synthetic biodegradable and biobased polymers: industrial aspects and technical products. Springer International Publishing, Cham, pp 269–304

Laskar N, Kumar U (2019) Plastics and microplastics: a threat to environment. Environ Technol Innov 14:100352

Law KL, Starr N, Siegler TR, Jambeck JR, Mallos NJ, Leonard GH (2020) The United States' contribution of plastic waste to land and ocean. Sci Adv 6(44):eabd0288

Marcharla E, Vinayagam S, Gnanasekaran L, Soto-Moscoso M, Chen WH, Thanigaivel S, Ganesan S (2024) Microplastics in marine ecosystems: a comprehensive review of biological and ecological implications and its mitigation approach using nanotechnology for the sustainable environment. Environ Res 256:119181

Mutha NH, Patel M, Premnath V (2006) Plastics materials flow analysis for India. Resour Conserv Recycl 47(3):222–244

New Bio-based Engineering Plastic "DURABIO" | Products | Mitsubishi Chemical Corporation (m-chemical.co.jp). https://www.m-chemical.co.jp/en/products/departments/mcc/pc/product/1201026_9368.html#:~:text=DURABIO%E2%84%A2%20is%20a%20bio-based%20polycarbonate%20resin%20derived%20mainly,transparency%2C%20excellent%20optical%20properties%2C%20and%20outstanding%20scratch%20resistance

Nyadjro ES, Webster JA, Boyer TP, Cebrian J, Collazo L, Kaltenberger G, Larsen K, Lau YH, Mickle P, Toft T, Wang Z (2023) The NOAA NCEI marine microplastics database. Sci Data 10(1):726

OECD (2023) Agricultural policy monitoring and evaluation 2023: overview. OECD Publishing. https://www.oecd.org/en/publications/2023-oecd-digital-government-index_1a89ed5e-en.html

Opara I, Lardier DT, Metzger I, Herrera A, Franklin L, Garcia-Reid P, Reid RJ (2020) "Bullets have no names": a qualitative exploration of community trauma among Black and Latinx youth. J Child Fam Stud 29:2117–2129

Paul S, Choudhury M, Deb U, Pegu R, Das S, Bhattacharya SS (2019) Assessing the ecological impacts of ageing on hazard potential of solid waste landfills: a green approach through vermitechnology. J Clean Prod 236:117643. https://doi.org/10.1016/j.jclepro.2019.117643

Pinheiro HT, MacDonald C, Santos RG, Ali R, Bobat A, Cresswell BJ, Francini-Filho R, Freitas R, Galbraith GF, Musembi P, Phelps TA (2023) Plastic pollution on the world's coral reefs. Nature 619(7969):311–316

Rahman MH, Bhoi PR (2021) An overview of non-biodegradable bioplastics. J Clean Prod 294:126218

Rajpal A, Choudhury M, Goswami S, Chakravorty A, Raghavan V (eds) (2024a) Waste management and treatment: advances and innovations, 1st edn. CRC Press. https://doi.org/10.1201/9781003258377(a)

Rajpal A, Choudhury M, Goswami S, Chakravorty A, Raghavan V (eds) (2024b) Waste management and treatment: advances and innovations, 1st edn. CRC Press. https://doi.org/10.1201/9781003258377(b)

Rana K (2022) This is how to ensure sustainable alternatives to plastic. World Economic Forum. https://www.weforum.org/stories/2022/03/are-we-replacing-plastic-with-more-energy-intensive-alternatives/

Raza ZA, Abid S, Banat IM (2018) Polyhydroxyalkanoates: characteristics, production, recent developments and applications. Int Biodeterior Biodegradation 126:45–56

Rosenboom JG, Langer R, Traverso G (2022) Bioplastics for a circular economy. Nat Rev Mater 7(2):117–137

Roy P, Choudhury M (2022) Bioeconomy and green plastic production. In: Baskar C, Ramakrishna S, Daniela La Rosa A (eds) Encyclopedia of green materials. Springer, Singapore. https://doi.org/10.1007/978-981-16-4921-9_189-1

Sarkar M, Upadhyay A, Pandey D, Sarkar C, Saha S (2023) Cellulose-based biodegradable polymers: synthesis, properties, and their applications. In: Biodegradable polymers and their emerging applications. Springer Nature, Singapore, pp 89–114

Shah M, Rajhans S, Pandya HA, Mankad AU (2021) Bioplastic for future: a review then and now. World J Adv Res Rev 9(2):056–067

Sharmila S, Teja, P.R., Gupta, D.V.C.G., Lakshmi, P.K., Kamalambigeswari, R. and Rebecca, L.J., 2021. Bioplastic production using corn starch with natural fillers and its SEM-EDS report. Plant Arch 21. https://doi.org/10.51470/PLANTARCHIVES.2021.v21.no1.040

Statista Research (2024) https://www.statista.com/topics/8744/bioplastics-industry-worldwide/topicOverview

Tekman MB, Walther BA, Peter C, Gutow L, Bergmann M (2022) Impacts of plastic pollution in the oceans on marine species, biodiversity and ecosystems, 1–221, WWF Germany, Berlin. https://doi.org/10.5281/zenodo.5898684

Tiso T, Narancic T, Wei R, Pollet E, Beagan N, Schröder K, Honak A, Jiang M, Kenny ST, Wierckx N, Perrin R (2021) Towards bio-upcycling of polyethylene terephthalate. Metab Eng 66:167–178

Tuuri EM, Leterme SC (2023) How plastic debris and associated chemicals impact the marine food web: a review. Environ Pollut 321:121156

Vieyra H, Molina-Romero JM, Calderón-Nájera JDD, Santana-Díaz A (2022) Engineering, recyclable, and biodegradable plastics in the automotive industry: a review. Polymers 14(16):3412

Zhao X, Wang Y, Chen X, Yu X, Li W, Zhang S, Meng X, Zhao ZM, Dong T, Anderson A, Aiyedun A (2023) Sustainable bioplastics derived from renewable natural resources for food packaging. Matter 6(1):97–127

Utilization of Micro-Nano Plastics Derived from Electronic Waste into Bitumen for Sustainable Environment

24

Arvind L. Chel, Akash Wani, Nischal C. Shetty, and Geetanjali Kaushik

24.1 Introduction

The concept of "microplastics" describes small fragments of plastic that have become a major concern for marine ecosystems, a sign of the worldwide influence of human activity. These particles, which are considered to be a risk to spacecraft borders, are produced when bigger plastic objects break down and transform into nanoplastics (NPs), which impact different ecosystems. The manufacture of plastics is expanding at a rapid rate worldwide, particularly in developing countries. This is a complicated issue that interacts with other environmental issues including carbon emissions and waste disposal. Trade has shifted as countries such as China are bringing electronics into the pollution narrative, which is gaining traction because of import limitations and the potential need to modify international waste management practices (Herat and Agamuthu 2012). A significant amount of microplastics can be associated with electronic waste. The improper handling and disposal of electronic devices, which frequently occur in developing countries that serve as landfills, causes mechanical degradation that leads to the accumulation of microplastics along with heavy metals and other dangerous materials, deteriorating the issue and making the area contaminated. An interdisciplinary strategy is needed to address the issue of microplastic pollution and to lessen its negative effects. This will help to promote an integrated effort toward sustainable environmental well-being (Awasthi and Li 2017). Globally, the amount of electronic waste has increased significantly as a result of technological innovation.

Despite the numerous benefits of incorporating e-waste-derived micro-nano plastics (MPs) into bitumen, several knowledge gaps remain. Those are as follows: (i) long-term durability, (ii) environmental impact, (iii) toxicity and health risks, (iv) regulatory framework, and (v) public perception and acceptance. The long-term durability and performance of bitumen-containing e-waste MPs under various environmental conditions is crucial for ensuring the sustainability of road infrastructure. To assess the overall sustainability of this technique, a thorough evaluation of

A. L. Chel · A. Wani
Department of Mechanical Engineering, Jawaharlal Nehru Engineering College, MGM University, Maharashtra, India
e-mail: arvindchel@alumni.iitd.ac.in

N. C. Shetty
Central Pollution Control Board Regional Directorate Pune, Maharashtra, India
e-mail: nischal.cpcb@nic.in

G. Kaushik (✉)
Department of Civil Engineering, Hi Tech Institute of Technology, Maharashtra, India

S. Goswami et al. (eds.), *Micro-Nano Plastics Exposure, Environmental Degradation and Public Health Crisis*, Sustainable Development Goals Series, https://doi.org/10.1007/978-981-96-5047-7_24

the environmental implications of incorporating electronic waste MPs into bitumen is required. This evaluation must take into account the possible leaching of MPs into the environment and their effects on ecosystems. While initial toxicity assessments may be conducted, further research is needed to assess the hazards associated with exposure to e-waste MPs in bitumen, especially for road workers and nearby residents. The development of clear regulations and standards governing the use of e-waste MPs in bitumen is essential to ensure safety, quality, and environmental compliance. Understanding public perceptions of using recycled materials in road construction, particularly those derived from e-waste, is crucial for gaining public support and acceptance. By addressing these study gaps and achieving the proposed objectives, one can better understand the potential benefits and challenges of utilizing e-waste MPs in bitumen, paving the way for its wider adoption and contributing to a more sustainable future. Converting electronic waste-derived MPs into bitumen is a sustainable and environmentally friendly practice that offers numerous benefits, including resource conservation, reduced pollution, improved road quality, economic advantages, and technological advancements.

24.2 Composition of E-Waste and Generation Status

Global Electronic Waste Monitor 2024 reported that global e-waste generation will increase to 62 million tons in 2022. This figure is projected to further increase up to 82 million tons by 2030. The rapid growth in e-waste generation is primarily driven by increased consumption of electronic devices, short product life, technological advancements, and inadequate waste management infrastructure.

Electronic waste is a complex mixture of various materials, including metals such as copper, aluminum, iron, gold, silver, platinum, and rare earth elements (Sachdeva and Sharma 2020). Specifically, e-waste comprises 30% organic substances such as plastics and additives, 30% ceramics such as silica, and 40% inorganic materials such as metals (Fig. 24.1). Plastics in EEE often contain brominated flame retardants, frequently in combination with the flame-quenching synergist Sb_2O_3, leading to the prevalent use of Br and Sb for their identification (Lakusic 2018). Concerns arise with inadequate e-waste management owing to the leakage of flame retardants and other hazardous compounds, including metals. E-waste products that are non-reusable pose challenges for solid waste management. These non-reusable e-waste materials are typically disposed of in landfills, hindering bio-degradation and consequently impacting the environment (Lakusic 2018).

24.3 Microplastics Sources Delivered from E-Waste

Discarded electronic items, including computers, TVs, and cell phones, are referred to as electronic waste. Recovery of metals and rare earth elements is typically associated with recycling and utilizing e-waste (Agarwal et al. 2023). Plastic which encases electronic equipment, contains chemicals such as flame retardants. The plastic components of broken electronics add to the overall amount of plastic waste in the environment. Uncontrolled recycling and open incineration are two improper disposal techniques that discharge fibers and fragments into the environment (Azhdarpour et al. 2016; Shaaban et al. 2024). These processes change macroplastics into microplastics, which exacerbates the global crisis of microplastic contamination. The countries with the biggest global output of e-waste are shown in Table 24.1.

24.4 Environmental Impact and E-Waste

There are several ways that e-waste-derived microplastics might find their way into the environment. According to Kumar et al. (2017), landfills serve as reservoirs for microplastics that

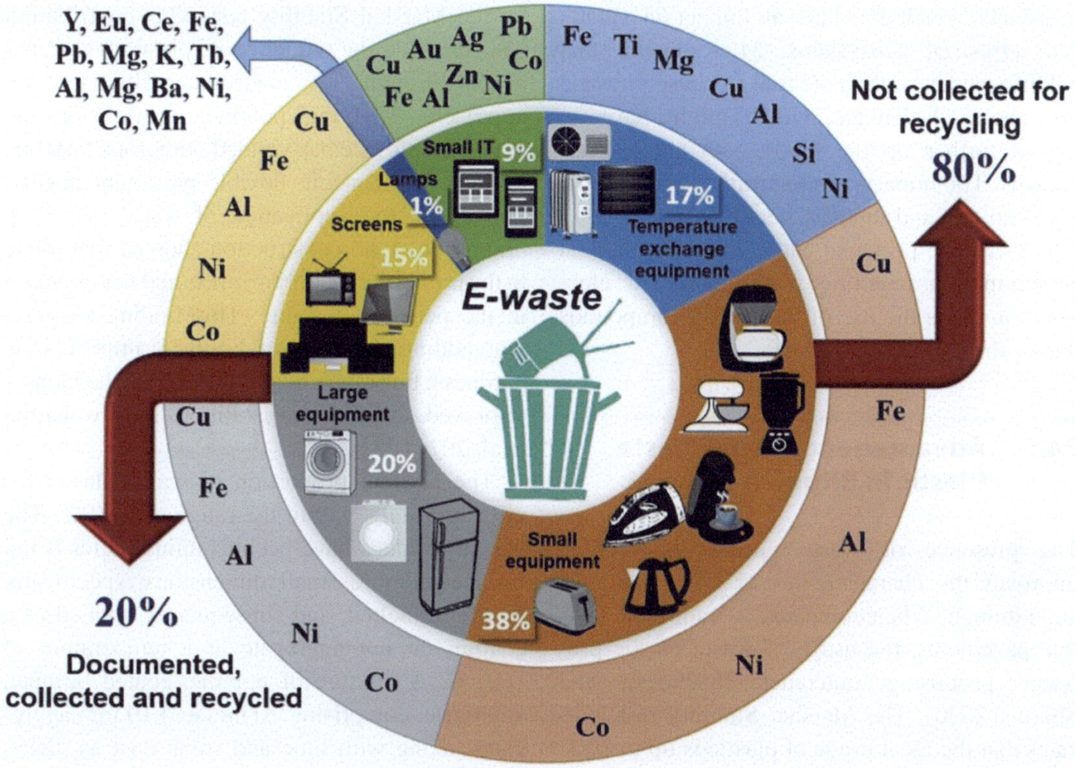

Fig. 24.1 Schematic representation of e-waste-created microplastics, their types and categorization, the existing management strategies, and their elemental compositions. (Reproduced with permission) (Bahadoran et al. 2022)

Table 24.1 E-waste-producing nations throughout the world, including their e-waste production in kilotons (Kt) and recycling rates

Sr. no.	Country	E-waste produced (Kt)	Recycling rate
1	China	10,129	16%
2	USA	6918	15%
3	India	3230	1%
4	Japan	2569	22%
5	Brazil	2143	0%
6	Russia	1631	6%
7	Indonesia	1618	NA
8	Germany	1607	52%
9	UK	1598	57%
10	France	1362	56%

gradually break down and leak into the ground and water (Kumar et al. 2017). Informally conducted electronic waste recycling processes, particularly in underdeveloped countries, can entail burning plastic parts in order to recover precious metals. As a result, plastic particles are released into the sky, where they may land in nearby soils and water sources or travel farther due to wind and water currents (Awasthi and Li 2017). Microplastics derived from electronic waste have a number of negative environmental effects. These microplastics have the ability to absorb harmful substances and release them into the environment, such as brominated flame retardants, which may have an impact on aquatic and terrestrial ecosystems. Microplastics that wildlife eats have the potential to cause physical and chemical damage, which might have an impact farther up the food chain and expose humans. The impacts of electronic wastes on the environment and human health are shown in Fig. 24.2. The principal cause of environmental contamination resulting from outmoded electronic items is the use of hazardous compounds during the production process.

24.5 Advantages of Using Waste Plastic in Bitumen

The presence of waste plastic in bitumen improves the characteristics of both aggregate and bitumen. When compared to traditional flexible pavements, the usage of waste plastic produces promising outcomes (Sachdeva and Sharma 2020). The Marshal Stability test indicates that the ideal usage of plastic is up to 10%. This improves the quality of flexible pavement while also considerably reducing electronic waste, promoting an environment-friendly approach. Surface coating on the aggregate surface by polymer has resulted in several benefits, all of which contribute to the quality of flexible pavement (Kumar et al. 2017). This demonstrates that the addition of polymer boosts the bitumen's hardness, offering confidence in the technological improvements. This might be attributed to the polymer bitumen blend's superior binding properties. The addition of plastic waste to bitumen enhanced its softening point. The larger the proportion of plastic waste used, the higher the softening point. The chemical composition of the additional polymers may impact the softening point. The inclusion of discarded plastic changes the characteristics of bitumen.

The inclusion of waste plastic in bitumen improves the properties of aggregate and bitumen. Using waste plastic in flexible pavements produces better results than regular flexible pavements (Geyer 2020). The optimal utilization of plastic can be achieved up to 10%, as determined by the Marshal Stability test. This contribution has enhanced the reduction of plastic waste disposal through an environmentally friendly approach. Applying polymer coating on the aggregate surface has yielded numerous benefits, ultimately enhancing flexible pavement quality. Studies on the application of waste plastic in bituminous road construction showed that a rise in the proportion of polymer resulted in a decrease in the penetration value. This finding suggests that adding polymer to bitumen improves its hardness; this effect may be related to the blend's improved binding capabilities (Vishwakarma et al. 2022).

The optimal proportion of waste plastic for utilization falls within the range of 5–10%. The plastic residues have been examined after being fragmented into small dimensions, specifically 2–4 mm, melted, and subsequently applied as a coating on hot aggregate at a temperature of 160 °C. A mixture of polymer-coated bitumen aggregate, comprising 20 mm and 10 mm aggregates, along with lime and stone dust as fillers, and various grades of bitumen such as 60/70 and 80/100-grade bitumen, has been used in the construction of several roads (Sachdeva and Sharma 2020). The shredded waste plastic is put to use. Roads built using shredded plastic waste outlasted those prepared with regular asphalt mixes in terms of lifetime. Because plastic has adhesive properties that increase strength and endurance to support larger loads, roads made from a combination of waste plastic function better than conventional ones. Plastic-bitumen roads have a life of about 10 years, whereas a normal "highway quality" road possesses a life of 5 years (Manjunath 2016).

Despite the potential marginally higher expense of constructing plastic roads in contrast to conventional methods, the substantial benefits in terms of longevity and environmental impact make this a forward-thinking and optimistic approach. It is anticipated that shortly, we will witness the establishment of robust, long-lasting, and environmentally friendly roads, alleviating the environment from various forms of plastic waste (Hake et al. 2020).

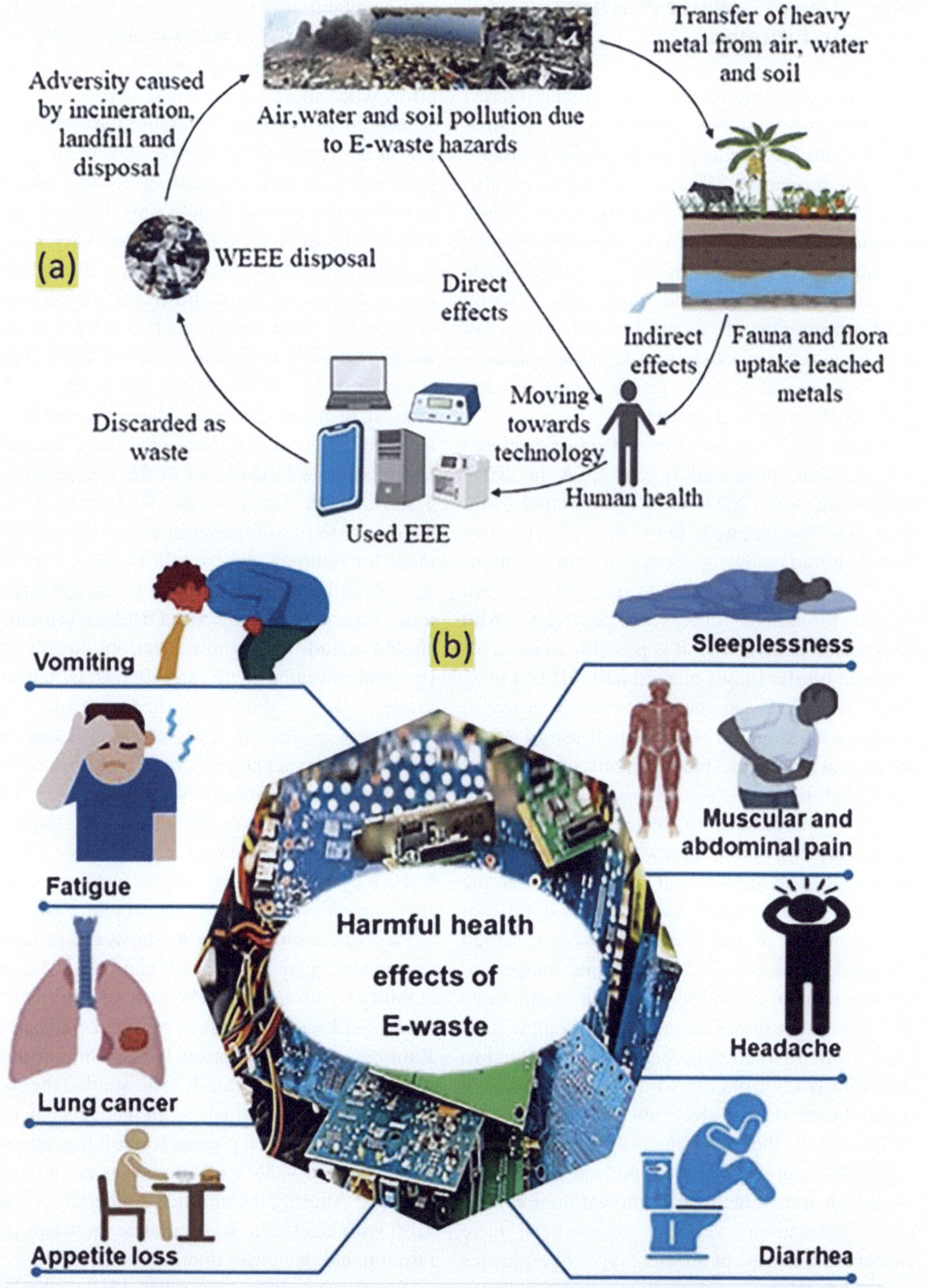

Fig. 24.2 Effect of (**a**) e-waste on the environment and (**b**) human health. (Reproduced with permission) (Twagirayezu et al. 2023)

24.6 Use of E-Waste Plastics in Bitumen

In modern times, there has been growing interest in using waste materials to build roads, emphasizing doing so while preserving the desirable concrete properties. The US government also encourages the use of eco-friendly materials and sustainable technologies to increase the lifespan of roads (Ma et al. 2021). In all, 8.3 billion metric tons of plastic were manufactured in the USA by 2017. Because people do not know that they can recycle or reuse plastics, plastic waste management has become a major problem very quickly (Abd El-Rahman et al. 2018).

About 35.7 million tons of plastic-related e-waste were produced in the USA in 2018, accounting for 12.2% of all municipal solid waste (MSW) created. Only 8.7% of plastics manufactured may be recycled; the remaining 75.6% and 15.7% end up in landfills or are burned to recover energy, respectively (Abd El-Rahman et al. 2018). It is possible to use a lot of waste plastic in hot mix asphalt (HMA) mixtures. To create sustainable roadways, research is still being done on the potential use of waste plastics as aggregate replacements or modifiers for asphalt binders.

The advancement of technology and economic growth had a noteworthy influence on many industries and social lifestyles. Over the past 20 years, there have been significant improvements in the electronics sector, which has made it necessary for electronic gadgets to receive regular upgrades to remain competitive and satisfy consumer demands. When new gadgets hit the market, people are more inclined to throw away their old electronics, contributing to a significant rise in electronic waste produced. Glass, metals, thermosetting polymers, and other materials comprise a sizable portion of e-waste. Based on their chemical composition for concrete applications, several researchers have examined the usage of different types of e-plastics in cement mixers, such as high-impact polystyrene, polyamides (PA), acrylonitrile butadiene styrene (ABS), and polycarbonate (PC). Bitumen is the binding ingredient in asphalt formation, and adaptable asphalt is the most common type used in India (Abd El-Rahman et al. 2018).

Bitumen and aggregate both have better properties when more waste plastic is present. When compared to traditional flexible pavements, the use of waste plastic in flexible pavements yields positive outcomes. According to the Marshal Stability test, plastic should only be used up to 10% of the time. As an environmentally beneficial method, this has increased the value of limiting the disposal of plastic trash. The ideal bitumen content has been determined to be 5.5%, with waste materials replacing 7.5% of this bitumen (Dombe et al. 2020). There has been a lot of interest in using electronic waste as a practical substitute for conventional materials, including aggregate in a DBM layer of flexible pavement. The Marshall Stability test determines if electronic waste plastic material is a viable replacement for bitumen and fly ash. In all control mix designs utilizing mineral aggregates, the bituminous mix composition for DBM is critical. It should include aggregates, electronic waste plastic, and bitumen with an 80/100 penetration grade. The ductility test result was 70.54 cm > 75 cm, the softening value was 48.25 > 55, the penetration value was 90 > 80/100, and the specific gravity was 1.03–1.00 g/cm^3. Bitumen may be strengthened, and the durability of pavement improved by adding waste plastic, 5–10% by weight. According to further research, the proportion of plastic electronic waste over 7.5% reduces binder stability; however, replacing coarse aggregate with fly ash can boost Marshall stability by up to 25%. The best stability is seen when 4–12% of fly ash is substituted (Shahan-ur-Rahman 2009). To lessen the accumulation of electronic waste, Vikram R et al. studied the recycling of electrical trash from gadgets such as radios, TVs, and cell phones for building applications. Currently, 80% of bitumen is employed as a binder. Altering the bitumen content by 5% and 10% with electronic waste plastic increases penetration and decreases ductility values.

There were difficulties when 15% of the bitumen in the bituminous mix was swapped out for electronic waste plastic. To address the issue of environmental waste, Sabarai Mani et al. (2017)

investigated the production of bitumen by employing fly ash and electronic waste for road building. This approach enhances stability, reduces bitumen consumption, and promotes an eco-friendly strategy by partially substituting bitumen with electronic waste and fly ash. The study reveals that when electronic waste plastic powder is gradually added in different percentages, the strength of the bitumen increases. A mixture of 5–10% of electronic waste plastic is more robust. It contains good strength and can avoid potholes in the pavement; 20% of electronic waste plastic is inappropriate for pavement, and 15% of electronic waste plastic in the bituminous region is weaker than 5% and 10% of e-waste powder in the mix design. For improved strength, the electronic waste plastic powder can be mixed with normal bitumen in pavement applications. Better strength was demonstrated in tests using electronic waste plastic powder to replace 5%, 10%, and 15% of bitumen. As a result, 10% replacement may be utilized for road pavement. Bitumen that has been altered by adding waste plastic in a weight proportion of around 5–10% enhances the strength and longevity of pavement (Shahan-ur-Rahman 2009). Bitumen's rheological characteristics are altered and its quality is improved by blending synthetic polymers such as plastics and rubbers. The pavement performs better and lasts longer due to this enhanced bitumen. Due to its higher values of Marshall Stability and Marshall Coefficient, the bitumen and polymer combination makes a good alternative material for smooth pavement construction (Vasudevan et al. 2005).

According to Kokare et al. (2021), modified bituminous concrete (BC) mixes with enhanced strength and stability qualities are produced by adding 8.0% weight of processed plastic (Kokare et al. 2021). Aging-resistant asphalt and recycled low-density polyethylene (LDPE) combine to safeguard the building. According to Devi et al. (2013), the use of plastics improves mixing characteristics and aids in void reduction while postponing pothole development. According to Huang et al. (2009), 15–30% of the aggregates may be swapped out for suitably recycled LDPE particles, which also have a good impact on the durability and functionality of asphalt pavements (Huang et al. 2009). There are no special methods needed for mixing asphalt with recycled LDPE; standard tools and procedures can be used. Using recycled electronic plastic particles in pavements presents a desirable way to control the amount of coarse aggregate needed. It is also a useful strategy for managing trash that is related to the release of plastics that are not biodegradable into the environment. Recycled plastic pellets were utilized in bituminous mixtures to partially replace aggregate (Zoorob and Suparma 2000).

When making CPU and keyboard housings, acrylonitrile butadiene styrene (ABS), a more sophisticated and superior plastic, is used instead of metal. Apart from ABS plastics, high-impact polystyrene, polyphenylene oxide (PPO), and other resins are commonly utilized in computers and televisions. The primary source of environmental contamination from outdated electronic items is the usage of toxic elements in their production. Electronic waste comprises over a thousand different types of toxins and chemicals, which can pose a major threat to human health and the environment. Landfilling is the most popular way to manage electronic waste because it is general knowledge that landfills leak. Over their lifetime, even the greatest "state of the art" ones are not completely sealed, thus some chemical and metal leaching will still happen. For more established dumpsites or those with laxer regulations, the situation is far worse. It is also necessary to take into account the vaporization of metallic and dimethylene mercury.

24.7 Bitumen Production Mechanism from E-Waste-Derived MPs

Figure 24.3 shows the wet process method used to prepare e-waste modified bitumen (e-WMB) using e-waste slurry. Using this method, a suitable solvent is added in a proportional amount to a beaker containing a known amount of e-waste material (e-WM) to create the e-WM slurry. The e-WM slurry is then combined with the basic

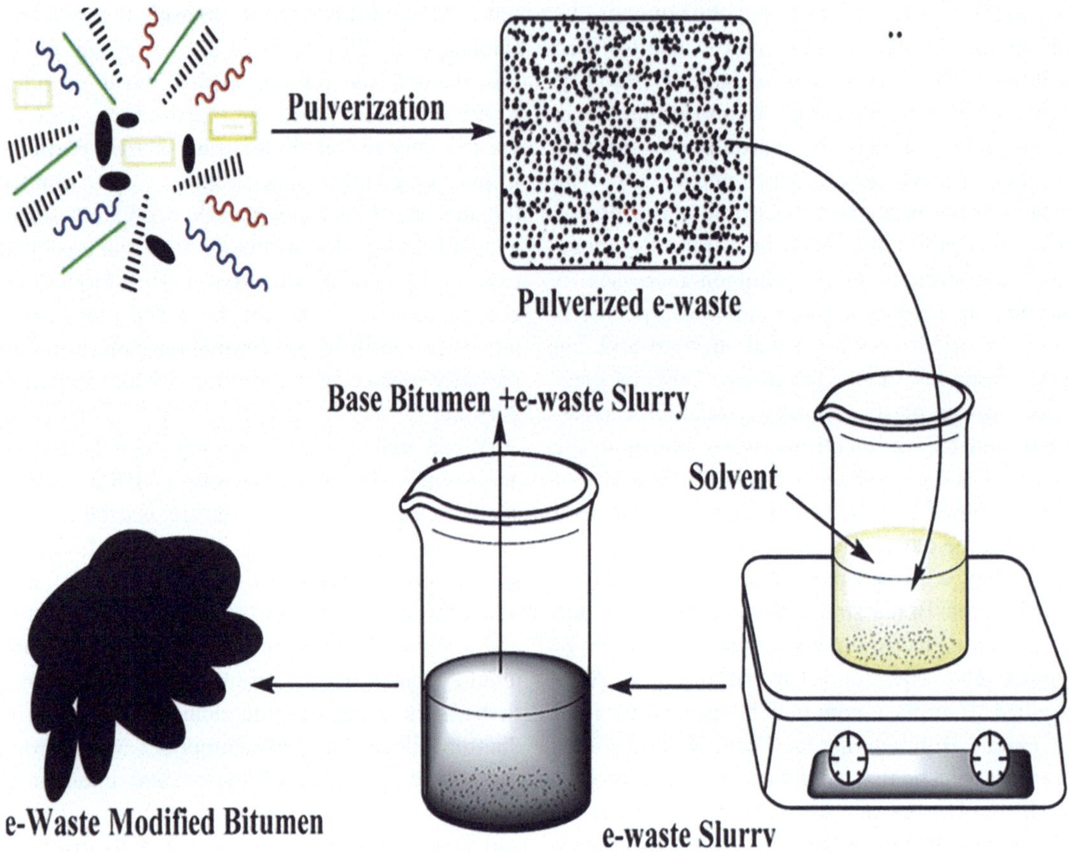

Fig. 24.3 Mechanism of the wet method used to prepare e-WMB from e-waste slurry. (Reproduced with permission) (Kumar et al. 2023)

bitumen. Acetone, benzene, toluene, and trichloroethylene were used separately as olefinic, aliphatic, and aromatic solvents to create a homogenous e-WM slurry. To cut down on the amount of excess solvent, the homogeneous slurry was concentrated in an oven after mixing. The concentrated e-WM slurry was also added to a 500 ml round-bottom glass assembly that included a thermowell jacket, a stirrer, and a container that held a predetermined amount of bitumen feedstock. After reaching the desired reaction temperature, the mixture is stirred with a mechanical stirrer for an ideal reaction duration of 90 minutes. After cooling the reaction mixture to 15 °C room temperature, the final product was analyzed to establish bitumen grade using preliminary testing.

24.8 Plastic Waste Utilization in Concrete/Road Construction

Dombe et al. (2020) developed bituminous mixes using plastic and electrical waste. Bitumen was substituted with waste plastic to the tune of 4.5–6% of the total quantity applied. Shredded electronic waste was used in place of some of the aggregates (7.5%, 10%, 12.5%, and 15% by volume of the mold). The bitumen's melting point was boosted by the use of plastic, which made the asphalt flexible all winter long. In addition, the shreds of plastic waste helped to maintain the state of the road. When 6.5% waste plastic was added, the bitumen's penetration value dropped by 6.68% and its softening point rose by 8.60%.

Surprisingly, bitumen's other properties remained unchanged. The specific gravity of aggregate covered with 7% waste plastic rose by 2.88%. Concurrently, the abrasion value, loss, and crushing value all decreased by 3–4% (Dombe et al. 2020).

Similarly, there may be drawbacks to adding plastic and electronic waste to concrete in addition to advantages. For instance, adding trash at different percentages throughout the course of a 28-day testing period caused the compressive strength to drop from 18.55 to 10.72 N/mm^2. Additionally, the split tensile strength dropped from 2.137 to 1.91 N/mm^2, and the flexural strength dropped from 3.14 to 2.74 N/mm^2. These results underscore the significance of taking a balanced approach when thinking about adding plastic and technology waste to concrete.

According to research, when 21.5% electronic waste replaces fine particles in M30 concrete, the concrete's compressive strength is reduced by 52.98% for concentrations of 0 to 21.5% (i.e., 7.5%, 14%, and 21.5%) (Damal et al. 2015). Another study discovered that adding 30% electronic waste decreased the concrete's slump value from 128 mm (control) to 75 mm (30% electronic waste) and its compressive strength from 47.18 to 22.15 N/mm^2 after 28 days of curing. Additionally, the inclusion of waste plastic during the curing period caused the split tensile strength to decrease from 4.9 to 3.8 N/mm^2 and the flexural strength to decrease from 4.35 to 2.5 N/mm^2 (Manjunath 2016). Another study looked at replacing natural aggregate in concrete with plastic waste. Around 60 cubes, 60 cylinders, and 40 prisms were cast in order to evaluate the compressive strength, split tensile strength, and flexural strength. Specifically, at weight percentages of 10%, 15%, and 20%, plastic fine (PF) aggregate was substituted for fine aggregates, while at weight percentages of 15%, 20%, and 25%, plastic coarse (PC) aggregate was substituted for coarse aggregates. Furthermore, 0.3% of steel fiber was added after the weight of cement. Concrete that had waste plastic added to it lost between 9 and 17% of its compressive strength.

Azhdarpour et al. (2016) demonstrated that inserting plastic particles into newly produced concrete has a substantial influence on many physical and strength-related parameters. Specifically, when the quantity of plastic particles rose, physical characteristics such as density and sonic velocity dramatically dropped. The specimens' compressive, tensile, and flexural strengths were raised when 5–10% of the tiny particles in the concrete were replaced with an equivalent number of polyethylene terephthalate (PET) fragments. The results of the investigation, however, indicated that the concrete's strength-related properties were considerably reduced when more than 10% of the particles were replaced. Therefore, it was suggested that replacing tiny particles in concrete specimens with PET fragments may enhance their strength-related properties, provided that the replacement rate remained below 10%. The inclusion of plastic in the concrete mix resulted in a drop in the fresh and dry densities of the concrete, according to measurements. Out of all the samples, the ones containing 30% plastic particles exhibited the biggest loss in density, at 9%.

24.9 Future Prospects and Recommendations

The growing amount of electronic waste poses a serious threat to the environment. Micro-nano plastics (MNPs) formed from electronic waste are particularly dangerous to human health and ecosystems because of their small size and tenacity. Adding these MNPs to bitumen during road building is one possible remedy that provides a creative and sustainable alternative. Bitumen containing micro-nanoplastics from electronic waste offers a viable path toward the creation of sustainable infrastructure. Through careful evaluation of important factors and thorough study, this strategy may be used to maximize possible benefits while reducing environmental dangers. A more circular economy and a sustainable future may be paved with the help of industries, academics, and policymakers working together. Assessments of toxicity and characterization, compatibility and performance, environmental impact, and standards and regulations are important factors to take into account when using micro-nanoplastics made from electronic waste in bitumen for a sustainable environment.

24.10 Conclusion

The world is producing enormous amounts of e-waste. It is a major reason for concern due to its negative effects on the ecosystem. Electronic waste recycling would improve public health, the environment, and the economy. The use of plastic in bitumen is a useful substitute for handling electronic waste. Research that has been published in the literature indicates that the quality of asphalt pavements might be improved by adding e-waste plastics to the materials used to make them. Therefore, more extensive research on this kind of topic is needed to address the rising problem of e-waste and its handling in a sustainable manner.

References

Abd El-Rahman AMM, El-Shafie M, Mohammedy MM, Abo-Shanab ZL (2018) Enhancing the performance of blown asphalt binder using waste EVA copolymer (WEVA). Egypt J Pet 27(4):513–521. https://doi.org/10.1016/j.ejpe.2017.08.002

Agarwal S, Darbar S, Saha S, Choudhury M, Singh RP (2023) E-waste management using different cost-effective, eco-friendly biological techniques: An overview. In: Waste management and resource recycling in the developing world. Elsevier, pp 205–235. https://doi.org/10.1016/B978-0-323-90463-6.00026-9

Awasthi AK, Li J (2017) An overview of the potential of eco-friendly hybrid strategy for metal recycling from WEEE. Resour Conserv Recycl 126:228–239. https://doi.org/10.1016/j.resconrec.2017.07.014

Azhdarpour AM, Nikoudel MR, Taheri M (2016) The effect of using polyethylene terephthalate particles on physical and strength-related properties of concrete; a laboratory evaluation. Constr Build Mater 109:55–62. https://doi.org/10.1016/j.conbuildmat.2016.01.056

Bahadoran A, De Lile JR, Masudy-Panah S, Sadeghi B, Li J, Sabzalian MH, Ramakrishna S, Liu Q, Cavaliere P, Gopinathan A (2022) Photocatalytic materials obtained from E-waste recycling: review, techniques, critique, and update. J Manuf Mater Process 6(4):69. https://doi.org/10.3390/jmmp6040069

Damal VS, Londhe SS, Mane AB (2015) Utilization of electronic waste plastic in concrete. Int J Eng Res Appl. 5(4):35–38

Dombe S, Tapase AB, Ghugal YM, Konnur BA, Akshay P (2020) Investigation on the use of E-waste and waste plastic in road construction. In: Badawy S, Chen D-H (eds) Recent developments in pavement engineering. Springer International Publishing, pp 85–99. https://doi.org/10.1007/978-3-030-34196-1_6

Geyer R (2020) Production, use, and fate of synthetic polymers. In: Plastic waste and recycling. Elsevier, pp 13–32. https://doi.org/10.1016/B978-0-12-817880-5.00002-5

Hake SL, Damgir RM, Awsarmal PR (2020) Utilization of plastic waste in bitumen mixes for flexible pavement. Transport Res Proc 48:3779–3785. https://doi.org/10.1016/j.trpro.2020.08.041

Herat S, Agamuthu P (2012) E-waste: a problem or an opportunity? Review of issues, challenges and solutions in Asian countries. Waste Manag Res 30(11):1113–1129. https://doi.org/10.1177/0734242X12453378

Huang K, Guo J, Xu Z (2009) Recycling of waste printed circuit boards: A review of current technologies and treatment status in China. J Hazard Mater 164(2–3):399–408. https://doi.org/10.1016/j.jhazmat.2008.08.051

Kokare AB, Salunkhe AS, Limkar CR (2021) Utilization of waste plastic for bituminous mix for improved performance of roads. In: Pawar PM, Balasubramaniam R, Ronge BP, Salunkhe SB, Vibhute AS, Melinamath B (eds) Techno-societal 2020. Springer International Publishing, pp 259–265. https://doi.org/10.1007/978-3-030-69925-3_26

Kumar A, Holuszko M, Espinosa DCR (2017) E-waste: an overview on generation, collection, legislation and recycling practices. Resour Conserv Recycl 122:32–42. https://doi.org/10.1016/j.resconrec.2017.01.018

Kumar K, Kumar A, Srivastava M, Chandra H, Singh RK (2023) Study of physical and rheological properties of modified bitumen prepared by blending non-metallicpart of e-waste. Int J Adhes Adhes 127:103516. https://doi.org/10.1016/j.ijadhadh.2023.103516

Lakusic S (2018) Use of e-plastic waste in bituminous pavements. J Croatian Associat Civil Eng 70(07):607–615. https://doi.org/10.14256/JCE.1375.2015

Ma Y, Zhou H, Jiang X, Polaczyk P, Xiao R, Zhang M, Huang B (2021) The utilization of waste plastics in asphalt pavements: a review. Cleaner Mater 2:100031. https://doi.org/10.1016/j.clema.2021.100031

Manjunath BTA (2016) Partial replacement of E-plastic waste as coarse-aggregate in concrete. Procedia Environ Sci 35:731–739. https://doi.org/10.1016/j.proenv.2016.07.079

Sabarai Mani S, Seeni Mohamed I, Nataraj Prabu P, Rajkumar M (2017) Experimental study on bituminous pavement by using ewaste and fly-ash. Int Res J Eng Sci 3:131–138

Sachdeva A, Sharma U (2020) An overview of utilization of E-waste plastic in road construction. In: Kanwar VS, Shukla SK (eds) Sustainable civil engineering practices, vol 72. Springer Singapore, pp 101–109. https://doi.org/10.1007/978-981-15-3677-9_11

Shaaban M, Wang X-L, Song P, Hou X, Wei Z (2024) Microplastic pollution and e-waste: Unraveling sources, mechanisms, and impacts across environments. Curr Opin Green Sustain Chem 46:100891. https://doi.org/10.1016/j.cogsc.2024.100891

Shahan-ur-Rahman A (2009) Use of waste plastic in construction of flexible pavement. New Building Materials & Construction World

Twagirayezu G, Uwimana A, Kui H, Birame CS, Irumva O, Nizeyimana JC, Cheng H (2023) Towards a sustainable and green approach of electrical and electronic waste management in Rwanda: a critical review. Environ Sci Pollut Res 30(32):77959–77980. https://doi.org/10.1007/s11356-023-27910-5

Vasudevan R, Rajasekaran S, Samanta S, Saravanavel S (2005) Utilization of waste plastics in construction of flexible pavements (Reuse of waste plastics—a path—breaking initiative). Global Plastics Environmental Conference 2005: GPEC 2005 – Creating Sustainability for the Environment, pp 315–381

Vishwakarma S, Kumar V, Arya S, Tembhare M, Rahul, Dutta D, Kumar S (2022) E-waste in information and communication technology sector: existing scenario, management schemes and initiatives. Environ Technol Innov 27:102797. https://doi.org/10.1016/j.eti.2022.102797

Zoorob SE, Suparma LB (2000) Laboratory design and investigation of the properties of continuously graded Asphaltic concrete containing recycled plastics aggregate replacement (Plastiphalt). Cem Concr Compos 22(4):233–242. https://doi.org/10.1016/S0958-9465(00)00026-3

Plastic Pollution Impacting Soil Microflora: A Study Toward Agricultural Productivity

25

Dipayan Das, Deep Deb, and Payel Paul

25.1 Introduction

Plastic pollution, a global environmental crisis, has long been recognized for its detrimental effects on marine ecosystems (Thushari and Senevirathna 2020). However, its impact on terrestrial environments, particularly soil ecosystems, has only recently begun to receive the attention it deserves (Gaur et al. 2024). Among the various components of soil, microflora, which includes bacteria, fungi, and archaea, plays a pivotal role in maintaining soil health and ecosystem functioning (Odelade and Babalola 2019). These microorganisms are responsible for nutrient cycling, organic matter decomposition, and maintaining soil structure (Wu et al. 2024). The intrusion of plastic, especially microplastics and nanoplastics, into soils poses a significant threat to these critical microbial communities, with far-reaching consequences for soil health, agricultural productivity, and ecosystem resilience (Seo et al. 2024).

Plastics in soil originate from various sources, including agricultural practices, industrial waste, and urban runoff. In agriculture, the use of plastic mulches, packaging materials, and slow-release fertilizers has led to an accumulation of plastic debris in soils (Russo et al. 2023). Additionally, the fragmentation of larger plastic items into microplastics, which are less than 5 millimeters in size, and nanoplastics, which are even smaller, has exacerbated this issue. These tiny plastic particles can persist in the soil for extended periods, potentially altering the physical and chemical properties of the soil environment (Azeem et al. 2021).

One of the primary concerns regarding plastic pollution in soil is its impact on soil microflora. The introduction of plastics into soil environments can disrupt the delicate balance of microbial communities, leading to shifts in their composition and function (Sajjad et al. 2022). For instance, some studies have shown that the presence of microplastics can lead to a reduction in the abundance of beneficial microorganisms, such as nitrogen-fixing bacteria and mycorrhizal fungi. These organisms are crucial for nutrient cycling and plant health, and their decline could lead to reduced soil fertility and lower crop yields (Aralappanavar et al. 2024).

Authors Deep Deb and Payel Paul have equally contributed to this chapter.

D. Das (✉)
Department of Microbiology, Royal School of Bio-Sciences, The Assam Royal Global University, Guwahati, Assam, India

D. Deb
Department of Biotechnology, University of North Bengal, Siliguri, West Bengal, India

P. Paul
Department of Botany, North Bengal St. Xavier's College, Rajganj, West Bengal, India

S. Goswami et al. (eds.), *Micro-Nano Plastics Exposure, Environmental Degradation and Public Health Crisis*, Sustainable Development Goals Series, https://doi.org/10.1007/978-981-96-5047-7_25

Moreover, plastics in soil can act as vectors for harmful chemicals and pathogens, further threatening soil microflora. Many plastics contain additives, such as phthalates and bisphenol A (BPA), which can leach into the soil and have toxic effects on microorganisms (Sobhani et al. 2021). Plastics can also absorb and accumulate environmental contaminants, including heavy metals and persistent organic pollutants (POPs), which may then be released into the soil (Verla et al. 2019). These contaminants can have detrimental effects on soil microflora, potentially leading to the proliferation of harmful microbial species while suppressing beneficial ones (Okoye et al. 2022).

The physical presence of plastics in soil can also have a direct impact on soil microflora by altering the soil's physical structure. Microplastics and nanoplastics can change the porosity and water-holding capacity of the soil, affecting the availability of water and nutrients to microorganisms (Wang et al. 2023). This can lead to changes in microbial activity and the breakdown of organic matter, further disrupting soil health. Additionally, the presence of plastics can create microhabitats that favor certain microbial species over others, leading to imbalances in the microbial community (Lear et al. 2021).

The consequences of plastic pollution on soil microflora extend beyond the immediate soil environment. As soil microflora play a key role in nutrient cycling, their disruption can have cascading effects on plant health and agricultural productivity (Adomako et al. 2022). For instance, a decline in nitrogen-fixing bacteria could lead to reduced nitrogen availability for plants, resulting in lower crop yields and increased reliance on chemical fertilizers (Li et al. 2022a). Furthermore, changes in soil microflora could impact the ability of soils to sequester carbon, potentially exacerbating climate change (Wu et al. 2024).

Addressing the issue of plastic pollution in soil requires a multifaceted approach. Reducing the use of plastics in agriculture, improving waste management practices, and promoting the development of biodegradable alternatives are essential steps in mitigating the impact of plastic pollution on soil microflora. Further research is needed to better understand how plastics affect soil microorganisms and to develop strategies for restoring and maintaining healthy soil microbial communities.

Thus, the impact of plastic pollution on soil microflora is a growing concern that has significant implications for soil health, agricultural productivity, and ecosystem resilience. The disruption of microbial communities in soil due to plastic contamination could lead to reduced soil fertility, lower crop yields, and increased environmental pollution. As awareness of this issue grows, it is crucial to take action to protect soil ecosystems and the vital services they provide.

25.2 Overview of Plastic Pollution in Terrestrial Ecosystems

Plastic pollution has emerged as a critical environmental issue, capturing global attention primarily due to its overwhelming presence in marine environments. However, terrestrial ecosystems, which include forests, grasslands, wetlands, and agricultural lands, are equally, if not more, affected by the proliferation of plastic waste. Unlike marine ecosystems, where the consequences of plastic pollution are often visible and immediate, the effects of plastic accumulation in terrestrial environments can be more insidious and long-lasting.

25.2.1 Sources of Plastic Pollution in Soil

Plastic pollution in terrestrial ecosystems originates from multiple sources, with human activities being the primary contributors. One significant source is the improper disposal of plastic waste (Rai et al. 2023). Plastics discarded in landfills or as litter often find their way into natural landscapes through wind, rain, and runoff (Kibria et al. 2023). Urban areas, in particular, contribute substantially to plastic pollution, as inadequate waste management systems allow plastic debris to escape into the environment.

Once in the environment, plastics can travel long distances, eventually accumulating in soils, rivers, and forests (Kumar et al. 2021).

Agricultural practices are another major source of plastic pollution in soil. The use of plastic products in agriculture has increased dramatically over the past few decades, driven by the demand for more efficient farming techniques. Plastics are used extensively in the form of mulching films, greenhouse covers, irrigation pipes, and packaging materials for fertilizers and pesticides (Hofmann et al. 2023). These plastics, often termed "agroplastics," are designed to enhance crop yields and reduce water usage. However, their widespread use has led to the unintended consequence of plastic accumulation in agricultural soils. Over time, these plastics degrade into smaller fragments, contributing to the growing problem of microplastic contamination in soils (Lakhiar et al. 2024).

Industrial activities also contribute significantly to plastic pollution in terrestrial ecosystems. Factories and manufacturing plants often generate plastic waste as byproducts of production processes. In some cases, this waste is improperly managed, leading to its release into the surrounding environment (Mihai et al. 2022). Additionally, the transportation and handling of raw materials and finished products can result in the spillage of plastic pellets, which are small, pre-production plastic particles used in the manufacture of plastic goods. These pellets, also known as "nurdles," are a significant source of microplastic pollution in terrestrial and aquatic environments (Karlsson et al. 2018).

25.2.2 Types of Plastics Found in Soil

Plastics found in terrestrial ecosystems are varied, reflecting the broad spectrum of plastic products used both in daily life and across industries. The most frequently found plastics in soils include polyethylene (PE), polypropylene (PP), polystyrene (PS), polyvinyl chloride (PVC), and polyethylene terephthalate (PET). Polyethylene (PE) is the most widely produced plastic globally and is commonly used in packaging materials, plastic bags, and agricultural films. Due to its widespread use, PE is frequently found in soils, particularly in agricultural areas where it is used as mulch or in greenhouses. PE is highly resistant to degradation, making it a persistent pollutant in the environment (Tiago et al. 2023) (Table 25.1).

Polypropylene (PP) is another common plastic, often used in the production of containers, automotive parts, and textiles. Like PE, PP is resistant to degradation and can accumulate in soils, contributing to microplastic pollution. PP is often found in soils near urban areas, where it is used in various consumer products and packaging (Alsabri et al. 2022).

Polystyrene (PS), widely used in packaging, disposable cups, and insulation materials, is also a common plastic pollutant in soils. PS is lightweight and brittle, making it prone to breaking down into smaller fragments that can easily be transported by wind and water into terrestrial ecosystems (Hwang et al. 2020).

Polyvinyl chloride (PVC) is used in construction materials, plumbing pipes, and medical devices. While less common in soils compared to PE and PP, PVC can still be found in areas where it is used extensively in construction or discarded improperly. PVC contains harmful additives, such as plasticizers and stabilizers, which can leach into the soil and pose a risk to soil health and the organisms that inhabit it (Kudzin et al. 2023).

Polyethylene terephthalate (PET) is primarily used in the production of beverage bottles and food packaging. PET is highly durable and resistant to environmental degradation, making it a significant contributor to plastic pollution in terrestrial ecosystems. PET can be found in soils, particularly in areas near landfills or where waste management practices are inadequate (Dhaka et al. 2022).

In addition to these common plastics, soils can also contain various additives and chemical compounds used in the manufacture of plastic products. These additives, such as plasticizers, flame retardants, and stabilizers, can leach into the soil and have adverse effects on soil chemistry and biology.

Table 25.1 Types of plastics, their characteristics, uses, and their impact on the environment

Type of plastic	Characteristics	Common uses	Impact on the environment	References
Polyethylene terephthalate (PET or PETE)	Lightweight, strong, transparent, and resistant to water	Beverage bottles, food packaging, polyester fibers	Recyclable but often ends up in landfills or oceans; can take hundreds of years to decompose. May release toxic chemicals when breaking down	Andreasi Bassi et al. (2022)
High-density polyethylene (HDPE)	Strong, durable, resistant to impact and chemicals	Milk jugs, detergent bottles, pipes, plastic bags	Recyclable, but HDPE waste in landfills can persist for centuries. Microplastics from HDPE can accumulate in marine environments	Mermertaş and Germirli Babuna (2019)
Polyvinyl chloride (PVC)	Flexible or rigid, durable, and resistant to chemicals	Pipes, cables, vinyl flooring, medical tubing	Difficult to recycle; releases toxic chemicals during production and when burned. Can leach harmful additives into the environment	Kudzin et al. (2023)
Low-density polyethylene (LDPE)	Flexible, lightweight, and moisture-resistant	Plastic bags, squeezable bottles, food wrap	Less commonly recycled; contributes significantly to plastic pollution, especially in marine environments	Abbasi et al. (2023)
Polypropylene (PP)	Tough, heat-resistant, and durable	Food containers, bottle caps, automotive parts, straws	Recyclable but not widely collected. PP waste accumulates in landfills and oceans, contributing to microplastic pollution	Alsabri et al. (2022)
Polystyrene (PS)	Lightweight, insulating, and easily molded	Disposable cups, food packaging (Styrofoam), insulation	Very difficult to recycle; breaks into microplastics that are harmful to marine life. PS can release toxic chemicals like styrene	Hwang et al. (2020)
Polycarbonate (PC) and other plastics	Strong, transparent, and impact-resistant	Water bottles, electronics, automotive parts	Some types (like polycarbonate) can leach toxic chemicals such as bisphenol A (BPA). Difficult to recycle, and often ends up in landfills	Zhou et al. (2023)

25.2.3 Extent of Plastic Pollution in Agricultural Soils

Agricultural soils are particularly vulnerable to plastic pollution due to the extensive use of plastics in modern farming practices (Sa'adu and Farsang 2023). The use of plastic mulch, in particular, has become a widespread practice in agriculture, as it helps conserve soil moisture, suppress weeds, and increase crop yields (Samphire et al. 2023). However, the benefits of plastic mulch come at a significant environmental cost, as these films often break down into small fragments that accumulate in the soil (Amare and Desta 2021).

An increasing amount of research is revealing the disturbing scale of plastic pollution in agricultural soils. Studies indicate that microplastics are now found in soils worldwide, with concentrations varying from a few particles to several thousand particles per kilogram of soil. In some instances, microplastic levels in agricultural soils surpass those found in marine sediments, highlighting the gravity of the issue (de Souza Machado et al. 2018) (Table 25.2).

The accumulation of plastics in agricultural soils can have far-reaching consequences for soil health and agricultural productivity (Mongil-Manso et al. 2023). Microplastics can alter the physical and chemical properties of soils, affect-

Table 25.2 Impact of plastic pollution on agriculture

Impact area	Description	Consequences	References
Soil quality degradation	Microplastics accumulate in soil through plastic mulch, fertilizers, or litter	Reduction in soil fertility and structure Impaired water retention and aeration Disruption of soil microbial activity, affecting nutrient cycling	Li et al. (2022b)
Plant growth inhibition	Microplastics interfere with soil-root interactions and nutrient uptake	Lower crop yields due to reduced nutrient absorption Stunted plant growth and development	Li et al. (2023b)
Water retention	Plastics in soil alter its capacity to retain moisture	Increased drought stress on crops due to poor water retention Difficulty in irrigating fields effectively	Wang et al. (2023)
Chemical contamination	Plastics can absorb and leach toxic chemicals (e.g., pesticides, herbicides, and heavy metals) into the soil and plants	Bioaccumulation of harmful chemicals in crops, leading to food safety risks Potential contamination of groundwater	Sajjad et al. (2022)
Microbial activity disruption	Microplastics disrupt the natural habitat of soil microorganisms, essential for decomposing organic matter and nutrient cycling	Reduced microbial diversity and functionality, affecting plant health Imbalance in nutrient availability for crops, leading to soil degradation over time	Chen et al. (2024)
Physical soil obstruction	Larger plastic fragments create physical barriers in the soil	Obstruction of root growth and penetration Reduced soil aeration, affecting oxygen availability for plant roots	Bodor et al. (2024)
Plastic mulch effects	While plastic mulch is used to prevent weed growth and conserve moisture, improper disposal leads to pollution	Breakdown of mulch into microplastics, contaminating the soil Negative impact on soil organic matter and long-term soil health	Somanathan et al. (2022)
Pest and disease acceleration	Plastics can alter the micro-environment in the soil, affecting the balance of pests and pathogens	Increased risk of pest and disease outbreaks due to changes in the micro-environment, possibly requiring increased pesticide use	de Souza Machado et al. (2019)
Livestock health	Animals grazing in areas with plastic-contaminated soils or pastures can ingest plastic	Health risks to livestock (blockages, malnutrition) Contaminated animal products entering the food chain, posing risks to human health	Urli et al. (2023)
Irrigation systems blockage	Plastic waste can clog irrigation canals and systems, reducing water availability for crops	Reduced irrigation efficiency, leading to water shortages for crops Higher costs of maintenance and cleaning of irrigation infrastructure	Kibria et al. (2023)
Long-term soil toxicity	Persistent plastic particles can degrade into smaller particles, remaining in the soil for centuries	Long-term contamination of agricultural lands Difficulties in reversing the degradation process, making land unusable for agriculture in severe cases	Cai et al. (2023)
Impact on aquatic agriculture	Plastic waste from agricultural runoff can pollute nearby water bodies, affecting fish farming and aquaculture	Fish and aquatic organisms ingest plastic, leading to bioaccumulation of toxins Reduced productivity of aquaculture farms, impacting food security and economic stability in affected regions	Iheanacho et al. (2023)

ing their structure, permeability, and water retention capacity. These changes can disrupt the balance of nutrients and microorganisms in the soil, leading to reduced soil fertility and compromised crop growth. Moreover, the presence of microplastics in the soil can impede the movement of water and air through the soil, further exacerbating the negative effects on plant growth (Wang et al. 2023).

Plastic pollution in agricultural soils not only affects soil health but also raises concerns about potential risks to human health (Fig. 25.1). Microplastics present in the soil can be absorbed by crops, introducing contaminants into the food chain. Although the full scope of health risks linked to microplastic ingestion is still being studied, increasing evidence suggests that long-term exposure to microplastics and the chemicals they carry may pose significant health threats to humans (Cverenkárová et al. 2021).

25.3 Soil Microflora: The Foundation of Soil Health

The microflora in soil, including bacteria, fungi, archaea, and other microorganisms, are fundamental to soil health and essential for the functioning of terrestrial ecosystems. They drive key processes that sustain soil fertility, enhance plant growth, and uphold the integrity of the ecosystem as a whole.

25.3.1 Importance of Soil Microflora

Soil microfloras are essential to the health and productivity of soil ecosystems. These microorganisms are responsible for numerous biochemical processes that sustain the soil environment and their ability to support plant and animal life. Some of the most critical roles played by soil microflora include nutrient cycling, organic matter decomposition, soil structure formation, and plant-microbe interactions (Wang et al. 2024b).

25.3.1.1 Nutrient Cycling

Soil microflora plays a key role in nutrient cycling, breaking down organic matter to release essential nutrients like nitrogen, phosphorus, and sulfur. Bacteria and fungi facilitate this process, making nutrients available for plant uptake. Nitrogen-fixing bacteria and archaea convert atmospheric nitrogen into a usable form for plants, while other microorganisms solubilize phosphorus, ensuring its absorption. These nutrient cycles are crucial for maintaining soil fertility and providing plants with the nutrients necessary for growth and survival (Adomako et al. 2022).

25.3.1.2 Organic Matter Decomposition

Soil microflora is also crucial in the decomposition of organic matter (Spohn 2016). As plants and animals die, their remains are broken down by soil microorganisms into simpler organic compounds, which eventually form humus. Humus is a stable organic component of soil that enhances soil structure, increases its water-holding capacity, and provides a reservoir of nutrients that can be slowly released to plants (Maier 2015). The decomposition process carried out by soil microflora is essential for the recycling of organic matter, preventing the accumulation of plant and animal debris in ecosystems, and maintaining the long-term fertility of soils (Cobaugh et al. 2015).

25.3.1.3 Soil Structure Formation

Soil microorganisms play a crucial role in forming and stabilizing soil structure. They produce compounds that bind soil particles into aggregates, enhancing soil aeration, water infiltration, and root penetration (Oades 1993). Fungi, in particular, play a significant role in soil aggregation by producing hyphae, which act as a physical binding agent for soil particles (Angulo et al. 2024). Well-aggregated soils are more resistant to erosion, have better water retention, and provide a more favorable environment for root growth and microbial activity (Lehmann et al. 2020).

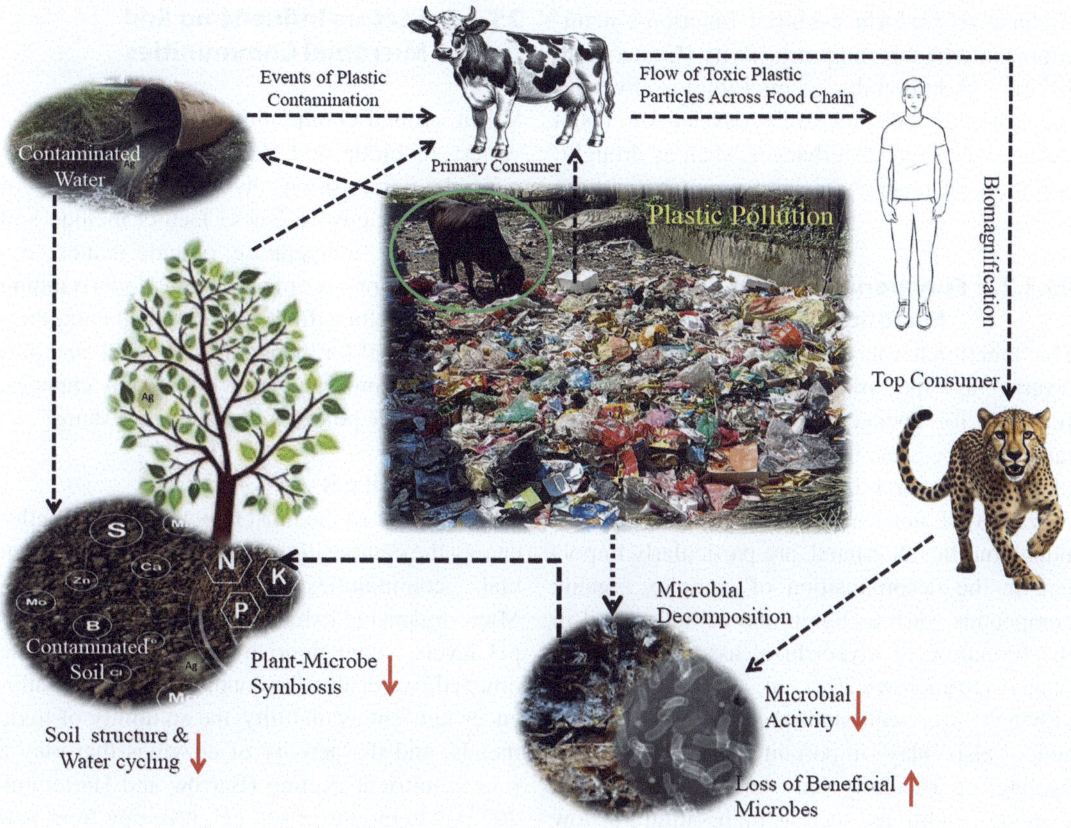

Fig. 25.1 Impact of plastic pollution on ecological food chain

25.3.1.4 Plant-Microbe Interactions

Soil microflora engages in various interactions with plants that are critical for plant health and productivity (Wu et al. 2009). Certain soil microorganisms, like mycorrhizal fungi, form symbiotic relationships with plants, improving nutrient and water absorption in exchange for carbohydrates. Similarly, rhizobia bacteria create nodules on legume roots to fix atmospheric nitrogen, benefiting the plant in return for nutrients. These interactions are crucial for plant growth, especially in nutrient-poor soils (Dolatabadian 2020).

25.3.2 Diversity and Functionality of Soil Microflora

The diversity of soil microflora is astounding, with a single gram of soil estimated to contain thousands to millions of microbial species. This diversity is not only a reflection of the vast array of microorganisms presents but also of the numerous ecological functions they perform. The functional diversity of soil microflora is crucial for the resilience and stability of soil ecosystems, enabling them to adapt to environmental changes and stressors.

25.3.2.1 Microbial Diversity and Ecosystem Resilience

The immense diversity of soil microflora is one of the key factors that contribute to the resilience of soil ecosystems. Different microorganisms specialize in different ecological functions, such as decomposing specific types of organic matter, fixing nitrogen, or solubilizing phosphorus (Philippot et al. 2021). This functional redundancy ensures that even if some microbial species are lost or their activity is reduced due to environmental stress, other microorganisms can continue to perform essential functions, maintaining soil health and productivity (Custer et al. 2022). Additionally, microbial diversity enhances the ability of soil ecosystems to resist and recover from disturbances, such as drought, pollution, or changes in land use (Che and Jin 2024).

25.3.2.2 Functional Roles of Soil Microflora

The functional roles of soil microflora are as diverse as the microorganisms themselves. Bacteria, for instance, are involved in a wide range of processes, including the decomposition of organic matter, nitrogen fixation, and the solubilization of nutrients (de Souza et al. 2015). Fungi, on the other hand, are particularly important in the decomposition of complex organic compounds, such as lignin and cellulose, and in the formation of mycorrhizal associations with plants (Komorowicz et al. 2023). Archaea, although less well-studied than bacteria and fungi, also play important roles in nutrient cycling, particularly in environments with extreme conditions, such as high salinity or low oxygen levels (Jung et al. 2020). The collective activity of these microorganisms ensures the continuous cycling of nutrients, the maintenance of soil structure, and the support of plant growth (Suman et al. 2022).

25.3.2.3 Soil Microflora and Ecosystem Services

The diversity and functionality of soil microflora are closely tied to the ecosystem services that provide benefits to humans from the environment (Romero et al. 2023). Healthy and functioning soil microbial communities provide a broad array of ecosystem services, such as soil formation, nutrient cycling, water purification, carbon sequestration, and the regulation of greenhouse gases (Singavarapu et al., 2023). By supporting these services, soil microflora play a crucial role in sustaining agricultural productivity, maintaining ecosystem stability, and mitigating climate change (Antoszewski et al. 2022).

25.3.3 Factors Influencing Soil Microbial Communities

Soil microbial communities are influenced by a variety of biotic and abiotic factors that determine the composition, diversity, and activity of soil microorganisms. These factors include soil pH, moisture, temperature, organic matter content, and the presence of pollutants. Understanding how these factors affect soil microbial communities is essential for managing soil health and predicting the impacts of environmental changes, such as plastic pollution, on soil ecosystems.

25.3.3.1 Soil pH

Soil pH is one of the most important factors influencing the composition and activity of soil microbial communities (Cao et al. 2016). Microorganisms exhibit varying preferences for pH levels; some flourish in acidic soils, while others thrive in alkaline conditions. Soil pH influences nutrient availability, the solubility of toxic metals, and the activity of enzymes that play a role in nutrient cycling (Barrow and Hartemink 2023). Alterations in soil pH, resulting from natural processes or human actions, can cause changes in the composition of microbial communities, which may impact the functioning of soil ecosystems (Wu et al. 2017).

25.3.3.2 Soil Moisture and Temperature

Soil moisture and temperature are critical factors that influence microbial activity and diversity (Frindte et al. 2019). Microorganisms require water for their metabolic processes, and soil moisture levels can affect the availability of nutrients, the diffusion of gases, and the mobility of microorganisms (Bian et al. 2022). Temperature, on the other hand, affects the rate of microbial metabolic processes, with most soil microorganisms having an optimal temperature range for growth and activity (Stres et al. 2008). Extreme temperatures, whether hot or cold, can inhibit microbial activity and reduce diversity, leading to changes in soil microbial communities and their ecological functions (Nedwell 1999).

25.3.3.3 Organic Matter Content

The availability of organic matter is a key determinant of soil microbial diversity and activity. Organic matter provides a source of energy and nutrients for soil microorganisms, supporting their growth and metabolic processes. The decomposition of organic matter by soil microflora is essential for nutrient cycling and the formation of soil structure (Prescott et al. 2019). Soils with high organic matter content tend to have more diverse and active microbial communities, while soils with low organic matter content may have reduced microbial diversity and functionality (Hao et al. 2021).

25.3.3.4 Presence of Pollutants

The introduction of pollutants, such as plastics, into the soil can have significant impacts on soil microbial communities. Plastics, particularly microplastics, can alter the physical and chemical environment of the soil, affecting microbial activity and diversity (Li et al. 2023a). Microplastics can change soil structure, reduce water infiltration, and alter the availability of nutrients, all of which can affect the composition and activity of soil microorganisms (Omidoyin and Jho 2023). Furthermore, plastics have the ability to adsorb and accumulate toxic chemicals, including heavy metals and persistent organic pollutants, which can further disrupt microbial communities and their ecological roles (de Souza Machado et al. 2019).

25.4 Impact of Plastic Pollution on Soil Microflora

Plastic pollution in agricultural soils is a rising environmental issue. Unlike ocean pollution, its impact on terrestrial ecosystems, particularly on soil microflora, is just starting to be recognized. Plastics persist for centuries, altering soil's physical, chemical, and biological properties. This can disrupt microbial communities, which are essential for nutrient cycling, organic matter decomposition, and plant growth. The long-term presence of plastics threatens both soil health and broader ecosystem functions.

25.4.1 Physical Effects of Plastics on Soil Microbes

Plastics in soil can physically alter the habitat of microorganisms by affecting soil structure, porosity, and water retention. Soil is a dynamic matrix where microorganisms reside in tiny pores and aggregates, which provide them with the necessary conditions for survival, such as access to water, nutrients, and oxygen. However, the introduction of plastic particles into the soil can disrupt these aggregates, leading to changes in soil structure and porosity (Kublik et al. 2022).

Larger plastic particles, such as plastic bags, bottles, and agricultural films, can block the natural pathways for air and water flow within the soil. This leads to poor aeration, reduced water infiltration, and altered water retention capacities (Bodor et al. 2024). In turn, these changes can create microenvironments within the soil that are less favorable for certain microbial species. For instance, anaerobic conditions may develop in areas where plastic blocks water flow, favoring anaerobic microbes while suppressing aerobic ones. As a result, the composition of the soil microbial community may shift, with certain groups of microorganisms becoming dominant while others decline (Hofmann et al. 2023).

Microplastics (particles smaller than 5 mm), which are increasingly found in soils due to the fragmentation of larger plastics, also pose a problem. These tiny particles can physically interact with soil particles, leading to compaction and reduced pore space (Osman et al. 2023). This compaction limits the movement of soil fauna, such as earthworms, which play a role in mixing soil and enhancing microbial activity. Reduced soil porosity also restricts gas exchange between the soil and atmosphere, leading to oxygen-deprived conditions that can negatively affect the metabolism of aerobic microorganisms. Overall, the physical presence of plastics in the soil can disrupt the delicate balance of soil microflora, with potential implications for soil health and fertility (Lakhiar et al. 2024).

25.4.2 Chemical Interactions Between Plastics and Soil Microbes

In addition to their physical effects, plastics can also introduce harmful chemicals into the soil environment, further impacting soil microbes. Many plastic products contain additives such as plasticizers (e.g., phthalates), stabilizers, and flame retardants, which are added to improve the performance and durability of plastics (Hahladakis et al. 2018). Over time, these additives can leach out of plastic debris and enter the soil. Some of these chemicals, such as bisphenol A (BPA) and phthalates, are known endocrine disruptors and can be toxic to soil microorganisms (Lucas et al. 2022).

The release of these chemicals into the soil can inhibit microbial growth and activity, leading to a reduction in microbial biomass and diversity. Studies have shown that phthalates can negatively affect the respiration and enzymatic activity of soil bacteria (Li et al. 2024), while BPA has been shown to disrupt the cell membranes of certain microbial species (Cao et al. 2023). Consequently, the overall metabolic activity of the soil microbial community may decline, which could negatively impact essential soil processes like nutrient cycling and the decomposition of organic matter (Zaborowska et al. 2021).

Microplastics can adsorb and concentrate on various environmental pollutants, including heavy metals and persistent organic pollutants (POPs) that may already exist in the soil. These toxins can adhere to the surfaces of plastic particles, forming a concentrated source of contaminants that soil microorganisms can ingest. The presence of microplastics can enhance the bioavailability of these pollutants to soil microbes, intensifying their toxic effects. (Guo et al. 2020). Heavy metals such as cadmium and lead can interfere with microbial enzymes and DNA, leading to reduced microbial growth and reproduction (Balali-Mood et al. 2021). Similarly, POPs such as polychlorinated biphenyls (PCBs) can disrupt microbial metabolism and hinder the breakdown of organic matter in the soil (Leigh et al. 2006).

Furthermore, the degradation of plastics in the soil, although slow, can release smaller plastic fragments and nanoparticles that may have distinct chemical properties compared to larger plastic particles. These smaller particles have a larger surface area-to-volume ratio, which may enhance their ability to adsorb pollutants and interact with microbial cells. As a result, the chemical interactions between plastics, soil pollutants, and microbes can create a complex and harmful environment for soil microorganisms, potentially leading to shifts in microbial community structure and function (Sutkar et al. 2023).

25.4.3 Biological Responses of Soil Microflora to Plastics

The presence of plastics in the soil can elicit various biological responses from soil microorganisms, depending on the type, size, and concentration of plastics, as well as the specific characteristics of the microbial community. One of the most significant biological responses is the alteration of microbial community structure. Some microbial species may be more tolerant to the presence of plastics, while others may be inhibited or outcompeted. Certain bacteria and fungi have been found to colonize plastic surfaces and even degrade plastic polymers, albeit at a slow rate (Yang et al. 2023). These microorganisms, known as plastic-degrading microbes, may gain a competitive advantage in soils contaminated with plastics, leading to changes in community composition (Cai et al. 2023).

The ability of some microbes to degrade plastics has led to an interest in their potential use for bioremediation of plastic-contaminated soils (Zeenat et al. 2021). However, the degradation of plastics by soil microbes is often incomplete, resulting in the accumulation of microplastics and other breakdown products (Ru et al. 2020). Moreover, the metabolic pathways involved in plastic degradation are not well understood, and the long-term effects of plastic degradation on soil microbial communities and soil health are still uncertain (Viljakainen and Hug 2021). While plastic-degrading microbes may play a role in mitigating plastic pollution in soils, their activity alone is unlikely to fully offset the negative impacts of plastics on soil microflora.

Beyond influencing community structure, plastics can impact the metabolic activity and functional diversity of soil microorganisms. Their presence may modify the rates of crucial microbial processes, including carbon and nitrogen cycling (Rillig et al. 2021). Some studies have shown that plastics can reduce microbial respiration rates, leading to lower carbon dioxide emissions from soil (Li et al. 2023a). This may be due to the physical obstruction of soil pores by plastic particles, which limits oxygen availability and slows down the microbial breakdown of organic matter. Similarly, plastics may interfere with nitrogen cycling by affecting the activity of nitrifying and denitrifying bacteria, which play a crucial role in converting nitrogen into forms that are usable by plants (Seeley et al. 2020).

The net effect of these biological responses is that plastic pollution can lead to changes in soil fertility and ecosystem functioning. Reduced microbial activity may result in slower decomposition of organic matter, leading to a buildup of organic residues in the soil (Zhang et al. 2022a). This can affect the availability of nutrients for plants and alter the overall nutrient balance in the soil. In agricultural soils, where healthy microbial communities are essential for crop productivity, plastic pollution may have particularly detrimental effects on soil fertility and plant growth (Mészáros et al. 2023).

25.4.4 Case Study: Microplastic Pollution and Soil Microbial Diversity

Microbial communities in soil are vital for various ecosystem functions, including nutrient cycling, organic matter decomposition, and plant growth promotion. Microplastics in soil can physically, chemically, and biologically interact with these microbial communities, leading to changes in microbial diversity, community structure, and functional capabilities. Case studies are crucial for examining specific instances where microplastic pollution has been shown to affect soil microbial diversity and explore the potential consequences of these changes on soil health and ecosystem functioning.

China investigated the effects of microplastic pollution on microbial diversity in agricultural soils (Tang 2023). The researchers applied varying concentrations of polyethylene microplastics (a common type of plastic) to soil samples and monitored changes in microbial community composition over time. The study revealed that microplastics had a significant impact on soil microbial diversity, with higher concentrations of microplastics leading to a reduction in overall microbial richness (Liu et al. 2024).

One of the key findings was that certain groups of bacteria, particularly those involved in nitrogen cycling, were negatively affected by the presence of microplastics. The abundance of nitrifying bacteria, which convert ammonium to nitrate, decreased in soils with high microplastic concentrations. This reduction in nitrifying bacteria could have significant implications for soil fertility, as nitrogen is a critical nutrient for plant growth (Wang et al., 2024a, b). In contrast, some bacterial groups, such as those capable of degrading plastic polymers, showed increased abundance in the presence of microplastics. These plastic-degrading microbes, while potentially beneficial for breaking down plastic particles, may not perform other essential functions for soil health (Yang et al. 2023).

Microplastic pollution could lead to a loss of functional diversity in soil microbial communities, with potential consequences for nutrient cycling and soil productivity. The selective pressure exerted by microplastics may favor certain microbial species at the expense of others, leading to a shift in community structure and a decline in overall microbial diversity.

25.5 Consequences of Altered Soil Microflora for Agricultural Productivity

Soil microflora, which includes bacteria, fungi, archaea, and other microorganisms, play a vital role in sustaining the health of agricultural ecosystems. These microorganisms are involved in essential processes such as nutrient cycling, organic matter decomposition, and maintaining soil structure. However, plastic pollution in soils, particularly the presence of microplastics, can disrupt soil microbial communities, leading to significant consequences for agricultural productivity.

25.5.1 Impact on Soil Fertility and Nutrient Cycling

Soil fertility is fundamentally linked to the activities of soil microbes. Microorganisms are responsible for breaking down organic matter, which in turn releases essential nutrients such as nitrogen, phosphorus, and potassium. These nutrients are critical for plant growth and overall crop productivity. When plastic particles accumulate in soil, they can negatively impact the structure and diversity of microbial communities, which in turn affects the efficiency of these microbial processes (Li et al. 2023a).

Microbial decomposers, particularly bacteria and fungi, are central to nutrient cycling. They break down plant residues and organic matter into simpler compounds, making nutrients available to plants. Plastic pollution, particularly in the form of microplastics, can diminish microbial biomass and activity by forming physical barriers in the soil that restrict access to organic matter. These microplastics can obstruct soil pores, hindering the movement of air and water vital for microbial respiration and metabolism. Consequently, microbial activity decreases, resulting in slower rates of organic matter decomposition and nutrient release (Li et al. 2023a).

Moreover, plastic pollution can disrupt essential microbial processes such as nitrogen fixation and nitrification, which are vital for transforming atmospheric nitrogen into usable forms for plants. The presence of plastic particles may hinder nitrogen-fixing bacteria that have symbiotic relationships with leguminous plants, ultimately decreasing nitrogen availability in the soil. This decline in nutrient cycling can adversely affect soil fertility, leading to lower crop yields and deteriorating soil health over time (Kim et al. 2023).

25.5.2 Effects on Plant-Microbe Interactions

Plant-microbe interactions are critical for agricultural productivity, as many plants rely on symbiotic relationships with microorganisms to enhance nutrient uptake, boost disease resistance, and improve overall resilience (Sharma et al. 2021). Mycorrhizal fungi form symbiotic associations with plant roots, extending their reach into the soil to help plants absorb nutrients such as phosphorus and water more efficiently (Bonfante and Genre 2010). Similarly, nitrogen-fixing bacteria colonize the roots of certain plants, helping convert atmospheric nitrogen into usable forms (Santi et al. 2013).

Plastic pollution can disrupt these beneficial plant-microbe interactions in several ways. Microplastics and chemical additives released from plastic materials can alter the microbial communities in the rhizosphere (the soil zone around plant roots), potentially reducing the abundance and functionality of beneficial microbes (Trojan et al. 2024). For instance, the chemical compounds leached from plastics, such as phthalates and bisphenol A (BPA), can be toxic to soil microorganisms, inhibiting their growth and activity (Rybczyńska-Tkaczyk et al. 2023). This inhibition can weaken the symbiotic relationships between plants and beneficial microbes, reducing the efficiency of nutrient uptake and leaving plants more vulnerable to nutrient deficiencies.

In addition, the reduction in microbial diversity caused by plastic pollution may lead to a decline in the population of microorganisms that help protect plants from soilborne pathogens (de Souza Machado et al. 2019). Beneficial microbes often outcompete harmful ones for resources and space, acting as a natural defense system for plants. If plastic pollution diminishes the abundance of these protective microbes, crops may become more susceptible to diseases, which can further reduce productivity and resilience. Thus, the disruption of plant-microbe interactions due to plastic pollution can have far-reaching consequences for crop health and agricultural yields.

25.5.3 Impact on Soil Structure and Water Retention

Soil structure plays a crucial role in maintaining adequate water retention, root penetration, and aeration, all of which are essential for healthy plant growth. Well-structured soil contains aggregates, which are clusters of soil particles bound together by organic matter, microorganisms, and fungal hyphae. These aggregates create pore spaces that allow water to infiltrate the soil, air to circulate, and roots to penetrate deeply (Prescott et al. 2019).

The presence of plastic particles in soil can disrupt soil structure by affecting the formation and stability of soil aggregates. Large plastic debris, such as agricultural films, can obstruct water flow and reduce soil porosity, leading to compaction and poor drainage. Microplastics, in particular, can lodge between soil particles, reducing pore spaces and altering the soil's water-holding capacity. This disruption in soil structure can lead to waterlogging in some areas while causing drought-like conditions in others (Bodor et al. 2024).

Changes in water retention and drainage directly affect plant growth. In soils where water retention is compromised, plants may struggle to access the moisture they need, especially in regions with limited rainfall or poor irrigation (Barbera et al. 2013). On the other hand, waterlogged soils can suffocate plant roots and lead to root diseases. Poor soil aeration due to compaction or blocked pores limits the oxygen available to plant roots and soil microbes, reducing overall soil productivity (Hogarth 2017).

Altered soil structure also makes it harder for plant roots to penetrate the soil, affecting root development and the plant's ability to access nutrients and water from deeper soil layers. Over time, these physical changes to the soil can lead to declines in plant health, reduced crop yields, and an increased need for artificial inputs such as fertilizers and irrigation.

25.5.4 Long-Term Implications for Sustainable Agriculture

The accumulation of plastic in agricultural soils poses a significant challenge to the long-term sustainability of global food systems (Lakhiar et al. 2024). As plastic pollution continues to build up in soils, the negative effects on soil microflora and soil health may become more severe, leading to declines in agricultural productivity. These long-term consequences could force farmers to rely more heavily on chemical inputs, such as synthetic fertilizers and pesticides, to compensate for the reduced efficiency of natural soil processes (Wanner 2021).

Increased reliance on chemical inputs is problematic for several reasons. First, excessive use of fertilizers can lead to nutrient imbalances in the soil, causing long-term degradation of soil quality (Lin et al. 2019). Second, the overuse of pesticides can harm non-target organisms, including beneficial soil microbes, further exacerbating the decline in soil biodiversity (Kaur et al. 2024). Moreover, the production and application of chemical inputs have environmental costs, contributing to greenhouse gas emissions, water pollution, and biodiversity loss.

The long-term presence of plastics in soil also raises concerns about the persistence of plastic-associated pollutants. Microplastics have the ability to adsorb and accumulate toxic chemicals, including heavy metals and persistent organic pollutants (POPs), from their environment. These pollutants can persist in the soil for decades, creating continuous risks to soil health, plant productivity, and human health throughout the food chain (Rafa et al. 2024).

For sustainable agriculture to be achieved, addressing plastic pollution in agricultural soils is essential. Strategies such as reducing plastic use in farming, promoting the adoption of biodegradable alternatives, and improving waste management practices must be prioritized (Hofmann et al. 2023). Additionally, further research is needed to better understand the long-term impacts of plastic pollution on soil ecosystems and to develop innovative solutions for mitigating these effects.

25.6 Mitigation Strategies and Future Directions

The detrimental effects of plastic pollution on soil ecosystems, particularly soil microflora, pose a significant challenge to agricultural productivity and environmental sustainability. Microplastics can alter soil structure, hinder microbial processes, and disrupt plant-microbe interactions, all of which are vital for healthy soil functioning. To mitigate these impacts, various strategies must be implemented at multiple levels, from reducing plastic use in agriculture to encouraging research into sustainable alternatives.

25.6.1 Reducing Plastic Use in Agriculture

One of the most effective strategies for mitigating the impact of plastic pollution on soil microflora is to reduce the use of plastics in agricultural practices. Plastics are commonly used in agriculture for mulch films, irrigation tubing, seedling trays, and packaging. While these materials offer short-term benefits like improved moisture retention, weed control, and crop protection, they also contribute to long-term environmental problems when they break down into microplastics (Sa'adu and Farsang 2023).

A promising solution is the adoption of alternative materials such as biodegradable mulches. Unlike traditional plastic mulches, biodegradable mulches can break down naturally in the soil, reducing the accumulation of plastic particles over time (Li et al. 2014). These alternatives are typically made from materials like plant starches, cellulose, or other organic compounds that are compatible with soil microbial processes (Zhang et al. 2022b). However, it is essential to ensure that these biodegradable materials do not release harmful chemicals during degradation and that they effectively fulfill their agricultural functions (Sintim et al. 2020).

Additionally, promoting more sustainable farming practices that minimize plastic waste is crucial. Techniques such as conservation agriculture, which focuses on minimal soil disturbance and maintaining organic soil cover, can reduce the need for plastic mulching and other synthetic materials (Evode et al. 2021). Furthermore, improved waste management practices are necessary to prevent plastic leakage into the environment. Farmers should be encouraged to recycle plastic waste, use longer-lasting and reusable materials, and adopt practices that minimize single-use plastics in farming (Hopewell et al. 2009).

Educating farmers and agricultural stakeholders about the long-term environmental costs of plastic pollution and providing them with accessible alternatives can drive the reduction of plastic use in agriculture. Incentive programs that reward farmers for using sustainable materials or implementing waste management practices can further encourage the transition toward a low-plastic agricultural system.

25.6.2 Promoting Soil Health Through Organic Amendments

A further approach to reduce the detrimental effects of plastic pollution on soil microflora is to enhance soil health by using organic amendments like compost, manure, and biochar. These amendments can help counteract the harmful impacts of plastic by increasing the organic matter in the soil and improving its structure, creating a more conducive environment for microbial growth and activity.

Compost is abundant in nutrients and organic matter, enhancing the physical properties of soil, such as its water retention and aeration. Additionally, it boosts microbial activity by providing a consistent source of carbon and nutrients for microorganisms (Ho et al. 2022). By boosting microbial diversity and activity, compost can counteract some of the negative impacts of microplastics on soil ecosystems.

Biochar, a carbon-rich substance created through the pyrolysis of organic materials, has demonstrated benefits such as improving soil structure, enhancing nutrient retention, and boosting microbial diversity. Additionally, it has the ability to bind pollutants, thereby decreasing their bioavailability and toxicity in the soil. Applying biochar to soils contaminated with plastic may help immobilize microplastics and protect soil microbial communities from disruption (Palansooriya et al. 2022).

Both compost and biochar can contribute to a more resilient soil system that is better able to withstand the physical and chemical stresses caused by plastic pollution (Leogrande et al. 2024). By improving soil structure, organic amendments can promote better water infiltration and root penetration, reducing the need for plastic-based solutions like mulches (Cooper et al. 2020). In this way, promoting soil health through organic amendments not only helps mitigate the effects of plastic pollution but also contributes to the overall sustainability of agricultural systems.

25.6.3 Research and Innovation in Plastic Alternatives

Research and innovation in the development of biodegradable plastics and sustainable alternatives to conventional plastic materials are crucial for reducing the environmental impact of plastic pollution in soils. While biodegradable plastics are already available, there is still significant room for improvement in terms of their environmental safety, functionality, and cost-effectiveness.

Biodegradable plastics that are specifically designed for agricultural use must be able to degrade fully and safely in soil without leaving harmful residues or microplastics (Filiciotto and Rothenberg 2021). Advances in materials science, including the use of bio-based polymers derived from renewable resources such as plant starches, proteins, or microbial fermentation products, can help develop more sustainable plastic alternatives (Babu et al. 2013). Furthermore, innovations in microbial ecology may lead to the discovery of microorganisms that can efficiently degrade plastic materials, offering the potential for bioengineering new solutions to plastic waste (Yang et al. 2023).

In addition to biodegradable plastics, other innovations such as edible coatings for seeds, natural fiber mulches, and plant-based films can further reduce the reliance on traditional plastic materials in agriculture (Atiwesh et al. 2021). These alternatives should be rigorously tested to ensure that they perform well in different agricultural contexts and do not introduce unintended environmental problems.

Collaborative research efforts between material scientists, microbiologists, and agricultural experts are essential to drive innovation in this area. Government funding and private-sector investment in sustainable plastic alternatives can accelerate the development of new technologies and bring them to market faster, reducing the environmental footprint of agricultural practices.

25.6.4 Policy Interventions and Public Awareness

Tackling plastic pollution in agricultural soils necessitates coordinated policy efforts. Governments and regulatory bodies can implement measures that promote sustainable farming, reduce plastic usage, and support research on plastic's impact on soil ecosystems. Policies might include regulations limiting non-biodegradable plastics in agriculture, establishing standards for disposal and recycling, and offering financial incentives for sustainable practices. Funding for research on plastic alternatives and microplastics is essential for evidence-based policies. Public awareness campaigns are crucial for educating farmers, agricultural workers, and consumers about the risks of plastic pollution and the benefits of sustainable practices, ultimately driving demand for eco-friendly products.

25.7 Conclusion

Plastic pollution is a growing threat to soil health and agricultural productivity. The accumulation of plastics in soils, particularly microplastics, can have profound effects on soil microflora, disrupting key ecological processes and reducing soil fertility. These changes have significant implications for the sustainability of agricultural systems, particularly in the face of increasing global food demand.

Addressing the issue of plastic pollution in soils requires a multifaceted approach, including reducing plastic use in agriculture, promoting soil health through organic amendments, and advancing research into sustainable materials. Additionally, policy interventions and public awareness are critical for driving the necessary changes at both the local and global levels.

The future of agriculture depends on the health of our soils, and protecting soil microflora from the impacts of plastic pollution is essential for ensuring the long-term sustainability of food production systems. By taking proactive steps to mitigate plastic pollution and promote soil health, we can safeguard the productivity of agricultural lands and contribute to the overall resilience of global food systems.

Acknowledgments D. Das, D. Deb, and P. Paul would like to thank The Assam Royal Global University, University of North Bengal, and North Bengal St. Xavier's College, respectively, for continuous support and motivation and for providing the necessary infrastructure to conduct the present review.

References

Abbasi T, Jaafarzadeh Haghighi Fard N, Madadizadeh F et al (2023) Environmental impact assessment of low-density polyethylene and polyethylene terephthalate containers using a life cycle assessment technique. J Polym Environ 31:3493–3508. https://doi.org/10.1007/s10924-023-02806-0

Adomako MO, Roiloa S, Yu F-H (2022) Potential roles of soil microorganisms in regulating the effect of soil nutrient heterogeneity on plant performance. Microorganisms 10:2399. https://doi.org/10.3390/microorganisms10122399

Alsabri A, Tahir F, Al-Ghamdi SG (2022) Environmental impacts of polypropylene (PP) production and prospects of its recycling in the GCC region. Mater Today Proc 56:2245–2251. https://doi.org/10.1016/j.matpr.2021.11.574

Amare G, Desta B (2021) Coloured plastic mulches: impact on soil properties and crop productivity. Chem Biol Technol Agric 8:4. https://doi.org/10.1186/s40538-020-00201-8

Andreasi Bassi S, Tonini D, Saveyn H, Astrup TF (2022) Environmental and socioeconomic impacts of poly(ethylene terephthalate) (PET) packaging management strategies in the EU. Environ Sci Technol 56:501–511. https://doi.org/10.1021/acs.est.1c00761

Angulo V, Bleichrodt R-J, Dijksterhuis J et al (2024) Enhancement of soil aggregation and physical properties through fungal amendments under varying moisture conditions. Environ Microbiol 26:e16627. https://doi.org/10.1111/1462-2920.16627

Antoszewski M, Mierek-Adamska A, Dąbrowska GB (2022) The importance of microorganisms for sustainable agriculture—a review. Meta 12:1100. https://doi.org/10.3390/metabo12111100

Aralappanavar VK, Mukhopadhyay R, Yu Y et al (2024) Effects of microplastics on soil microorganisms and microbial functions in nutrients and carbon cycling—a review. Sci Total Environ 924:171435. https://doi.org/10.1016/j.scitotenv.2024.171435

Atiwesh G, Mikhael A, Parrish CC et al (2021) Environmental impact of bioplastic use: a review. Heliyon 7:e07918. https://doi.org/10.1016/j.heliyon.2021.e07918

Azeem I, Adeel M, Ahmad MA et al (2021) Uptake and accumulation of Nano/microplastics in plants: a critical review. Nanomaterials (Basel) 11:2935. https://doi.org/10.3390/nano11112935

Babu RP, O'Connor K, Seeram R (2013) Current progress on bio-based polymers and their future trends. Prog Biomater 2:8. https://doi.org/10.1186/2194-0517-2-8

Balali-Mood M, Naseri K, Tahergorabi Z et al (2021) Toxic mechanisms of five heavy metals: mercury, Lead, chromium, cadmium, and arsenic. Front Pharmacol 12:643972. https://doi.org/10.3389/fphar.2021.643972

Barbera AC, Maucieri C, Cavallaro V et al (2013) Effects of spreading olive mill wastewater on soil properties and crops, a review. Agric Water Manag 119:43–53. https://doi.org/10.1016/j.agwat.2012.12.009

Barrow NJ, Hartemink AE (2023) The effects of pH on nutrient availability depend on both soils and plants. Plant Soil 487:21–37. https://doi.org/10.1007/s11104-023-05960-5

Bian H, Li C, Zhu J et al (2022) Soil moisture affects the rapid response of microbes to labile organic C addition. Front Ecol Evol 10. https://doi.org/10.3389/fevo.2022.857185

Bodor A, Feigl G, Kolossa B et al (2024) Soils in distress: the impacts and ecological risks of (micro)plastic pollution in the terrestrial environment. Ecotoxicol Environ Saf 269:115807. https://doi.org/10.1016/j.ecoenv.2023.115807

Bonfante P, Genre A (2010) Mechanisms underlying beneficial plant–fungus interactions in mycorrhizal symbiosis. Nat Commun 1:48. https://doi.org/10.1038/ncomms1046

Cai Z, Li M, Zhu Z et al (2023) Biological degradation of plastics and microplastics: a recent perspective on associated mechanisms and influencing factors. Microorganisms 11:1661. https://doi.org/10.3390/microorganisms11071661

Cao H, Chen R, Wang L et al (2016) Soil pH, total phosphorus, climate and distance are the major factors influencing microbial activity at a regional spatial scale. Sci Rep 6:25815. https://doi.org/10.1038/srep25815

Cao H, Li Z, Chen B et al (2023) From environmental toxicants to antibacterial agents: BPA analogues selectively inhibit the growth of gram-positive bacteria by disturbing biosynthesis of the Cell Wall. Environ Health 1:291–299. https://doi.org/10.1021/envhealth.3c00092

Che Y, Jin G (2024) Plant–soil microbial diversity and structural attributes jointly dominate the multifunctionality of the temperate forest. Ecol Indic 166:112282. https://doi.org/10.1016/j.ecolind.2024.112282

Chen Y, Li Y, Liang X et al (2024) Effects of microplastics on soil carbon pool and terrestrial plant performance. Carbon Res 3:37. https://doi.org/10.1007/s44246-024-00124-1

Cobaugh KL, Schaeffer SM, DeBruyn JM (2015) Functional and structural succession of soil microbial communities below decomposing human cadavers. PLoS One 10:e0130201. https://doi.org/10.1371/journal.pone.0130201

Cooper J, Greenberg I, Ludwig B et al (2020) Effect of biochar and compost on soil properties and organic matter in aggregate size fractions under field conditions. Agric Ecosyst Environ 295:106882. https://doi.org/10.1016/j.agee.2020.106882

Custer GF, Bresciani L, Dini-Andreote F (2022) Ecological and evolutionary implications of microbial dispersal. Front Microbiol 13. https://doi.org/10.3389/fmicb.2022.855859

Cverenkárová K, Valachovičová M, Mackuľak T et al (2021) Microplastics in the food chain. Life (Basel) 11:1349. https://doi.org/10.3390/life11121349

de Souza Machado AA, Lau CW, Till J et al (2018) Impacts of microplastics on the soil biophysical environment. Environ Sci Technol 52:9656–9665. https://doi.org/10.1021/acs.est.8b02212

de Souza Machado AA, Lau CW, Kloas W et al (2019) Microplastics can change soil properties and affect plant performance. Environ Sci Technol 53:6044–6052. https://doi.org/10.1021/acs.est.9b01339

de Souza R, Ambrosini A, Passaglia LMP (2015) Plant growth-promoting bacteria as inoculants in agricultural soils. Genet Mol Biol 38:401–419. https://doi.org/10.1590/S1415-475738420150053

Dhaka V, Singh S, Anil AG et al (2022) Occurrence, toxicity ajnd remediation of polyethylene terephthalate plastics. A review. Environ Chem Lett 20:1777–1800. https://doi.org/10.1007/s10311-021-01384-8

Dolatabadian A (2020) Plant–microbe interaction. Biology (Basel) 10:15. https://doi.org/10.3390/biology10010015

Evode N, Qamar SA, Bilal M et al (2021) Plastic waste and its management strategies for environmental sustainability. Case Stud Chem Environ Eng 4:100142. https://doi.org/10.1016/j.cscee.2021.100142

Filiciotto L, Rothenberg G (2021) Biodegradable plastics: standards, policies, and impacts. ChemSusChem 14:56–72. https://doi.org/10.1002/cssc.202002044

Frindte K, Pape R, Werner K et al (2019) Temperature and soil moisture control microbial community composition in an arctic–alpine ecosystem along elevational and micro-topographic gradients. ISME J 13:2031–2043. https://doi.org/10.1038/s41396-019-0409-9

Gaur N, Sharma S, Yadav N (2024) Chapter 2: Environmental pollution. In: Garg VK, Yadav A, Mohan C et al (eds) Green chemistry approaches to environmental sustainability. Elsevier, pp 23–41

Guo J-J, Huang X-P, Xiang L et al (2020) Source, migration and toxicology of microplastics in soil. Environ Int 137:105263. https://doi.org/10.1016/j.envint.2019.105263

Hahladakis JN, Velis CA, Weber R et al (2018) An overview of chemical additives present in plastics: migration, release, fate and environmental impact during their use, disposal and recycling. J Hazard Mater 344:179–199. https://doi.org/10.1016/j.jhazmat.2017.10.014

Hao J, Chai YN, Lopes LD et al (2021) The effects of soil depth on the structure of microbial communities in agricultural soils in Iowa (United States). Appl Environ Microbiol 87:e02673–e02620. https://doi.org/10.1128/AEM.02673-20

Ho TTK, Tra VT, Le TH et al (2022) Compost to improve sustainable soil cultivation and crop productivity.

Case Stud Chem Environ Eng 6:100211. https://doi.org/10.1016/j.cscee.2022.100211

Hofmann T, Ghoshal S, Tufenkji N et al (2023) Plastics can be used more sustainably in agriculture. Commun Earth Environ 4:1–11. https://doi.org/10.1038/s43247-023-00982-4

Hogarth PJ (2017) Mangrove ecosystems☆. In: Reference module in life sciences. Elsevier

Hopewell J, Dvorak R, Kosior E (2009) Plastics recycling: challenges and opportunities. Philos Trans R Soc Lond Ser B Biol Sci 364:2115–2126. https://doi.org/10.1098/rstb.2008.0311

Hwang J, Choi D, Han S et al (2020) Potential toxicity of polystyrene microplastic particles. Sci Rep 10:7391. https://doi.org/10.1038/s41598-020-64464-9

Iheanacho S, Ogbu M, Bhuyan MS, Ogunji J (2023) Microplastic pollution: an emerging contaminant in aquaculture. Aquacult Fish 8:603–616. https://doi.org/10.1016/j.aaf.2023.01.007

Jung J, Kim J-S, Taffner J et al (2020) Archaea, tiny helpers of land plants. Comput Struct Biotechnol J 18:2494–2500. https://doi.org/10.1016/j.csbj.2020.09.005

Karlsson TM, Arneborg L, Broström G et al (2018) The unaccountability case of plastic pellet pollution. Mar Pollut Bull 129:52–60. https://doi.org/10.1016/j.marpolbul.2018.01.041

Kaur R, Choudhary D, Bali S et al (2024) Pesticides: an alarming detrimental to health and environment. Sci Total Environ 915:170113. https://doi.org/10.1016/j.scitotenv.2024.170113

Kibria MG, Masuk NI, Safayet R et al (2023) Plastic waste: challenges and opportunities to mitigate pollution and effective management. Int J Environ Res 17:20. https://doi.org/10.1007/s41742-023-00507-z

Kim K, Song I-G, Yoon H, Park J-W (2023) Sub-micron microplastics affect nitrogen cycling by altering microbial abundance and activities in a soil-legume system. J Hazard Mater 460:132504. https://doi.org/10.1016/j.jhazmat.2023.132504

Komorowicz M, Janiszewska-Latterini D, Przybylska-Balcerek A, Stuper-Szablewska K (2023) Fungal biotransformation of hazardous organic compounds in wood waste. Molecules 28:4823. https://doi.org/10.3390/molecules28124823

Kublik S, Gschwendtner S, Magritsch T et al (2022) Microplastics in soil induce a new microbial habitat, with consequences for bulk soil microbiomes. Front Environ Sci 10. https://doi.org/10.3389/fenvs.2022.989267

Kudzin MH, Piwowarska D, Festinger N, Chruściel JJ (2023) Risks associated with the presence of polyvinyl chloride in the environment and methods for its disposal and utilization. Materials (Basel) 17:173. https://doi.org/10.3390/ma17010173

Kumar R, Verma A, Shome A et al (2021) Impacts of plastic pollution on ecosystem services, sustainable development goals, and need to focus on circular economy and policy interventions. Sustain For 13:9963. https://doi.org/10.3390/su13179963

Lakhiar IA, Yan H, Zhang J et al (2024) Plastic pollution in agriculture as a threat to food security, the ecosystem, and the environment: an overview. Agronomy 14:548. https://doi.org/10.3390/agronomy14030548

Lear G, Kingsbury JM, Franchini S et al (2021) Plastics and the microbiome: impacts and solutions. Environ Microbiome 16:2. https://doi.org/10.1186/s40793-020-00371-w

Lehmann A, Zheng W, Ryo M et al (2020) Fungal traits important for soil aggregation. Front Microbiol 10:2904. https://doi.org/10.3389/fmicb.2019.02904

Leigh MB, Prouzová P, Macková M et al (2006) Polychlorinated biphenyl (PCB)-degrading bacteria associated with trees in a PCB-contaminated site. Appl Environ Microbiol 72:2331–2342. https://doi.org/10.1128/AEM.72.4.2331-2342.2006

Leogrande R, Vitti C, Castellini M et al (2024) Residual effect of compost and biochar amendment on soil chemical, biological, and physical properties and durum wheat response. Agronomy 14:749. https://doi.org/10.3390/agronomy14040749

Li C, Moore-Kucera J, Lee J et al (2014) Effects of biodegradable mulch on soil quality. Appl Soil Ecol 79:59–69. https://doi.org/10.1016/j.apsoil.2014.02.012

Li C, Li X, Yang Y et al (2022a) Degradation reduces the diversity of nitrogen-fixing bacteria in the alpine wetland on the Qinghai-Tibet plateau. Front Plant Sci 13:939762. https://doi.org/10.3389/fpls.2022.939762

Li S, Ding F, Flury M et al (2022b) Macro- and microplastic accumulation in soil after 32 years of plastic film mulching. Environ Pollut 300:118945. https://doi.org/10.1016/j.envpol.2022.118945

Li Y, Hou Y, Hou Q et al (2023a) Soil microbial community parameters affected by microplastics and other plastic residues. Front Microbiol 14. https://doi.org/10.3389/fmicb.2023.1258606

Li Z, Yang Y, Chen X et al (2023b) A discussion of microplastics in soil and risks for ecosystems and food chains. Chemosphere 313:137637. https://doi.org/10.1016/j.chemosphere.2022.137637

Li X, Jiang N, Zhang J et al (2024) Soil health hazards of di(2-ethylhexyl) phthalate: new perspectives on earthworms from different ecological niches DNA damage, gut microbial disruption and soil enzyme changes. J Hazard Mater 467:133700. https://doi.org/10.1016/j.jhazmat.2024.133700

Lin W, Lin M, Zhou H et al (2019) The effects of chemical and organic fertilizer usage on rhizosphere soil in tea orchards. PLoS One 14:e0217018. https://doi.org/10.1371/journal.pone.0217018

Liu Z, Wen J, Liu Z et al (2024) Polyethylene microplastics alter soil microbial community assembly and ecosystem multifunctionality. Environ Int 183:108360. https://doi.org/10.1016/j.envint.2023.108360

Lucas A, Herrmann S, Lucas M (2022) The role of endocrine-disrupting phthalates and bisphenols in cardiometabolic disease: the evidence is mounting. Curr Opin Endocrinol Diabetes Obes 29:87–94. https://doi.org/10.1097/MED.0000000000000712

Maier RM (2015) Chapter 16: Biogeochemical cycling. In: Pepper IL, Gerba CP, Gentry TJ (eds) Environmental microbiology, 3rd edn. Academic, San Diego, pp 339–373

Mermertaş M, Germirli Babuna F (2019) Life cycle environmental impact analysis of HDPE packaging materials for different disposal options. In: Balkaya N, Guneysu S (eds) Recycling and reuse approaches for better sustainability. Springer International Publishing, Cham, pp 55–61

Mészáros E, Bodor A, Kovács E et al (2023) Impacts of plastics on plant development: recent advances and future research directions. Plants (Basel) 12:3282. https://doi.org/10.3390/plants12183282

Mihai F-C, Gündoğdu S, Markley LA et al (2022) Plastic pollution, waste management issues, and circular economy opportunities in rural communities. Sustain For 14:20. https://doi.org/10.3390/su14010020

Mongil-Manso J, Jiménez-Ballesta R, Trujillo-González JM et al (2023) A comprehensive review of plastics in agricultural soils: a case study of Castilla y León (Spain) farmlands. Land 12:1888. https://doi.org/10.3390/land12101888

Nedwell DB (1999) Effect of low temperature on microbial growth: lowered affinity for substrates limits growth at low temperature. FEMS Microbiol Ecol 30:101–111. https://doi.org/10.1111/j.1574-6941.1999.tb00639.x

Oades JM (1993) The role of biology in the formation, stabilization and degradation of soil structure. Geoderma 56:377–400. https://doi.org/10.1016/0016-7061(93)90123-3

Odelade KA, Babalola OO (2019) Bacteria, fungi and archaea domains in Rhizospheric soil and their effects in enhancing agricultural productivity. Int J Environ Res Public Health 16:3873. https://doi.org/10.3390/ijerph16203873

Okoye CO, Addey CI, Oderinde O et al (2022) Toxic chemicals and persistent organic pollutants associated with micro-and Nanoplastics pollution. Chem Eng J Adv 11:100310. https://doi.org/10.1016/j.ceja.2022.100310

Omidoyin KC, Jho EH (2023) Effect of microplastics on soil microbial community and microbial degradation of microplastics in soil: a review. Environ Eng Res 28. https://doi.org/10.4491/eer.2022.716

Osman AI, Hosny M, Eltaweil AS et al (2023) Microplastic sources, formation, toxicity and remediation: a review. Environ Chem Lett 21:2129–2169. https://doi.org/10.1007/s10311-023-01593-3

Palansooriya KN, Sang MK, Igalavithana AD et al (2022) Biochar alters chemical and microbial properties of microplastic-contaminated soil. Environ Res 209: 112807. https://doi.org/10.1016/j.envres.2022.112807

Philippot L, Griffiths BS, Langenheder S (2021) Microbial community resilience across ecosystems and multiple disturbances. Microbiol Mol Biol Rev 85:e00026–e00020. https://doi.org/10.1128/MMBR.00026-20

Prescott CE, Frouz J, Grayston SJ et al (2019) Chapter 13: Rehabilitating forest soils after disturbance. In: Busse M, Giardina CP, Morris DM, Page-Dumroese DS (eds) Developments in soil science. Elsevier, pp 309–343

Rafa N, Ahmed B, Zohora F et al (2024) Microplastics as carriers of toxic pollutants: source, transport, and toxicological effects. Environ Pollut 343:123190. https://doi.org/10.1016/j.envpol.2023.123190

Rai M, Pant G, Pant K et al (2023) Microplastic pollution in terrestrial ecosystems and its interaction with other soil pollutants: a potential threat to soil ecosystem sustainability. Resources 12:67. https://doi.org/10.3390/resources12060067

Rillig MC, Leifheit E, Lehmann J (2021) Microplastic effects on carbon cycling processes in soils. PLoS Biol 19:e3001130. https://doi.org/10.1371/journal.pbio.3001130

Romero F, Hilfiker S, Edlinger A et al (2023) Soil microbial biodiversity promotes crop productivity and agro-ecosystem functioning in experimental microcosms. Sci Total Environ 885:163683. https://doi.org/10.1016/j.scitotenv.2023.163683

Ru J, Huo Y, Yang Y (2020) Microbial degradation and valorization of plastic wastes. Front Microbiol 11. https://doi.org/10.3389/fmicb.2020.00442

Russo M, Oliva M, Hussain MI, Muscolo A (2023) The hidden impacts of micro/nanoplastics on soil, crop and human health. J Agric Food Res 14:100870. https://doi.org/10.1016/j.jafr.2023.100870

Rybczyńska-Tkaczyk K, Skóra B, Szychowski KA (2023) Toxicity of bisphenol A (BPA) and its derivatives in divers biological models with the assessment of molecular mechanisms of toxicity. Environ Sci Pollut Res 30:75126–75140. https://doi.org/10.1007/s11356-023-27747-y

Sa'adu I, Farsang A (2023) Plastic contamination in agricultural soils: a review. Environ Sci Eur 35:13. https://doi.org/10.1186/s12302-023-00720-9

Sajjad M, Huang Q, Khan S et al (2022) Microplastics in the soil environment: a critical review. Environ Technol Innov 27:102408. https://doi.org/10.1016/j.eti.2022.102408

Samphire M, Chadwick DR, Jones DL (2023) Biodegradable plastic mulch films increase yield and promote nitrogen use efficiency in organic horticulture. Front Agron 5. https://doi.org/10.3389/fagro.2023.1141608

Santi C, Bogusz D, Franche C (2013) Biological nitrogen fixation in non-legume plants. Ann Bot 111:743–767. https://doi.org/10.1093/aob/mct048

Seeley ME, Song B, Passie R, Hale RC (2020) Microplastics affect sedimentary microbial communities and nitrogen cycling. Nat Commun 11:2372. https://doi.org/10.1038/s41467-020-16235-3

Seo Y, Zhou Z, Lai Y et al (2024) Micro- and nanoplastics in agricultural soils: assessing impacts and navigating mitigation. Sci Total Environ 931:172951. https://doi.org/10.1016/j.scitotenv.2024.172951

Sharma P, Kumar T, Yadav M et al (2021) Plant-microbe interactions for the sustainable agriculture and food security. Plant Gene 28:100325. https://doi.org/10.1016/j.plgene.2021.100325

Singavarapu B, Du J, Beugnon R et al (2023) Functional potential of soil microbial communities and their subcommunities varies with tree mycorrhizal type and tree diversity. Microbiol Spectr 11:e04578-22. https://doi.org/10.1128/spectrum.04578-22

Sintim HY, Bary AI, Hayes DG et al (2020) In situ degradation of biodegradable plastic mulch films in compost

and agricultural soils. Sci Total Environ 727:138668. https://doi.org/10.1016/j.scitotenv.2020.138668
Sobhani Z, Fang C, Naidu R, Megharaj M (2021) Microplastics as a vector of toxic chemicals in soil: enhanced uptake of perfluorooctane sulfonate and perfluorooctanoic acid by earthworms through sorption and reproductive toxicity. Environ Technol Innov 22:101476. https://doi.org/10.1016/j.eti.2021.101476
Somanathan H, Sathasivam R, Sivaram S et al (2022) An update on polyethylene and biodegradable plastic mulch films and their impact on the environment. Chemosphere 307:135839. https://doi.org/10.1016/j.chemosphere.2022.135839
Spohn M (2016) Element cycling as driven by stoichiometric homeostasis of soil microorganisms. Basic Appl Ecol 17:471–478. https://doi.org/10.1016/j.baae.2016.05.003
Stres B, Danevčič T, Pal L et al (2008) Influence of temperature and soil water content on bacterial, archaeal and denitrifying microbial communities in drained fen grassland soil microcosms. FEMS Microbiol Ecol 66:110–122. https://doi.org/10.1111/j.1574-6941.2008.00555.x
Suman J, Rakshit A, Ogireddy SD et al (2022) Microbiome as a key player in sustainable agriculture and human health. Front Soil Sci 2. https://doi.org/10.3389/fsoil.2022.821589
Sutkar PR, Gadewar RD, Dhulap VP (2023) Recent trends in degradation of microplastics in the environment: a state-of-the-art review. J Hazard Mater Adv 11:100343. https://doi.org/10.1016/j.hazadv.2023.100343
Tang KHD (2023) Microplastics in agricultural soils in China: sources, impacts and solutions. Environ Pollut 322:121235. https://doi.org/10.1016/j.envpol.2023.121235
Thushari GGN, Senevirathna JDM (2020) Plastic pollution in the marine environment. Heliyon 6:e04709. https://doi.org/10.1016/j.heliyon.2020.e04709
Tiago GAO, Mariquito A, Martins-Dias S, Marques AC (2023) The problem of polyethylene waste—recent attempts for its mitigation. Sci Total Environ 892:164629. https://doi.org/10.1016/j.scitotenv.2023.164629
Trojan M, Koutný M, Brtnický M et al (2024) The interaction of microplastics and microbioplastics with soil and a comparison of their potential to spread pathogens. Appl Sci 14:4643. https://doi.org/10.3390/app14114643
Urli S, Corte Pause F, Crociati M et al (2023) Impact of microplastics and Nanoplastics on livestock health: an emerging risk for reproductive efficiency. Animals (Basel) 13:1132. https://doi.org/10.3390/ani13071132
Verla AW, Enyoh CE, Verla EN, Nwarnorh KO (2019) Microplastic–toxic chemical interaction: a review study on quantified levels, mechanism and implication. SN Appl Sci 1:1400. https://doi.org/10.1007/s42452-019-1352-0
Viljakainen VR, Hug LA (2021) New approaches for the characterization of plastic-associated microbial communities and the discovery of plastic-degrading microorganisms and enzymes. Comput Struct Biotechnol J 19:6191–6200. https://doi.org/10.1016/j.csbj.2021.11.023
Wang Z, Li W, Li W et al (2023) Effects of microplastics on the water characteristic curve of soils with different textures. Chemosphere 317:137762. https://doi.org/10.1016/j.chemosphere.2023.137762
Wang S, Wang W, Rong S et al (2024a) Key factors and mechanisms of microplastics' effects on soil nitrogen transformation: a review. Soil Environ Health 2:100101. https://doi.org/10.1016/j.seh.2024.100101
Wang X, Chi Y, Song S (2024b) Important soil microbiota's effects on plants and soils: a comprehensive 30-year systematic literature review. Front Microbiol 15. https://doi.org/10.3389/fmicb.2024.1347745
Wanner P (2021) Plastic in agricultural soils—a global risk for groundwater systems and drinking water supplies?—a review. Chemosphere 264:128453. https://doi.org/10.1016/j.chemosphere.2020.128453
Wu CH, Bernard SM, Andersen GL, Chen W (2009) Developing microbe–plant interactions for applications in plant-growth promotion and disease control, production of useful compounds, remediation and carbon sequestration. Microb Biotechnol 2:428–440. https://doi.org/10.1111/j.1751-7915.2009.00109.x
Wu Y, Zeng J, Zhu Q et al (2017) pH is the primary determinant of the bacterial community structure in agricultural soils impacted by polycyclic aromatic hydrocarbon pollution. Sci Rep 7:40093. https://doi.org/10.1038/srep40093
Wu H, Cui H, Fu C et al (2024) Unveiling the crucial role of soil microorganisms in carbon cycling: a review. Sci Total Environ 909:168627. https://doi.org/10.1016/j.scitotenv.2023.168627
Yang X-G, Wen P-P, Yang Y-F et al (2023) Plastic biodegradation by in vitro environmental microorganisms and in vivo gut microorganisms of insects. Front Microbiol 13:1001750. https://doi.org/10.3389/fmicb.2022.1001750
Zaborowska M, Wyszkowska J, Borowik A, Kucharski J (2021) Bisphenol A—A dangerous pollutant distorting the biological properties of soil. Int J Mol Sci 22. https://doi.org/10.3390/ijms222312753
Zeenat EA, Bukhari DA et al (2021) Plastics degradation by microbes: a sustainable approach. J King Saud Univ Sci 33:101538. https://doi.org/10.1016/j.jksus.2021.101538
Zhang J, Ren S, Xu W et al (2022a) Effects of plastic residues and microplastics on soil ecosystems: a global meta-analysis. J Hazard Mater 435:129065. https://doi.org/10.1016/j.jhazmat.2022.129065
Zhang M, Xue Y, Jin T et al (2022b) Effect of long-term biodegradable film mulch on soil physicochemical and microbial properties. Toxics 10:129. https://doi.org/10.3390/toxics10030129
Zhou X, Zhai Y, Ren K et al (2023) Life cycle assessment of polycarbonate production: proposed optimization toward sustainability. Resour Conserv Recycl 189:106765. https://doi.org/10.1016/j.resconrec.2022.106765

Remediation Techniques for Soil Contaminated with Toxic Materials from E-Waste-Associated Microplastics

26

Suchitra Mitra, Mohammad Abdul Kader, R. K. Saran, and Sofia Lazar

Abstract

E-waste associated microplastics contamination of soil is a growing environmental concern because it contains toxic materials that can leach into the soil when improperly disposed or recycled. These toxic materials include heavy metals, flame retardants, and organic pollutants. Hazardous compounds can pose significant health risks to humans when they are improperly managed and released into the environmental matrices through air, water, and soil pathways. Developing countries need proper technological infrastructure and resources to manage the huge amounts of e-waste associated microplastics they. As a result, over 90% of e-waste associated microplastics in under-developed and developing countries relies on rudimentary and primitive methods. This improper treatment of e-waste associated microplastics releases hazardous pollutants into the environment, which causes serious health problems for humans. It is important to study the properties of these hazardous compounds of e-waste associated microplastics in order to develop effective remediation techniques. It is an overview of the sources and types of e-waste associated microplastics contaminants, as well as the different remediation techniques that can be used to mitigate soil contamination. The chapter also discusses the advantages and limitations of physical, chemical, and biological soil remediation techniques.

Keywords

Remediation · Soil contaminants · Heavy metals · Hazardous compounds

26.1 Introduction

E-waste is any electronic device that is no longer in use or is broken which includes a wide range of items, such as computers, televisions, cell phones, appliances, and batteries (Priya et al. 2023). It contains a number of hazardous

S. Mitra (✉)
Indian Institute of Science Education and Research, Kolkata, West Bengal, India

M. A. Kader
Center for Irrigation and Water Management, Rural Development Academy (RDA), Bogura, Bangladesh
e-mail: kader@rda.gov.bd

R. K. Saran
Department of Microbiology, Maharaja Ganga Singh University, Bikaner, Rajasthan, India

S. Lazar
Institute for Water Research, Rhodes University, Grahamstown, South Africa
e-mail: s.lazar@ensh.dz

S. Goswami et al. (eds.), *Micro-Nano Plastics Exposure, Environmental Degradation and Public Health Crisis*, Sustainable Development Goals Series, https://doi.org/10.1007/978-981-96-5047-7_26

substances, mainly plastics and associated toxic materials such as heavy metals, lead, cadmium, and mercury that are disposed of improperly, and contaminate the soil ecosystem. The plastics and associated toxic materials in e-waste can leach into the soil and groundwater, contaminating the soil and water supply for miles around. Soil contamination by plastics and associated toxic materials can have several negative impacts on human health and the environment (Kanwar et al. 2020). Hazardous substances in e-waste-associated microplastics can accumulate in the body over time and cause a variety of health problems, including neurological damage, kidney failure, and cancer (Hou et al. 2020). In addition, early lead poisoning in children may cause developmental and learning disabilities. Moreover, plastics and associated toxic materials in e-waste can harm plants and animals, and they can also pollute water footprints. Soils in areas with e-waste can get polluted with harmful substances like toxic metals and organic compounds. These include heavy metals, PAHs, and polybrominated diphenyl ethers. It is crucial to handle e-waste-associated microplastics properly to protect our environment and our health (Zhang et al. 2012). Improper e-waste-associated microplastics management causes serious environmental and human health problems because of the toxic substances associated. This issue has a disproportionate impact on a number of Asian countries (Yu et al. 2022). Therefore, it is critical to conduct regular monitoring of plastics and associated toxic materials in the environment due to their detrimental effects on human health, vegetation, animals, soil, and water.

26.2 E-Waste-Associated Plastic Contaminants in the Environment

E-waste, from a wide range of electrical and electronic equipment, exhibits a complex composition. However, e-waste comprises various other components, resulting in a diverse mix of hazardous and nonhazardous materials. Approximately, steel accounts for 50% of e-waste, while copper, aluminum, and other metals make up 13%, and plastics contribute to 21% (ETC/SCP). Plastics constitute a substantial portion of e-waste. Key types include polycarbonate (PC), acrylonitrile butadiene styrene (ABS), polystyrene (PS), polyethylene (PE), and polyvinyl chloride (PVC). These plastics are often combined with additives like flame retardants, plasticizers, and stabilizers, which improve functionality but pose environmental risks. For instance, brominated flame retardants (BFRs) in e-waste plastics are known for their persistence, bioaccumulation, and toxicity (Tian et al. 2011). E-waste consists of a multitude of substances, and approximately 1% of its total weight is made up of hazardous and toxic metals. Among these metals are valuable ones like Au (0.1%), Ag (0.2%), and even Pd (0.005%). Recoverable precious metals significantly support recycling infrastructure, whereas some other heavy metals, i.e., lead, nickel, and various plastics have postrecovery value (Cui and Zhang 2008). Microplastics derived from e-waste plastics are a significant concern due to their incorporation of toxic materials. These microplastics often carry hazardous additives such as brominated flame retardants (BFRs), heavy metals (e.g., lead, cadmium, and mercury), and phthalates used as plasticizers. These substances are designed to enhance the durability, flexibility, or fire resistance of plastics but pose severe environmental and health risks. When e-waste plastics degrade into microplastics, these toxic materials can leach into the soil and water, contaminating ecosystems (Geibig and Socolof 2004). For instance, BFRs are persistent organic pollutants known for their bioaccumulation potential and endocrine-disrupting effects, while heavy metals can impair soil microbial activity and harm plant growth. Additionally, phthalates have been linked to reproductive and developmental toxicity. The small size of microplastics increases their mobility, facilitating the transfer of these toxic substances through the food chain, potentially impacting human health (Fig. 26.1). Consequently, the presence of such toxic materials in e-waste-associated microplastics underscores the need for proper e-waste management

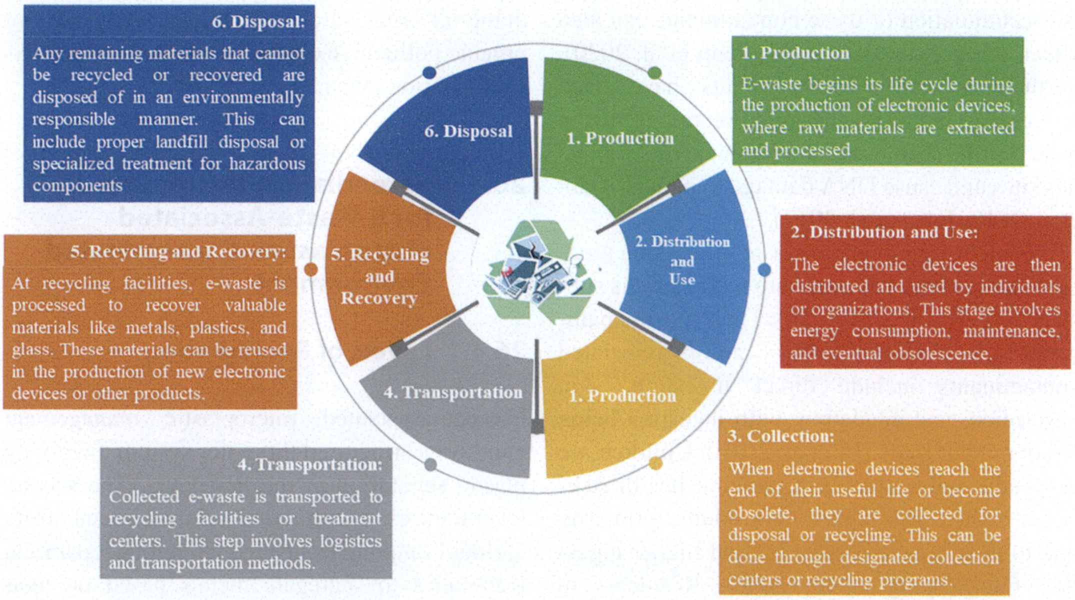

Fig. 26.1 Lifecycle assessment of e-wastes

and mitigation strategies to protect environmental and public health.

26.3 Impact of E-Waste-Associated Microplastics on Health and Environment

When e-waste-associated microplastics are not recycled properly and are illegally disposed of, it can cause extensive environmental contamination which can have significant implications for both the soil ecosystem and human health. The contamination, emanating from e-waste-associated microplastics, affects various environmental mediums, including soil, air, dust, and water, thereby impacting ecosystems and communities. This contamination is underpinned by a multitude of biogeochemical, physicochemical, geomorphological, and pedological factors (Kit et al. 2013). One major consequence of these activities is soil contamination. E-waste-associated microplastics release toxic substances into the environment, and these harmful substances can persist, posing risks to both the soil and cultivated crops. These pollutants exhibit extended environmental longevity, heightening exposure risks. The consequences extend to aquatic ecosystems, as e-waste-associated microplastics can infiltrate aquatic systems through leaching and runoff, which pollute the groundwater and surface water. Regions near e-waste recycling sites have reported high levels of pollutants in sediment and aquatic organisms, including PBDEs and PCBs (Wang et al. 2011a, b; Wu et al. 2008). Moreover, water sources have been contaminated with heavy metals like lead, surpassing safe drinking water thresholds (Wang and Guo 2006). These pollutants pose a severe threat to aquatic ecosystems, as they can accumulate in aquatic organisms, possibly entering the food chain and endangering health (Luo et al. 2011). Open burning practices associated with e-waste-associated plastics have contributed significantly to air pollution, engendering health risks among nearby populations (Chen et al. 2009). Studies have identified the accumulation of POPs like PBDEs and PCBs in soil and plants within e-waste-associated microplastics-affected areas (Liu et al. 2008; Shi et al. 2019). Consequently, this can lead to increased concentrations of these contaminants in crops and vegetables, potentially exposing humans to food intake (Fu et al. 2008; Liang et al. 2017).

Bioaccumulation of these contaminants can also affect wildlife (Li et al. 2019; Peng et al. 2020). Furthermore, these contaminants have been observed to reduce the germination rate of rice in paddy soils near e-waste-associated microplastics sites and cause DNA damage to plants (Alabi et al. 2012; Iqbal et al. 2017).

People residing near e-waste processing areas face elevated risks of exposure to hazardous substances (Awasthi et al. 2018). The exposure pathways for microplastics and associated contaminants include direct ingestion, skin absorption, and inhalation, with ingestion being the dominant route (Li et al. 2016). Children are especially vulnerable, experiencing health risks up to eight times higher than adults, primarily due to their smaller body size and higher ingestion rate (Leung et al. 2008). Residents in e-waste-associated microplastics recycling areas are at risk of heavy metal exposure. The health risks associated with these exposures are both carcinogenic and noncarcinogenic. Some heavy metals, such as Cr, Cu, Cd, and Pb, have noncarcinogenic effects, with hazard quotients (HQ) exceeding 1 (Zheng et al. 2013). Lead (Pb) has been identified as a significant carcinogenic risk (Xue et al. 2012). As consistently observed in all monitored samples, the lifetime carcinogenic risk of these metals is as follows: Cr > Ni > Pb > Cd (Zheng et al. 2013). Furthermore, residents in e-waste-associated plastics recycling areas can be exposed to polychlorinated biphenyls (PCBs) through dietary and nondietary routes, including inhalation and dermal exposure. Awareness should be raised about these risks and work toward implementing proper e-waste-associated plastics management practices to protect the health and well-being of pregnant women and their babies (Song and Li 2015). Neonates born to mothers in e-waste-associated microplastics-affected areas may have higher rates of stillbirth, low birth weight, low infant lengths, and lower Apgar scores compared to neonates from control sites (Xu et al. 2012). PCBs and dioxins from e-waste-associated microplastics exposure during the perinatal stage can increase the risk of infectious and allergic diseases in infants (Potera 2000). Moreover, children living near e-waste dumping sites are susceptible to persistent organic pollutant exposure, which can affect thyroid function (Wang et al. 2011a, b).

26.4 Remediation Techniques for E-Waste-Associated Microplastics-Contaminated Environment

26.4.1 Levels of Remediation

E-waste-associated microplastic management employs a structured three-tier system, involving plastic separation at the first level. The second level focuses on extracting valuable metals from various components, using meticulous separation techniques to segregate metals based on their physical properties. Efficient magnetic and eddy current separation in e-waste-associated microplastics management requires preparatory steps, including milling, hammering, shredding, and material screening. To separate valuable metals such as copper and aluminum from highly residual valuable metals (e.g., silver and gold), these processes employ electrostatic, gravimetric, and eddy current technologies. The effectiveness of metal recovery from these technologies relies on multiple factors including particle size, shape, feeding rate, and operational optimization. Notably, eddy current separation is most efficient with granular metals larger than 5 mm, while materials smaller than 2 mm are typically subjected to electrostatic precipitation for copper recovery. However, electrostatic separation is less efficient at sizes below 100 μm compared to column air separators. However, the third level of treatment depends on residual fractions input, technologies used, and final output. E-waste-associated microplastics feedstock undergoes processes to generate heat energy and concentrate metals, in addition to eddy current and magnetic separation. Developed nations utilize diverse techniques for recycling materials like plastics, and ferrous and nonferrous metals, yielding highly efficient and substantial results. For instance, magnetic separation can recover approximately 90–95% of ferrous metals from e-waste-associated microplastics, while eddy current separation attains over 90% recovery for nonferrous metals (MoEF 2010).

26.4.2 Remediation Techniques for E-Waste-Associated Microplastics-Contaminated Soil

Soil contamination with toxic materials associated with electronic waste (e-waste-associated microplastics) is a rising environmental concern (Yu et al. 2022). To address this issue, various remediation techniques have been developed. In this article, we will discuss several remediation methods for e-waste-associated microplastics-contaminated soil, including phytoremediation, bioremediation, chemical treatment, electrokinetic remediation, soil washing, thermal desorption, encapsulation and capping, landfill disposal, and in situ remediation (Fig. 26.2).

Remediation techniques for addressing contamination and pollutants can be broadly classified into three main categories: physical, chemical, and biological approaches. These approaches have evolved over time, offering diverse strategies to address environmental issues. The specific techniques are also distinguished by their location of application which are namely:

- *In situ*: The "in situ" method allows for the treatment of pollution right at the site where it is located. This implies that there is no requirement to relocate or dig up the soil, which can be a more efficient and cost-effective method. The treatment is conducted immediately at the location where the pollution exists, eliminating the need for soil excavation.
- *On site*: The "on site" technique involves removing the soil from the polluted area and processing it right there. This allows for careful monitoring and control throughout the process.
- *Ex situ*: With the "ex situ" method, the remediation process takes place at a separate location, away from the polluted area. This involves excavating the contaminated soil and transporting it to a processing plant for treatment.

Phytoremediation

Phytoremediation is an environmentally friendly technique for soil remediation that utilizes plants to remove or break down contaminants. This technique involves employing plants to extract or degrade pollutants from the soil (Raffa et al. 2021). While phytoremediation is a relatively new technology, it holds promise as a cost-effective approach to cleaning up contaminated soil (Awa and Hadibarata 2020). There are different methods within phytoremediation:

- Phytoextraction: This method involves the cultivation of specific plants known as hyperaccumulators in contaminated soil. These hyperaccumulators absorb and accumulate heavy metals from the soil.
- Phytostabilization: This technique uses plants to immobilize contaminants, preventing them from leaching into groundwater. It is especially effective for sites with high-metal contamination.

Bioremediation

Bioremediation utilizes microorganisms or plants to degrade or detoxify contaminants in the soil. Two primary methods of bioremediation include the following:

- Biostimulation involves adding nutrients like nitrogen and phosphorus to enhance the activity of indigenous microorganisms that can degrade contaminants.
- In bioaugmentation, specific microorganisms or plants with the ability to detoxify contaminants are introduced into the soil. These microorganisms can be naturally occurring or genetically engineered.

Physicochemical Treatments

Several chemical treatment methods are employed in soil remediation, each targeting the modification of contaminant mobility and bioavailability. Chemical stabilization relies on sub-

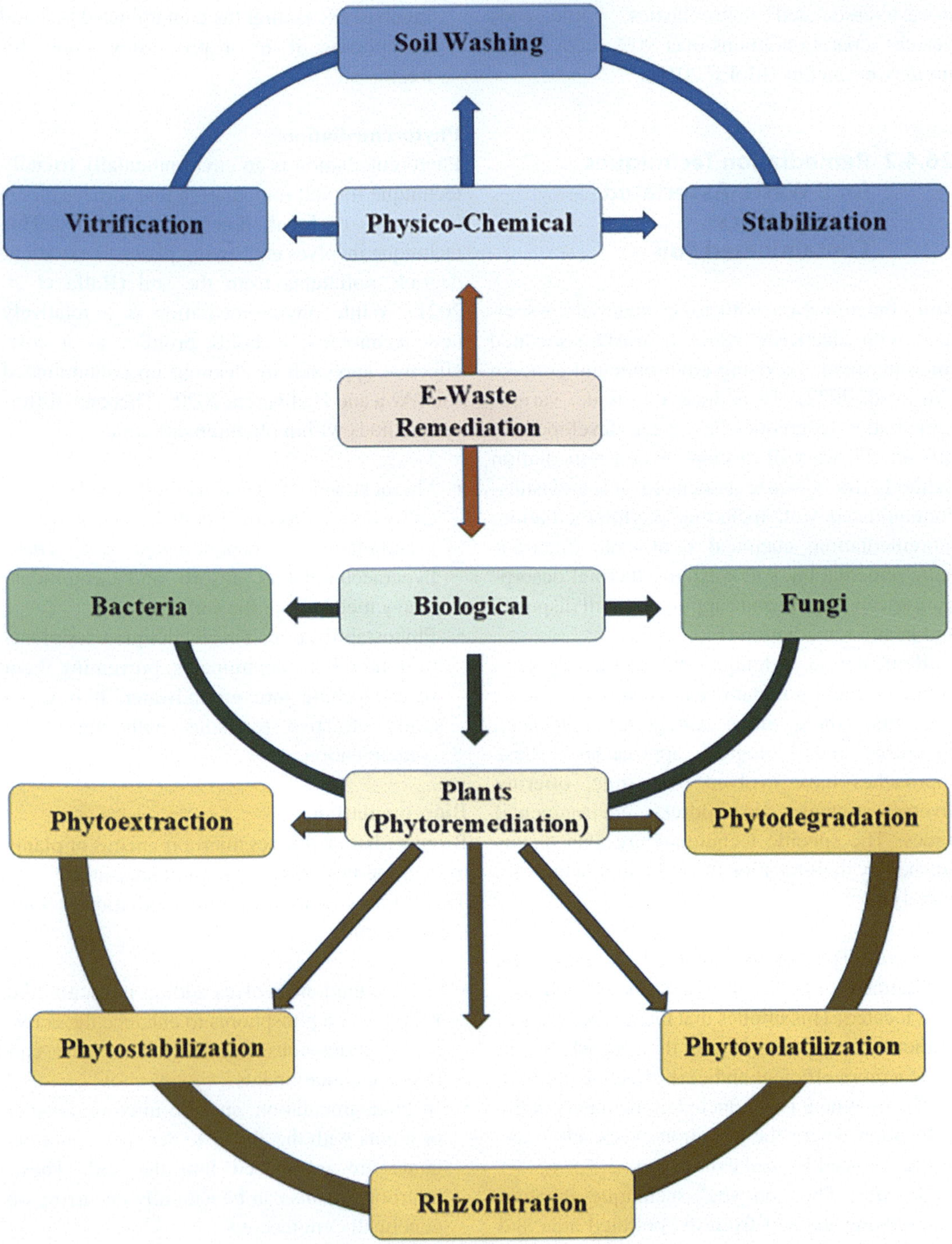

Fig. 26.2 Available techniques to remediate contaminants from e-waste-associated microplastics (Gunarathne et al. 2019)

stances such as lime, cement, and phosphate-based compounds to bind with contaminants and reduce their mobility (Oladoye et al. 2022). Meanwhile, chemical oxidation/reduction utilizes chemical agents like hydrogen peroxide or sodium persulfate to transform contaminants into lower toxic forms (Oladoye et al. 2022). Chemical immobilization involves the addition of passivators to the soil. This method is known for its cost-effectiveness and suitability for broad-scale in situ soil remediation, with particular effectiveness in preventing food chain pollution (Oladoye et al. 2022). Carbon black (CB) and modified carbon black (MCB) nanoparticles have gained traction due to their extensive surface areas and potent adsorption and chelating capabilities, rendering them valuable tools in remediating metal-contaminated soil. Electrokinetic remediation, on the other hand, relies on an electrical field to mobilize and transport contaminants within the soil. Mechanisms like electroosmosis and electrophoresis are harnessed to move contaminants toward electrodes for removal. This approach is highly effective for heavy metals and ionizable contaminants across diverse soil types and is advantageous for preserving site integrity by avoiding the need to remove contaminated soil (Awa and Hadibarata 2020; Yan et al. 2020). Soil washing is yet another method of using water/chemical solutions to remove and eliminate contaminants from the soil. Chelating agents and surfactants aid in solubilizing and mobilizing contaminants, with subsequent treatment of the leachate to remove or immobilize contaminants (Oladoye et al. 2022; Priya et al. 2023). Thermal desorption utilizes heat to vaporize contaminants from the soil, with high-temperature desorption suitable for heavily contaminated soils and low-temperature desorption for less contaminated soils (Paz-Ferreiro et al. 2018). The physical extraction of microplastics from contaminated soils is another critical aspect of remediation efforts. Techniques such as flotation using higher-density solutions, like sodium chloride, have proven effective in recovering microplastics from soil matrices (Thakur et al. 2023). This method capitalizes on the differences in density between microplastics and soil particles, allowing for the selective removal of plastic contaminants. Furthermore, the integration of filtration and membrane technologies can enhance the efficiency of microplastic extraction, although challenges remain in optimizing these systems for smaller microplastic particles (Namasivayam and Avinash 2023; Sharma et al. 2023).

26.5 Policies, Awareness, and Future Research

It is super important to come up with effective ways to clean up the environment and deal with e-waste-associated microplastics contamination. The following strategies can be used to solve the remediations of e-waste-associated microplastics contaminated in soil and water environments. One cool strategy is to use a combination of biological and physical methods, like high temperatures and ultrasonication, along with soil washing using an oil-water solvent system. This approach has been proven to be really effective in getting rid of PBDE and heavy metal contamination in mixed-contaminated sites (Ye et al. 2015). It is awesome to see that after using this method, the microbiological counts, biomass C, N, and functioning diversity all increase (Salam and Varma 2019). Cleaning up e-waste-associated microplastics sites is a complex issue that involves different people and factors like economic development, education, and environmental awareness. They are doing things like tracking e-waste-associated microplastics globally and running projects in places like Nigeria and Ethiopia (Lin et al. 2022). It is really great to see everyone coming together to tackle this environmental challenge.

26.6 Conclusion

Electronic waste (e-waste-associated microplastics) has emerged as a pressing environmental concern due to the hazardous materials it contains, which can lead to soil contamination. This chapter concludes that a comprehensive approach is necessary to effectively address soil contamination caused by e-waste-associated microplastics-associated toxic materials. This includes the implementation of stringent regulations and policies governing e-waste-associated microplastics management and disposal, responsible recycling practices, safe e-waste-associated microplastics disposal, public awareness campaigns, eco-friendly product design, and international collaborations. Furthermore, public awareness campaigns are essential for reducing the amount of contaminated soil. Educating people about the dangers of e-waste-associated microplastics and the importance of disposing of it properly is a crucial step in solving this problem. Manufacturers can contribute to reducing e-waste-associated microplastics contamination by prioritizing eco-friendly design, utilizing materials that are less harmful when discarded and easier to recycle. A shift toward more sustainable product design can significantly decrease the environmental impact of e-waste-associated microplastics. E-waste-associated microplastic is a global issue that transcends borders, and international collaboration is necessary to develop unified strategies for its management. Contaminant type and concentration, soil characteristics, site conditions, and economic considerations all play a role in determining the most suitable approach. The field of e-waste-associated microplastics soil remediation is continuously evolving, and future research is essential. Addressing soil contamination caused by e-waste-associated microplastics requires a coordinated effort from governments, businesses, researchers, and the public.

References

Alabi OA, Bakare AA, Xu X, Li B, Zhang Y, Huo X (2012) Comparative evaluation of environmental contamination and DNA damage induced by electronic-waste in Nigeria and China. Sci Total Environ 423:62–72

Awa SH, Hadibarata T (2020) Removal of heavy metals in contaminated soil by phytoremediation mechanism: a review. Water Air Soil Pollut 231(2):47

Awasthi AK, Wang M, Awasthi MK, Wang Z, Li J (2018) Environmental pollution and human body burden from improper recycling of e-waste associated microplastics in China: a short-review. Environ Pollut 243:1310–1316

Chen D, Bi X, Zhao J, Chen L, Tan J, Mai B, ... Wong M (2009) Pollution characterization and diurnal variation of PBDEs in the atmosphere of an E-waste dismantling region. Environ Pollut 157(3):1051–1057

Cui J, Zhang L (2008) Metallurgical recovery of metals from electronic waste: a review. J Hazard Mater 158(2–3):228–256

Fu J, Zhou Q, Liu J, Liu W, Wang T, Zhang Q, Jiang G (2008) High levels of heavy metals in rice (Oryzasativa L.) from a typical E-waste-associated microplastics recycling area in southeast China and its potential risk to human health. Chemosphere 71(7):1269–1275

Geibig JR, Socolof ML (2004) Life-cycle impacts of lead and lead-free solder use in reflow soldering of electronics manufacturing.

Gunarathne V, Rajapaksha AU, Vithanage M, Adassooriya N, Cooray A, Liyanage S, ... Ok YS (2019) Heavy metal dissolution mechanisms from electrical industrial sludge. Sci Total Environ 696:133922

Hou D, O'Connor D, Igalavithana AD, Alessi DS, Luo J, Tsang DC et al (2020) Metal contamination and bioremediation of agricultural soils for food safety and sustainability. Nat Rev Earth Environ 1(7):366–381

Iqbal M, Syed JH, Breivik K, Chaudhry MJI, Li J, Zhang G, Malik RN (2017) E-waste associated microplastics driven pollution in Pakistan: the first evidence of environmental and human exposure to flame retardants (FRs) in Karachi City. Environ Sci Technol 51(23):13895–13905

Kanwar VS, Sharma A, Srivastav AL, Rani L (2020) Phytoremediation of toxic metals present in soil and water environment: a critical review. Environ Sci Pollut Res 27:44835–44860

Kit BK, Fakhouri TH, Park S, Nielsen SJ, Ogden CL (2013) Trends in sugar-sweetened beverage consumption among youth and adults in the United States: 1999–2010. Am J Clin Nutr 98(1):180–188

Leung MC, Williams PL, Benedetto A, Au C, Helmcke KJ, Aschner M, Meyer JN (2008) Caenorhabditis elegans: an emerging model in biomedical and environmental toxicology. Toxicol Sci 106(1):5–28

Li F, Zhang J, Huang J, Huang D, Yang J, Song Y, Zeng G (2016) Heavy metals in road dust from Xiandao District, Changsha City, China: characteristics, health risk assessment, and integrated source identification. Environ Sci Pollut Res 23:13100–13113

Li H, Hu Y, Sun Y, De Silva AO, Muir DC, Wang W et al (2019) Bioaccumulation and translocation of tetrabromobisphenol A and hexabromocyclododecanes in mangrove plants from a national nature reserve of Shenzhen City, South China. Environ Int 129:239–246

Liang Y, Yi X, Dang Z, Wang Q, Luo H, Tang J (2017) Heavy metal contamination and health risk assessment in the vicinity of a tailing pond in Guangdong, China. Int J Environ Res Public Health 14(12):1557

Lin S, Chen XW, Cai Z, Shi J, Fu J, Jiang G, Wong MH (2022) Remediation of emerging contaminated sites due to uncontrolled e-waste associated microplastics recycling. Chem Eng J 430:133169

Liu H, Zhou Q, Wang Y, Zhang Q, Cai Z, Jiang G (2008) E-waste associated microplastics recycling induced polybrominated diphenyl ethers, polychlorinated biphenyls, polychlorinated dibenzo-p-dioxins and dibenzo-furans pollution in the ambient environment. Environ Int 34(1):67–72

Luo C, Liu C, Wang Y, Liu X, Li F, Zhang G, Li X (2011) Heavy metal contamination in soils and vegetables near an e-waste associated microplastics processing site, south China. J Hazard Mater 186(1):481–490

Ministry of Environment and Forests (2010) E-waste associated microplastics (management and handling) rules 2010. Ministry of Environment and Forests, New Delhi. http://moef.nic.in/downloads/rules-and-regulations/1035e_eng.pdf. Retrieved January 2012

Namasivayam SKR, Avinash GP (2023) Review of green technologies for the removal of microplastics from diverse environmental sources. Environ Qual Manag 33:449–465. https://doi.org/10.1002/tqem.22131

Oladoye PO, Olowe OM, Asemoloye MD (2022) Phytoremediation technology and food security impacts of heavy metal contaminated soils: a review of literature. Chemosphere 288:132555

Paz-Ferreiro J, Gascó G, Méndez A, Reichman SM (2018) Soil pollution and remediation. Int J Environ Res Public Health 15(8):1657

Peng X, Zhu Z, Xiong S, Fan Y, Chen G, Tang C (2020) Tissue distribution, growth dilution, and species-specific bioaccumulation of organic ultraviolet absorbents in wildlife freshwater fish in the Pearl River catchment, China. Environ Toxicol Chem 39(2):343–351

Potera C (2000) Exposure now sickness later early exposure to PCBs and dioxins may increase some childhood diseases. Environ Health Perspect 108(12):A574–A575

Priya AK, Muruganandam M, Ali SS, Kornaros M (2023) Clean-up of heavy metals from contaminated soil by phytoremediation: a multidisciplinary and eco-friendly approach. Toxics 11(5):422

Raffa CM, Chiampo F, Shanthakumar S (2021) Remediation of metal/metalloid-polluted soils: a short review. Appl Sci 11(9):4134

Salam MD, Varma A (2019) A review on impact of e-waste associated microplastics on soil microbial community and ecosystem function. Pollution 5(4):761–774

Sharma S, Sharma V, Chatterjee S (2023) Contribution of plastic and microplastic to global climate change and their conjoining impacts on the environment – a review. Sci Total Environ 875:162627

Shi J, Xiang L, Luan H, Wei Y, Ren H, Chen P (2019) The health concern of polychlorinated biphenyls (PCBs) in a notorious e-waste associated microplastics recycling site. Ecotoxicol Environ Saf 186:109817

Song Q, Li J (2015) A review on human health consequences of metals exposure to e-waste associated microplastics in China. Environ Pollut 196:450–461

Thakur B, Singh J, Singh J, Angmo D, Vig AP (2023) Biodegradation of different types of microplastics: molecular mechanism and degradation efficiency. Sci Total Environ 877:162912

Tian M, Chen SJ, Wang J, Zheng XB, Luo XJ, Mai BX (2011) Brominated flame retardants in the atmosphere of E-waste and rural sites in southern China: seasonal variation, temperature dependence, and gas-particle partitioning. Environ Sci Technol 45(20):8819–8825. https://doi.org/10.1021/es202284p

Wang JP, Guo XK (2006) Impact of electronic wastes recycling on environmental quality. Biomed Environ Sci 19(2):137

Wang H, Han M, Yang S, Chen Y, Liu Q, Ke S (2011a) Urinary heavy metal levels and relevant factors among people exposed to e-waste associated microplastics dismantling. Environ Int 37(1):80–85

Wang Y, Luo C, Li J, Yin H, Li X, Zhang G (2011b) Characterization of PBDEs in soils and vegetations near an e-waste associated microplastics recycling site in South China. Environ Pollut 159(10):2443–2448

Wu JP, Luo XJ, Zhang Y, Luo Y, Chen SJ, Mai BX, Yang ZY (2008) Bioaccumulation of polybrominated diphenyl ethers (PBDEs) and polychlorinated biphenyls (PCBs) in wild aquatic species from an electronic waste (e-waste associated microplastics) recycling site in South China. Environ Int 34(8):1109–1113

Xu X, Yang H, Chen A, Zhou Y, Wu K, Liu J, ... Huo X (2012) Birth outcomes related to informal e-waste recycling in Guiyu, China. Reprod Toxicol 33(1):94–98

Xue ZJ, Liu SQ, Liu YL, Yan YL (2012) Health risk assessment of heavy metals for edible parts of vegetables grown in sewage-irrigated soils in suburbs of Baoding City, China. Environ Monit Assess 184:3503–3513

Yan A, Wang Y, Tan SN, Mohd Yusof ML, Ghosh S, Chen Z (2020) Phytoremediation: a promising approach for revegetation of heavy metal-polluted land. Front Plant Sci 11:359

Ye M, Sun M, Wan J, Fang G, Li H, Hu F, ... Kengara FO (2015) Enhanced soil washing process for the remediation of PBDEs/Pb/Cd-contaminated electronic waste site with carboxymethyl chitosan in a sunflower oil–water solvent system and microbial augmentation. Environ Sci Pollut Res 22:2687–2698

Yu H, Ma J, Chen F, Zhang Q, Wang Y, Bian Z (2022) Effective remediation of electronic waste contaminated soil by the combination of metal immobilization and phytoremediation. J Environ Chem Eng 10(3):107410

Zhang WH, Ying-Xin WU, Simonnot MO (2012) Soil contamination due to e-waste associated microplastics disposal and recycling activities: a review with special focus on China. Pedosphere 22(4):434–455

Zheng J, Chen KH, Yan X, Chen SJ, Hu GC, Peng XW et al (2013) Heavy metals in food, house dust, and water from an e-waste associated microplastics recycling area in South China and the potential risk to human health. Ecotoxicol Environ Saf 96:205–212

Facemask and PPE (COVID-19)-Associated Micro- and Nanoplastic Wastes—Study of Cause and Effect

27

N. Yogeesh

27.1 Introduction

The COVID-19 pandemic, caused by the SARS-CoV-2 virus, necessitated widespread use of personal protective equipment (PPE) to mitigate viral transmission. Facemasks, gloves, and other PPE items became essential for healthcare workers and the general public, leading to an unprecedented surge in production and consumption globally. According to Shakil et al. (2020), an estimated 129 billion facemasks and 65 billion gloves were used monthly during the pandemic's peak. While these materials played a crucial role in curbing infection rates, improper disposal and lack of adequate recycling systems contributed significantly to environmental pollution (Prata et al. 2020). The lightweight and durable properties of PPE components, such as polypropylene and polyethylene, facilitate their dispersion into ecosystems, where they degrade into micro- and nanoplastics. These particles pose long-term risks to terrestrial and aquatic environments, potentially entering food chains and impacting biodiversity (De-la-Torre et al. 2021). The urgency of addressing the environmental implications of this surge in PPE usage highlights the need for sustainable alternatives and improved waste management practices. The improper disposal of personal protective equipment (PPE) during the COVID-19 pandemic has resulted in significant environmental challenges. Discarded facemasks, gloves, and other single-use PPE items have been found in terrestrial and aquatic ecosystems, contributing to an alarming increase in plastic pollution. Studies estimate that nearly 1.6 million tonnes of plastic waste per day were generated globally during the pandemic, with a substantial portion attributed to PPE (Benson et al. 2021). These materials, primarily composed of non-biodegradable polymers like polypropylene, degrade into microplastics and nanoplastics, persisting in the environment for decades (Patrício Silva et al. 2021). In aquatic ecosystems, such waste poses severe threats to marine organisms through ingestion, entanglement, and habitat disruption. Moreover, the leaching of hazardous additives from PPE can contaminate soil and water, affecting biodiversity and ecosystem health (Ammendolia et al. 2021). Despite the critical role PPE played in controlling COVID-19 transmission, its environmental footprint underscores the need for urgent intervention through effective waste management, recycling innovations, and the development of sustainable alternatives.

This chapter aims to provide a concise exploration of the environmental challenges posed by

N. Yogeesh (✉)
Department of Mathematics, Government First Grade College, Tumkur, Karnataka, India

S. Goswami et al. (eds.), *Micro-Nano Plastics Exposure, Environmental Degradation and Public Health Crisis*, Sustainable Development Goals Series, https://doi.org/10.1007/978-981-96-5047-7_27

micro- and nanoplastic wastes derived from the widespread use of facemasks and PPE during the COVID-19 pandemic. It seeks to unravel the lifecycle of these materials, from their composition and degradation into micro- and nanoplastics to their dispersion in various ecosystems. By examining the accumulation of these pollutants in soil, water, and air, the chapter evaluates their impact on biodiversity, food safety, and public health. Furthermore, it integrates global data and case studies to highlight pollution hotspots and the scale of the issue. The chapter also focuses on mitigation strategies, offering insights into sustainable alternatives, effective recycling techniques, and policy frameworks to address this emerging environmental crisis. Through this multidisciplinary approach, the chapter aspires to foster awareness, encourage policy action, and stimulate innovation toward sustainable solutions for managing PPE-associated plastic pollution.

27.2 Composition and Sources of Micro- and Nanoplastic Wastes

Facemasks and personal protective equipment (PPE) are predominantly made from synthetic polymers such as polypropylene, polyethylene, and polycarbonate, which are favored for their lightweight, durability, and cost-effectiveness (Aragaw 2020). These polymers, however, are non-biodegradable and susceptible to fragmentation under environmental stressors such as ultraviolet radiation, mechanical abrasion, and microbial activity. As these materials degrade, they release microplastics (particles <5 mm) and nanoplastics (particles <1 μm), which disperse into terrestrial and aquatic ecosystems (De-la-Torre and Aragaw 2021). Facemasks, for instance, consist of multiple layers, including a melt-blown filter and spun-bonded fabrics, which degrade into fibers that contribute significantly to microplastic waste in marine environments (Patrício Silva et al. 2021). Gloves and gowns, often made from polyethylene or nitrile, similarly fragment into microscopic particles. The improper disposal of these materials, combined with their widespread use during the pandemic, has intensified their environmental footprint, creating persistent sources of pollution with long-term ecological and health implications. Understanding the composition and degradation pathways of PPE materials is crucial for developing targeted solutions to mitigate their environmental impact.

The degradation of facemasks and personal protective equipment (PPE) into micro- and nanoplastic wastes occurs through physical, chemical, and biological processes, driven by environmental conditions. PPE materials, primarily composed of thermoplastics such as polypropylene, degrade via photodegradation, mechanical fragmentation, and oxidative weathering, leading to the formation of microscopic particles (Yang et al. 2021). Photodegradation, caused by ultraviolet (UV) radiation, breaks polymer chains into smaller fragments, often initiating the formation of microplastics in exposed environments such as landfills and surface waters (Patrício Silva et al. 2021). Mechanical abrasion, exacerbated by wind and water currents, further reduces the particle size, facilitating their transformation into nanoplastics. Additionally, biofilm formation on PPE surfaces accelerates biodegradation, albeit at a slower rate, through microbial activity that weakens polymer integrity (Aragaw 2020). The multilayered structure of facemasks, including melt-blown and spun-bonded fabrics, increases the release of fibers under environmental stress. These degradation pathways not only contribute to widespread pollution in ecosystems but also enhance the potential for these particles to infiltrate food chains and affect human health. Understanding these pathways is essential for mitigating the environmental and public health risks associated with PPE-related plastic pollution.

27.3 Biotic Degradation Mechanisms of Plastics

27.3.1 PPE-Driven Plastic Pollution in Aquatic Ecosystems

The widespread use and improper disposal of personal protective equipment (PPE) during the

COVID-19 pandemic have significantly contributed to plastic pollution in aquatic ecosystems. PPE items, such as facemasks and gloves, often end up in water bodies due to poor waste management systems, littering, and stormwater runoff (De-la-Torre et al. 2021). Once in aquatic environments, these materials degrade into micro- and nanoplastics, releasing fibers and particles that are ingested by marine organisms, including fish, crustaceans, and plankton (Ardusso et al. 2021). The physical and chemical characteristics of PPE, such as buoyancy and the presence of additives, allow them to persist and travel across freshwater and marine ecosystems, accumulating in remote and biodiverse regions (Ammendolia et al. 2021). These pollutants can disrupt food chains, affect organismal health through ingestion and entanglement, and potentially transfer toxic substances. For instance, polypropylene facemasks have been identified as a major source of plastic fibers in coastal waters, contributing to significant biodiversity loss and ecological imbalance (Prata et al. 2020). Addressing PPE-driven aquatic pollution requires urgent attention to improve disposal practices and promote the adoption of biodegradable alternatives.

27.3.2 Soil Contamination and Terrestrial Impact

The improper disposal of personal protective equipment (PPE), particularly facemasks and gloves, has led to significant plastic pollution in terrestrial environments. These materials, primarily made from non-biodegradable polymers such as polypropylene and polyethylene, can persist in soils for decades, contributing to the accumulation of microplastics (Patrício Silva et al. 2021). When PPE items are discarded in landfills or left in public spaces, they break down through weathering processes like wind erosion, UV radiation, and mechanical abrasion, releasing smaller plastic particles into the soil (Prata et al. 2020). These microplastics, which can range in size from a few micrometers to several millimeters, contaminate the soil and affect its quality by altering physical properties such as porosity and water retention (Wang et al. 2019). Moreover, the leaching of harmful chemicals, including additives used in the production of PPE, may disrupt soil microbial communities and negatively impact plant growth (Benson et al. 2021). The terrestrial spread of PPE-driven plastic pollution also increases the risk of these particles entering the food chain through direct ingestion by animals or through contaminated plants, highlighting the need for comprehensive waste management strategies and the development of eco-friendly alternatives.

27.3.3 Airborne Micro- and Nanoplastic Particles

The rapid increase in the use of personal protective equipment (PPE) during the COVID-19 pandemic has not only led to land and water pollution but also contributed to airborne plastic pollution. Facemasks, gloves, and other single-use PPE items, when exposed to environmental stressors like wind, UV radiation, and mechanical disturbances, can release microplastics and nanoplastics into the atmosphere (Patrício Silva et al. 2021). These tiny plastic particles, which are often a result of the degradation of PPE materials like polypropylene and polyethylene, can be transported over long distances by wind currents, eventually settling in remote regions (Prata et al. 2020). Airborne microplastics pose significant concerns as they can be inhaled by both humans and animals, potentially leading to respiratory issues and other health problems (De-la-Torre et al. 2021). Studies have shown that facemasks, when improperly discarded or degraded in situ, contribute to the generation of microplastic fibers that become suspended in the air (Benson et al. 2021). The accumulation of these particles in the atmosphere highlights the need for effective waste management systems and the development of biodegradable or recyclable alternatives to mitigate the environmental and health risks associated with PPE-driven plastic pollution.

27.3.4 Impact on Ecosystems and Public Health

PPE-driven plastic pollution, particularly from facemasks and gloves, poses significant threats to ecosystems and public health through bioaccumulation and food chain disruptions. As PPE materials such as polypropylene and polyethylene degrade into microplastics and nanoplastics, they are ingested by a variety of marine and terrestrial organisms, ranging from plankton to fish, and even larger animals (Prata et al. 2020). This ingestion of plastic particles can lead to bioaccumulation, where these toxic substances accumulate in the tissues of organisms, potentially impacting growth, reproduction, and overall health (De-la-Torre et al. 2021). As the microplastics move up the food chain, they can affect human health, either directly through consumption of contaminated seafood or indirectly through soil contamination affecting agriculture (Patrício Silva et al. 2021). The release of harmful additives, such as plasticizers and flame retardants, from PPE items further exacerbates the risk, as these chemicals can leach into ecosystems and disrupt hormonal and immune systems in wildlife and humans alike (Benson et al. 2021). This contamination cycle underscores the urgent need for sustainable solutions to manage PPE waste, as the long-term consequences of microplastic pollution on both ecosystems and human health remain a significant concern.

27.4 Case Studies and Global Data: Analysis of PPE Waste Hotspots

The global surge in PPE usage during the COVID-19 pandemic has led to significant plastic pollution, particularly in regions with high population density and inadequate waste management infrastructure. Case studies from countries such as India, the Philippines, and Italy have highlighted PPE waste hotspots, where improper disposal practices and the lack of effective waste management systems have exacerbated environmental pollution (De-la-Torre et al. 2021). In coastal regions, such as Southeast Asia, large quantities of PPE have been found along shorelines, contributing to the accumulation of microplastics in marine environments (Prata et al. 2020). In densely populated urban areas, PPE waste is often discarded in public spaces or improperly disposed of, resulting in high concentrations of plastics in landfills and municipal waste streams (Benson et al. 2021). Additionally, data from the United Nations Environment Programme (UNEP) estimates that over 129 billion face masks and 65 billion gloves were used globally each month during the peak of the pandemic, with a significant proportion of this waste entering natural ecosystems (Patrício Silva et al. 2021). These case studies and global data emphasize the need for comprehensive waste management strategies, including the adoption of sustainable alternatives and more robust recycling programs to address PPE-driven plastic pollution in identified hotspots.

- *India*—India, with its large population and rapidly urbanizing landscape, has struggled to manage the enormous surge in PPE waste during the pandemic. A case study in New Delhi, one of the most populated cities in India, revealed that improper disposal of PPE, including masks, gloves, and face shields, contributed to a considerable increase in plastic waste (Cubas et al. 2023). A significant proportion of this waste found its way into open landfills and rivers due to the lack of proper segregation and disposal systems. The Yamuna River, one of the most polluted rivers in India, was reported to have significant amounts of PPE waste, further contributing to the microplastic contamination in the water. Additionally, many rural and urban areas experienced insufficient waste collection and processing systems, which meant that a large portion of PPE waste was either incinerated improperly or dumped in the open environment (Patrício Silva et al. 2021). The absence of a formal collection and recycling framework, along with a lack of public awareness about the risks of PPE pollution, intensified the challenge.

- *The Philippines*—The Philippines, an archipelago with an already strained waste management infrastructure, also became a hotspot for PPE-driven plastic pollution. The country has experienced a notable increase in PPE waste during the pandemic, with millions of face masks and gloves entering the waste stream. Research has highlighted the surge in waste within coastal communities and informal settlements in cities like Manila, where waste management systems were not designed to handle such high volumes of pandemic-related materials (Ardusso et al. 2021). A study conducted by the Ocean Conservancy in 2020 found that discarded PPE, including masks and face shields, contributed significantly to marine plastic pollution, particularly in coastal areas where most of the population lives. These items often end up in landfills, or worse, in waterways and oceans, where they fragment into microplastics. Additionally, many residents in these areas resorted to improper disposal methods, such as throwing PPE into streets or rivers, thereby exacerbating the issue. The Philippines, already dealing with high levels of marine pollution from plastic waste, faced the added pressure of PPE waste, which further strained the local environment and aquatic biodiversity (Prata et al. 2020).
- *Italy*—Italy, one of the first countries in Europe to face a severe COVID-19 outbreak, confronted significant challenges with PPE waste, particularly in cities such as Rome and Milan. With the rapid increase in mask and glove usage, waste management systems became overwhelmed, especially during the peak of the pandemic. A report by the Italian Ministry of Environment in 2021 estimated that the country had seen a dramatic rise in medical waste, most of which was not segregated properly (De-la-Torre et al. 2021). The improper disposal of PPE led to high volumes of plastic waste in public spaces, including parks, beaches, and along city streets. In particular, large quantities of PPE waste were found in the Mediterranean Sea, impacting marine ecosystems. Italy's waste management infrastructure, already challenged by high volumes of non-medical plastic waste, struggled to cope with the sudden influx of PPE waste. In addition to municipal waste challenges, the recycling and treatment of PPE waste were limited due to the contamination risks associated with handling medical materials. This created further problems in managing the waste in an environmentally sound manner, contributing to the accumulation of microplastics in the environment.

27.5 Mitigation Strategies and Policy Recommendations

To mitigate the environmental impacts of PPE-driven plastic pollution, it is essential to transition toward sustainable alternatives to conventional PPE materials. One promising approach is the development of biodegradable or compostable PPE options made from natural fibers, such as cotton, hemp, or biodegradable plastics derived from renewable sources like cornstarch or algae (Patrício Silva et al. 2021). These alternatives can reduce the persistence of PPE in the environment and decrease the release of microplastics into ecosystems. Additionally, enhancing recycling and waste management systems is crucial for minimizing the environmental footprint of PPE. Policies that mandate the recycling of PPE materials, alongside the implementation of extended producer responsibility (EPR) schemes, can incentivize manufacturers to design products that are easier to recycle and less harmful to the environment (Prata et al. 2020). Public awareness campaigns are also necessary to educate individuals about proper PPE disposal methods and the benefits of using reusable options, such as fabric masks or face shields (Benson et al. 2021). Governments should also prioritize research into alternative materials and technologies for PPE production while integrating environmental considerations into public health guidelines. By combining these strategies, it is possible to reduce the burden of PPE pollution on both the environment and public health. Effective recycling and waste management techniques are crucial to mitigate the environmental impact of PPE waste generated during the COVID-19 pandemic. However, the heterogeneity of PPE materials, such as polypropylene, polyethylene, and nitrile, presents significant challenges for recycling. Traditional recycling processes are not equipped to handle contaminated or mixed medical wastes, which include masks, gloves, and face shields, due to hygiene concerns and material complexity (Prata et al. 2020). One promising approach is the development of specialized recycling technologies tailored for PPE materials. These include advanced pyrolysis and chemical recycling methods, which break down PPE plastics into reusable monomers or energy sources, thereby reducing landfill disposal and microplastic generation (Benson et al. 2021). Furthermore, some researchers are exploring the feasibility of integrating PPE waste into composite materials for construction or infrastructure projects, providing a second life for otherwise discarded items (Patrício Silva et al. 2021). To improve waste management, countries are encouraged to adopt extended producer responsibility (EPR) frameworks, which hold manufacturers accountable for the end-of-life disposal of their products. Public health agencies can also promote the use of reusable PPE options, such as washable face shields and cloth masks, thereby reducing the demand for single-use items. Effective waste management and recycling policies, combined with innovative technological solutions, are key to reducing the environmental footprint of PPE waste.

27.6 Conclusion and Future Directions

The COVID-19 pandemic has significantly increased the global reliance on personal protective equipment (PPE), leading to an alarming surge in plastic pollution. Improper disposal and inadequate waste management systems in countries like India, the Philippines, and Italy have exacerbated the environmental impact of PPE, contributing to land, water, and air pollution, with long-term consequences for ecosystems and public health (Prata et al. 2020). The degradation of PPE materials into microplastics poses severe risks of bioaccumulation in the food chain, threatening biodiversity and human health (Patrício Silva et al. 2021). Addressing these challenges requires urgent and innovative solutions, including the development of sustainable alternatives to conventional PPE materials, such as biodegradable plastics and reusable PPE options (Benson et al. 2021). Recycling and waste management techniques, including advanced chemical recycling and pyrolysis, offer promising solutions to mitigate the environmental footprint of PPE waste (Patrício Silva et al. 2021). Additionally, the implementation of extended producer responsibility (EPR) schemes and public awareness campaigns can significantly improve waste disposal practices and reduce the accumulation of PPE waste in ecosystems. By embracing these innovative solutions and policies, we can mitigate the environmental impact of PPE pollution and work toward a more sustainable future.

References

Ammendolia J, Saturno J, Brooks AL, Jacobs S, Jambeck JR (2021) An emerging source of plastic pollution: environmental presence of plastic personal protective equipment (PPE) debris related to COVID-19 in a metropolitan city. Environ Pollut 269:116160. https://doi.org/10.1016/j.envpol.2020.116160

Aragaw TA (2020) Surgical face masks as a potential source for microplastic pollution in the COVID-19 scenario. Mar Pollut Bull 159:111517. https://doi.org/10.1016/j.marpolbul.2020.111517

Ardusso M, Forero-López AD, Buzzi NS, Spetter CV, Fernández-Severini MD (2021) COVID-19 pandemic repercussions on plastic and antiviral polymeric textile causing pollution on beaches and coasts of South America. Sci Total Environ 763:144365. https://doi.org/10.1016/j.scitotenv.2020.144365

Benson NU, Bassey DE, Palanisami T (2021) COVID pollution: impact of COVID-19 pandemic on global plastic waste footprint. Heliyon 7(2):e06343., ISSN 2405-8440. https://doi.org/10.1016/j.heliyon.2021.e06343

Cubas ALV, Moecke EHS, Provin AP, Dutra ARA, Machado MM, Gouveia IC (2023) The impacts of plastic waste from personal protective equipment used during the COVID-19 pandemic. Polymers (Basel) 15(15):3151. https://doi.org/10.3390/polym15153151. PMID: 37571045; PMCID: PMC10421242

De-la-Torre GE, Aragaw TA (2021) What we need to know about PPE associated with the COVID-19 pandemic in the marine environment. Mar Pollut Bull 163:111879. https://doi.org/10.1016/j.marpolbul.2020.111879

De-la-Torre GE, Pizarro-Ortega CI, Dioses-Salinas DC, Ammendolia J, Okoffo ED (2021) Investigating the current status of COVID-19 related plastics and their potential impact on human health. Curr Opin Toxicol 27:47–53. https://doi.org/10.1016/j.cotox.2021.08.002

Patrício Silva AL, Prata JC, Walker TR, Duarte AC, Ouyang W, Barcelò D, Rocha-Santos T (2021) Increased plastic pollution due to COVID-19 pandemic: challenges and recommendations. Chem Eng J 405:126683. https://doi.org/10.1016/j.cej.2020.126683

Prata JC, Silva ALP, Walker TR, Duarte AC, Rocha-Santos T (2020) COVID-19 pandemic repercussions on the use and management of plastics. Environ Sci Technol 54(13):7760–7765. https://doi.org/10.1021/acs.est.0c02178

Shakil MH, Munim ZH, Tasnia M, Sarowar S (2020) COVID-19 and the environment: a critical review and research agenda. Sci Total Environ 745:141022., ISSN 0048-9697. https://doi.org/10.1016/j.scitotenv.2020.141022

Wang J, Liu X, Li Y, Powell T, Wang X, Wang G, Zhang P (2019) Microplastics as contaminants in the soil environment: a mini-review. Sci Total Environ 691:848–857. https://doi.org/10.1016/j.scitotenv.2019.07.209

Yang H, Chen G, Wang J (2021) Microplastics in the marine environment: sources, fates, impacts and microbial degradation. Toxics 9(2):41. https://doi.org/10.3390/toxics9020041. PMID: 33671786; PMCID: PMC7927104

Micro- and Nanoplastics Impact on Food Safety and Security

28

Neelesh Kumar Maurya, Neha Kumari,
Latika Yadav, Poonam Maurya, and Sanjay Kumar

28.1 Introduction

Plastic has emerged as one of the most significant challenges to the environment that we are currently confronted with. About 3.5 million metric tons per year are being produced as garbage from plastic in India. It is anticipated that the total amount of waste, including municipal solid waste, waste made of plastic, and waste from automobiles, will be three times as high by the year 2025 (Padmalosan et al. 2023). Recycling only accounts for a small fraction of plastic usage. A clarion appeal for various activities for managing plastic trash in the country is required due to the enormous volumes of plastic garbage leaking out. In this light, let us examine the problems that are related to waste plastic and the potential solutions. Plastic, which is durable, nonreactive, and lightweight, provides various benefits to businesses, customers, and other links in the social chain (Greene et al. 2022; Ragasri and Sabumon 2023). This is entirely due to the fact that it is both inexpensive and adaptable. Plastics are an important part of the healthcare business because they help keep things sanitary. Single-use plastics characterize each and every syringe and surgical instrument. In the automotive industry, it has enabled a significant decrease in vehicle weight, which has led to a reduction in fuel consumption and, as a consequence, the influence that automobiles have on the environment. Helmets made of plastic protect our heads from harm (Silva et al. 2021; Pironti et al. 2021).

28.2 Microplastics: A Brief Recall

Given How Pervasive Plastic Has Become From our toothbrushes to debit cards—India and the rest of the world are having trouble disposing of the vast versatilities of plastic waste it generates. Plastic waste is not being collected

N. K. Maurya (✉)
Department of Nutrition and Dietetics, School of Allied Health Science, Sharda University, Greater Noida, India

N. Kumari
Centre of Advanced Study in Botany, Institute of Science, Banaras Hindu University, Varanasi, India

L. Yadav
Department of Home Science, Vijay Singh Pathik Government Post Graduate Degree College (Affiliated to Maa Shakumbhari University, Punwarka, Saharanpur), Kairana, Shamli, Uttar Pradesh, India

P. Maurya
Shambhunath Institute of Pharmacy, Prayagraj, Uttar Pradesh, India

S. Kumar
Division of Nephrology, Mayo Clinic, Rochester, MN, USA
e-mail: sanjay.kumar7@sharda.ac.in

S. Goswami et al. (eds.), *Micro-Nano Plastics Exposure, Environmental Degradation and Public Health Crisis*, Sustainable Development Goals Series, https://doi.org/10.1007/978-981-96-5047-7_28

at the rate of slightly more than 10,000 tons/day (Chow et al. 2023).

Unsustainable Packaging The packaging business in India is the country's largest consumer of plastics. An unsustainable approach to packaging in India is expected to result in a loss of over 133 billion dollars' worth of the value of plastic material over the course of the next 10 years, according to a report that will be published in 2020 (FICCI Report 2023). Packaging that is not sustainable is universal packaging and uses plastic that is only used once. Even while it is limited to large cities, the surge in popularity of online retail and food delivery apps is one factor leading to an increase in the amount of waste plastic (Kumar et al. 2023). According to recent reports, the two largest online delivery firms in India, Swiggy and Zomato, deliver over 28 million orders every month. E-commerce businesses have also come under fire for their wasteful usage of plastic packaging materials. Destabilizing the Food Chain Polluting plastics, such as plankton, can affect some of the smallest organisms on the planet. When these organisms become toxic owing to microplastic ingestion, they may adversely affect the entire ecosystem, including humans who consume them as a food source. Microplastics can cause damage to animals' livers, reproductive systems, and gastrointestinal tracts; larger fragments, such as plastic bags and straws, can suffocate and starve marine life (Silva et al. 2021). Larger items, such as these, can also have a direct impact on the blue economy. Research that revealed the presence of microplastics in 90% of the sample presented water was published by the World Health Organization in 2018, and it was stunning. Humans inhale microplastic particles present in the air, highlighting an emerging environmental and public healthconcern. We also breathe in plastic as a residue resource waste management, which includes burning trash outdoors in the open air. Human hormonal imbalances caused by plastic toxicity can affect pregnancy, delivery, and reproduction (Greene et al. 2022; Ragasri and Sabumon 2023; Kumar et al. 2023).

28.3 Food, Food Safety, and Security

One of the most fundamental requirements for living is the consumption of food. Food contains the nutrients needed for the development, maintenance, and repair of bodily tissues as well as the control of essential functions (Chow et al. 2023). Material is required by an organism's body to support growth, repair, and other crucial processes and supply energy. This material is most commonly of proteins, carbohydrates, and fats. It is possible to determine whether or not a food is "balanced" by determining whether or not it comprises a range of food kinds in the appropriate amounts and proportions. This guarantees that the body obtains the required quantity of calories, proteins, minerals, vitamins, and other nutrients. It also leaves room for a small supply of additional nutrients that can assist the body in getting through the brief period of reduced calorie consumption. In addition, a diet that is considered to be balanced should contain bioactive phytochemicals such as dietary fiber, antioxidants, and nutraceuticals, all of which are advantageous to one's health in a variety of different ways. In a diet that is thought to be balanced, the proportion of calories that come from carbohydrates, proteins, and fats should, respectively, be between 60% and 70%, 10% and 12%, and 20% and 25% while monitoring persons weight and keeping a healthy energy level (Chen et al. 2022; Ramos-Jiménez et al. 2023). A diet that contains nutrients will not only help a person feel better in general but will also provide a person with more energy and make it simpler for the person to manage stress. The bulk of social and cultural events cannot function without providing some form of sustenance. In addition to positively impacting, one's health, it also can strengthen the bonds of friendship and community between individuals. Food security and safety are two interdependent components that support an environmentally conscious future. In order to achieve sustainability over the long term, this paper will contend that the objectives of food security and safety must be aligned, and the trade-offs between these three objectives must be controlled carefully and supported by data. Additionally, this paper will argue that achieving sustainability requires aligning the goals of food safety and security (Ramos-Jiménez et al. 2023). We require creative solutions for our future food security and sustainability in order to achieve the Sustainable Development Goals (SDG), which include the eradication of hunger and poverty, sustainable land use, clean water, climate change mitigation, responsible production and consumption, and sustainable life on land and in the water that does not compromise food safety (Silva et al. 2021; Pironti et al. 2021). It is possible to achieve sustainability and food security through a variety of methods, such as reducing the amount of food lost or wasted, increasing the number of plant-based foods one consumes, or recycling food and food products. The trade-offs between food security and safety are challenging. For instance, building nutrient-recycling circular food production systems could result in a cycle of pathogen accumulation. This presents a challenge when attempting to balance food safety and security. In order to attain food security, the methods and techniques that are used must not only be sustainable but also correspond with food safety and public health standards (Bai et al. 2023).

28.4 Micro-Nanoplastics in Food and Associated Products

The microplastic compounds found in food include those that are purposefully added by manufacturers, such as fillers and stabilizers, and those that collect as byproducts, such as residues and impurities (Domenech et al. 2023; Kumar et al. 2023). The following are examples of frequent types of microplastics found in food:

1. Epoxy resins and polycarbonate plastics, which are utilized in many consumer goods, are produced using BPA. Manufacturers frequently utilize polycarbonate plastics to create containers for storing food and beverages, including water bottles.
2. Dioxin is a pollutant that can be regarded as an ubiquitously persistent environmental pollutant in the ecosystem and is a result of the production of herbicides and the bleaching of paper.
3. Phthalates are found in many different forms of food packaging. They are responsible for giving plastics more flexible, transparent, and long-lasting properties.
4. Polyethylene and polypropylene are the two polymers that are most commonly found in food as well as in the environment. These two plastics are also responsible for the light-weight and long-lasting nature of packaging.
5. BPA and BPF are examples of microplastics that can be detected in food, along with mono-(3-carboxy propyl), mono-(carboxy isononyl), and mono-(carboxyisoctyl).

28.5 The Presence of Plastic in the Food Chain as well as the Accumulation of Pollutants in Living Organisms

When plastic gets into the environment, it has a tendency to attach to other contaminants that are already there. Through a process referred to as bioaccumulation, the associated toxins from plastic that makes its way through the food chain can also make their way through the food chain and accumulate in the fat and tissue of animals. In addition, throughout the manufacturing process, chemicals are frequently added to plastic to endow it with certain properties that are desired. Even when the plastic is inside an animal's body, these chemicals can leach out of the plastic and cause harm (as Table 28.1 explained) (Bai et al. 2023).

28.5.1 Diffusion of Toxins Along the Food

Toxins like PCBs and dioxins are not the only contaminants that can make their way up the food chain; plastic is not even the primary culprit. Compared to the exposure that occurs through the animals' regular diet, plastic's contribution to

Table 28.1 Occurrence of microplastics in other food items (Liebezeit and Liebezeit 2014; Iñiguez et al. 2017; Karami et al. 2017; Kosuth et al. 2018)

Food items	Levels of mp	Size range	Site
Salts	17; 94% 1–10 particles/kg of salt	>149 μm	Australia, France, Iran, Japan, Malaysia, New Zealand, Portugal, and South Africa from local supermarkets
	21; 100% 50–280 particles/kg of salt	10–3500 μm	Spanish salt producers
Beer	24; 100% 2–79 fibers/L, 12–109 fragments/L, 2–66 granules/L	Not specified	Germany from local supermarkets
	12; 100% 0–14.3 particles/L	100–5000 μm	USA purchased from Minneapolis, Duluth, Alpena, Michigan, and Rochester (liquor stores, breweries)
Sugar	5; 100% 217 ± 123 fibers/kg of sugar 32 ± 7 fragments/kg of sugar	10–20 μm	From local supermarkets
Honey	19; 100% 166 ± 147 fibers/kg of honey 9 ± 9 fragments/kg of honey	10–20 μm	Germany, France, Italy, Spain, and Mexico from local supermarkets or producers
Drinking water Mineral water	38; 100% 2–44 particles/L in single-use plastic bottles	1–500 μm	Grocery stores in Germany

the bioaccumulation of poisons is quite minimal. Animals which excrete plastic that they have swallowed may actually be doing their bodies a favor by cleaning themselves since the toxins in their bodies have linked themselves to the plastic (Menéndez-Pedriza et al. 2022).

28.5.2 Recycled Plastic Wastes, Black Plastics, and Food Safety

Plastic is recycled from black electrical equipment housing. Additionally, Plymouth University found bromine, antimony, and lead in it. We should think about black plastics. It is utilized in disposable cutlery, coffee cup lids, and food packaging trays. Consumers should avoid it at all costs. Black plastics are unrecyclable. They could also contain unsafe levels of hazardous substances including heavy metals and flame retardants. Black plastic pollution is very harmful. The most widely used plastic is black. Again, existing plastic sorting machinery cannot detect black plastics (Guo et al. 2022), making black plastic recycling problematic. Low recycling rates mean lots of black plastic ends up in the environment as garbage. Over time, they produce more microplastics than white plastics. Future research and development are needed to analyze and control black microplastics' environmental/health risks: (1) reduce the manufacturing of black plastic products and ban single-use food containers and mulch; (2) create a new detectable and harmless black pigment; (3) switch to more permanent substitutes and materials; (4) create a new recycling system involving new separation sorting methods (Hossain et al. 2021); (5) establish a comparable and harmonized sampling and analytical approach for the amount of environmental microplastics and nanoplastics; (6) establish a standard reporting dossier containing complete information on microplastic characteristics, special chemical information; (7) assess ecological toxicity and human health efficacy (As, plasticizers, flame retardants, black dyes) leaked from black plastic items; and (8) collaborate with management, policymakers, industry, researchers, and the public to avoid and limit leakage (Yuan et al. 2022).

28.5.3 Toxic Components from Plastic Wastes in Vegetables, Eggs, and Meat

Plastic waste toxins are detected in vegetables, milk, eggs, chicken, fish, and meat. Currently, the presence of nanoplastics in foods is a growing concern. Small plastic particles under 5 mm are called microplastics. Primary microplastics, like microbeads in personal care items, are minuscule. Secondary microplastics, on the other hand, begin as plastic bags and disintegrate into microplastic particles. Microplastics in the marine environment have been studied extensively (Ugoeze et al. 2021; Lin et al. 2022). Plastic garbage can pollute our waters. It "breaks down" into tiny particles and creates microplastic. Sunlight's UV radiation degrades this. Light and buoyant microplastics can travel far on ocean currents. They are found on beaches, in arctic sea ice, and in deep-sea sediments. Fish, seagulls, and other marine animals mistake microplastics for food. Once consumed by aquatic creatures, they may enter the human food chain (Menéndez-Pedriza et al. 2022). Multiple factors impact the migration of plastic components from packaging materials into food, including temperature, contact time, food composition, and the physicochemical properties of the polymers. Most polymers used in packaging materials can alter under external circumstances, releasing chemical components. UV radiation, high temperatures, and pH variations cause polymer-specific physicochemical changes (Ugoeze et al. 2021; Lin et al. 2022). The quality of the plastic material, food properties, and contact time all affect plastic component migration. Water concentrations increased two aldehydes and antimony (Sb), but in vitro bioassays showed no increase in water extracts' cytotoxicity, genotoxicity, or estrogenic activity. Food cooked in polythene bags released lead (Pb), cadmium (Cd), chromium (Cr), and other heavy metals at 7%, 16%, 8%, and 48%, respectively. Cobalt (Co), respectively, resulting in food concentrations of up to 121, 12, 9.5, and 15 ppm after 5 h exposure at 95 °C, mainly on the food surface (Ebrahimi et al. 2022).

28.5.4 Microplastics in Various Foods

Italian researchers found microplastic in fruits and vegetables. Apples and carrots contain the most microplastics. However, pears, broccoli, lettuce, potatoes, radishes, and turnips contain microplastics. Plant roots absorb microplastic-contaminated water, contaminating fruits and vegetables. Microplastic enters plant shoots, forming edible parts. Fruit is more contaminated than vegetables. Vegetable plants have shallower, less developed root systems than fruit trees, which absorb more water (Ugoeze et al. 2021; Lin et al. 2022). A list of MO levels in different food items has been given in Table 28.1.

28.5.4.1 Salts

Researchers found microplastics in 90% of global table salt brands. If so, human seasonings may include microplastics, even if the dish does not. Salt contains microplastics worldwide (Zhang et al. 2022). Thus, no salt is microplastic-free. Rock salt has the fewest microplastics, followed by lake and sea salt. Salt has the most microplastics (Zhang et al. 2020; Guo et al. 2022; Udovicki et al. 2022).

28.5.4.2 Tea

Microplastics are found in food and beverages like tea. Tea has microplastics. Heat-sealed polypropylene plastic prevents teabags from separating. Some manufacturers use mesh "temples" instead of paper tea bags. The good news is that several teabag companies are phasing out the plastic. My plastic teabag alternatives site has an up-to-date list (Zhang et al. 2020; Sewwandi et al. 2023).

28.5.4.3 Bottled Water

Drinking water contains microplastics. If this makes a person want to avoid tap water, bottled water includes even more microplastics. Researchers found that bottled water has twice as much plastic as tap water. Drinking bottled or tap water is the biggest contributor to worldwide plastic usage. Bottled water is poisoned for unknown reasons (Guo et al. 2022; Udovicki et al. 2022). Source water may be polluted. However, plastic polymers used to make bottles and caps are more likely to have caused it (Sewwandi et al. 2023).

28.5.4.4 Beer

Researchers found microplastic in beer. The study examined many drinks. Despite finding microplastics in most liquids, beer had the most.

Even though beer and tap water had similar particle counts, this does not always mean the water supply is problematic. Another study found significant differences in microplastic contamination in two beers made in the same city using the same municipal water supply. This suggests that beermaking may produce microplastics (Guo et al. 2022; Udovicki et al. 2022).

28.5.4.5 Rice

The rice inside paper or plastic containers was the same quality. Precooked rice, such as microwaveable packets, has four times more plastic. This suggests avoiding quick rice and making your own. Researchers tried various methods to reduce rice microplastic. Shaking rice in its package did not impact its plastic content. They found gold at the top. Prewashing rice minimizes plastic contamination by 20–40% (Guo et al. 2022; Udovicki et al. 2022).

28.5.4.6 Seafood

Microplastic in fish and shellfish is expected. Zoologists think fish ingested microplastics from the water or seafloor or ate food that had already eaten them. Commercial fish may receive fishmeal with microplastics. However, seafood varies (Guo et al. 2022; Udovicki et al. 2022). Mussels, oysters, shrimp, prawns, and other filter-feeding marine species may include more microplastic because they start their lives in an animal's stomach, shellfish particularly (Guo et al. 2022; Udovicki et al. 2022).

28.5.4.7 Honey

My most disturbing microplastic-food revelation was that honey could include microplastic. Bees are among the most intriguing and smart creatures. Bees look fuzzy due to their hundreds of tiny hairs. While the bee flies, these little hairs charge positively. Air friction against the bee contributes to this. A positive-charged bee attracts negatively charged pollen grains to its landing place. Bees normally adhere these grains to their hair. People were probably wondering why microplastic matters. Bee hairs can retain more than pollen. Positively charged hairs attract other particles. Bees increasingly seem to prefer flying plastic over plant debris or dust. The hive turns this plastic into honey (Guo et al. 2022; Udovicki et al. 2022).

28.6 How Grassroots Education, Awareness, and Stringent Global Food Safety Guidelines Work: FDA, CDC, and WHO Roles in Plastic Waste Food Safety Issues

Due to their larger volume-to-surface area ratio, microplastics were the biggest plastic waste problem. Microplastics enter the ocean through improper waste management and dumping. Microplastics like plastic pellets are utilized to make larger things, but they can also be made by mechanically disintegrating larger plastics. Plastic ropes break down into microfibers. Toothpaste and face-wash microbeads emit huge amounts of microplastics into the environment. Recent research suggests that microplastics, specifically microbeads, microplastic fibers, and degraded microplastics, may be more harmful to aquatic environments than microplastics. However, studies and regulations to regulate plastic pollution are still lacking (Rose et al. 2023).

28.6.1 National Recycling Strategy

The EPA reiterated its goal to increase US recycling rates to 50% by 2030 in November 2021. The Strategy includes strategic objectives and initiatives to improve municipal solid waste recycling in the USA and make it more dependable and affordable (McCauley and Jestratijevic 2023).

28.6.2 Environmental Protection Agency (EPA)

Recycling has been a part of the Environmental Protection Agency (EPA) decades-long implementation of the Resource Conservation and

Recovery Act (RCRA) and the more recent Sustainable Materials Management (SMM) approach, which aims to lessen the environmental impacts of materials throughout their lifecycle. EPA publishes a yearly report on municipal solid waste (MSW) generation, recycling, composting, burning with energy recovery, and landfilling called Advancing Sustainable Materials Management: Facts and Figures. The article includes economic, material, management, and MSW generation indicators (Hossain et al. 2021).

28.6.3 Waste Wise Program

To increase material, use and reuse throughout their life cycles, the EPA collaborates with organizations that serve businesses, governments, and the general public. Partners demonstrate waste reduction, environmental stewardship, and sustainable materials management in their business model, including waste-handling operations. Through stakeholder involvement, Trash Free Waters assists US and foreign communities in eradicating land-based marine litter sources. More than 200 US location-based projects exist. The Trash Free Waters International Implementation Guide offers detailed instructions for including all interested parties in decisions regarding land-based marine litter at the federal, state, and local levels (Onyena et al. 2021).

28.6.4 Decision-Makers in Developing Countries

Because of the low costs of financing, operation, and maintenance, open dumping is the predominant form of solid waste management (SWM) in the majority of less developed nations, including Africa and the Middle East, China, East Asia, Oceania, and Europe. This nonengineered method lacks liners, gas collection, and leachate collection and treatment, contaminating air, water, and soil. Poor domestic solid waste management causes biological, psychological, ergonomic, and noncommunicable disorders. Pollution breeds insects, rats, and flies. These biological vectors gradually induce diarrhea, dysentery, gastrointestinal difficulties, worm infestations, food poisoning, dengue fever, cholera, leptospirosis, bacterial infections, skin, nose, and eye irritations, and respiratory symptoms. Landfill methane, carbon dioxide, sulfur dioxide, and nitrogen dioxide induce inflammation, bronchoconstriction, and immune cell destruction. Inhaling residual hydrogen chloride and hydrogen fluoride causes coughing, chest tightness, and shortness of breath (VaraPrasad et al. 2022).

28.6.5 United States Department of Energy (DOE)

The Plastics Innovation Challenge, inaugurated in 2018, coordinates DOE's many plastics recycling, degrading, upcycling, and circularity activities. The program aims to cut greenhouse gas emissions by 50%, save 50% of energy, treat 90% of plastics, and use 75% carbon. Research projects like Energy Frontier Research Centers solve fundamental science problems, public-private partnerships like the BOTTLE Consortia (bottle.org) and the REMADE Institute reduce plastic waste, and the Better Plants Program works with industrial partners to reduce plastic waste (Milbrandt et al. 2022).

28.6.6 United States Department of Agriculture (USDA)

The USDA supports bioplastics research, development, and consumer and buyer education. Scientists from the Agricultural Research Service converted crop-derived sugars into the chemicals needed to create nylon polymers sustainably by using waste products from cornstalk, straw, and sugarcane bagasse as well as sustainable enzymes. The USDA's BioPreferred Program promotes biobased products through federal procurement, certification, and labeling. The USDA Certified Biobased Product designation promotes bioproduct awareness and sales (Golden et al. 2015).

28.6.7 United States Department of the Interior (DOI)

The Bureau of Safety and Environmental Enforcement (BSEE) requires offshore energy companies to train all Outer Continental Shelf (OCS) staff annually and apply marine debris reduction best practices (Renchen et al. 2021).

28.6.8 The National Park Service (NPS)

The National Park Service (NPS) collaborates with partners to promote conservation and outdoor recreation in the USA and abroad. The National Park Service works with local, state, and federal partners on beach cleanups and educational programs to teach visitors about plastic pollution and marine debris and how their choices and actions may assist Ocean Plastics (National Park Service) (nps.gov); 180 coastal national wildlife refuges and five marine national monuments are under the USFWS's management. The USFWS, partners, and many volunteers clear reefs and beaches of plastic litter. The USFWS removes fishing nets and plastic waste from Papahanaumokuakea Marine National Monument alongside NOAA and others (more than 950 metric tons since 1996) (Baldwin et al. 2020).

28.6.9 The United States Food and Drug Administration (FDA)

Recycled Plastics for Food Packaging: The FDA works with producers to guarantee safe food contact with recycled plastics. The program helps keep plastic food contact products out of landfills and the ocean by safely recycling high-quality plastics (Baldwin et al. 2020).

28.6.10 National Aeronautics and Space Administration (NASA)

NASA supported an experiment to test satellite remote sensing's ability to identify microplastics in our oceans, a crucial function of the PACE mission. When plastic trash in the ocean degrades due to sunlight and ocean wave motion, ocean currents can carry microplastics hundreds or thousands of miles away, making them hard to track and remove. NASA's Cyclone Global Navigation Satellite System is helping researchers track ocean microplastics (CYGNSS) (Nakayama and Osako 2023).

28.6.11 The NIST (NIST)

NIST assists the USA in developing documented standards, reference materials, and reference data and convenes different stakeholders on US manufacturing and technological innovation needs. NIST also studies polymer thermal and mechanical properties during recycling, including depolymerization and degradation in manufacturing and natural environments.

28.6.12 NOAA Marine Debris Program (NOAA)

Marine debris harms the US economy, marine ecosystem, and navigation safety; hence, the NOAA Marine Debris Program was created in 2006. Through outreach and education, the Program promotes community-driven marine debris removal projects and forms partnerships worldwide to reduce marine debris.

28.6.13 National Science Foundation (NSF)

The National Science Foundation funds fundamental research on recycling, polymer chemistry and physics, alternative materials, sustainable and circular processes, the fate and effects of plastic materials lost to the environment, pollution reduction, control systems, and remediation (NSF) (www.state.gov).

28.6.14 International State Department of the United States

Environmental Cooperation under CAFTA-DR and Panama: Under the terms of the US-Dominican Republic-Central America Free Trade Agreement (CAFTA), the Bureau of Oceans and International Environmental and Scientific Affairs (OES) of the State Department is working with the Environmental Protection Agency (EPA) to improve solid waste management capabilities with Central American nations and the Dominican Republic, including addressing land-based sources of ocean plastic pollution (www.state.gov).

28.6.15 West Africa's Plastic Waste Management

The International Union for Conservation of Nature (IUCN) awarded the OES Office of Environmental Quality a grant to develop waste management capabilities and stop the flow of plastic trash into Senegal's marine environment. ASEAN's plastic pollution problem and integrated solid waste management strategies for sustainable local and regional solutions were covered at a session of the US-Singapore Third Country Training Program 2021 (www.state.gov).

28.6.16 Global Ghost Gear Initiative (GGGI)

Throughout 2017, the State Department gave the Ocean Conservancy and GGGI about $1.5 million to fight marine waste, including ghost gear, in the Caribbean. The State Department awarded the Ocean Conservancy about $1 million to help Vietnam fight marine garbage. The USA will join GGGI in 2020. The APEC Marine Debris Management and Innovation (MDMI) Sub-Fund was established by the USA in 2018. The State Department was able to increase the number of APEC members by donating $800,000. Each year, nations can apply for grants from the MDMI for projects that reduce marine plastic pollution (www.state.gov).

28.6.17 Other Recent State Department Programs Include All Hands-on Deck

Accelerating Initiatives to Reduce Ocean Plastic in Vietnam, Social Mobilization for Marine Waste Management in Vietnam, and a Community-Based Marine Litter Reduction Program led by the Vietnam government. To combat marine debris, data, creative local solutions, and outreach are used (www.state.gov)

28.6.18 The United States Agency for International Development (USAID)

USAID's main worldwide program, Clean Cities, Blue Ocean, implements the Save Our Seas 2.0 Act and reduces the estimated 11 million tons of plastics that enter the ocean each year. USAID's Municipal Waste Recycling Program reduced land-based ocean plastic pollution in Indonesia, the Philippines, Sri Lanka, and Vietnam from 2016 to 2021. Through financing and technical help, the initiative presented innovative and scalable ocean plastic pollution reduction strategies, many of which are currently being scaled by Clean Cities, Blue Ocean, local governments, and others (Agnelli and Tortora 2022).

28.6.19 Private Sector Partnerships

In 2019, USAID partnered with Circulate Capital to leverage over $100 million in a private-sector investment plan led by Circulate Capital and funded by global firms. USAID provides a $35 million, 50% loan portfolio guarantee to encourage private capital investment in South and Southeast Asia's recycling value chain through the DFC (Agnelli and Tortora 2022).

28.6.20 Nongovernment Organizations (NGO)

NGOs worldwide arrange these to promote awareness of marine litter and remove harmful materials that disintegrate into microplastics. The International Coastal Cleanup (ICC), led by the US NGO Ocean Conservancy, is one of the main operational organizers of these activities, contributing financially and socially. The ICC organized a September weekend litter assessment and beach cleanup in 70 countries; 790,000 people collected 18 million pounds of rubbish from 25,000 coastal miles in 2016. These actions may affect marine plastic pollution or the environment. The cost-effectiveness of plastic pollution prevention strategies versus reactive actions like beach cleanups is unknown. It bans single-use plastic items on July 1, 2022. Extended Producer Responsibility (EPR), which requires manufacturers to collect and recycle their products after use, promotes circularity. Asia's first plastics pact is India's. The Plastics Pact will unite stakeholders to reduce, reuse, and recycle plastics across the value chain. The mascot, Prakriti, promotes minor lifestyle improvements that aid the environment (Eisfeld-Pierantonio et al. 2022).

28.6.21 Project REPLAN

Khadi and Village Industries Commission (KVIC) launched Project REPLAN, which stands for Reducing Plastic in Nature, which aims to decrease the use of plastic bags by providing an environmentally friendly substitute. Hotspots identifying plastic leakage hotspots from plastic manufacturing, consumption, and disposal can help governments create successful plastic regulations. Items made of plastic can be swapped out for nonplastic, recyclable, or biodegradable alternatives. Product designers can help to find reusable design items and alternatives to single-use plastic by promoting oxo-biodegradable plastics, which disintegrate faster owing to UV radiation and heat. Bacteria may eat plastic waste since it is so ubiquitous. Japanese plastic-eating bacteria were engineered to decompose polyester polymers (food packaging and plastic bottles) (www.pib.gov.in).

28.6.22 National Dashboard on Single-Use Plastic Elimination and Plastic Waste Management

On World Environment Day in June 2022, India began a statewide awareness campaign about single-use plastics. To empower citizens to monitor the sale, use, and manufacturing of SUP in their region and address the plastic catastrophe, a smartphone app for single-use plastics grievance redressal was developed (Peng et al. 2021).

28.6.23 Plastic Waste Management Amendment Rules, 2022

It limits the production, importation, stocking, distribution, sale, and use of numerous single-use plastic products as of July 1, 2022. Additionally, it has implemented Extended Producer Responsibility (EPR), which embraces circularity by mandating that product manufacturers collect and dispose of their items after their useful lives (Peng et al. 2021). It is the first of its kind in Asia. In order to minimize, reuse, and recycle plastics throughout the material's value chain, stakeholders will come together under the auspices of the Plastics Pact, an audacious and cooperative initiative (Peng et al. 2021).

28.6.24 Mascot "Prakriti"

To raise public awareness about simple changes that can be made in one's lifestyle to help the environment. The Union Environment Minister, Bhupendra Yadav, introduced a mascot that would later be named "Prakriti" on April 5, 2022. The purpose of introducing this mascot was to raise awareness among the country's people about the relatively few adjustments that can lead to an improved environment for the children and grandchildren of today (Peng et al. 2021).

28.7 Future Prospects and Conclusion

The UN Economic and Social Commission for Asia and the Pacific project aims to end tourism-related plastic pollution by 2025, with various government organizations working together to promote 100% reusable, recyclable, or compostable plastic packaging. In 2022, 124 countries, including India, agreed to draft a legally binding agreement to address the entire cycle of plastics, from manufacture to disposal. In July 2019, plastic shopping bags were banned in 68 countries with varying enforcement. The EU Single-Use Plastics Directive became effective in July 2021. Developing new, standardized techniques to isolate, characterize, and quantify nanoplastics in drinking and seawater, sediments, ecosystems, and living things is necessary for successful plastic policies. Recognizing plastic leakage hotspots in manufacture, use, and disposal can assist governments in developing successful plastic policies. Product designers can help discover single-use plastic alternatives and reusable design goods. Future prospects include increased plastic production and use, fragmentation of larger plastics due to environmental factors like sunlight and wave action, microplastics in wastewater treatment plants, potential improvements in detection methods, and the development of mitigation strategies.

Microplastic contamination poses a growing global problem that poses a potential threat to food safety and security. The long-term effects of microplastic ingestion on human health are still unknown, but they could be significant. There is an urgent need to develop and implement strategies to reduce microplastic pollution at its source. Additional considerations include public awareness campaigns to educate consumers about the issue of microplastic contamination, adaptation of regulatory frameworks to address challenges posed by microplastics, and further research to understand the full extent of microplastic contamination in the food chain. Further investigation into the potential human health effects of microplastic ingestion is crucial, and the development of cost-effective methods for removing microplastics from food is crucial.

References

Agnelli A, Tortora P (2022) The role of development cooperation in tackling plastic pollution: key trends, instruments, and opportunities to scale up action. OECD Publishing

Bai C, Zhou H, Sarkis J (2023) Evaluating Industry 4.0 technology and sustainable development goals – a social perspective. Int J Prod Res 12:1–21

Baldwin AK, Spanjer AR, Rosen MR, Thom T (2020) Microplastics in Lake Mead national recreation area, USA: occurrence and biological uptake. PLoS One 15(5):e0228896

Chen LK, Arai H, Assantachai P et al (2022) Roles of nutrition in muscle health of community-dwelling older adults: evidence-based expert consensus from Asian Working Group for Sarcopenia. J Cachexia Sarcopenia Muscle 13(3):1653–1672

Chow J, Perez-Garcia P, Dierkes R, Streit WR (2023) Microbial enzymes will offer limited solutions to the global plastic pollution crisis. Microb Biotechnol 16(2):195–217

Domenech J, Annangi B, Marcos R et al (2023) Insights into the potential carcinogenicity of micro-and nano-plastics. Mutat Res Rev Mutat Res 2:108453

Ebrahimi P, Abbasi S, Pashaei R et al (2022) Investigating impact of physicochemical properties of microplastics on human health: a short bibliometric analysis and review. Chemosphere 289:133146

Eisfeld-Pierantonio SM, Pierantonio N, Simmonds MP (2022) The impact of marine debris on cetaceans with consideration of plastics generated by the COVID-19 pandemic. Environ Pollut 5:118967

FICCI Report (2023) India as the world's largest plastic polluter. Available at: drishtiias.com. Accessed 25 Mar 2023

Golden JS, Handfield RB, Daystar J, McConnell TE (2015) An economic impact analysis of the US bio-based products industry: a report to the Congress of the United States of America. Ind Biotechnol 11(4):201–209

Greene J, Skolnik CL, Merritt MW (2022) How medicine becomes trash: disposability in health care. Lancet 400(10360):1298–1299

Guo X, Lin H, Xu S, He L (2022) Recent advances in spectroscopic techniques for the analysis of microplastics in food. J Agric Food Chem 70(5):1410–1422

Hossain S, Rahman MA, Chowdhury MA et al (2021) Plastic pollution in Bangladesh: a review on current status emphasizing the impacts on environment and public health. Environ Eng Res 26(6):210–231

Iñiguez ME, Conesa JA, Fullana A (2017) Microplastics in Spanish table salt. Sci Rep 7:8620

Karami A, Golieskardi A, Keong Choo C et al (2017) The presence of microplastics in commercial salts from different countries. Sci Rep 7:46173

Kosuth M, Mason SA, Wattenberg EV (2018) Anthropogenic contamination of tap water, beer, and sea salt. PLoS One 13(4):e0194970

Kumar V, Singh E, Singh S et al (2023) Micro-and nano-plastics (MNPs) as emerging pollutants in groundwater: environmental impact, potential risks, limitations, and way forward towards sustainable management. Chem Eng J 459:141568

Liebezeit G, Liebezeit E (2014) Synthetic particles as contaminants in German beers. Food Addit Contam Part A 31(9):1574–1578. https://doi.org/10.1080/19440049.2014.945099

Lin S, Ali MU, Zheng C et al (2022) Toxic chemicals from uncontrolled e-waste recycling: exposure, body burden, health impact. J Hazard Mater 426:127

McCauley E, Jestratijevic I (2023) Exploring the business case for textile-to-textile recycling using post-consumer waste in the US: challenges and opportunities. Sustainability 15(2):1473

Menéndez-Pedriza A, Jaumot J, Bedia C (2022) Lipidomic analysis of polyethylene microplastics and polychlorinated biphenyls on human hepatoma cells. J Hazard Mater 421:126777

Milbrandt A, Coney K, Badgett A, Beckham GT (2022) Quantification and evaluation of plastic waste in the United States. Resour Conserv Recycl 183:106363

Nakayama T, Osako M (2023) Development of a process-based eco-hydrology model for evaluating macro-and micro-plastics dynamics. Ecol Model 476:110243

Onyena AP, Aniche DC, Ogbolu BO et al (2021) Governance strategies for mitigating microplastic pollution in the marine environment: a review. Microplastics 1(1):15–46

Padmalosan P, Vanitha S, Kumar VS et al (2023) Investigation on waste materials in clay brick production. Mater Today Proc. https://doi.org/10.1016/j.matpr.2023.01.238

Peng Y, Wu P, Schartup AT et al (2021) Plastic waste release caused by COVID-19 and its fate in the global ocean. Proc Natl Acad Sci 118(47):e2111530118

Pironti C, Ricciardi M, Motta O et al (2021) Microplastics in the environment: intake, human exposure, and toxicological effects. Toxics 9(9):224

Ragasri S, Sabumon PC (2023) A critical review on slaughterhouse waste management and framing sustainable practices in managing slaughterhouse waste in India. J Environ Manag 327:116823. https://doi.org/10.1016/j.jenvman.2022.116823

Ramos-Jiménez A, Zavala-Lira RA, Moreno-Brito V, González-Rodríguez E (2023) FAT/CD36 participation in human skeletal muscle lipid metabolism: a systematic review. J Clin Med 12(1):318. https://doi.org/10.3390/jcm12010318

Renchen GF, Butler CB, Matthews TR (2021) Marine debris knows no boundaries: characteristics of debris accumulation in marine protected areas of the Florida Keys. Mar Pollut Bull 173:112957. https://doi.org/10.1016/j.marpolbul.2021.112957

Rose PK, Jain M, Kataria N, Sahoo PK, Garg VK, Yadav A (2023) Microplastics in multimedia environment: a systematic review on its fate, transport, quantification, health risk, and remedial measures. Groundw Sustain Dev 20:100889. https://doi.org/10.1016/j.gsd.2023.100889

Sewwandi M, Wijesekara H, Rajapaksha AU, Soysa S, Vithanage M (2023) Microplastics and plastics-associated contaminants in food and beverages; global trends, concentrations, and human exposure. Environ Pollut 317:120747. https://doi.org/10.1016/j.envpol.2023.120747

Silva GC, Galleguillos Madrid FM, Hernández D, Pincheira G, Peralta AK, Urrestarazu Gavilán M, Vergara-Carmona V, Fuentes-Peñailillo F (2021) Microplastics and their effect in horticultural crops: food safety and plant stress. Agronomy 11(8):1528. https://doi.org/10.3390/agronomy11081528

Udovicki B, Andjelkovic M, Cirkovic-Velickovic T, Rajkovic A (2022) Microplastics in food: scoping review on health effects, occurrence, and human exposure. Int J Food Contam 9(1):1–6. https://doi.org/10.1186/s40550-022-00092-7

Ugoeze KC, Amogu EO, Oluigbo KE, Nwachukwu N (2021) Environmental and public health impacts of plastic wastes due to healthcare and food products packages: a review. J Environ Sci Public Health 5(1):1–31. https://doi.org/10.26502/jesph.96120128

VaraPrasad K, Mondal A, Cheela V, Dubey BK (2022) Occurrences, impacts and characterization of microplastics in terrestrial ecosystem to aid policy. Curr Opin Environ Sci Health 27:100361. https://doi.org/10.1016/j.coesh.2022.100361

Yuan Z, Nag R, Cummins E (2022) Human health concerns regarding microplastics in the aquatic environment: from marine to food systems. Sci Total Environ 823:153730. https://doi.org/10.1016/j.scitotenv.2022.153730

Zhang Q, Xu EG, Li J, Chen Q, Ma L, Zeng EY, Shi H (2020) A review of microplastics in table salt, drinking water, and air: direct human exposure. Environ Sci Technol 54(7):3740–3751. https://doi.org/10.1021/acs.est.9b04535

Zhang J, Ren S, Xu W, Liang C, Li J, Zhang H, Li Y, Liu X, Jones DL, Chadwick DR, Zhang F, Wang K (2022) Effects of plastic residues and microplastics on soil ecosystems: a global meta-analysis. J Hazard Mater 435:129065. https://doi.org/10.1016/j.jhazmat.2022.129065

Systematic Assessment of Ecological and Fiscal Productivity Impact of Micro-Nano Technological Innovation in Plastics Waste Deterioration and Prevention

29

X. Agnello J Naveen, P. Jaisankar, Kandiah Pakeerathan, S. Suseela, S. Saratha, C. Parameswari, and S. Alagendran

Abstract

This research explains the holistic characteristic of Nano micro-plastic Exposure, degradation and microbial parameters by using different medium dilution factors to convert polymers into useful plastics which can be used for other factors depending upon the concept of 3R systems Degradation of Plastics using engineering analysis and concepts. It has been used in two combination systems microbial + nanotechnology. And learn on the effects and change towards the assimilation properties this is an exploratory research based on the factors, variables, and review depending on secondary resources. Factors like capital cost + fixed cost + variable cost + cost-benefit analysis and their advantage in

X. A. J. Naveen (✉)
Assistant Professor of Environmental Science, Department of Crop Management, Thanthai Roever Institute of Agriculture and Rural Development (TRIARD), Perambalur, Tamil Nadu, India

P. Jaisankar
Department of Horticulture, Pushkaram College of Agriculture Sciences, Pudukkottai, Tamil Nadu, India

K. Pakeerathan
Department of Agricultural Biology, The University of Jaffna, Kilinochchi, Sri Lanka
e-mail: pakeerathank@univ.jfn.ac.lk

S. Suseela
Assistant Professor of Physics, Department of Science and Humanities, Roever Engineering College (REC), Perambalur, Tamil Nadu, India

S. Saratha
Assistant Professor of Biochemistry, Department of Crop Management, Thanthai Roever Institute of Agriculture and Rural Development (TRIARD), Perambalur, Tamil Nadu, India

C. Parameswari
Assistant Professor of Soil Science and Agricultural Chemistry, Department of Crop Management, Thanthai Roever Institute of Agriculture and Rural Development (TRIARD), Perambalur, Tamil Nadu, India

S. Alagendran
Associate Professor of Biochemistry, Division of Crop Management, Dhanalaksmi Srinivasan Agriculture College, Perambalur, Tamil Nadu, India

S. Goswami et al. (eds.), *Micro-Nano Plastics Exposure, Environmental Degradation and Public Health Crisis*, Sustainable Development Goals Series, https://doi.org/10.1007/978-981-96-5047-7_29

waste and energy management. The result gives a valid classification of applications and technological upgrades on their reliability and feasibility in recycling the plastic waste and circular economics present in the standard operating procedure. The method used is exploratory research investigated through SWOT analysis with an interview schedule. The study envisages every important practical concept like clean development mechanism intervention with a mixed raw material analysis and resilient circular economy on resource substitutions; this has an importance on developing the relationship between the institutional research and development of waste management platform, monitoring, and assessment, Clean and green production, bioremedial technologies Waste-to-energy, Organic biomass, Bio-fuels, Biochemicals, composting. The risk assessment, and life cycle analysis (LCA) in Supply Chain and cradle to grave concept. Environmental decision support system towards future technologies with energy, environmental, safety and quality assurance audits.

Keywords

Nanotechnology · Plastics · Micro-plastics · Synthetic technology · Circular economics · Microbial fragmentation · SWOT analysis · Microbes · Economic variables

29.1 Introduction

Pollution from plastic is a major source of pollution that must be addressed because it is non-biodegradable. Plastics can be classified into Polyethylene, Polypropylene, Poly-methyl Methacrylate, Nylon (Polyamide), Teflon (Polytetrafluoroethylene), and Polyurethane. It has been identified as having different effects on management perception depending upon the range of the polymer managing its properties, including poisonous, fragmented, and highly inflammable and also toxic, such as a carcinogenic agent, and it can adsorb, absorb, and sediment due to its breaking ability, and it can also be easily activated in all forms, liquid, solid, air, and in plant life. Based on the solid material, the particles change into matter from matter, and depending on the microgram to milligram, it can affect soil microbes. Biotic and abiotic compounds like human health and can be chemically and physically visible for years; these can be traded through circular economics with the global and Indian communities. In the past 2 years, people have become aware of using plastics and plastic products' environmental day has the theme of "Beating the plastics" or solution for plastics via two methods, using the plastics we can employ on the application platform. Using recyclable or synthesized can be reduced. Economic cost-benefit analysis and calculating resources durability, mobility, carcinogenicity, shifts in use application, LCA concept, bio-remediation, and phytoremediation are methods (Nalladiyil and Babu 2024) using macrophytes can reduce the risk of biomagnification on water (Gao et al. 2024). It is a critical component in improving technology and lowering prices nowadays. This sector has been a major source of income. Because of the trade-off, it has become a critical commodity, with a cutoff point causing significant externalizing of their knowledge. The circular plastic business model helps comprehend difficulties like plastic. Reuse, recycling, reduction, re-engineering, industrial plastic ecology, and energy from waste co-processing materials are all reliant on raw material-producing goods, the ability, and the outcome of polymer segmentation. Polymer governance is built on a structural basis of material supply chain, logistics, training, and mixed raw materials. Transportation of materials are to be reported as per the regulator norm by giving proper. Circular economy (CE) uses a variety of holistic methods to promote economic growth, including sustainability as a strategic policy, managerial business trends, informational, technical, and organizational influence (Ruban 2012), product stewardship, and eco-plastic design can be sustainable options. The Sustainable Development Goals, or SDGs, created the 2030 policy of "zero waste" from plastic products (Keith et al. 2024), and the theme for 2023s June 5 Environmental Day was Solution to Plastics. The goals of plastic economy techniques, in particular, will help us achieve Goal 12.

Specifically, circular economy strategies will help achieve Goal 12. Plastic manufacturing technologies in industries include absorbing rubbish into landfills or burning it, extending the life cycle, and achieving ecological balance. Customer education and stakeholder satisfaction are critical for updating and publishing on plastic usage and the client feedback on plastic life cycle waste is unknown (Jiao et al. 2024; Piao et al. 2024); several aspects of plastic plays a vital role it has its negative and positive like economic, environmental, social, and technological viewpoints. By 2050, 90 billion tonnes of plastics (Sana et al. 2024) can be changed and re-engineered into 3R logistic system, i.e., affordable, from end-to-end material with increased producer accountability (also called EPR) (Bhaskar et al. 2024; Ganguly and Chakraborty 2024). The analyses highlight the interplay between CE and plastics management processes, which are "the most advanced technology" systems for humanity.

29.2 Material and Methods

The study has been reviewed on 25 recent research and review papers about plastic waste management and advanced concepts of the circular economy, utilizing four criteria to discuss the technical advancements concept: (1) Basic level, (2) Secondary level, (3) Policy and regulation settings, (4) The combination and interaction were completed utilizing four separate elements (Nanotechnology). From this variable and factors are revived and given an outcome by the exploratory research on SWOT (strengths, weaknesses, opportunities, and threats analysis) (Henry et al. 2024; Bentil et al. 2024).

29.3 A Worldwide and Indian Outlook for Plastics Waste Management in the Circular Waste Economy

By 2026, worldwide plastic waste management is predicted to total 10 billion metric tons (Macheca et al. 2024). The sustainability economy that is utilized in incorporating the product into the end-of-life system is a combined business model that has existed since the beginning of the twentieth century (Table 29.1). Material waste drives creative thinking, with new concepts, capacities, and effectiveness being evaluated regularly. From the year 1970 until the year 2000, lawmakers developed programs in varying proportions favoring various manufacturers. At this time, other things to consider are the amount of plastics, toxic waste materials, the procedure, the fee that the tax has established in its orientation, and international agencies such as the US Environmental Protection Agency. Organization for Economic Co-operation and Development (OECD), European Union (EU), introduced the Plastic Emissions Directive. Best Available Technology (Zeri 2013) introduced best available technology and circular economy (CE), United Nations Environmental Protection (the UN Environmental Protection Agency), Southeast Asia, Europe, and Africa countries to develop with the regulatory standard operating system. Additionally, there are now categorized technologies on the market for converting solid/hazardous waste into functional/recyclable materials. This regulatory method cannot be feasible in reducing pollution since plastic trash has solids; semi-solids are complicated physical, chemical, and biological in the application of treatment processes. To limit the incorporation of new technologies in the 70s, the EPA developed the rule with a nanotechnology firm that integrated equipment in abatement by employing hazardous/solid waste pollution prevention. The present technology provides three treatment methods: physical segregation, heavy metal removal, and physical processes, which mainly serve to remove solids through reduction and waste (Varshney et al. 2022; Alkhawaldeh and Al-Rousan 2022). Figure 29.1 shows the many uses for handling plastic trash, including recycling, reuse, re-engineering, and remodeling.

29.4 Discussion of the Different Plastic Disposal Concepts

A road map, diffusion, supply chain, material logistics, institutional structure, and private-public partnerships are all necessary for micro-

Table 29.1 Plastics waste applications and their conceptual framework

Plastic scraps visions	Concept orientation	Reviewing and season
Plastics sustainable economy	Plastic trash can be recycled and reduced in certain ways using the manufacturing material of the raw polymers	Lee et al. (2024)
Reuse and reduce of solid waste	Non-hazardous waste can be recycled, and reused, plastic waste	Debele et al. (2024)
Reuse and reduce hazardous waste	Non-biodegradable, dumping plastic waste at safe disposal, being combustible, leachate, extremely exploding, hazardous, detrimental, risky, gene-toxic, infectious, or lethal having thermal heating methods, bio-magnification, bio-accumulation	Pilapitiya and Ratnayake (2024), Bentil et al. (2024)
Reuse	Reuse polymer depends upon the raw material specification	Yusuf et al. (2024)
Recycle	Plastics can be recycled again depending upon the raw material and standard operating procedure	Santos et al. (2024)
Refuse	Ban the use of plastic or change over to fully cloth material; it is 100% eco-friendly. Awareness is important to the public	Shen and Worrell (2024)
Reduce	Plastics raw material thickness can be changed into partial bio-degradable and non-degradable material again if it gets fragmented into soil, water gets polluted	Naveen et al. (2022b)
Re-engineering economy	Changing the total manufacturing of plastics into synthetic material using green chemistry, value-added products	Naveen et al. (2021b)
Sanitary waste dump	Waste disposal facilities cannot recycle garbage placed under oxygen-deprived circumstances; however, if it has a biological component, it may be utilized as the energy of 50% of their costs	Sajid (2025)
Incineration	Trash is burnt in a setting of additional oxygen (air), producing vapor and particles	Crîstiu et al. (2024)
Digestion without air	Natural and renewable resources may be recovered and utilized (vermicomposting)	Su et al. (2024)
Combustion	Ignition is the biochemical destruction of substances that decompose in the absence of air, producing hydrogen, carbon monoxide, hydrocarbon carbon dioxide	Mishra et al. (2024)
A combination of basic materials	Separation of toxic and harmless substances to refurbish and utilize them	Schade et al. (2024)
Ecology of industry	Anything that may be recycled and used again as scrap might be considered waste	Naveen et al. (2021a, b)
Garbage derived from power	Co-processing recycles a co-generation plant with the self-sufficiency of industrial raw material	Naveen et al. (2020)
Supply chain management	Recycling and reusing the co-processing have raw for business management, creating demand	Naveen et al. (2020)
Circular economy	Production, consumption, and disposal of plastics, designed to minimize waste	Jefferson (2019)
Thermoplastics	Extended producer responsibility	Phung et al. (2024)
Nano-plastics	Fragmentation of larger to smaller particles	Vohl et al. (2024)
Thermo-setting polymers	Irreversible hardening	Tang and Li (2024)
Compostable bio-plastics	Bioplastics	Nizamuddin et al. (2024)
Micro-plastics	Toxic recycling	Saif et al. (2024)
Bio-based bio-plastics	Carbon footprint reduction	Deng et al. (2024)
Photo-oxidative degradation	Photoinduced recycling	Michelas et al. (2024)
Ozone degradation	Photocatalyst	Hu and Hu (2024)

(continued)

Table 29.1 (continued)

Plastic scraps visions	Concept orientation	Reviewing and season
Mechanochemical degradation	Mechano-chemical reaction based	Aydonat et al. (2024)
Catalytic degradation	Oxidative catalysis and electrochemical	Liu et al. (2024b)
Aerobic bio-degradation	Semi-degradation and nutrient	Chiemchaisri et al. (2024)
Anaerobic bio-degradation	Recycling via anaerobic digestion and biochemical process	Belone et al. (2024)
Polymer degradation by microorganisms	Additives of enzymes	Aguiar et al. (2024)
Bio-deterioration	Synthetic plastics, photodegradation, and biodeterioration	Gu et al. (2024)
Bio-fragmentation	Landfill, recycling	Shabbir and Hussain (2024)
Mineralization	Low molecules	Ruffino and Zanetti (2024)
Assimilation	Analyzing the physical, chemical, and biological properties	Liu et al. (2024a, b)

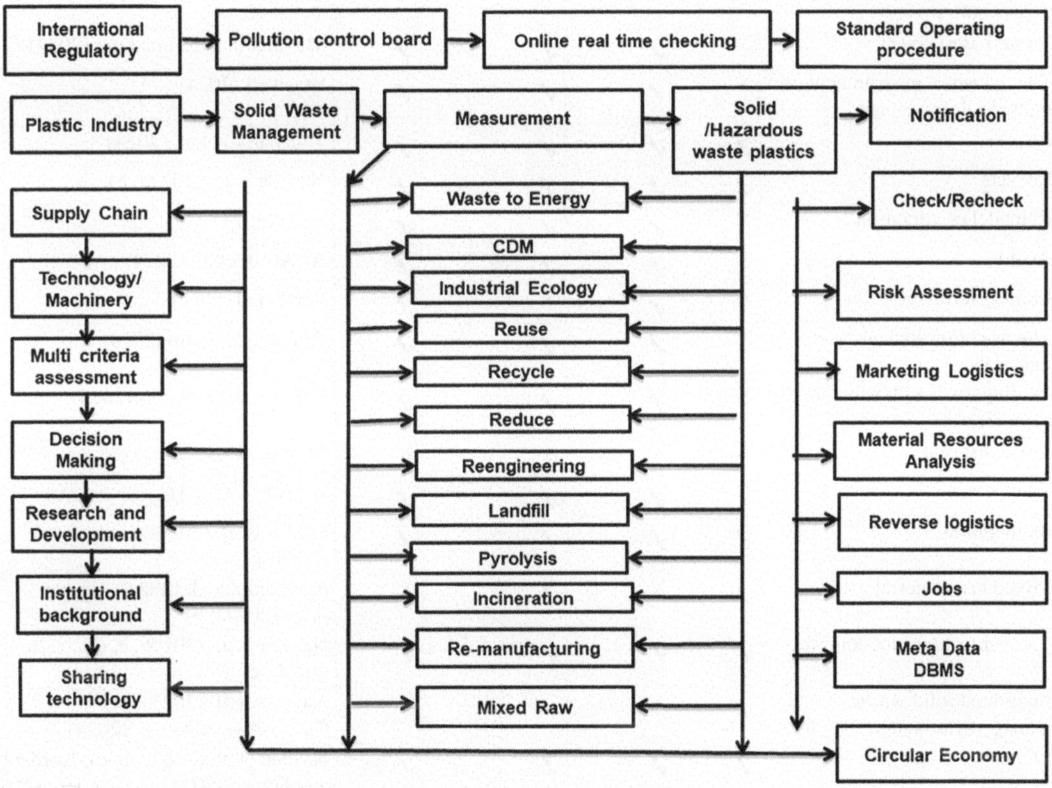

Fig. 29.1 Hierarchy of plastic waste management system

plastics since manufacturing input/output capability is essential to building the sector's consistency (Table 29.2). These have different logic and math for businesses that outsource; medium-sized and small enterprises face financial challenges that prevent them from sustaining their plastics sector, while massive sectors in India benefited from regulations and subsidies

Table 29.2 Application process of recycling, reducing, and reusing plastics waste in India

Technical of plastic 3R idea (useful methods)	Secondary data analysis				
	Strength	Weakness	Opportunities	Threats	Critical gap
1. *Basic level*					
Approach	✓	✓	✓	✓	Dewika et al. (2024)
Spreading the ideas	X	X	X	X	Naveen et al. (2015)
Procurement	✓	✓	✓	X	Naveen et al. (2019b, c, d)
Assessment	✓	✓	✓	✓	Gan et al. (2024)
Fiscal accounting	✓	✓	✓	X	Hossain (2024)
Requirement	✓	✓	✓	✓	Thacharodi et al. (2024)
Equipment/gadgets	✓	✓	✓	X	Todd et al. (2024)
Research and engineering	✓	✓	✓	X	Cao et al. (2024)
Operations in reversal	✓	✓	✓	✓	Farida et al. (2024)
Economics for supplies	X	X	✓	✓	McIvor et al. (2024)
Government and non-government cooperation	X	✓	X	✓	Kurniawan et al. (2024)
2. *Secondary level*					
Waste-based power	X	✓	✓	✓	Mirsalami and Mirsalami (2024)
The industrial environment or mutual reinforcement	X	✓	✓	✓	Braswell (2024)
Reuse	✓	✓	✓	✓	Thacharodi et al. (2024)
Recycle	✓	✓	✓	✓	Naveen et al. (2022a, b)
Remodel or alteration	X	✓	✓	✓	Stutz (2024)
Avoid	X	✓	✓	✓	Al-Abedi et al. (2024)
Making fertilizer	X	✓	X	✓	Wahl et al. (2024)
The use of biological materials	X	✓	✓	✓	Palit and Hussain (2024)
Trading trash with valuable resources	X	✓	✓	✓	Belioka and Achilias (2024)
Toxic disposal	✓	X	X	X	Muis et al. (2024)
Pyrolysis	✓	✓	✓	✓	Kim et al. (2024)
Incineration	✓	✓	✓	✓	Crîstiu et al. (2024); Naveen et al. (2024b)
Mixed raw material	✓	X	✓	X	Stallkamp et al. (2024); Naveen et al. (2024a, b)
Cleaner tech/production	✓	✓	✓	✓	Naveen et al. (2019b, c, d, 2022b, 2023b, c)
Integrated solid waste management system	✓	X	✓	X	Naveen et al. (2023a, b, c); Castro-Amoedo et al. (2024)
LCA	✓	✓	✓	✓	Naveen et al. (2023a, b, c); Nordahl and Scown (2024)
Hazardous risk assessment	✓	✓	✓	✓	Naveen et al. (2023a, b, c); Ghobakhloo et al. (2024)
Assessment of natural resources	✓	✓	✓	✓	Hahladakis et al. (2024)

(continued)

Table 29.2 (continued)

Technical of plastic 3R idea (useful methods)	Secondary data analysis				
	Strength	Weakness	Opportunities	Threats	Critical gap
Procedure for standard operating	✓	✓	✓	✓	Merrington (2024)
Multiple variables evaluation	✓	✓	✓	✓	Schultz et al. (2024)
Participants	✓	✓	✓	✓	Desole et al. (2024)
3. *Law, levy*					
Regulatory framework	✓	✓	X	✓	Srinivasan et al. (2022); Santos et al. (2024)
Public funding support	X	✓	X	✓	Naveen et al. (2019a, b, c, d); Abdul Rahuman et al. (2022)
4. *Model concept*					
Method of computational models	✓	✓	✓	✓	Appavu and Naveen (2023); Sandoval-Reyes et al. (2024)
Ecological, power, and manufacturing assessment	X	X	X	X	Naveen et al. (2019c)
Setting up a facility for instruction and education	✓	✓	✓	✓	Naveen and Arivoli (2021); AlQaraleh et al. (2024)
Monitoring	✓	X	✓	X	Tripathi (2024)
Decision-making	✓	X	✓	X	Castro-Amoedo et al. (2024)
Reporting	✓	✓	✓	✓	Schade et al. (2024)

S, W, O, T—Strength, Weakness, Opportunity, Threats, Plastics waste management, ✓—Merits, X—Demerits, Hazardous Risk Assessment
Note: 1. B*—Basic level, 2. S*—Strength, 3. LL*—Law and levy, 4. MC*—Model concept (Refer to Table 29.3)

obtained by the industries in technological advancement, solid waste database systems, and cooperative work. The plastics industry's ability to calculate raw material consumption necessitates additional consumption. Micro-plastic degradation is the breakdown of plastic materials into tiny pieces by a range of mechanisms, such as mechanical, thermal, and photo-degradation. The type of goods is determined by their quality. The plastic concepts from investigation and technology: the majority of the nation's waste disposal processes (state-of-the-art technology) and harmful plastics are connected to micro-nanotechnology, which can help a business operate in a sustainable economy. Massive, medium-sized, and small enterprises can apply these technologies, depending on raw material consumption, daily output material capacity, and revenue quality, greatly enhancing the handling and breakdown of plastic trash. This is an objective assessment of the economic and environmental effects of this emerging technology.

29.4.1 Overview of Micro-Nano Technology in Plastic Degradation

Micro-nano technology refers to the use of extremely small materials or particles (ranging from nanometers to micrometers) to enhance chemical, physical, and biological processes. In plastic waste management, these technologies can be applied to accelerate degradation, improve recyclability, and reduce the environmental impact of plastics.

29.4.2 Environmental Efficacy Implications

Accelerated degradation: Traditional plastic can take hundreds of years to degrade, but micro-nano technologies can significantly speed up the process. Nanomaterials such as metal oxide nanoparticles (e.g., TiO_2, ZnO) or bio-nanocomposites can enhance photo-degradation or biodegradation of plastics by catalyzing or facilitating breakdown mechanisms at a molecular level.

29.4.3 Targeted Degradation

By designing micro-nanoparticles that target specific polymers, plastics can be broken down selectively without generating harmful by-products. This helps prevent micro-plastic pollution, a significant environmental concern, as traditional degradation often results in micro-plastics that persist in ecosystems.

29.4.4 Reduction of Landfill Dependency

By integrating these technologies into waste management practices, the reliance on landfills can be reduced. This has the potential to conserve land and reduce pollution from leachate and greenhouse gas emissions.

29.4.5 Reduced Eco-toxicity

Some nanomaterials may help reduce the toxicity of degraded plastic particles by neutralizing harmful chemicals released during the degradation process, such as bisphenol A (BPA) or phthalates.

29.4.6 Challenges and Risks

The introduction of nanoparticles into ecosystems could pose new risks, such as the potential toxicity of certain nanoparticles themselves. The long-term effects of nano-particles on organisms and ecosystems are not fully understood, and some could bio-accumulate, potentially causing harm to wildlife.

29.5 Economic Efficacy Implications

29.5.1 The Economy Rehabilitation Methods

Micro-nanotechnology can reduce recycling costs by boosting the effectiveness of organizing, segregation, and cleansing activities. Tiny particles, for instance, may be applied in detectors to autonomously comprehend and organize different kinds of polymers, conserving period and energy.

Enhancing retrieved product excellence management is integrating the field of nanotechnology to reuse and recycling procedures may lead to better-utilized plastics which can clash with virgin plastics in vitality and durability. The following improves the financial feasibility of recycled plastic products and fosters enterprises to use feasible financial concepts.

29.5.2 Business: The Expense and Level

One economic impediment is the high initial cost of developing and deploying micro-nano equipment on a large scale. Nevertheless, improvements in nano-manufacture methods are gradually cutting these costs, establishing the gadgets further reasonable to consumption handling sectors.

29.5.3 Demand for Trained Employees and Job Creation

The application of micro-nanotechnologies in the disposal of waste may lead to the creation of positions in industries including recycling, processing of waste, and nanotechnology research. However, it would increase the need for training and education courses since it would require qualified employees who could manage these intricate operations.

29.5.4 Marketing Potential

The development of polymers that decompose with nanoparticles creates new market opportunities, especially for environmentally aware consumers and businesses. Because of their environmental credentials, these materials might be more expensive, encouraging the use of polyethylene improved by nanotechnology. Despite the great promise of micro-nanotechnology, a thorough evaluation must also take safety and legislation into account.

29.5.5 Laws

The use of nanoparticles in environmental applications is not regulated in many nations. To guarantee that technology advancements do not unintentionally harm people and the environment, exact requirements and testing procedures are essential.

29.5.6 LCA Assessment

To ensure that these advancements are truly environmentally friendly, a thorough life cycle assessment (LCA) of micro-nano methods to plastic degradation must be carried out. This means assessing potential emissions, any unanticipated ecological effects, and the energy and material inputs required for nano-manufacturing.

29.5.7 Future Directions and Research Gaps

Nano-materials Development: More studies are needed on eco-friendly, biodegradable, or even self-degrading nanoparticles that can decay in the environment without damage.

Regulation and Legislation: The government and international organizations must develop policies that encourage the safe application of micro-nanotechnology in the handling of waste. This involves encouraging innovation while ensuring safety requirements are met.

Collaboration Among Those who are Stakeholders: Academics, industry, and regulators must work together to bridge the gap between research and real-world applications. Public-private collaborations may hasten the research and development of these innovations. Plastic is also called solid/hazardous trash and is classified into 88 varieties of harmful waste plastics (Karthikeyan et al. 2018); 58 industrial facilities owned by the Central Pollution Control Board are situated in desirable areas for plastic co-processing trash (Neves et al. 2020). The Resource Conservation and Recovery Act can help to manufacture, recover, repurpose, reflect, restore, repair, revise, reuse, minimize, recycle, redesign, decline, refuel, and remanufacture to a standard process (Farmand et al. 2022). This comprises overall commercial scraps, merchandise recoveries like ecological sustainability and renewable energy, and material recover. Future events 8R systems are to develop high tech knowledge gain an to improvise (Kolesnik and Merkulina 2021; Appavu and Naveen, 2023). Hazard Waste Prevention Guidelines and Site Visits for the Practical Managing of Toxic The combination of processes and Combustible. These issues are relevant to both large and small companies (Ghosh et al. 2022). Industry suggestions have been offered by the Ecological Training Units, Resources Recovery Set, and Program Goals Series. Conventional Hazardous Material Storage, Treatment, and Disposal Facilities, Toxic Waste Management Company System, and Rehabilitation, Storage, and Destruction Facilities, Environmental information system, cleaner technology in SMEs, waste minimization circle, waste reduction demonstration in small industries project, comprehensive industry documents, and total quality management system cleaner manufacturing in India (National Sustainable Development Centers establish database management system, with Innovation Showing Center for Conservation of Energies, and Technological The absorption of the regulation granted by Organizations legislation) (Naveen et al. 2022a, b). The purpose of this regulation is to ensure that the financial statements are accurate; conducting an audit for the green rating project involves classifying industrial units according to their production footprint and reduction of carbon emissions. Cleaner technology is one of the reasons. A key component of Green Chemistry and Green Innovation is energy conservation. Examples of this include Plasma Pyrolysis as Technology, 50 norms of operation, and the present bio-remediation of trash disposal sites (Lang et al. 2022).

29.6 Details for the Company, as well as Green Economy on Microplastic Waste Handling System

India and European countries are encouraged to work together on pollution reduction through the European Union Joint Environmental Education Program. Contributing educational institutions in the study included two institutions throughout Europe. The Environment safeguarding Learning and Investigation established at a University of a Quarter in the Netherlands, as well as Indian Institutes, the Indian Institute of Technology in Bombay, India, the Plastic Pollution Studies and Training Institute in the Indian city of Hyderabad, and the Sweden Institution's worldwide Institute of Industry—based on the atmosphere initiatives.

The International Center of Industrial Environment Assessment at Lund Academic oversees the project, which is sponsored by the European Commission Council. Academic success in preventive environmental management, the creation of study units in India to train students alongside business legislators, the creation of application tools like datasets, publications, and handbooks, and the establishment of 28 continuous points for communication regarding the topic between India and European countries are often the program's principal objectives. The Center Buildings Research Facility, the Central Leather Research Institute, the Energy and Resources Institute, the National Accreditation Board for Testing and Calibration Laboratories accredited laboratory, the National Sustainable Energy, the Council of Research in Science and Industry, the InterMinisterial Group Technical Expert Committee from the Indian Institute of Technology, the National Sustainability Research Facility (Shiuly et al. 2022), and the Central Leather Research Institute are additional research organizations. In 2003, they launched the Better Production website, which is connected to a monthly magazine and is updated and published on the webpage. Sixty numbers were sent as instructions. According to Ahmad et al. (2022), India has embraced the Gujarat Green Production Center-Environmental Management System (Naveen et al. 2022a) model, which links the nation's sustainable economy. To encourage the development and adoption of pilot/demonstration projects by small and medium-sized enterprises and collaboration sectors, the Clean Plastic Technology cells division of India has a system called the Research and Development on Circular economies and waste Source Reduction Demonstration Project. Twenty-four industries have been selected by the Ministry to get the newest technological advancements. The plan includes to improvise 25 finished projects, 6 fresh ones, and 18 ongoing efforts. small to mid-sized companies. The Center for the Advancement of Science, the Pollution Control Board, the Center for Mining and Sustainability, the Environment Protection Training and Research Institute, the Indian Institute of Chemical Techs, and numerous other government and private organizations are helping CE create Written agreements of Understanding on training and providing technical upgrades, such as Terra Green (Luthra et al. 2022; Arivoli et al. 2021a). Research projects were funded by the National Institute of Mechanical Engineering Technology, the Technical Showing Center, the Ministry of Power, the State Pollution Control Board, the Energy Research Facility, the Council for Science, Information, and the Environment, the National Environmental Engineering Research Institute, New Tech, and other commercial organizations (Arivoli et al. 2021b; Arivoli and Naveen 2022; Malik et al. 2022).

29.7 Energy Conservation (Co-generation) in Recycling Micro-Plastic and Finance are Examples of Circular Economies

Every industry's certification procedure has been strict (extremely polluting) according to the Microplastics (Energy conservation with Technology Absorption) law, which requires an audit to finish the circular economy process and checks financial/production input and output. In India's large-to-medium-sized industry, the red industrial category has a strict and mandatory regulatory strategy for the direction of every business (energy, recycling, co-generation, co-processing, and reusing). With concepts like reverse and material logistics and the demand and surplus data-driven notion in determining energy conservation activities, supply chain management has reinforced capacity building for clean development.

If a market-based mechanism under the Energy Efficiency system is envisioned using a static policy of Perform-Achieve-Trade, a hybrid approach database management system is produced. According to Naveen et al. (2022a, b), it examines eight distinct industries that produce solid waste for self-sufficient energy consumption, dry solid state material power consumption of raw material co-processing systems, and thermal, power, aluminum, cement, fertilizer, iron, and steel. These sectors comprise 40% of India's total energy logistics, making them the country's main drivers (Rajaparthiban et al. 2022). Utilizing millions of tons of recycled plastic garbage is the aim. The Department of Non-Conventional (Renewable) states that energy sources are linked to the grid, and other industries are also investing in solar and wind power supplies (Singh et al. 2022). For cogeneration energy production, microplastics are further separated into second and third polymer energy by gasifier, combustion, sludge, and other industrial waste (station capacity of 30 MW). Thermal co-processing (sludge) will utilize the electricity provided by such co-generation units to meet captive needs. Business process wastes such as those from the manufacturing of steel and marble slurry waste should be investigated for co-processing or co-incineration. Dolo char is used as an alternative fuel and raw material blended with coal in atmospheric fluidized bed burning, circulating fluidized bed combustion boilers, and power plants with captives. The state's electrical grid may receive the extra electricity produced. Conditions about inputs, raw materials, production capacity, technical advancements, training facilities, and the yearly mean increase in power generation are characteristics of all industries, whether small, medium, or big (Rao et al. 2022; Naveen et al. 2022a, b). Through nine experimental initiatives in the areas of publishing, training, seminars, public affairs, and research and innovation, the International Development Program has contributed up to 10% of the total. Foreign investors were given access to set up a global framework of new subsidiaries with an MoU on the scrap to the developmentt of cost-cutting renewables like waste-to-energy (Table 29.3).

29.8 Future Aspect and Significances

Plastic Solid waste reduction through Economics of sustainability also entails deciding amongst legislative options and socioeconomic incentives/disincentives. It is critical to evaluate how the sustainability economy evolves when pollution levels fall, as well as the function of other factors such as production level (or company size) and the impact of such expense trend on the economy of the firm in question. The option is between economics and technology that has made a significant influence in the reduction of plastic waste. Approaches may have to be chosen based on their effectiveness; the relative cost that society incurs to the operating price of the instruments involved in the price functions for CE may differ between hazardous and solid waste management, nonetheless if an approach appears to be more effective, polluting firms will successfully try to avoid CE to ensure that it can save

Table 29.3 Findings of applications of microplastics using SWOT analysis

Strength	*Weakness*
B*: Updating and enhancing fresh innovations from both local and global CE, establishing several research and development collaborations to identify new business ventures, logistics of supply chains, commercial partnerships, and government grants. S*: Emerging entrepreneurs are beneficial in plastic sectors such as pollution and mixed raw materials in the CE supply chain, where material resources are crucial. LL*: The governmental entity and the business community are hammering out an agreement. The government has provided numerous subsidies in order to improve the institutional basis, and there was a lot of notice by 6R on energy from garbage. MC*:The evaluation should be reported by three consultants CPCB, the scientific center, and a third party suggest enhancing the quality of the production and output waste handling through monitoring; it helps to increase organizing improvement, review, and decision-making in plastic CE.	B*: Economic constraints make updating, upgrading, servicing, and maintaining a time-consuming operation, and the presence of medium- and small-scale firms weakens the supply chain. Large-scale businesses are few, but they have their material logistics for reuse as well as recycling. S*: The majority of polluting companies recycle and repurpose banned materials. The cost problem creates a trade-off idea in small and medium-sized enterprises. LL*: Pooled or combined is used to enforce strict rules. However, there are still no established authority or quality certification, and the funding for SMEs is quite little. There are benefits to large-scale industries. MC*: SMEs face high costs, are unable to meet targets owing to societal standards and have time delays, yet large-scale industries benefit.
Opportunities	*Threats*
B*: Developing the CE is a connected enterprise with a vast material supply network that combines market value with the need for initial supplies. S*: In India, energy is very vital. Composting is a vital business; raw and mixed materials are important ideas, like co-processing; and many co-generation facilities give zero energy waste. The transport and distribution of raw materials will be highly sought out due to the size and diversity of SMEs. **LL***: Increase the number of incentives and exclusions for the use of environmentally friendly technologies and raise the caliber of firms' output. **MC***: Inspections can guarantee a database management system assessment and preservation of the company's CE reputation.	**B***: Plastic particles also depend on the manufacturing, availability, and logistics of materials. With several financial constraints, SMEs are reluctant to update the oversight policy; yet, major firms carry out their own logistical evaluations. S*: CE for integrated systems is practiced by SMEs. Although they still concentrate on landfills, pyrolysis, and burning, raw material production has expanded on a bigger scale, and the CE supplier network is experiencing growth. LL*: Large-scale businesses, like CE, have a proper inspection, while SMEs do not have a competent inspection. MC*: Due to limited time, SMEs lack a proper CE audit system, and the CE audit procedure is expensive and time-consuming.

money. For any industry to manage solid waste material with regulation (standard operating procedure), the cost of nano-microplastic is a significant economic consideration. It provides details on material logistics, recycling locations, subsidies, private government partnerships with financial assistance, and how to modify the gradual reduction of technological complexity and periodically upgrade it. It also covers classroom instruction, instructional background, and the development of new Memorandums of Understanding. In order to provide effective business alternatives, research and development, notice, ratification of the regulation, modification of the new legislation, the pollution control board (PCB) format offered by economic viewpoint, and regular monitoring are all necessary for the management of plastics trash (the context of databases management system). Sima pro-software is an essential software in the Artifical intelligence and machine learning technology to develop in an research platform.

29.9 Conclusion

Sustainable economy and recycling plastics, recovering, and reduction policy entails a step-by-step rise including several stages of treatment, such as primary and secondary stages of treatment, with each set having its own extra capital cost, with the marginal cost being nearly constant throughout. Though pollution control agencies now enforce contamination regulations, enterprises must regard their solid/hazardous/low energy usage/recycling goods as overall quality assurance. Whatever are potentially harmful to society, the standards commonly utilized for standard-setting remain unclear. A financial analyst would think in terms of setting it at a point where the marginal cost of control equals the marginal benefit from avoiding expenses incurred by pollution prevention gadgets, in terms of whether normal or environmentally conscious technology can be used for performance, preservation of the environment, and benefit to society, along with employing cutting-edge technology.

References

Abdul Rahuman M, Srinivasan VM, Naveen XAJ, Ramadoss R et al (2022) Assessment of yield component traits in cowpea (Vigna ungiculata L. Walp) genotypes. Int J Pharm Res Appl 7:1431–1439

Aguiar MI, Sousa AF, Teixeira G, Tavares AP, Ferreira AM, Coutinho JA et al (2024) Enhancing plastic waste recycling: evaluating the impact of additives on the enzymatic polymer degradation, vol 429. Catal Today, p 114492

Ahmad Z, Aishwarya S, Amanullah MF et al (2022) Zero Waste Biorefinery. In: Zero Waste BiorefineryY. K. Nandabalan, V. K. Garg, N. K. Labhsetwar, & A. Singh (Eds.). Verlag: Springer Nature Singapore

Alkhawaldeh AA, Al-Rousan RZ (2022) Upgrading cyclic response of heat-damaged RC beam-column joints using CFRP sheets. Case Stud Constr Mater 17:e01699

Al-Abedi HJ, Smith JD, Al-Rubaye H, Shakor ZM, Erdem A, Ani PC (2024) Synergistic co-pyrolysis of corn stover and refuse-derived fuel with microplastics: kinetic and thermodynamic study. Biofuels 15:1197–1213

AlQaraleh L, Hajar HAA, Matarneh S et al (2024) Multi-criteria sustainability assessment of solid waste management in Jordan. J Environ Manag 366:121929

Appavu A, Naveen XAJ (2023) Phytoremediation influenced in dairy effluent water treatment using constructed wetland application with aquatic macrophytes: a literature study of several constructed wetland systems for dairy wastewater treatment. In: Multidisciplinary approaches to sustainable human development (pp. 255–282). IGI Gobal, Hershey

Arivoli A, Thangavelu S, Naveen XAJ (2021a) Influence of Sugar Mill Effluent. In: Bioremediation and Green Technologies: Sustainable Approaches to Mitigate Environmental Impacts (p. 271). Springer, Cham

Arivoli A, Naveen XAJ, Thangavelu S (2021b) Impact of sugar mill effluent on the quality of groundwater in Cuddalore district, Tamil Nadu, India. Bioremediation and Green Technologies: Sustainable Approaches to Mitigate Environmental Impacts (pp. 41–52). Springer, Cham

Arivoli A, Naveen XAJ (2022) Efficacy of bio-absorbent concept in textile effluent treatment technology using low-cost materials by implementing Banana bark and Orange peel. In: IoT for Sustainable Smart Cities and Society (pp. 301–317). Cham: Springer International Publishing

Aydonat S, Hergesell AH, Seitzinger CL, Lennarz R, Chang G, Sievers C, Göstl R et al (2024) Leveraging mechanochemistry for sustainable polymer degradation. Polym J 56:249

Belioka MP, Achilias D (2024) Sustainable management and valorization of microplastics. In: From waste to wealth. Springer, Singapore

Belone MCL, Brosens D, Kokko M, Sarlin E et al (2024) Effects of mesophilic and thermophilic anaerobic digestion of sewage sludge on different polymers: perspectives on the potential of the treatment to degrade microplastics. Sci Total Environ 907:168014

Bentil J, Braimah S, Obeng AF (2024) Solid waste management in higher educational institution: an investigation using the SWOT analysis and the circular economy principle perspective. Circ Econ Sustain 4:1249

Bhaskar K, Radulovic V, Griffin JJ et al (2024) Challenges and opportunities of partnering with the informal sector. Center on Regulation and Markets working paper 11

Braswell CB (2024) Downstream dilemma: navigating microplastic's impact on freshwater symbiosis in the anthropocene. Doctoral dissertation, Virginia Tech

Cao Y, Bian J, Han Y, Liu J, Ma Y, Feng W et al (2024) Progress and prospects of microplastic biodegradation processes and mechanisms: a bibliometric analysis. Toxics 12(7):463

Castro-Amoedo R, Granacher J, Kantor I, Dahmen A, Barbosa-Povoa A, Maréchal F et al (2024) On the role of system integration in plastic waste management. Resour Conserv Recycl 201:107295

Chiemchaisri W, Chiemchaisri C, Muenmee S et al (2024) Biodegradation of plastic wastes under semi-aerobic condition with active methane oxidation activities and nutrient supply. Case Stud Chem Environ Eng 10:100809

Crîstiu D, d'Amore F, Bezzo F et al (2024) Economic and environmental optimisation of mixed plastic waste supply chains in Northern Italy comparing incineration and pyrolysis technologies. Comput Chem Eng 180:108503

Debele AD, Demeke S, Bekele T, Malimo M et al (2024) Recycling and reusing potential of disposable low-density polyethylene plastic waste for flexible paver tile construction for outdoor application. Heliyon 10:e29381

Deng Y, Zeng Z, Feng W, Liu J, Yang F et al (2024) Characteristics and migration dynamics of microplastics in agricultural soils. Agriculture 14:157

Desole MP, Gisario A, Fedele L, Barletta M et al (2024) Life cycle assessment (LCA) and multi criteria decision analysis (MCDA) of eco-friendly packaging for dairy products and fourth range. Procedia CIRP 122:927–932

Dewika M, Markandan K, Ruwaida JN, Sara YY, Deb A, Irfan NA, Khalid M et al (2024) Integrating the quintuple helix approach into atmospheric microplastics management policies for planetary health preservation. Sci Total Environ 954:176063

Farida Y, Siswanto N, Vanany I et al (2024) Reverse logistics toward a circular economy: consumer behavioral intention toward polyethylene terephthalate (PET) recycling in Indonesia. Case Stud Chem Environ Eng 10:100807

Farmand P, Nicolai H, Schumann C, Attili A, Berger L, Li T, ... Pitsch H (2022) Numerical investigation and assessment of flamelet-based models for the prediction of pulverized solid fuel homogeneous ignition and combustion. Combust Flame 235:111693

Gan M, Zhang Y, Shi P, Cui L, Zhang C, Guo J (2024) Occurrence, potential sources, and ecological risk assessment of microplastics in the inland river basins in Northern China. Mar Pollut 205:116656

Ganguly RK, Chakraborty SK (2024) Plastic waste management during and post Covid19 pandemic: challenges and strategies towards circular economy. Heliyon 10:e25613

Gao S, Zhang S, Feng Z, Lu J, Fu G, Yu W et al (2024) The bio-accumulation and magnification of microplastics under predator–prey isotopic relationships. J Hazard Mater 480:135896

Ghobakhloo S, Mostafaii GR, Khoshakhlagh AH, Moda HM, Gruszecka-Kosowska A et al (2024) Health risk assessment of heavy metals in exposed workers of municipal waste recycling facility in Iran. Chemosphere 346:140627

Ghosh UK, Islam MN, Siddiqui MN, Cao X, Khan MAR (2022) Proline, a multifaceted signalling molecule in plant responses to abiotic stress: understanding the physiological mechanisms. Plant Biol 24(2):227–239

Gu JD, Ford T, Mitton B, Mitchell R et al (2024) Research on biodeteriotration of plastics. Int Biodeterior Biodegradation 186:105699

Hahladakis JN, Iacovidou E, Gerassimidou S et al (2024) Plastic waste in a circular economy. In: Environmental materials and waste. Elsevier, Oxford

Henry NGEH, Tembo FM, Walde Z et al (2024) Analyzing the strengths, weaknesses, opportunities, and threats (SWOT) of plastic waste management in Cameroon. Int Res J Innov Eng Technol 8:158–165

Hossain MS (2024) People's attitudes regarding plastics and microplastics pollution: perceptions, behaviors, and policy implications. Mar Policy 165:106219

Hu J, Hu J (2024) Mineralization characteristics and behavior of polyethylene microplastics through ozone-based treatment. Chemosphere 349:140839

Jefferson M (2019) Whither plastics?—petrochemicals, plastics and sustainability in a garbage-riddled world. Energy Res Soc Sci 56:101229

Jiao H, Ali SS, Alsharbaty MHM, Elsamahy T, Abdelkarim E, Schagerl M et al (2024) A critical review on plastic waste life cycle assessment and management: challenges, research gaps, and future perspectives. Ecotoxicol Environ Saf 271:115942

Karthikeyan OP, Trably E, Mehariya S, Bernet N, Wong JW, Carrere H (2018) Pretreatment of food waste for methane and hydrogen recovery: a review. Bioresour Technol 249:1025–1039

Keith M, Koller M, Lackner M et al (2024) Carbon recycling of high value bioplastics: a route to a zero-waste future. Polymers 16:1621

Kim D, Han M, Kim N, Kim J, Jung SP et al (2024) Waste plastic pyrolysis industry: current status and prospects. J Korean Soc Environ 46:395

Kolesnik GV, Merkulina IA (2021) Waste management of production and consumption as an element of on the circular economy. In IOP Conference Series: Earth and Environmental Science (Vol. 666, No. 2, p. 022064). IOP Publishing

Kurniawan TA, Mohyuddin A, Othman MHD, Goh HH, Zhang D, Anouzla A et al (2024) Beyond surface: unveiling ecological and economic ramifications of microplastic pollution in the oceans. Water Environ Res 96:e11070

Lang C, Cheng G, Tu B, Han J (2022) Learning what not to segment: a new perspective on few-shot segmentation. In: Proceedings of the IEEE/CVF conference on computer vision and pattern recognition (pp. 8057–8067)

Lee S, Huang M, Lee J, Choi H, Jo IY, Na H et al (2024) Eco-friendly materials for a zero e-waste society: challenges and opportunities in engineering plastics. Adv Sustain Syst 8:2300428

Liu Q, Martinez-Villarreal S, Wang S, Tien NNT, Kammoun M, De Roover Q, Len C, Richel A et al (2024a) The role of plastic chemical recycling processes in a circular economy context. Chem Eng J 498:155227

Liu Y, Shi J, Jin H, Guo L et al (2024b) Chemical recycling methods for managing waste plastics: a review. Environ Chem Lett 22:149

Luthra S, Kumar A, Sharma M, Garza-Reyes JA, Kumar V (2022) An analysis of operational behavioural factors and circular economy practices in SMEs: An emerging economy perspective. J Bus Res 141:321–336

Macheca AD, Mutuma B, Adalima JL, Midheme E, Lúcas LH, Ochanda VK, Mhlanga SD et al (2024) Perspectives on plastic waste management: challenges and possible solutions to ensure its sustainable use. Recycling 9:77

Malik P, Patel K, Pinto C, Jaiswal R, Tirupathi R, Pillai S, Patel U (2022) Post-acute COVID-19 syndrome (PCS) and health-related quality of life (HRQoL)—A systematic review and meta-analysis. J Med Virol 94(1):253–262

McIvor MJ, McIlhagger AT, Archer E, Dooher T, Gault A, Golbang A et al (2024) A natural fibre supply chain in Northern Ireland. NICOM LLP and Ulster University, Belfast

Merrington A (2024) Recycling of plastics. In: Applied plastics engineering handbook. William Andrew Publishing, Norwich

Michelas M, Wimberger L, Boyer C et al (2024) A general approach for photo-oxidative degradation of various polymers. Macromol Rapid Commun 45:e2400358

Mirsalami SM, Mirsalami M (2024) Achieving optimal output of microplastic petroleum waste and improving energy and environmental indicators by optimizing the pyrolysis process and modifying the fuel composition. Fuel 376:132699

Mishra R, Shu CM, Gollakota AR, Pan SY et al (2024) Unveiling the potential of pyrolysis-gasification for hydrogen-rich syngas production from biomass and plastic waste. Energy Convers Manag 321:118997

Muis R, Fariz A, Darma R, Yunus S, Tasrief R, Rachman I, Matsumoto T et al (2024) Investigating the potential of landfilled plastic waste – a case study of Makassar landfill, Eastern Indonesia. Ecol Eng Environ Technol 25:185

Nalladiyil A, Babu GL (2024) Dumpsite remediation through landfill mining and rehabilitation: a circular economy perspective. J Indian Inst Sci 104:489

Naveen XAJ, Arivoli A (2021) Corporate social responsibility in environmental protection of specific sugar factory in Tamil Nadu: a case study. In: The new horizon & emerging challenges in post COVID-19 business scenario, vol 1. PG & Research Department of Commerce, Tiruchengode, pp 232–249

Naveen XAJ, Ravichandran M, Balamurugan J et al (2015) Clean technology and its efficacy: strategies of environmental management. J Environ Soc Sci 2:1–10

Naveen XAJ, Arivoli A et al (2019a) Case study to understand the benefits of cleaner technology by the assessment of water, air and soil quality around the sugar industry in Erode, Tamil Nadu. Int J Sci Technol Res 8:2070–2075. ISSN 2277-8616

Naveen XAJ, Arivoli A, Boopathi S, Balamurugan J et al (2019b) A review on cleaner-technology in pollution control at ultra-red industry in Tamil Nadu, India. Int J Sci Technol Res 8:1892–1898. ISSN 2277-8616

Naveen XAJ, Boopahti S, Arivoli A et al (2019c) Efficacy of cleaner technology in wastewater treatment and energy conservation using (UASB-R) anaerobic digester at bagasse-based pulp and paper mill in Tamil Nadu. Res Rev Int J Multidiscip 6:563–577

Naveen XAJ, Boopathi S, Arivoli A et al (2019d) Efficacy of cleaner technology in recycling wastewater treatment process using add-on applications at sugar industry in Tamil Nadu. Poll Res 38:S38–S46

Naveen XAJ, Boopathi S, Arivoli A, Balamurugan J et al (2020) Win-Win strategy of cleaner technology-environmental and economic scores of sugar industry in Tamil Nadu. Indian J Environ Prot 40:186–191

Naveen XAJ, Boopathi S, Arivoli A (2021a) Case experience on WIN WIN approach towards industrial pollution control-economic and environmental efficacy of cleaner technology in select cement industry in Tamil Nadu. In: Bioremediation and green technologies. Springer, Cham

Naveen XAJ, Boopathi S, Arivoli A, Kannan A (2021b) Economic and environmental benefits of cleaner technology in industrial pollution control: case study of select sugar industry in Tamil Nadu. In: Bioremediation and green technologies. Springer, Cham

Naveen XAJ, Arivoli A, Boopathi S, Rajkumar V (2022a) A case study on evaluation of energy management system by implication of advanced technology in typical cement factory Tamil Nadu, India. In: Smart technologies for energy and environmental sustainability. Springer, Cham

Naveen XAJ, Arivoli A, Malarmannan VC, Boopathi S (2022b) A review on cleaner technology for sustainable industrial practices: global and Indian scenario genesis of technologies to control pollution. In: Computational modelling in industry 4.0. Springer, Singapore

Naveen XAJ, Devi MP, Parameswari C (2023a) Comprehensive environmental management system report in specific sugar mill, India: a case study using SWOT analysis–road map and diffusion of EMS, classification of the environmental management system. In: Management, technology, and economic growth in smart and sustainable cities. IGI Global, Hershey

Naveen XAJ, Arivoli A, Jaisankar P, Parameshwari C et al (2023b) Monitoring air and soil parameters around the proximity of a typical cement plant in India to determine the efficacy of clean tech used in production. Int J All Res Educ Sci Methods 11:1405

Naveen XAJ, Wahab K, Srinivasan VM, Jaishankar P, Abdul Rahuman M, Ramadoss R, Vishnu L et al (2023c) Industrial solid/hazardous waste recovery and management in India: a conceptual nexus of circular economy. Int J Pharm Res Appl 8:492–513

Naveen XAJ, Arivoli A, Boopathi S, Rajkumar V, Devi MP (2024a) Assessment and monitoring of environmental quality of the specific sugar industry in Tamil Nadu using GIS monitoring systems. In: Sustainable innovation for industry 6.0. IGI Global, Hershey

Naveen XAJ, Boopathi S, Arivoli A, Wahab K, Rahuman A, Srinivasan VM, Ramadoss R (2024b) Clean and green applications towards sustainable development: a case study in select sugar distillery and cement industries in Tamil Nadu. In: Digital transformation for improved industry and supply chain performance. IGI Global, Hershey

Neves A, Godina R, Azevedo SG, Matias JC (2020) A comprehensive review of industrial symbiosis. J Clean Prod 247:119113

Nizamuddin S, Baloch AJ, Chen C, Arif M, Mubarak NM et al (2024) Bio-based plastics, biodegradable plastics, and compostable plastics: biodegradation mechanism, biodegradability standards and environmental stratagem. Int Biodeterior Biodegradation 195:105887

Nordahl SL, Scown CD (2024) Recommendations for life-cycle assessment of recyclable plastics in a circular economy. Chem Sci 15:9397

Palit S, Hussain CM (2024) Microplastics generated from bioplastics—a far-reaching. In: Bioplastics for sustainability: manufacture, technologies, and environment. Elsevier, Amsterdam

Phung NHT, Lien LT, Nguyen Ha KB, Tran LM, Nguyen-Viet HA, Anh Duc P, Trinh BS et al (2024) Effects of chronic exposure to microplastics and microplastics associated with polychlorinated biphenyl 153 on Daphnia magna. Hum Ecol Risk Assess Int J 30:22

Piao Z, Agyei Boakye AA, Yao Y et al (2024) Environmental impacts of biodegradable microplastics. Nat Chem Eng 1:661

Pilapitiya PNT, Ratnayake AS (2024) The world of plastic waste: a review. Cleaner Mater 11:100220

Rajaparthiban J, Suresh G, Ganesamoorthy R et al (2022) Effect of polyurethane ratio on mechanical behavior of banana fiber/polyurethane-vinylester matrix composites. Matéria (Rio de Janeiro) 27:e13182

Rao NS, Sunitha B, Das R, Kumar BA (2022) Monitoring the causes of pollution using groundwater quality and chemistry before and after the monsoon. Phys Chem Earth (Pt A, B,C) 128:103228

Ruban A (2012) Life cycle assessment of plastic bag production. Master thesis in Sustainable Development

Ruffino B, Zanetti M (2024) Is the assimilation to a solid recovered fuel a viable solution for automobile shredder residues' management? Environ Res 247:118131

Saif M, Blay-Roger R, Zeeshan M, Bobadilla LF, Reina TR, Nawaz MA, Odriozola JA et al (2024) Navigating the legislative interventions, challenges, and opportunities in revolutionizing textile upcycling/recycling processes for a circular economy. ACS Sustain Resour Manag 1:2338

Sajid M (2025) Recent trends in sanitary landfilling and future prospects. In: Waste management and treatment. CRC Press, Boca Raton

Sana A, Mahmood A, Ibrahim M et al (2024) Recovery of resources and circular economy from biomass-derived waste through aerobic and anaerobic digestion-based technique. In: Nanomaterials in biomass conversion. Woodhead Publishing, Cambridge, MA

Sandoval-Reyes M, He R, Semeano R, Ferrão P (2024) Mathematical optimization of waste management systems: methodological review and perspectives for application. Waste Manag 174:630

Santos G, Esmizadeh E, Riahinezhad M et al (2024) Recycling construction, renovation, and demolition plastic waste: review of the status quo, challenges and opportunities. J Polym Environ 32:479

Schade A, Melzer M, Zimmermann S, Schwarz T, Stoewe K, Kuhn H et al (2024) Plastic waste recycling—a chemical recycling perspective. ACS Sustain Chem Eng 12:12270

Schultz FC, Valentinov V, Kirchherr J, Reinhardt RJ, Pies I et al (2024) Stakeholder governance to facilitate collaboration for a systemic circular economy transition: a qualitative study in the European chemicals and plastics industry. Bus Strateg Environ 33:2173

Shabbir A, Hussain N (2024) Biodegradation of plastic waste: mechanisms, perspectives, and challenges. In: Plastic waste management: methods and applications. Wiley-VCH, Weinheim

Shiuly A, Hazra T, Sau D, Maji D (2022) Performance and optimisation study of waste plastic aggregate based sustainable concrete–A machine learning approach. Clean Waste Syst 2:100014

Shen L, Worrell E (2024) Plastic recycling. In: Handbook of recycling. Elsevier, Amsterdam

Singh T, Poterba T, Curtis D, Akil H, Al Eissa M, Barchas JD, ... Daly MJ (2022) Rare coding variants in ten genes confer substantial risk for schizophrenia. Nature 604(7906):509–516

Srinivasan VM, Abdul Rahuman M, Soban Babu G, Naveen XAJ et al (2022) Efficacy of Trichoderma against green gram root rot pathogen Macrophomina phaseolina. Int J Adv Res 10:324–328

Stallkamp C, Hennig M, Volk R, Stapf D, Schultmann F et al (2024) Pyrolysis of mixed engineering plastics: economic challenges for automotive plastic waste. Waste Manag 176:105

Stutz T (2024) Design of an HDPE bottle collection and pre-cleaning system for recycling in Blantyre, Malawi. Master's thesis, ETH Zurich

Su Z, He PJ, Lü F, Zhang H, Ren L, Lü K et al (2024) Certified compostable products and sustainable food waste management via laboratory-level simulation anaerobic digestion and postdigestion composting. ACS Sustain Chem Eng 12:4146

Tang KHD, Li R (2024) The effects of plastisphere on the physicochemical properties of microplastics. Bioprocess Biosyst Eng 48:1

Thacharodi A, Hassan S, Meenatchi R, Bhat MA, Hussain N, Arockiaraj J et al (2024) Mitigating microplastic pollution: a critical review on the effects, remediation, and utilization strategies of microplastics. J Environ Manag 351:119988

Todd PA, Yong CLX, Foo SH, Ying LSM, Ledet J et al (2024) Plastic-less equipment for sampling marine microplastics. Front Mar Sci 11:1345591

Tripathi NG (2024) Plastic waste crisis: how smart cities can lead the change. In: Plastic pollution: challenges and green solutions. Springer, Singapore

Varshney A, Dahiya P, Sharma A, Pandey R, Mohan S (2022) Fly ash application in soil for sustainable agriculture: an Indian overview. Energy Ecol Environ 7(4):340–357

Vohl S, Kristl M, Stergar J et al (2024) Harnessing magnetic nanoparticles for the effective removal of micro- and nanoplastics: a critical review. Nanomaterials 14(14):1179

Wahl A, Davranche M, Rabiller-Baudry M, Pédrot M, Khatib I, Labonne F, Gigault J et al (2024) Condition of composted microplastics after they have been buried for 30 years: vertical distribution in the soil and degree of degradation. J Hazard Mater 462:132686

Yusuf BO, Abdalla TA, Alahmari TS, Hassan R et al (2024) Adaptive reuse of waste plastic as binders in composites for sustainable construction. Cleaner Eng Technol 22:100812

Zeri A (2013) Deconstructing the industrial emissions directive's (2010/75/EU) regulatory standards: a tale of cautious optimism. UCLJLJ 2:173

The manufacturer's authorised representative in the EU is Springer Nature Customer Service Centre GmbH, Europaplatz 3, 69115 Heidelberg, Germany. If you have any concerns regarding our products, please contact ProductSafety@springernature.com

Printed and bound by CPI Group (UK) Ltd, Croydon, CR0 4YY
01/10/2026
02225936-0020